# ADVANCES IN NUMERICAL PARTIAL DIFFERENTIAL EQUATIONS AND OPTIMIZATION

PROCEEDINGS OF THE FIFTH
MEXICO-UNITED STATES WORKSHOP

# SIAM PROCEEDINGS SERIES LIST

Neustadt, L.W., Proceedings of the First International Congress on Programming and Control (1966)

Hull, T.E., Studies in Optimization (1970)

Day, R.H. & Robinson, S.M., Mathematical Topics in Economic Theory and Computation (1972)

Proschan, F. & Serfling, R.J., Reliability and Biometry: Statistical Analysis of Lifelength (1974)

Barlow, R.E., Reliability & Fault Tree Analysis: Theoretical & Applied Aspects of System Reliability & Safety Assessment (1975)

Fussell, J.B. & Burdick, G.R., Nuclear Systems Reliability Engineering and Risk Assessment (1977)

Duff, I.S. & Stewart, G.W., Sparse Matrix Proceedings 1978 (1979)

Holmes, P.J., New Approaches to Nonlinear Problems in Dynamics (1980)

Erisman, A.M., Neves, K.W. & Dwarakanath, M.H., Electric Power Problems: The Mathematical Challenge (1981)

Bednar, J.B., Redner, R.,Robinson, E. & Weglein, A., Conference on Inverse Scattering: Theory and Application (1983)

Voigt, R.G., Gottlieb, D. & Hussaini, M. Yousuff, Spectral Methods for Partial Differential Equations (1984)

Chandra, Jagdish, Chaos in Nonlinear Dynamical Systems (1984)

Santosa, Fadil, Symes, William W., Pao, Yih-Hsing & Holland, Charles, Inverse Problems of Acoustic and Elastic Waves (1984)

Gross, Kenneth I., Mathematical Methods in Energy Research (1984)

Babuska, I., Chandra, J. & Flaherty, J., Adaptive Computational Methods for Partial Differential Equations (1984)

Boggs, Paul T., Byrd, Richard H. & Schnabel, Robert B., Numerical Optimization 1984 (1985)

Angrand, F., Dervieux, A., Desideri, J.A. & Glowinski, R., Numerical Methods for Euler Equations of Fluid Dynamics (1985)

Wouk, Arthur, New Computing Environments: Parallel, Vector and Systolic (1986)

Fitzgibbon, William E., Mathematical and Computational Methods in Seismic Exploration and Reservoir Modeling (1986)

Drew, Donald A. & Flaherty, Joseph E., Mathematics Applied to Fluid Mechanics and Stability: Proceedings of a Conference Dedicated to R.C. DiPrima (1986)

Heath, Michael T., Hypercube Multiprocessors 1986 (1986)

Papanicolaou, George, Advances in Multiphase Flow and Related Problems (1987)

Wouk, Arthur, New Computing Environments: Microcomputers in Large-Scale Computing (1987)

Chandra, Jagdish & Srivastav, Ram, Constitutive Models of Deformation (1987)

Heath, Michael T., Hypercube Multiprocessors 1987 (1987)

Glowinski, R., Golub, G.H., Meurant, G.A. & Periaux, J., First International Conference on Domain Decomposition Methods for Partial Differential Equations (1988)

Salam, Fathi M.A. & Levi, Mark L., Dynamical Systems Approaches to Nonlinear Problems in Systems and Circuits (1988)

Datta, B., Johnson, C., Kaashoek, M., Plemmons, R. & Sontag, E., Linear Algebra in Signals, Systems and Control (1988)

Ringeisen, Richard D. & Roberts, Fred S., Applications of Discrete Mathematics (1988)

McKenna, James & Temam, Roger, ICIAM '87–Proceedings of the First International Conference on Industrial and Applied Mathematics (1988)

Rodrigue, Garry, Parallel Processing for Scientific Computing (1989)

Chan, Tony F., Meurant, Gerard, Periaux, Jacques & Widlund, Olof B., Domain Decomposition Methods (1989)

Caflish, Russel E., Mathematical Aspects of Vortex Dynamics (1989)

Wouk, Arthur, Parallel Processing and Medium-Scale Multiprocessors (1989)

Flaherty, Joseph E., Paslow, Pamela J., Shephard, Mark S. & Vasilakis, John D., Adaptive Methods for Partial Differential Equations (1989)

Kohn, Robert V. & Milton, Graeme W., Random Media and Composites (1989)

Mandel, Jan, McCormick, S.F., Dendy, J.E., Jr., Farhat, Charbel, Lonsdale, Guy, Parter, Seymour V., Ruge, John W. & Stüben, Klaus, Proceedings of the Fourth Copper Mountain Conference on Multigrid Methods (1989)

Colton, David, Ewing, Richard & Rundell, William, Inverse Problems in Partial Differential Equations (1990)

Chan, Tony F., Glowinski, Roland, Periaux, Jacques & Widlund, Olof B., Third International Symposium on Domain Decomposition Methods for Partial Differential Equations (1990)

Dongarra, Jack, Messina, Paul, Sorensen, Danny C. & Voigt, Robert G., Proceedings of the Fourth SIAM Conference on Parallel Processing for Scientific Computing (1990)

Glowinski, Roland & Lichnewsky, Alain, Computing Methods in Applied Sciences and Engineering (1990)

Coleman, Thomas F. & Li, Yuying, Large-Scale Numerical Optimization (1990)

Aggarwal, Alok, Borodin, Allan, Gabow, Harold, N., Galil Zvi, Karp, Richard M., Kleitman, Daniel J., Odlyzko, Andrew M., Pulleyblank, William R., Tardos, Éva & Vishkin, Uzi, Proceedings of the Second Annual ACM-SIAM Symposium on Discrete Algorithms (1991)

Cohen, Gary, Halpern, Laurence & Joly, Patrick, Mathematical and Numerical Aspects of Wave Propagation Phenomena (1991)

Gómez, S., Hennart, J. P., & Tapia, R. A., Advances in Numerical Partial Differential Equations and Optimization, Proceedings of the Fifth Mexico-United States Workshop (1991)

# ADVANCES IN NUMERICAL PARTIAL DIFFERENTIAL EQUATIONS AND OPTIMIZATION

PROCEEDINGS OF THE FIFTH
MEXICO-UNITED STATES WORKSHOP

Edited by  S. Gómez
IIMAS-UNAM

J. P. Hennart
IIMAS-UNAM

R. A. Tapia
Rice University

**siam.**
Society for Industrial and Applied Mathematics

Philadelphia

# ADVANCES IN NUMERICAL PARTIAL DIFFERENTIAL EQUATIONS AND OPTIMIZATION

PROCEEDINGS OF THE FIFTH MEXICO-UNITED STATES WORKSHOP

Proceedings of the Fifth Mexico-United States Workshop on Advances in Numerical Partial Differential Equations and Optimization held January 26, 1989, in Merida, Yucatan, Mexico.

**Library of Congress Cataloging-in-Publication Data**

Mexico-United States Workshop on Advances in Numerical Partial Differential Equations and Optimization (5th : 1989 : Mérida, Mexico)
    Advances in numerical partial differential equations and optimization : proceedings of the fifth Mexico-UnitedStates Workshop / edited by S. Gómez, J. P. Hennart, R. A. Tapia.
      p. cm.
    "Proceedings of the Fifth Mexico-United States Workshop on Advances in Numerical Partial Differential Equations and Optimization, Mérida, Mexico, 1989" --T.p. verso.
    Includes bibliographical references and index.
    ISBN 0-89871-269-6
    1. Differential equations, Partial--Numerical solutions--Congresses.  2. Mathematical optimization--Congresses.  I. Gómez, S. (Susana) II. Hennart, J. P. (Jean Pierre), 1942-   .
III. Tapia, Richard A.  IV. Title.
QA377.M49 1989                                          91-658
515'.353--dc20                                             CIP

All rights reserved. Printed in the United States of America. No part of this book may be reproduced, stored of transmitted in any manner without the written permission of the Publisher. For information, write the Society for Industrial and Applied Mathematics, 3600 University City Science Center, Philadelphia, Pa 19104-2688.

Copyright © 1991 by the Society for Industrial and Applied Mathematics.

# PREFACE

The Fifth Mexico-United States Workshop on Numerical Analysis was held January 26, 1989, in Merida, Yucatan, México. It was organized by the Institute for Research in Applied Mathematics of the National University of Mexico in collaboration with the Mathematical Sciences Department at Rice University, as were the previous workshops in 1978, 1979, 1981 and 1984. The program of this research workshop concentrated on the numerical aspects of three main areas, namely optimization, linear algebra, and differential equations, both ordinary and partial.

As were the third and fourth workshops, this one was supported by a generous grant from the Mexican National Council for Science and Technology and the U.S. National Science Foundation and was part of the Joint Scientific and Technical Cooperation Program existing between these two countries.

S. Gómez
Instituto de Investigaciones en Matematicas Aplicadas Y En Sistemas,
Universidad Nacional Autónoma de México, México.

J. P. Hennart
Instituto de Investigaciones en Matematicas Aplicadas Y En Sistemas,
Universidad Nacional Autónoma de México, México.

R. A. Tapia
Rice University, Houston, Texas

# LIST OF CONTRIBUTORS

**Gonzalo Alduncin,** Instituto de Geofisica, Universidad Nacional Autónoma de México, 04510-Mexico, D.F., México.

**G. D. Allen,** Department of Mathematics, Texas A&M University, College Station, TX 77843.

**Filippo Aluffi-Pentini,** Dipartmento di modelli Matematici per le scienze applicate Università di Roma "La Sapienza"-00161 Roma-Italy.

**Robin Anderson,** Numertrix, Ltd., Toronto, Ontario, Canada.

**S. Barrera P.,** Facultad de Ciencias, Universidad Nacional Autónoma de México, México.

**Emanuele Caglioti,** Istituto Nazionale di Alta Matematica "F.Severi" Piazzale Aldo Moro 5-00185 Roma-Italy.

**Gilberto Calvillo-Vives,** Instituto Politécnico Nacional and Banco de México.

**Jorge Carrera,** Instituto de Geofisica, Universidad Nacional Autónoma de México, 04510-Mexico, D.F., México.

**J. L. Castellanos N.,** ICIMAF, Academia de Ciencias, Cuba.

**Andrew R. Conn,** Department of Combinatorics and Optimization, University of Waterloo, Waterloo, Ontario N2L 3G1, Canada.

**J. D. F. Cosgrove,** Cameron University, Lawton, OK.

**Jane K. Cullum,** Mathematical Sciences Dept., IBM Research Division, T. J. Watson Research Center, Yorktown Heights, NY 10598.

**E. Del Valle,** Departamento de Ingeniería Nuclear, Escuela Superior de Física y Matemáticas del IPN, Unidad Profesional "Adolfo López Mateos", 07738 México, D. F., México.

**Ron S. Dembo,** Algorithms, Inc., Toronto, Ontario, Canada and the University of Toronto, Canada.

**J. C. Díaz,** Center for Parallel and Scientific Computing, University of Tulsa, Tulsa, OK.

**Bjorn Enquist,** Department of Mathematics, University of California, Los Angeles, CA 90024.

**R. E. Ewing,** Department of Mathematics, Chemical Engineering, and Petroleum Engineering, University of Wyoming, Laramie, Wyoming 82071.

**David M. Gay,** AT&T Bell Laboratories, Murray Hill, NJ 07974-2070.

**Cristina Gígola,** Departmento de Matemáticas, Escuela Superior de Físcia y Matemáticas, COFAA, IPN, México.

**Susana Gómez,** Instituto de Investigaciones en Matemáticas Aplicadas y en Sistemas, Universidad Nacional Autónoma de México, México.

**John D. Gonglewski,** U. S. Air Force's Weapons Laboratory, Optical Phased Array Branch, Kirtland Air Force Base, New Mexico 87117.

**Raphael T. Haftka,** Department of Aerospace and Ocean Engineering, Virginia Polytechnic Institute & State University, Blacksburg, VA 24061.

**J. P. Hennart,** Departamento de Métodos Matemáticos y Numéricos, Instituto de Investigaciones en Matemáticas Aplicadas y en Sistemas de la Universidad Nacional Autónoma de México, Apartado Postal 20-726,01000 México, D. F., México.

**Quyen Quang Huynh,** Department of Computer Sciences, Uppsala University, Sturegatan 4B 2TR, 75223 Uppsala, Sweden.

**P. G. Jacobs,** Department of Mathematics, University of Wyoming, Laramie, Wyoming 82071.

**Yuying Li,** Computer Science Department, Cornell University, Upson Hall, Ithaca, NY, 14853.

**Guillermo López-Mayo,** Universidad Autónoma de Puebla.

**Frederick H. Lutze,** Department of Aerospace and Ocean Engineering, Virginia Polytechnic Institute & State University, Blacksburg, VA 24061.

**C. G. Macedo, Jr.,** Center for Parallel and Scientific Computing, University of Tulsa, Tulsa, OK.

**Luciano Misici,** Dipartimento di Matematica e Fisica, Universitià di Camerino- 62032 Camerino(MC)-Italy.

**P. Nelson, Jr.,** Department of Nuclear Engineering Texas A&M University, College Station, TX 77843

**Alfredo Nicolás-Carrizosa,** Sección de Graduados, ESIME, IPN, México. Edif. 8, U.P. Zacatenco, México 14, D.F., México.

**R. B. Ojeda C.,** Facultad de Ciencias, Universidad Naional Autónoma de México, México.

**R. R. Parashkevov,** Department of Mathematics, University of Wyoming, Laramie, Wyoming 82071.

**A. Perez D.,** ICIMAF, Academia de Ciencias, Cuba.

**Raymond H. Plaut,** Department of Civil Engineering, Virginia Polytechnic Institute & State University, Blacksburg, VA 24061.

**R. T. Rockafellar,** Department of Mathematics, University of Washington GN-50, Seattle, WA 98195.

**D. Romero,** Instituto de Investigaciones en Matemáticas Aplicadas y en Sistemas, Universidad Nacional Autónoma de México, Av. Universidad s/n, Cuernavaca, Mor.(México).

**A. Sánchez-Flores,** Instituto de Investigaciones en Matemáticas Aplicadas y en Sistemas, Universidad Nacional Autónoma de México, Av. Universidad s/n, Cuernavaca, Mor. (México).

**J. Shen,** Department of Mathematics, University of Wyoming, Laramie, Wyoming 82071.

**Philip Y. Shin,** Department of Mechanical Engineering, Naval Postgraduate School, Monterey, CA 93940.

**J. Valdés,** Gerencia de Tecnología Reglamentación y Servicios, Comisión Nacional de Seguridad Nuclear y Savaguardias, 01030, México, D. F., México. Present address, Depto. de Energía Nuclear, Instituto de Investigaciones Eléctricas, Interior del Internado Palmira s/n, 62000 Cuernavaca, Morelos, (México).

**Homer F. Walker,** Department of Mathematics and Statistics, Utah State University, Logan, UT 84332-3900.

**Layne T. Watson,** Department of Computer Science, Virginia Polytechnic Institute & State University, Blacksburg, VA 24061

**Ralph A. Willoughby,** Mathematical Sciences Department, IBM Research Division, T. J. Watson Research Center, Yorktown Heights, NY 10598.

**Francesco Zurilli,** Dipartimento di Matematica "G. Castelnuovo" Università di Roma "La Sapienza"-00185 Roma-Italy.

# CONTENTS

1    An Efficient Linesearch for Convex Piecewise-Linear/Quadratic Functions
Ron S. Dembo and Robin Anderson

9    The One-Dimensional Space Allocation Problem: A Simulated Annealing Approach
D. Romero and A. Sánchez-Flores

17    Stopping Tests That Compute Optimal Solutions for Interior-Point Linear Programming Algorithms
David M. Gay

43    Polyhedral Invariants and a Representative Theorem in Projective Geometry
Gilberto Calvillo-Vives and Guillermo López-Mayo

50    Towards a Theory of Optimization in Projective Space
Gilberto Calvillo-Vives

60    Iterative Gradient-Newton Type Methods for Steady Shock Computations
Bjorn Enquist and Quyen Quang Huynh

76    Applications of Adaptive Grid-Refinement Methods
R. E. Ewing, P. G. Jacobs, R. R. Parashkevov, and J. Shen

101    Approximate Inverse Preconditioning for Nonsymmetric Sparse Systems
J. D. F. Cosgrove, J. C. Díaz, and C. G. Macedo, Jr.

112    On Generalized Finite Difference Methods for Approximating Solutions to Integral Equations
G. D. Allen and P. Nelson, Jr.

141    Discrete-Ordinates Equations in Slab Geometry: A Mixed-Hybrid Nodal Version of the Diffusion Synthetic Acceleration Method
E. Del Valle, J. P. Hennart, and J. Valdés

158    Upwind Finite Element Approximations of the Advection-Diffusion Problem
Gonzalo Alduncin and Jorge Carrera

171    Fast Elliptic Solvers Using Mixed-Hybrid Nodal Finite Elements
J. P. Hennart and E. Del Valle

185    A New Discrete Functional for Grid Generation
P. Barrera S., J. L. Castellanos N., R. B. Ojeda C., and A. Perez D.

198  Computing Eigenvalues of Large Matrices, Some Lanczos Algorithms and a Shift and Invert Strategy
Jane K. Cullum and Ralph A. Willoughby

247  Large-Scale Extended Linear-Quadratic Programming and Multistage Optimization
R. T. Rockafellar

262  A Finite Difference Approach to the Kuramoto-Sivashinsky Equation
Alfredo Nicolás-Carrizosa

273  A Numerical Method for the Three Dimensional Inverse Acoustic Scattering Problem with Incomplete Data
Filllippo Aluffi-Pentini, Emanuele Caglioti, Luciano Misici, and Francesco Zurilli

284  The Application of Globally Convergent Homotopy Methods to Nonlinear Optimization
Layne T. Watson, Raphael T. Haftka, Frederick H. Lutze, Raymond H. Plautt, and Philip Y. Shin

299  Relation Between the Regularization and the Multipliers Method for the Min-Max Problem
Cristina Gígola and Susana Gómez

320  Two Second Order Regularization Methods to Solve the Finite Min-Max Problem
Cristina Gígola and Susana Gómez

332  Quasi-Newton Methods for Minimum-Likelihood Estimation
John D. Gonglewski and Homer F. Walker

346  An Approach to Nonlinear $l_\infty$ Approximation
Andrew R. Conn and Yuying Li

# CHAPTER 1

An Efficient Linesearch for Convex Piecewise-Linear/Quadratic Functions

Ron S. Dembo*
Robin Anderson†

## Introduction

Functions that are piecewise linear/quadratic arise in optimization in a number of ways; for example, through the use of smoothed exact penalty functions[1], in statistics[2] and in stochastic optimization[3]. These functions, although smooth and continuously differentiable, pose significant problems to off-the-shelf optimization techniques. They often behave more like nondifferentiable functions than the continuous functions supposed by standard optimizers. This is because a single function may contain segments that are very small with high curvature and others that are large with low curvature. Thus a given function may either be very well or very poorly conditioned depending on the range under consideration. This can cause havoc in existing codes since they are based on methods that presuppose more predictable behavior.

Consider the following problem:

$$\text{Minimize} \quad f(x)$$

where  f: $R^n \rightarrow$ R  is a continuously-differentiable piecewise linear/quadratic function. We use  g(x) to denote the gradient of f(.) at x.

Since this is a smooth problem one may wish to apply standard methods to solve it. Most methods would proceed as follows:

---

*Algorithms, Inc., Toronto, Ontario, Canada and the University of Toronto, Canada.
†Numertrix, Ltd., Toronto, Ontario, Canada.

(*Optimality Criterion*)　　　WHILE x is not optimal　DO:

(*Search Direction Calculation*) Find a direction d such that $g(x)^T d < 0$;

(*Linesearch*)　　　　　　Minimize$_\alpha$　$f(x + \alpha d)$

(*Update*)　　　　　　　　$x \leftarrow x + \alpha^* d$

In this paper we focus on efficient methods for computing an exact solution to the linesearch problem. Under mild conditions on the direction d (see [4] for example), algorithms conforming to such a framework will then converge to a local minimum of the problem from an arbitrary starting point (global convergence).

The search direction subproblem itself may pose significant difficulties because of the potential radical changes in the conditioning of $f(x)$ over small intervals. An efficient procedure for treating this subproblem in the piecewise linear/quadratic case is still an open research question and will be examined in a subsequent paper.

## The Linesearch Subproblem

Before describing our method we need to make an additional restrictive assumption:

$f(x)$ *is continuously differentiable and convex.*

The nonconvex case is extremely interesting because of the wide variety of functions that may be approximated using combinations of piecewise concave and convex quadratics. This subject however is beyond the scope of this paper. Dropping the continuous differentiability assumption also leads to an important class of problems. The approach described in this paper is a precursor to methods for the nondifferentiable case.

We have therefore limited this study to the one dimensional, convex, continuously differentiable, piecewise linear/quadratic linesearch problem. This problem in its own right may pose many difficuties that are sure to confound existing methods. There may be thousands of segments of varying sizes whose boundaries are not known a priori. We only know whether or not the function is linear or quadratic by evaluating it at the point in question. The range over which it remains linear or quadratic is either given or may be computed.

## The Linesearch Subproblem

For the special case we are considering, it is possible to characterize the nature of the neigborhood of an optimal solution.

*Proposition :*   *An optimal solution occurs at the minimum of a quadratic segment.*

This observation is a consequence of the assumed convexity and differentiability of f(x).

As a general principle therefore, algorithms for this problem will work best in a neighborhood of a solution if they are based on a quadratic model. However, as is commonly the case in optimization, the neighborhood is problem specific and usually cannot be determined a priori. Using a quadratic model when far from a neighborhood of the solution may result in slow convergence. A good algorithm should be able to adapt to the nature of the function as can best be judged fron local information.

Our method attempts to build up a coarse approximation to the function based on local information. In general, when far from a minimum we view the function as piecewise linear. At each stage we assume that we know the function values at two distinct points that bracket the minimum. The *left hand point*, $\alpha_L$, has a negative gradient and the gradient at the *right hand point*, $\alpha_R$, is positive. We say that the left hand point is *quadratic/linear* if it lies on a quadratic/linear segment. By *the left hand line* we mean the line corresponding to the linear segment when the left point is linear. Similarly, *the right hand quadratic* corresponds to the quadratic approximation at a right hand point that is quadratic.

The linesearch algorithm has the following general form:

```
START with  [ α_L, α_R ]  bracketing the minimum.

COMPUTE a new approximation to the minimum  =  α̂;

WHILE    ∇_α f( x  +  α̂ d )  ≠  0

         If    ∇_α f( x  +  α̂ d )  >  0 ,   α_R  ←  α̂
         If    ∇_α f( x  +  α̂ d )  <  0 ,   α_L  ←  α̂ ;
```

## Computing a new approximation to the minimum, $\hat{\alpha}$

$\boxed{\text{CASE I}}$ *The left and right hand points are both linear;*

The function is approximated as a piecewise linear function and the next approximation to the minimum is taken to be the intersection of the left and right hand lines.

$\boxed{\text{CASE II}}$ *The left hand point is linear and the right hand point is quadratic (or vice versa);*

We know that the current left and right hand points are not in the neighborhood of a minimum. We therefore continue to treat the problem as if it were piecewise linear and our next approximation point is the intersection of the linear approximations at the two points.

An alternative (when the Hessian of the quadratic is available) is to compute the minimum of the right hand quadratic (approximated at the right hand point ) and, if it is to the right of the current left hand point, accept it as our new approximating point. Otherwise, take the intersection of the right hand quadratic and the left hand linear function as the new point.

The alternative involves more work but is likely to be a better approximation to the underlying piecewise quadratic function. A similar strategy could be followed if the left hand point was quadratic and the right hand point linear or if both were quadratic.

When the Hessian is not available or is expensive to compute relative to the gradient, then left and right hand linear approximations are used until both left and right hand points are quadratic (CASE III).

$\boxed{\text{CASE III}}$ *The left and right hand points are both quadratic;*

Here we have reason to suspect that we have bracketed the quadratic segment on which the minimum lies. We therefore fit a quadratic through the two points which will be exact in a neighborhood of the solution;

## An Application

An important specially-structured linear programming problem for which no exceptional algorithm exists is the linear network optimization problem with sideconstraints. Network linear optimization problems are solved extremely efficiently by specializations of the Simplex method. However, if only a few sideconstraints are added the cost of solution goes up substantially.

In [1] Dembo has proposed solving such problems using an algorithm based on successive smoothing of an exact penalty function. The formulation and the algorithm may be summarized as follows.

### Problem Definition

$$\underset{x \in X}{\text{Minimize}} \quad c^T x + \mu \sum_i | \sum_j s_{ij} x_j - \bar{s}_i |$$

where  $X \equiv \{ x \mid A x = b \; ; \; l \leq x \leq u \}$;
$\mu$ is a large penalty.

### Algorithm

WHILE not optimal DO:

(Search Dir.)  $\underset{x \in X}{\text{Minimize}} \quad c^T x + \mu \sum_i | \sum_j s_{ij} x_j - \bar{s}_i |_\epsilon$

$$\text{where } | y |_\epsilon \equiv \begin{cases} (y - \epsilon/2) & \text{if } y > \epsilon \\ (-y - \epsilon/2) & \text{if } y < -\epsilon \\ (y^2/2\epsilon) & \text{if } -\epsilon \leq y \leq \epsilon; \end{cases}$$

Let $\hat{x}$ be the solution;
Define  $d \equiv (\hat{x} - x)$;
Let $\bar{\alpha}$ be the maximum value of $\alpha$ such that $x + \alpha d \in X$;

(Linesearch)  $\underset{0 \leq \alpha \leq \bar{\alpha}}{\text{Minimize}} \quad c^T(x + \alpha d) + \mu \sum_i | \sum_j s_{ij}(x_j + \alpha_j) - \bar{s}_i |_\epsilon$

(Update)  $x \leftarrow x + \alpha^* d$ ;
reduce $\epsilon$;

## Computational Results

We have implemented the above algorithm for problems arising from short-term hydroelectric power scheduling[ 4 ]. These problems all have the form of linear network flow problems with roughly 20% sideconstraints. In practice they are all solved with standard linear programming codes. Even for such codes the sideconstraints add significantly to the solution time. For example, in experiments run using MINOS[ 5 ], the solution to the problem with sideconstraints required more than three times the computation time required for the same models without sideconstraints.

For the experiments considered here we studied the behavior of the linesearch, in the context of the algorithm described in the previous section, for the following three problems.

| Problem # Sideconstraints | Bounded Variables | Equality | Constraints |
|---|---|---|---|
| 1 (3-period) | 81 | 30 | 6 |
| 2 (24-period) | 696 | 240 | 48 |
| 3 (48-period) | 1398 | 480 | 96 |

The 24 and 48 period problems were 1 and 2 day horizon hydroelectric scheduling models with real data. The 3-period problem is derived from these problems for purposes of presentation.

The number of possible segments in any linesearch for these problems is given by

$$( \text{\# sideconstraints} * 3 )$$

since there are three segments corresponding to each penalty term. Thus it is clear that each linesearch in problem 3 faces a formidable task, namely it has to sort through 288 segments to find an exact solution.

definition of test problems , actual algorithm tested, numerical results, code, graphical results

### Numerical Results

A summary of the numerical results for problem 3 (288 segments) illustrates the typical behaviour of our linesearch algorithm. Note that in each case we find an exact minimum along the piecewise-linear-quadratic line. The linesearch number refers to the major iteration number of the smoothed exact-penalty algorithm described above (ie at each iteration an exact linesearch on a function with 288 segments is performed).

## Computational Results

| Linesearch Number | Number of trials (= no. of function evaluations) |
|---|---|
| 1 | 5 |
| 2 | 13 |
| 3 | 3 |
| 4 | 11 |
| 5 | 13 |
| 6 | 9 |
| 7 | 13 |
| 8 | 7 |
| 9 | 11 |
| 10 | 9 |

It is important to place these computational results in perspective. Firstly, it appears as if one can compute an exact minimum along the line in a number of steps (function evaluations) which is far less than the number of segments along the line. Secondly, an exact linesearch is performed at each major iteration in order to obtain an estimate of the Lagrange multipliers corresponding to the sideconstraints[ 1 ]. This information becomes particularly important for evaluating termination conditions and highlights one of the particularly important tradeoffs that are possible in the above algorithm for networks with sideconstraints. Either one does a crude linesearch at each iteration and then expends additional work on computing Lagrange multiplier estimates or one does an exact linesearch and gets the Lagrange multiplier estimates "free". The trade-off is the additional work done to do an exact linesearch versus the cost of computing Lagrange multiplier estimates. In the example above the cost of doing the exact linesearch appears to be minimal when compared with (say) a first-order least-squares estimate of the multipliers.

## An alternative linesearch

An important alternative in the smoothing algorithm is to use a projection method to maintain feasibility with respect to the box constraints, $l \leq x \leq u$, instead of limiting the stepsize as specified above.

Let $\bar{\alpha}$ be the maximum stepsize satisfying $A*(x+\alpha d) = b$;

Define $x(\alpha) \equiv [x + \alpha d]^{\#}$ where $[y]^{\#}$ is the projection of y onto $l \leq y \leq u$.

The projection linesearch below will then replace the linesearch in the algorithm described above.

$$(Linesearch) \quad \underset{0 \leq \alpha \leq \bar{\alpha}}{\text{Minimize}} \quad c^T(x(\alpha)) + \mu \sum_i |\sum_j s_{ij}(x_j(\alpha)) - \bar{s}_i|_\epsilon$$

We have not yet tested this alternative.

## References

1. R. S. Dembo, *Smoothed exact penalty algorithms for large-scale nonlinear optimization*, presented at the 13th International Mathematical Programming Symposium, Tokyo, August 1988.

2. R. T. Rockafellar and R. J-B. Wets, *A Lagrangian finite generation technique for solving linear-quadratic problems in stochastic programming*, Mathematical Programming Study 28(1986)63-93.

3. J. E. Dennis and R. Schnable, Numerical Methods for Unconstrained Optimization and Nonlinear Equations, Prentice-Hall Series in Computational Mathematics, 1983.

4. R. S. Dembo, A. Chiarri, J. Gomez and L. Paradinas, *Managing Hidroeléctrica Española's Hydroelectric Power System*, to appear in *INTERFACES* 20:1 January-February 1990, pp 115-135.

5. B. A. Murtagh and M. A. Saunders, "*MINOS 5.1 User's Guide*," 1987. Report SOL 77-9, Department of Operations Research, Stanford University, Stanford, California.

# CHAPTER 2

The One-Dimensional Space Allocation Problem:
A Simulated Annealing Approach

D. Romero*
A. Sánchez-Flores*

**Abstract.** Simulated annealing has recently proved to be a suitable approach to obtain approximate solutions of difficult optimization problems. In this paper we describe the implementation of heuristic methods incorporating in various ways the simulated annealing technique in order to solve the one-dimensional space allocation problem, an *NP*–complete combinatorial optimization problem where a linear ordering of $n$ objects is sought, and for which no general exact method has ever solved instances with $n \geq 20$.

**1. Introduction.** In the realm of combinatorial optimization, the one-dimensional space allocation problem (ODSAP) stands as a redoubtable one: given a set of $n$ one-dimensional objects of different length, a linear ordering of the objects is sought that minimizes the weighted distances between every pair of objects. As an example, suppose that the physical arrangement of rectangular rooms along one side of a corridor is to be designed (Beghin-Picavet and Hansen (1982), and Chan and Francis (1979)). The rooms have the same width, but not necessarily the same lengths, which are known. Suppose known, further, the number of people per unit of time, going from one room to another. Which is the sequence of the rooms that minimizes the total distance traveled by people?

A situation alike might arise in placement of files on a computer storage device (Grossman and Silverman (1973), Pratt (1972), and Troya and Vaquero (1981)), where we want to minimize the average seek time, or in classification of items along one scale, like books (Romero (1978)) or phonemes (Doyle (1988)), where an *similarity* degree between every pair of items is provided, and we want a linear ordering of the items such that similar items are "close" from each other.

---

*Instituto de Investigaciones en Matemáticas Aplicadas y en Sistemas, Universidad Nacional Autónoma de México, Av. Universidad s/n, Cuernavaca, Mor. (MEXICO)

The ODSAP is NP–complete (Garey et al. (1976)), thus a "good" (polynomial) algorithm for its optimal solution is very improbable to be found. Moreover, unlike some NP–complete problems (i.e., the traveling salesman problem), for which there exist algorithms to solve medium size instances (Padberg and Rinaldi (1987)), the practical scope of known, general exact algorithms for the ODSAP, like branch-and-bound (Simmons (1969)), or dynamic programming (Beghin-Picavet and Hansen (1982), and Picard and Queyranne (1981)), has never gone beyond $n = 18$. Ad hoc heuristics to deal with bigger problems are scarce: the Troya and Vaquero (1981) procedure behaves well with randomly generated data, but in some circumstances its application could seriously go astray; finally, Picard and Queyranne (1981) have suggested the combination of elementary interchange, swapping and insertion procedures. Despite the pessimism contributed by both theory and practice, a few polynomially solvable cases of the ODSAP have been found, such as the so-called linear ordering problem with independent destinations (Bergmans (1972), and Pratt (1972)), the all prominent matrix case (Romero (1978)), and the identical similarities case (Chan and Francis (1979), and Picard and Queyranne (1978)). On the other hand, simulated annealing (Černý (1985), and Kirkpatrick et al. (1983)) has recently proved to be a suitable approach to obtain approximate solutions of tough combinatorial optimization problems, like coloring graphs (Chams et al. (1987)), quadratic assigment (Burkard and Rendl (1984)), protein folding (Bohr and Brunak (1989)), and graph bipartition (Banavar et al. (1987)).

In this article we describe the development and implementation of heuristic methods that incorporate the simulated annealing concept in various ways. After introducing notation and formulation of the ODSAP in the next section, the simulated annealing technique is presented in section 3. The aim of section 4 is to report our computational experience. To end, our conclusions can be found in the last section.

**2. Notation and Formulation.** Let $N$ be a set of one-dimensional objects $1, \ldots, n$, with lengths $\ell_1, \ldots, \ell_n$, respectively. The symmetric matrix $A = \{A_{ij}\}$, with $A_{ij} \in \Re$ and $A_{ii} = 0$, for all $i, j \in N$, will be referred to as the *similarity matrix* of $N$. Let $\sigma = (\sigma_1, \ldots, \sigma_n)$ be an element of the set $P$ of permutations on $N$, and let $\sigma^{-1} = (\sigma_1^{-1}, \ldots, \sigma_n^{-1})$ denote the inverse of $\sigma$, i.e., $\sigma_j^{-1} = k \iff \sigma_k = j$. When the objects are juxtaposed according to $\sigma$, i,e., $\sigma_1\sigma_2\sigma_3\ldots\sigma_n$, the cost is $C(\sigma) = \sum_{i<j} A_{ij} D_{ij}(\sigma)$, where $D_{ij}(\sigma)$ is the sum of the lengths of the objects that lay between $i$ and $j$, i.e., if $\sigma_i^{-1} < \sigma_j^{-1}$ then $D_{ij}(\sigma) = \sum_{k=\sigma_i^{-1}+1}^{\sigma_j^{-1}-1} \ell_{\sigma_k}$, otherwise $D_{ij}(\sigma) = D_{ji}(\sigma)$. The ODSAP is then to find a minimum cost permutation in $P$.

When all the lengths are equal, the ODSAP is called the linear ordering problem (Adolphson and Hu (1973)), which is an NP–complete case (Garey et al. (1976)) of the quadratic assignment problem (Lawler (1963)), and whose name is shared (alas!) by a different combinatorial optimization problem (Grotschel et al. (1984)).

**3. Simulated Annealing.** In order to apply the simulated annealing concept to the ODSAP, we associate to every $p \in P$ a predefined *neighborhood* $H(p) \subset P$. For instance, $H(p)$ might be defined at the set of all solutions obtainable from

$p$ by interchanging any two objects. The method starts with a randomly chosen $p \in P$ and with a variable $t$ (called *temperature*) initialized to $t_0$. Given a current solution $p$, a *neighbor* $q \in H(p)$ is generated. If $\Delta := C(q) - C(p)$ is negative or if $x < exp(-\Delta/t)$ ($x$ is a random number uniformly distributed in $(0,1)$), then $q$ is accepted becoming the new current solution, otherwise no change is made. The process is iterated until a *steady state* is detected, then $t$ is decreased multiplying it by a positive *cooling factor* $\varphi < 1$ and the whole process is started anew. The algorithm stops when no changes occur during a complete cycle at constant $t$: a *local minimum* with respect to $H(p)$ has been reached.

Of course, the performance of an actual heuristic method based on this general simulated annealing scheme, will heavily depend on the selected cooling strategy and on the neighborhood definition.

**4. Implementation.** We have essentially tested variations of two general methods, $A$ and $B$, each with diverse strategies using integers $y, z \in N$ to obtain a neighbor $q$ of $p$, yielding a total of 56 different heuristics. With "interchange strategy" ($\alpha$), permutation $q$ is obtained from $p$ by interchanging the objects in positions $y$ and $z$. With "insertion-from strategy" ($\beta$), the object in position $z$ goes to position $y$ and objects between positions $y$ and $z$ (including the object in position $y$) are shifted one place towards position $z$. "Insertion-to strategy" ($\gamma$) is the same as $\beta$ with the role of positions $y$ and $z$ interchanged. Actually, $\beta$ and $\gamma$ strategies are different because $y$ consecutively takes all values from 1 to $n$, whereas $z$ is randomly generated. Finally, also the combined strategies $\alpha\beta$, $\alpha\gamma$, $\beta\gamma$ and $\alpha\beta\gamma$, were implemented, which essentially consist in alternating the simple strategies composing them.

METHOD $A$. A steady state test is made every time a set $U$ of $n$ accepted solutions is formed. Let $U_i$ be the last formed set of $n$ accepted solutions, and let $V$ be the average cost of the set $U_{i-1}$ of $n$ accepted solutions formed just before. A steady state has been attained whenever $||G| - |H|| < n/3$, where $G = \{p \mid p \in U_i, C(p) > V\}$, and $H = \{p \mid p \in U_i, C(p) < V\}$; in this case, the temperature is decreased multiplying it by $\varphi$. At a given iteration, if $j$ is the accumulated number of generated solutions, then $y = 1 + j \bmod N$. To generate $z$ this method uses the parameter $\xi$ and the *radius* $r$. The integer $z$ is then a uniform random number such that $|z - y| < r$. Let $\rho$ be the total availabe time for a run. The radius $r$ is initialized to $n$ and maintains this value for a time slot equal to $\xi\rho$; then $r$ is reduced to a half, maintaining this value for a time slot equal to the remaining available time multiplied by $\xi$, i.e., $\xi(\rho - \xi\rho)$; again the last calculated $r$ is reduced to a half, and maintains this value for a time slot equal to the remaining available time multiplied by $\xi$; and so on. Note thas as $r$ decreases the search becomes more "local". When implementing combined strategies, the strategy in use changes every time a solution is generated.

METHOD $B$. This method somewhat varies with method $A$. There is not a steady state test, as such. Instead, every $3n$ iterations the temperature is updated in a regulatory way, with possible increments. Consider the $i$th temperature updating with the same radius $r$. Let $T_r$ be the time slot available for this radius, and let $Y_i$

be the number of accepted solutions since the last temperature updating such that they could not be generated with the next and smaller value of $r$. Temperature then is controlled in such a way that $Y_i$ decreases linearly with respect to time having a value of zero at the end of the time slot, i.e., $t := t(1 + w/T_r - Y_i/Y_1)$, where $w$ is the remaining available time for radius $r$.

In order to improve the final solution in both methods we devised the 3-LOC local improvement procedure, with neighborhood defined as: $q \in H(p) \iff q$ is obtainable from $p$ by freely permuting any 3 *consecutive* objects.

For $n = 20, 30, 40, 50$, the heuristics were tested on five similarity matrices: the entries in the first two matrices were randomly generated in $(0, n)$ and $(0, n^2)$, respectively; the third matrix was euclidean, and the last two matrices were sparse (around 20% density). The length vectors were chosen trying not to follow any pattern, with integer entries between 1 and $n$. The time allowed to each run was a fixed quantity $\rho_n$.

Four variations of methods $A$ and $B$ were considered. The first three correspond to the $\xi$ values 0.6, 0.85, and 1.0 (denoted as $A_{0.6}$, $A_{0.85}$, $A_{1.0}$, $B_{0.6}$, $B_{0.85}$, and $B_{1.0}$, respectively). The fourth variation, $A_r$ ($B_r$), consist in running three times $A_{1.0}$ ($B_{1.0}$), each for a time equal to $\rho_n/3$, and taking the best solution found. In all variations of method $A$, $\varphi = 0.9$ was used. XT and AT models of a Printaform microcomputer, at their maximum velocity (10MHz), were used for $n = 20, 30$, and $n = 40, 50$, respectively. As our initial experiments showed that problems defined on sparse similarity matrices were more difficult, we studied them separately. In table I, which corresponds to non-sparse similarity matrices, the entries are obtained as follows. Each of the 56 heuristics was run 8 times on each of the matrices of sizes 20, 30 and 40, and 6 times for $n = 50$. From these 8 (6) runs we obtained the average cost and the percent error of this average with respect to the cost of the best known solution. Then, for each heuristic the corresponding entry is the average of the percent error of the twelve non-sparse matrices.

| METHOD | $\alpha$ | $\beta$ | $\gamma$ | $\alpha\beta$ | $\alpha\gamma$ | $\beta\gamma$ | $\alpha\beta\gamma$ | AVERAGE |
|---|---|---|---|---|---|---|---|---|
| $A_{0.6}$ | 1.50 | 1.00 | 0.89 | 1.26 | 1.02 | 1.13 | 1.18 | 1.14 |
| $A_{0.85}$ | 1.20 | 0.67 | 0.58 | 0.79 | 0.80 | 0.59 | 0.73 | 0.77 |
| $A_{1.0}$ | 0.72 | 0.47 | 0.35 | 0.59 | 0.58 | 0.48 | 0.56 | 0.54 |
| $A_r$ | 1.38 | 0.91 | 0.86 | 1.00 | 1.24 | 0.86 | 1.05 | 1.04 |
| $B_{0.6}$ | 0.63 | 0.45 | 0.59 | 0.64 | 0.57 | 0.41 | 0.55 | 0.55 |
| $B_{0.85}$ | 0.80 | 0.44 | 0.37 | 0.50 | 0.51 | 0.43 | 0.53 | 0.51 |
| $B_{1.0}$ | 0.69 | 0.40 | 0.35 | 0.46 | 0.43 | 0.44 | 0.39 | 0.45 |
| $B_r$ | 0.61 | 0.37 | 0.31 | 0.42 | 0.42 | 0.34 | 0.38 | 0.41 |

TABLE I. Percent error in non-sparse matrices. $\rho_{20} = 2$min, $\rho_{30} = 3$min, $\rho_{40} = 4$min, $\rho_{50} = 6$min.

Globally, method $B$ looked superior to method $A$. Among the variations of method $A$, $A_{1.0}$ was the only to behave well. Simulated annealing with strategy $\alpha$ had a poor performance. Nevertheless, when all the lengths are equal this strategy would become competitive, because of the easiness to compute the cost increment. In general, the best heuristics made good usage of 3-LOC; table II shows the percent error without using 3-LOC in the best heuristics of table I.

| $A_{1.0} - \gamma$ | $B_{1.0} - \gamma$ | $B_r - \gamma$ | $B_r - \beta\gamma$ |
|---|---|---|---|
| 1.82 | 1.37 | 1.93 | 2.24 |

TABLE II. Percent error of best strategies in table I without using 3-LOC.

Table III shows the performance of the best heuristics (BH) for varying $n$. Note that we have incorporated results of tests with $n = 60$. All the entries come from the same microcomputer *(AT)*.

| $n$ | 20 | 30 | 40 | 50 | 60 |
|---|---|---|---|---|---|
| BH | $B_r - \gamma$ | $A_{1.0} - \gamma$ | $B_{1.0} - \gamma$ | $B_{1.0} - \gamma$ | $B_{1.0} - \gamma$ |
| ERROR | 0.15 | 0.18 | 0.28 | 0.33 | 0.40 |

TABLE III. Percent error of the best heuristics. $\rho_{20} = 1\text{min } 47\text{sec}$, $\rho_{30} = 2\text{min } 40\text{sec}$, $\rho_{40} = 4\text{min}$, $\rho_{50} = 6\text{min}$, $\rho_{60} = 9\text{min}$.

We have also applied the best heuristics of each method to special cases, obtaining low errors from the true optimal solutions. Table IV shows the average results of three independent destinations cases. Our tests on 40-by-40 prominent matrices always obtained the exact optimal solutions.

| $A_{1.0} - \gamma$ | $B_r - \gamma$ | $B_{0.85} - \gamma$ | $B_{1.0} - \gamma$ |
|---|---|---|---|
| 0.13 | 0.06 | 0.07 | 0.05 |

TABLE IV. Percent error on independent destinations cases. $n = 50$.

Regarding sparse similarity matrices, the values of $\rho_n$ were $\rho_{20} = 4\text{min}$, $\rho_{30} = 6\text{min}$, $\rho_{40} = 8\text{min}$, $\rho_{50} = 12\text{min}$. Table V was computed very much like table I. Each heuristic was run 8, 7, 6, and 5 times on each of the matrices of size 20, 30, 40, and 50, respectively.

| METHOD | $\beta$ | $\gamma$ | $\alpha\beta$ | $\alpha\gamma$ | $\beta\gamma$ | $\alpha\beta\gamma$ |
|---|---|---|---|---|---|---|
| $A_{0.6}$ | | | 2.58 | 1.69 | 2.50 | 2.62 |
| $A_{0.85}$ | | 1.80 | 2.79 | | 2.39 | |
| $A_{1.0}$ | 1.40 | | 2.29 | | | 1.95 |
| $B_{0.85}$ | 2.43 | | | | 1.98 | |
| $B_r$ | | 1.95 | | | | 3.13 |

TABLE V. Percent error in sparse matrices. $\rho_{20} = $ 4min, $\rho_{30} = $ 6min, $\rho_{40} = $ 8min, $\rho_{50} = $ 12min.

From our initial results we found useless to compute all the entries, hence we worked only on the most promising heuristics. Note that although the allowed computing time doubles the allowed time for non-sparse matrices, the errors are much greater. Outstanding combinations to deal with sparsity are $A_{1.0} - \beta$, and $A_{0.6} - \alpha\gamma$, although other heuristics are also competitive. Globally, methods $A$ and $B$ are comparable.

In general, as $n$ grows the average error seems to increase rapidly with all the methods —for example, testing a few cases with some of the best heuristics, we obtained around 5% error for $n = 60$ ($\rho_{60} = $ 16min), and 10% for $n = 70$ ($\rho_{70} = $ 24min).

**6. Conclusion.** We have presented our computational experience with heuristics that incorporate the simulated annealing concept, for the solution of the ODSAP with both sparse and non-sparse matrices.

In general, the best performance was obtained when the simulated annealing temperature is a function both of the remaining available time for the run and of the present results.

We noted that sparsity of similarity matrices increased the difficulty: the error in all methods grows rapidly with the size of the instances.

We also observed the convenience to apply a local improving technique to the final solution obtained by any of the proposed methods.

When applied to special cases (independent destinations, all prominent matrices), our methods obtained average errors less than 0.1% from the true optimal solutions.

Finally, notwithstanding that simple interchange has been the basis of heuristics galore in combinatorial optimization, we found that, when simulated annealing is applied to the ODSAP, insertion strategies appear superior to interchange strategies.

To conclude, our computational results confirm the suitability of simulated annealing to deal with the ODSAP.

# REFERENCES

ADOLPHSON, D. and HU, T.C. (1973), Optimal Linear Ordering, SIAM *Journal of Applied Mathematics* **25**, 403–423.

BEGHIN-PICAVET, M. and HANSEN, P. (1982), Deux Problèmes d'Affectation non Linéaires, RAIRO *Recherche Opérationnelle* **16**, 263–276.

BANAVAR, J.R., SHERRINGTON, D., and SOURLAS, N. (1987), Graph Bipartitioning and Statistical Mechanics, *Journal of Physics* **A20**, L1–L7.

BERGMANS, P.P. (1972), Minimizing Expected Travel Time on Geometrical Patterns by Optimal Probability Rearrangements, *Inform. Control* **20**, 331–350.

BOHR, H. and BRUNAK, S. (1989), A Traveling Salesman Approach to Protein Conformation, *Complex Systems* **3**, 9–28.

BURKARD, R.E. and RENDL, F. (1984), A Thermodynamically Motivated Simulation Procedure for Combinatorial Optimization Problems, *European Journal of Operational Research* **17**, 169–174.

ČERNÝ, V. (1985), Thermodynamical Approach to the Traveling Salesman Problem: An Efficient Simulation Algorithm, *Journal of Optimization Theory and Applications* **45**, 41–51.

CHAMS, M., HERTZ, A., and DE WERRA, D. (1987), Some Experiments with Simulated Annealing for Coloring Graphs, *European Journal of Operational Research* **32**, 260–266.

CHAN, A.W. and FRANCIS, R.L. (1979), Some Layout Problems on the Line with Interdistance Constraints and Costs, *Operations Research* **27**, 952–971.

DOYLE, J. (1988), Classification by Ordering a (sparse) Matrix: a 'Simulated Annealing' Approach, *Appl. Math. Modelling* **12**, 86–94.

GAREY, M.R., JOHNSON, D.S., and STOCKMEYER, L. (1976), Some Simplified NP-Complete Graph Problems, *Theoretical Comp. Sci.* **1**, 237–267.

GROSSMAN, D.D. and SILVERMAN, H.F. (1973), Placement of Records on a Secondary Storage Device to Minimize Acces Time, *Journal of the Association for Computing Machinery* **20**, 429–438.

GROTSCHEL, M., JUNGER, M., and REINELT, G. (1984), A Cutting Plane Algorithm for the Linear Ordering Problem, *Operations Research* **32**, 1195–1220.

KIRKPATRICK, S., GELATT, JR., C.D., and VECCHI, M.P. (1983), Optimization by Simulated Annealing, *Science* **220**, 671–680.

LAWLER, E.L. (1963), The Quadratic Assignment Problem, *Management Science* **9**, 586–599.

PADBERG, M.W. and RINALDI, G. (1987), Optimization of a 532-City Traveling Salesman Problem by Branch and Cut, *Operations Research Letters* **6**, 1–8.

PICARD, J.C. and QUEYRANNE, M. (1978), On the One-Dimensional Space Allocation Problem, Rapport Technique EP78-R-48, École Polytechnique de Montréal, Canada.

PICARD, J.C. and QUEYRANNE, M. (1981), On the One-Dimensional Space Allocation Problem, *Operations Research* **29**, 371–391.

PRATT, V.R. (1972), An $N\log N$ Algorithm to Distribute $N$ Records in a Sequential Access File: In *Complexity of Computer Computations* R.E. Miller and J.W. Tatcher (eds.). Plenum Press, New York.

ROMERO, D. (1978), *Variations sur l'Effet Condorcet*, Thèse Doct. 3e Cycle, University of Grenoble, France.

SIMMONS, D.M. (1969), One-Dimensional Space Allocation: An Ordering Algorithm, *Operations Research* **17**, 812–826.

TROYA, J.M. and VAQUERO, A. (1981), An Approximation Algorithm for Reducing Expected Head Movement in Linear Storage Devices, *Information Processing Letters* **13**, 218–220.

# CHAPTER 3

## Stopping Tests that Compute Optimal Solutions for Interior-Point Linear Programming Algorithms

David M. Gay*

**Abstract.** Knowledge of which inequalities in a linear programming problem can be slack in an optimal solution should be enough to let one compute such a solution. When running interior-point algorithms for linear programming, one can estimate the set $S$ of slack inequalities. The same linear algebraic machinery needed for carrying out these algorithms can be used to check whether the estimate is the true $S$. Computational experience with a dual affine algorithm suggests that an optimal solution can often be computed with relatively little overhead.

**1. Introduction.** Karmarkar's famous paper [13] has provoked considerable interest in interior-point methods for linear and, more recently, general convex programming; see Todd [27] for an excellent overview of this work (which we will not attempt to survey here).

Interior-point linear programming methods produce a sequence of iterates that, under reasonable assumptions, converge to an optimal solution, but these methods never actually compute an optimal solution. From the viewpoint of complexity theory, this is not an issue, as many interior-point algorithms manage in work bounded by a polynomial function of the "size" of the problem to produce an iterate close enough to an optimal solution that it could be modified, with relatively little more work, into an optimal solution. Moreover, for many practical applications, it suffices to find a point that is merely close to optimal. Sometimes, however, one would like to have an optimal basis, e.g., for use in conventional sensitivity analysis. The question of how hard it is in practice to find an optimal solution is also interesting in its own right. Finally, after identifying an optimal solution, one can stop iterating.

---

*AT&T Bell Laboratories, 600 Mountain Avenue, Murray Hill, NJ 07974-2070, U.S.A.

This paper describes stopping tests that often find optimal solutions, even for highly degenerate problems. These tests use the same linear algebraic machinery generally needed to carry out interior-point algorithms, so modifying existing implementations to exploit the tests should be easy. The tests often add relatively little overhead, and they sometimes save iterations. The solution found is generally not basic (unless the problem has a nondegenerate solution), but it could be converted to a basic optimal solution by a procedure that Megiddo [19] has recently described.

Various other authors have proposed ways to compute optimal solutions in the context of interior-point algorithms for linear programming. Some, e.g., [16, 18, 24], resort to pivoting techniques (which might require many iterations on degenerate problems); others, e.g., [15, 26, 29], require computing components of a relevant projection operator (which could be expensive on sparse problems).

The next section establishes some notation and defines the problem to be solved. Section 3 defines a set $\tilde{B}$ that characterizes the solution, describes computing estimates $B$ of $\tilde{B}$, and gives rules for deciding when to check whether $B = \tilde{B}$. Sections 4 and 5 discuss checking whether $B = \tilde{B}$, §6 presents some computational experience, and §7 offers concluding discussion.

**2. The LP Problem.** Let $A \in \mathbb{R}^{m \times n}$, $b \in \mathbb{R}^m$, and $c \in \mathbb{R}^n$ be given. (As usual, $\mathbb{R}^k$ denotes the set of real vectors of $k$ components, $\mathbb{R}^{m \times n}$ denotes the set of real $m \times n$ matrices, superscript "T" denotes transpose, and $\mathbf{0}$ denotes a vector or matrix of zeros of dimension dictated by context.) We shall refer to the standard-form linear programming problem of finding $x \in \mathbb{R}^n$ to

(P) $$\text{minimize } c^T x \text{ subject to } Ax = b, \, x \geq \mathbf{0}$$

as the *primal* LP. As is well known, there is a corresponding *dual* LP, i.e., the problem of finding $y \in \mathbb{R}^m$ to

(D) $$\text{maximize } b^T y \text{ subject to } A^T y \leq c.$$

Assume for now that both (P) and (D) are *feasible*, i.e., that there exist $x \in \mathbb{R}^n$ and $y \in \mathbb{R}^m$ that satisfy the constraints of (P) and (D) ($Ax = b$, $x \geq \mathbf{0}$, and $A^T y \leq c$, with inequalities understood componentwise). Then, as again is well known, there exist optimal solutions $\tilde{x}$ for (P) and $\tilde{y}$ for (D), and they satisfy

(1a) $$A\tilde{x} = b$$

(1b) $$\tilde{x} \geq \mathbf{0}$$

(1c) $$\tilde{s} = c - A^T \tilde{y} \geq \mathbf{0}$$

(1d) $$c^T \tilde{x} = b^T \tilde{y}.$$

The *complementary slackness* condition is easily seen:

(2) $$\tilde{x}^T \tilde{s} = 0,$$

i.e., for $1 \leq i \leq n$,

either $\tilde{x}_i = 0$ and $\tilde{s}_i \geq 0$

or $\tilde{x}_i \geq 0$ and $\tilde{s}_i = 0$.

The triple $(\tilde{x}, \tilde{y}, \tilde{s})$ (with $\tilde{s} = c - A^T\tilde{y}$) is said to enjoy *strict complementarity* if $\tilde{x} + \tilde{s} > \mathbf{0}$, i.e., for each $i$, $1 \leq i \leq n$,

either $\tilde{x}_i = 0$ and $\tilde{s}_i > 0$

or $\tilde{x}_i > 0$ and $\tilde{s}_i = 0$.

The present work is concerned with finding strictly complementary primal and dual optimal solutions $\tilde{x}$ and $\tilde{y}$. By reasoning that we now sketch, the Farkas Lemma assures the existence of such solutions. Let $A_{\bullet j}$ denote the $j$th column of $A$. Suppose $(\hat{x}, \hat{y}, \hat{z})$ is an optimal triple having $\hat{x}_i = \hat{s}_i = 0$, and let $N_0 = \{k: \hat{s}_k = 0 \text{ and } k \neq i\}$ and $N_1 = \{k: \hat{x}_k > 0\} \subset N_0$. The Farkas Lemma tells us that either

1. there exists a $z \in \mathbb{R}^m$ such that $z^T A_{\bullet i} < 0$, $z^T A_{\bullet j} \leq 0$ for $j \in N_0$, and $z^T A_{\bullet j} = 0$ for $j \in N_1$

or else

2. for $j \in N_0$ there exist multipliers $\lambda_j$ with $\lambda_j \leq 0$ for $j \in N_0 \setminus N_1$, such that $A_{\bullet i} = \sum_{j \in N_0} \lambda_j A_{\bullet j}$.

In case 1, for all sufficiently small $\varepsilon > 0$, $\tilde{y} = \hat{y} + \varepsilon z$ is a dual optimal solution that has $c_i > \tilde{y}^T A_{\bullet i}$ and $c_j > \tilde{y}^T A_{\bullet j}$ for all $j$ with $c_j > \hat{y}^T A_{\bullet j}$. In case 2, the $w \in \mathbb{R}^n$ defined by $w_i = 1$, $w_j = -\lambda_j$ for $j \in N_0$, and $w_j = 0$ for other $j$ is a vector in the null space of $A$ having $w_i > 0$ and $w_j \geq 0$ whenever $\hat{x}_j = 0$, whence for all sufficiently small $\varepsilon > 0$, $\tilde{x} = \hat{x} + \varepsilon w$ is a primal optimal solution with $\tilde{x}_i > 0$ and $\tilde{x}_j > 0$ for all $j$ with $\hat{x}_j > 0$. Repeating this argument at most $n$ times, we obtain an optimal triple $(\tilde{x}, \tilde{y}, \tilde{s})$ that enjoys strict complementarity.

The primal and dual affine algorithms are the easiest interior-point algorithms to explain. Moreover, they work well in practice, even though they are suspected not to have polynomial complexity (cf Megiddo and Shub [20]). The primal affine algorithm works by making the linear change of variables that transforms the current strictly feasible iterate $x^k$ (i.e., $Ax^k = b$ with $x^k > \mathbf{0}$) into $e$, the vector of ones, computing the projected steepest-descent direction, moving a certain amount in this direction (but not so far as to lose strict feasibility), and translating the resulting point back to the original variables. This algorithm is most succinctly stated in its differential form:

$$(3) \qquad x' = -D_x P_{AD_x} D_x c,$$

where $D_x$ is the diagonal matrix

$$D_x = \text{diag}(x_1, \cdots, x_n)$$

having $x_i$ as its $i$th diagonal element, and

$$P_{AD_x} = I - D_x A^T (AD_x^2 A^T)^\dagger AD_x,$$

with $I$ denoting the identity matrix of appropriate dimension, and superscript $\dagger$ denoting the pseudo-inverse, so that $P_{AD_x}$ projects orthogonally onto the null space of $AD_x$. Integrating (3) by Euler's method, we obtain the iteration

$$x^{k+1} = x^k - \alpha_k D_{x^k} P_{AD_{x^k}} D_{x^k} c,$$

where $\alpha_k$ is a step-length parameter chosen to keep $x^{k+1}$ strictly feasible.

One can derive the dual affine algorithm by transforming (D) into standard form and applying (3); see, e.g., [10] for details. In differential form, the resulting algorithm is

(4) $$y' = (AD_{1/s}^2 A^T)^\dagger b,$$

where $s = c - A^T y > 0$ and $D_{1/s}$ is the diagonal matrix having $(s_i)^{-1}$ as its $i$th diagonal element.

The primal and dual affine algorithms share linear algebraic needs with all the other recent interior-point algorithms of which I am aware: one must solve systems of linear equations involving $\overline{AD}^2 A^T$ for some diagonal matrix $\overline{D}$, or, equivalently (mathematically), one must solve a linear least-squares problem involving $\overline{AD}$. The strategies presented below rely on this linear algebraic machinery.

**3. Stopping Tests.** For each $i$ with $1 \leq i \leq n$, the optimality conditions for (P), (D) imply that if $\tilde{s}_i > 0$ for some optimal solution, then $\tilde{x}_i = 0$ for all optimal solutions, and, conversely, if $\tilde{x}_i > 0$ for some optimal solution, then $\tilde{s}_i = 0$ for all optimal solutions. Since (P), (D) have a strictly complementary optimal triple $(\tilde{x}, \tilde{y}, \tilde{s})$, there is a unique $\tilde{B} \subset \{1, 2, \cdots, n\}$ such that $\tilde{x}_i > 0$ for $i \in \tilde{B}$ and $\tilde{s}_i > 0$ for $i \in \tilde{N} = \{1, 2, \cdots, n\} \setminus \tilde{B}$. At the 1985 Asilomar workshop on linear programming, Narendra Karmarkar remarked that one might regard $\tilde{B}$ as *the* solution to (P), (D); given $\tilde{B}$, one should be able to compute a suitable $(\tilde{x}, \tilde{y}, \tilde{s})$.

Our stopping tests proceed by computing predictions $B$ of $\tilde{B}$ and trying quickly to check whether $B = \tilde{B}$. An optimal triple $(\tilde{x}, \tilde{y}, \tilde{s})$ results if the check succeeds.

Consider first the choice of $B$. Many interior-point algorithms have the property that if there exists an optimal triple $(\tilde{x}, \tilde{y}, \tilde{s})$ in which $\tilde{x}_i > 0$, then the iterates $(x^k, y^k, s^k)$ satisfy $\lim_{k \to \infty} x_i^k > 0$ and, alternatively, if there exists an optimal triple in which $\tilde{s}_i > 0$, then $\lim_{k \to \infty} s_i^k > 0$. Thus these algorithms are such that

$$\tilde{B} = \{1 \leq i \leq n: \lim_{k \to \infty} \frac{x_i^k}{s_i^k} = +\infty\}$$

$$= \{1 \leq i \leq n: \lim_{k \to \infty} \frac{s_i^k}{x_i^k} = 0\}.$$

For such algorithms, it seems natural to base an estimate $B$ of $\tilde{B}$ on the values $\frac{s_i^k}{x_i^k}$.

Let $(x, y, s) = (x^k, y^k, s^k)$ denote the current iterate (with, as usual, $s = c - A^T y$). We have just argued that for many algorithms, choosing

$$B = B^{(k)} = B_\tau \equiv \{i: \frac{s_i}{x_i} \leq \tau\} \tag{5}$$

for some fixed tolerance $\tau > 0$ (e.g., $\tau = 1$) should eventually result in $B^{(k)} = \tilde{B}$. Unfortunately, $B_\tau$ is scale-dependent: if we changed $A$ to $\bar{A} = AZ$, $c$ to $\bar{c} = Zc$, and $x$ to $\bar{x} = Z^{-1}x$ for some positive-definite diagonal matrix $Z$, then $B_\tau$ could change. For a particular choice of $\tau$, a fiend could choose $Z$ to make the first $k$ for which $B^{(k)} = \tilde{B}$ arbitrarily large. Initial equilibration of $A$ is a good way to reduce the effects of scale on (5). The method of Curtis and Reid [4] works well (and, as implemented in the Harwell Library routine MC19A, is fast); the computing reported in §6 uses this equilibration method.

One possible drawback to (5) is that it involves both primal and dual variables. Some interior-point algorithms, such as Karmarkar's original projective algorithm (when applied to a problem in his canonical form), recur only primal or only dual variables. Others, such as projective algorithms that use duality as suggested by Todd and Burrell [28] to bound the optimal objective value, only maintain strictly feasible primal ($x > 0$) or strictly feasible dual ($s > 0$) variables. Karmarkar [14] has suggested a choice of $B$ (meant for "dual" algorithms) that only involves the dual slack vector $s$:

$$B = B^{(k)} = \{i: s_i \leq \sqrt{\bar{s} \cdot \bar{h}}, 1 \leq i \leq n\}, \tag{6}$$

where

$$\bar{s} = \frac{1}{n} \sum_{i=1}^{n} s_i$$

and

$$\bar{h} = \left[\frac{1}{n} \sum_{i=1}^{n} s_i^{-1}\right]^{-1},$$

i.e., $B$ is the set of $i$ for which $s_i$ is at most the geometric mean of the arithmetic and harmonic means of the components of $s$. This choice also invites initial equilibration of $A$. Many interior-point algorithms stay close to the path of centers, i.e., the iterates satisfy $x_i^k s_i^k \approx \lambda = \lambda_k$ (with $\lambda_k \to 0$ as $k \to \infty$). For such algorithms, the arithmetic mean satisfies $\bar{s} \approx \tilde{s}$, and the harmonic mean satisfies $\bar{h} \approx \lambda/\tilde{x}$, where $\tilde{s}$ and $\tilde{x}$ are the arithmetic means of (the components of) $\tilde{s}$ and $\tilde{x}$, respectively. Thus $\sqrt{\bar{s} \cdot \bar{h}} \approx \lambda^{\frac{1}{2}} \sqrt{\tilde{s}/\tilde{x}}$, so (6) should pick $B$ correctly for relatively large $\lambda$ values.

Our stopping tests use the operators $(AD(B)^2 A^T)^\dagger$ and $P_{AD(B)}$, where $D(B)$ is the diagonal matrix having

$$D(B)_{i,i} = \begin{cases} D_{i,i} & \text{if } i \in B \\ 0 & \text{otherwise} \end{cases}.$$

During iterations in which we try to find an optimal solution, i.e., carry out the stopping tests, it is convenient to use $D(B) = D^{(k)}(B^{(k)})$ in place of $D^{(k)}$ (when computing factorizations and search directions). If we fail to find an optimal solution, it is sometimes best to start a new iteration (i.e., to recompute the operators using $D^{(k)}$ rather than $D^{(k)}(B^{(k)})$). In the computing of §6, we did so when the consistency check (8), described below, failed.

Attempting to compute an optimal solution and failing costs time. A simple way to prevent premature effort at trying to verify the optimality of a choice of $B$ is to wait until

$$(7) \qquad \frac{\min\{s_i: s_i \geq \sqrt{\overline{s \cdot h}}\}}{\max\{s_i: s_i \leq \sqrt{\overline{s \cdot h}}\}} > \mu_{sep}$$

for several iterations in a row (where $\mu_{sep}$ is an appropriate tolerance) before using $D^{(k)}(B^{(k)})$ in place of $D^{(k)}$. In the computing reported below, we only used $D^{(k)}(B^{(k)})$ when (7) held at least twice in succession for $\mu_{sep} = 50$. Moreover, after a failure to confirm optimality, we increased $\mu_{sep}$ by the left-hand side of (7).

Another time-saving device is to record the most recent failed choice of $B$ and to reject the current $B$ if it has not changed.

After deciding to use $D^{(k)}(B^{(k)})$ rather than $D^{(k)}$, it is helpful to carry out an initial consistency check to see whether $B = B^{(k)}$ is obviously wrong. In the computing reported below, we use the following subiteration, with $D = D^{(k)}$, to check whether $b$ lies in the column space of $AD(B)$:

(8a) $\qquad x^0 := f\ell\left[D(B)^2 A^T (AD(B)^2 A^T)^\dagger b\right];$

$\qquad j := 0;$

(8b) $\qquad$ while $\|b - Ax^j\|_\infty > \varepsilon_{cc} a^T |x^j|$ do

(8c) $\qquad\qquad x^{j+1} := f\ell\left[x^j + D(B)^2 A^T (AD(B)^2 A^T)^\dagger (b - Ax^j)\right];$

$\qquad\qquad j := j + 1;$

$\qquad\qquad$ if $\|b - Ax^j\|_\infty > \frac{1}{2}\|b - Ax^{j-1}\|_\infty$

$\qquad\qquad$ then stop subiteration (8) and reject $B$.

In (8b), $a$ is the vector of $\ell_\infty$ column norms of $A$, i.e.,

(8d) $\qquad a_i = \|A_{\bullet i}\|_\infty,$

and $|x^j|$ is the vector whose $i$th component is $|x_i^j|$. The floating-point result notation $f\ell(\cdot)$ in (8a) and (8c) emphasizes that we are dealing with an approximation of $(\cdot)$. In the computing reported below, the consistency-check tolerance $\varepsilon_{cc}$ is $1000 \cdot (m+n) \cdot \varepsilon_{mach}$, where $\varepsilon_{mach}$ is the unit roundoff (about $10^{-16}$ in the computing of §6). This choice of $\varepsilon_{cc}$ is obviously crude; ideally, $\varepsilon_{cc}$ would involve an estimate of the condition of $AD(B)^2 A^T$, i.e., the ratio of largest to smallest *positive* eigenvalues of $AD(B)^2A^T$.

Note that the computationally expensive part of (8a) is available for free from (4) if one is executing the dual affine algorithm (or is otherwise computing the right-hand side of (4)).

It is sometimes useful to seek an optimal solution even when (7) does not hold (or has not held long enough). The following alternative often proved worthwhile in the computing reported below. In this computing, a dual feasible $y^k$ is always available, and a primal feasible $x^k$ is sometimes available. If a feasible $x^k$ is available, let $\delta^k = c^T x^k - b^T y^k$, i.e., let $\delta^k$ be the duality gap. Otherwise, let $\delta^k = b^T(y^{k+1} - y^k)$ be the dual-function improvement achieved in the current iteration. If

(9a) $$\delta^k \leq \max\{\tau_{abs}, \tau_{rel} \cdot |b^T y^{k+1}|\}$$

and

(9b) $$\max\{\min\{\left|\frac{\hat{x}_i}{s_i}\right|, \left|\frac{s_i}{\hat{x}_i}\right|\}: 1 \leq i \leq n\} \leq \tau_{rat},$$

in which $\hat{x} = D_{1/s}^2(AD_{1/s}^2 A^T)^\dagger b$, then compute $B$ by (5) (with $x := \hat{x}$ and $\tau := 1$) and check whether it is optimal, using the tests described in §4 and §5. The computing of §6 used $\tau_{abs} = \tau_{rel} = 10^{-7}$ and $\tau_{rat} = 10^{-1}$.

**4. Dual Optimality Check.** If $B$ is optimal, i.e., $B = \tilde{B}$, then for any dual optimal solution $\tilde{y}$, we must have

$$D(B)A^T\tilde{y} = D(B)c,$$

where $D$ is any positive-definite diagonal matrix. This implies

$$AD(B)^2 A^T \tilde{y} = AD(B)^2 c,$$

which suggests computing

(10) $$\hat{y} = f\ell\left[(AD(B)^2 A^T)^\dagger AD(B)^2 c\right]$$

as an approximation to $\tilde{y}$. In the computing reported below, we used the current scale matrix $D_{1/s}$ as $D$ and applied the following iterative refinement to (10):

$\hat{y}^0 := y^k$ (the current iterate)

$j := 0$

loop

(11a) $$\hat{y}^{j+1} := f\ell\left[\hat{y}^j + (AD(B)^2 A^T)^\dagger AD(B)^2(c - A^T \hat{y}^j)\right]$$

$j := j + 1$

$\hat{s}^j := c - A^T \hat{y}^j$

(11b) $\quad\sigma^j := \max\{\text{if } i \in B \text{ then } |\hat{s}_i| \text{ else } -\hat{s}_i : 1 \leq i \leq n\}$

(11c) $\quad$ if $\sigma^j \leq \varepsilon_{df}$
$\qquad$ then accept $y^j$ as a suitable approximation to $\tilde{y}$
$\qquad$ otherwise, if $j \geq 2$ and $\sigma^j \geq \frac{1}{2}\sigma^{j-1}$
$\qquad$ then reject $B$.

We used $\varepsilon_{df} = 10(m + n)\varepsilon_{mach}$ in (11c). It was occasionally necessary to iterate (11) more than once when accepting a $B$ as $\tilde{B}$. Two iterations of (11) were often required when $B$ was rejected.

**5. Primal Optimality Check.** If $B = \tilde{B}$, then any optimal solution $\tilde{x}$ of (P) satisfies $\sum_{i \in B} A_{\bullet i}\tilde{x}_i = b$, so if $D = D^{(k)}$ is the current (positive-definite, diagonal) scale matrix, then $b$ is in the column space of $AD(B)$. Thus we must have

$$AD(B)^2 A^T (AD(B)^2 A^T)^\dagger b = b,$$

which suggests using the consistency-check iteration (8) to try to compute $x$ with $Ax = b$ and $x_i = 0$ for $i \in N = \{1, 2, \cdots, n\} \setminus B$. To enforce $x_i > 0$ for $i \in B$, we modify (8) to restrict the step length in (8c) and, if more than one iteration of the while loop in (8) is necessary, to recompute $D$. Thus we arrive at iteration (12):

(12a) $\quad x^0 := f\ell\left[D(B)^2 A^T (AD(B)^2 A^T)^\dagger b\right];$

(12b) $\quad$ modify $x^0$ to enforce $x_i^0 > 0$ for $i \in B$, as discussed in §6;
$\qquad j := 0;$

(12c) $\quad$ while $\|b - Ax^j\|_\infty > \varepsilon_{cc} a^T |x^j|$ do

(12d) $\quad\quad \delta^j := f\ell\left[D(B)^2 A^T (AD(B)^2 A^T)^\dagger (b - Ax^j)\right];$

(12e) $\quad\quad \lambda_{ls} := \dfrac{(b - Ax^j)^T A\delta^j}{\|A\delta^j\|_2^2};$

(12f) $\quad\quad \lambda_{wall} := \varepsilon_{wall} \cdot \min\{\dfrac{x_i^j}{\delta_i^j} : \delta_i^j > 0, i \in B\};$

(12g) $\quad\quad x^{j+1} := x^j + \min\{\lambda_{ls}, \lambda_{wall}\} \cdot \delta^j;$
$\qquad\qquad j := j + 1;$
$\qquad\qquad$ if $\|b - Ax^j\|_\infty > \frac{1}{2}\|b - Ax^{j-1}\|_\infty$
$\qquad\qquad$ then stop iteration (12) and reject $B$;

(12h) $\quad\quad D := \text{diag}(x_1^j, \cdots, x_n^j).$

In (12c), $a$ is again given by (8d). A trivial induction shows that (12) preserves the properties $D_{ii} > 0$ and $x_i^j > 0$ for $i \in B$, and $x_i^j = 0$ and $\delta_i^j = 0$ for $i \in N$.

The $\lambda_{ls}$ given by (12e) is the step length that minimizes the residual in the least-squares sense; if (12d) were exact, then $\lambda_{ls}$ would be 1. The computing reported below uses $\varepsilon_{wall} = 0.9$ in (12f).

It is sometimes helpful to iteratively refine $\delta^j$ in (12d). We do so with an iteration similar to (8).

An alternative to (12) that sometimes saves time is based on the problem

(13a)
$$\text{maximize} \sum_{i \in B} \log(x_i)$$

subject to

(13b)
$$\sum_{i \in B} A_{\bullet i} x_i = b.$$

(The solution of (13) is sometimes called a "center" or "the analytic center"; see [27] for references and discussion of centers.) The first-order necessary (KKT) conditions for (13) are

(14)
$$\begin{bmatrix} D_{x_B}^{-2} & A_B^T \\ A_B & 0 \end{bmatrix} \begin{bmatrix} x_B \\ z \end{bmatrix} = \begin{bmatrix} 0 \\ b \end{bmatrix},$$

in which $A_B$ denotes the matrix whose columns are $\{A_{\bullet i}: i \in B\}$, $x_B$ similarly denotes the vector whose components are $\{x_i: i \in B\}$, and $D_{x_B}$ is the diagonal matrix whose diagonal elements are the components of $x_B$. Block elimination in (14) reveals that the optimal solution $\tilde{x}_B$ of (13) satisfies

$$\tilde{x}_B = D_{x_B}^2 A_B^T (A_B D_{x_B}^2 A_B^T)^\dagger b.$$

Newton's method applied to (14) leads to the iteration

(15)
$$x_B^{j+1} = x_B^j + \alpha_j \left[ D_{x_B}^2 A_B^T (A_B D_{x_B}^2 A_B^T)^\dagger (b - 2A_B x_B) + x_B \right],$$

in which $\alpha_j \in (0, \infty)$ is a step-length parameter (which would be 1 for the undamped Newton iteration). Here, again, it is sometimes helpful to iteratively refine the step direction $\Delta x_B$ to ensure that it satisfies $A_B \Delta x_B = b - Ax_B$.

An analogous alternative to (11) would be to solve the following problem:

$$\text{maximize} \sum_{i \in N} \log(s_i)$$

subject to

$$A_B^T y = c_B$$

and

$$s_N = c_N - A_N^T y.$$

Unfortunately, this leads to solving equations involving $A_B^T (A_N D_{1/s_N} A_N^T)^\dagger A_B$, which departs from using the operators required for most current interior-point algorithms.

## 6. Computational Experience.
We experimented with a dual affine algorithm, i.e., an iteration, based on (4), of the form

$$(16a) \qquad y^{j+1} := y^j + \alpha_j (AD_{1/s}^2 A^T)^\dagger b$$

in which

$$(16b) \qquad s = s^j = c - A^T y^j.$$

(We also experimented with the dual projective algorithm described in [9]; although it sometimes required fewer iterations, it was seldom if ever faster, so we refrain here from discussing it further.)

Unfortunately, many details affect implementations based on (16) and thus affect the computational results described below. For the record, we now summarize our handling of these details, while making no claims about the "best" approach.

As in [1], we chose $\alpha_j = 0.9$ for the first ten iterations and $\alpha_j = 0.99$ thereafter.

We deal with bounds as suggested in [9]. Thus if $A$ in the primal LP (P) has the form

$$(17) \qquad A = \begin{bmatrix} \hat{A} & 0 \\ I & I \end{bmatrix}$$

and if $D$ (or $D := D(B)$) is correspondingly partitioned as

$$D = \begin{bmatrix} D_\ell & 0 \\ 0 & D_u \end{bmatrix},$$

then we factor $AD^2 A^T$ as

$$(18a) \qquad \hat{L}^T \begin{bmatrix} \hat{A} D_\ell^2 D_u^2 (D_\ell^2 + D_u^2)^{-1} \hat{A}^T & 0 \\ 0 & D_\ell^2 + D_u^2 \end{bmatrix} \hat{L},$$

where

$$(18b) \qquad \hat{L} = \begin{bmatrix} I & 0 \\ (D_\ell^2 + D_u^2)^{-1} D_\ell^2 \hat{A}^T & I \end{bmatrix},$$

so that the main computational burden is in factoring (or otherwise solving equations involving) $\hat{A} D_\ell^2 D_u^2 (D_\ell^2 + D_u^2)^{-1} \hat{A}^T$; since $\hat{D} = D_\ell D_u (D_\ell^2 + D_u^2)^{-\frac{1}{2}}$ is a diagonal matrix, this burden again involves a matrix of the form $A\bar{D}^2 A^T$, where $\bar{D}$ is a diagonal matrix. It is convenient to treat all problems as though $A$ had the form (17), replacing the factors in (18) by their limiting values as components of $D_u$ corresponding to variables without bounds tend to $+\infty$. By computing vectors of the diagonal elements of the diagonal matrices in (18) once per iteration, we arrange that inner loops are unaffected by the presence or absence of bounds.

Our execution of (12b) involved (17): let $\hat{x}$ denote the $\hat{n} = \frac{1}{2} n$ components of $x$

that correspond to $\hat{A}$, and let $u$ denote the vector of upper bounds on $x$, so that $0 \leq \hat{x}_i \leq u_i \leq +\infty$. Let $\mu = 0.001 \cdot \min\{u_i: 1 \leq i \leq \hat{n}\}$ if this yields $\mu < +\infty$, and let $\mu = 1$ otherwise. Finally, let $y^{(b)}$ denote the components of the dual vector $y$ that correspond to the identity matrices in (17). We carried out (12b) as follows:

If $u_i < +\infty$
  then if $i + \hat{n} \in N$
    then $x_i^0 := u_i$;
    else if $x_i^0 \leq 0$
      then $x_i^0 := \mu$
      else $x_i^0 := \min\{x_i^0, 0.99 \cdot u_i\}$;
  else if $x_i^0 \leq 0$
    then $x_i^0 := \mu$.

We used two schemes for choosing a feasible $y^0$; each worked better than the other on some problems. Both are variants of the much despised "big M" method. In the first, henceforth called the "individual-bounding scheme", we impose a temporary upper bound on each variable that has upper bound $+\infty$; in the second, henceforth called the "sum-bounding scheme", we impose a temporary upper bound on the sum of all the variables. In both schemes, $y_i^0 = 0$ for the $i$ corresponding to $\hat{A}$ in (17), and

$$y_{i+\hat{m}}^0 = \begin{cases} 0 & \text{if } u_i = +\infty \text{ and (sum-bounding or } c_i > 0) \\ -c_i & \text{if } c_i > 0 \\ 2c_i & \text{if } c_i < 0 \\ \tau \|A_{\bullet i}\|_\infty & \text{if } c_i = 0 \end{cases}$$

where $\hat{m} = m - \hat{n}$, so the components $\{y_{i+\hat{m}}: 1 \leq i \leq \hat{n}\}$ correspond to the block of identity matrices in (17), and where $\tau = -\max\{|c_i|/\|\hat{A}_{\bullet i}\|_\infty\}$ for the individual-bounding scheme and $\tau = -1$ for the sum-bounding scheme. We increase the temporary bounds a limited number of times (until the bounds reach $10^{10}$) if they threaten to become tight, and we eliminate them once we have a $y$ that would be strictly feasible without them.

For completeness, we note that the temporary bounding scheme can detect infeasibilities, at least when the limit on temporary bounds is large enough. We claim to detect primal infeasibility if ever we obtain a dual objective value that exceeds the largest value permitted by the limiting temporary bound value, and we detect dual infeasibility by solving a second linear programming problem if the first problem turns out to have an optimal solution in which one of the temporary bounds is tight. Such detection of infeasibility is not an issue in the computing reported below.

Rounding errors sometimes caused $s = b - A^T y$ to wander outside the strictly positive orthant. With the sum-bounding scheme, it was then occasionally helpful to reduce the component of $y$ corresponding to the sum-bounding row by $10 \cdot |\min\{s_i\}|$.

To operate with $(AD^2A^T)^\dagger$ or $(AD(B)^2A^T)^\dagger$, we computed a Cholesky factorization $LL^T$ of $Q\hat{A}\hat{D}^2\hat{A}^T Q^T$, where $Q$ is a permutation matrix chosen by the variant of the minimum degree algorithm implemented by the PORT routine SPMOR. We used one of the Cholesky factorization variants described in [11] (the *alist* variant when the *alist* was less then $10^6$ words long and the column-dispatch variant otherwise) with the following modifications when a pivot was negligible ($\varepsilon_{sing}$ times its original value; we used $\varepsilon_{sing} = 10^{-15}$ in the computing reported here): we zero the corresponding column of $L$, and, for the column-dispatch scheme, we do not dispatch the column. Negligible pivots are particularly likely when we use $D(B)$ in place of $D$ in the stopping tests.

It is sometimes helpful to augment (16) with "centering" steps, i.e., steps that increase

$$(19) \qquad \sum_{i=1}^{n} \log(c_i - A_{\bullet i}^T y)$$

while not decreasing the linear objective $b^T y$. Such steps are of theoretical interest, as they can convert the affine algorithm into one with a polynomial-time complexity bound [3]; sometimes they are even worthwhile in practice. Reasoning like that behind (4) (see, e.g., the dual-algorithm section of [10]) gives search direction $-(AD_{1/s}^2 A^T)^\dagger A s^{-1}$ for (19), where $(s^{-1})_i = (s_i)^{-1}$. This reasoning, sketched in Appendix A, suggests that the constraint that $b^T y$ not be increased should be enforced, if necessary, by an adjustment in the dual-affine search direction. The augmented (16) thus has the form

$$(20a) \qquad p^j := (AD_{1/s}^2 A^T)^\dagger b,$$

where, again,

$$(20b) \qquad s := s^j = c - A^T y^j;$$

$$(20c) \qquad y^{j+\frac{1}{2}} := y^j + \alpha_j p^j;$$

$$(20d) \qquad q^j := (AD_{1/s}^2 A^T)^\dagger A(c - A^T y^{j+\frac{1}{2}})^{-1};$$

if $b^T q^j < 0$

$$(20e) \qquad \text{then } q^j := q^j - \left[\frac{b^T q^j}{b^T p^j}\right] p^j;$$

$$(20f) \qquad y^{j+1} := y^{j+\frac{1}{2}} + \beta_j q^j,$$

with $\beta_j$ chosen to approximately maximize (19) in the direction $q^j$. Note that (20d) uses the scale matrix $D_{1/s}$ from the start of the current iteration; in principle, it might be desirable to use an updated scaling matrix in (20d), but that would probably cost more time than it saved.

We experimented with the *netlib* collection of linear programming test problems (see [5] and [8]), as the version available from netlib@research.att.com stood on 31 October 1989. Thus we attempted to solve 83 problems, whose sizes and estimated degrees of degeneracy are shown in Appendix B.

Our computations were carried out on an SGI 4D/240S with 64MB memory and

SMD disks. We used a single processor (a 25MHz MIPS R3000). Compilation was with a combination of `cc -g` and `lcc`, where `cc` is the vendor-supplied C compiler and `lcc` is an experimental C compiler by Chris Fraser and Dave Hanson — see [7]. (The relatively small amount of Fortran source code, e.g. for the routines `MC19A` and `SPMOR` mentioned above, was converted to C by *f2c* [6]).

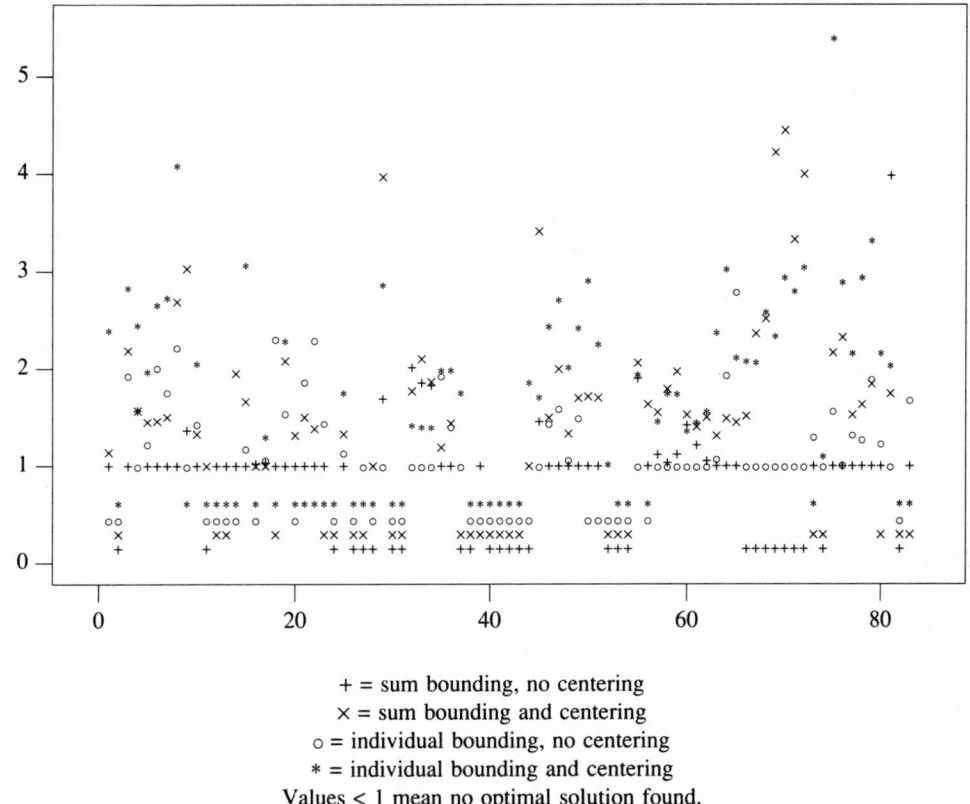

+ = sum bounding, no centering
× = sum bounding and centering
o = individual bounding, no centering
∗ = individual bounding and centering
Values < 1 mean no optimal solution found.

**Figure 1. Relative times among variants that found optimal solutions.**

Figures 1–4 summarize our experience with four variants of the dual affine algorithm: the sum-bounding and individual-bounding choices of $y^0$ described above, both with and without centering (i.e., iterations (20) and (16)). In all figures, the *x*-axis indicates the test problems as they are numbered when their names are sorted alphabetically. Some numbers behind the figures are shown in Appendix B.

The run times include reading a binary representation of the problem from which the fixed columns and (possibly resulting) empty rows had been removed and slack and surplus variables had been added. The computing reported here involved no other preprocessing.

Figure 1 shows relative solution times for those problems for which our stopping tests delivered optimal solutions. The times are relative to the time taken by the variant that solved the problem most quickly; values less than one signify cases where the tests failed to deliver an optimal solution. As is plain from Figure 1, none of the four variants is clearly superior to the others. Table 1 summarizes, for each variant, the number of problems for which the tests delivered an optimal solution. The final "net optimal" line shows number of problems for which at least one dual affine variant stopped with an optimal solution, i.e., about 5/6 of the then current *netlib* collection.

| Centering | Bounding | Optimal Solutions |
|---|---|---|
| no | sum | 56 |
| yes | sum | 58 |
| no | indiv. | 56 |
| yes | indiv. | 53 |
| net optimal | | 70 |

**Table 1. Counts of optimal solutions found**

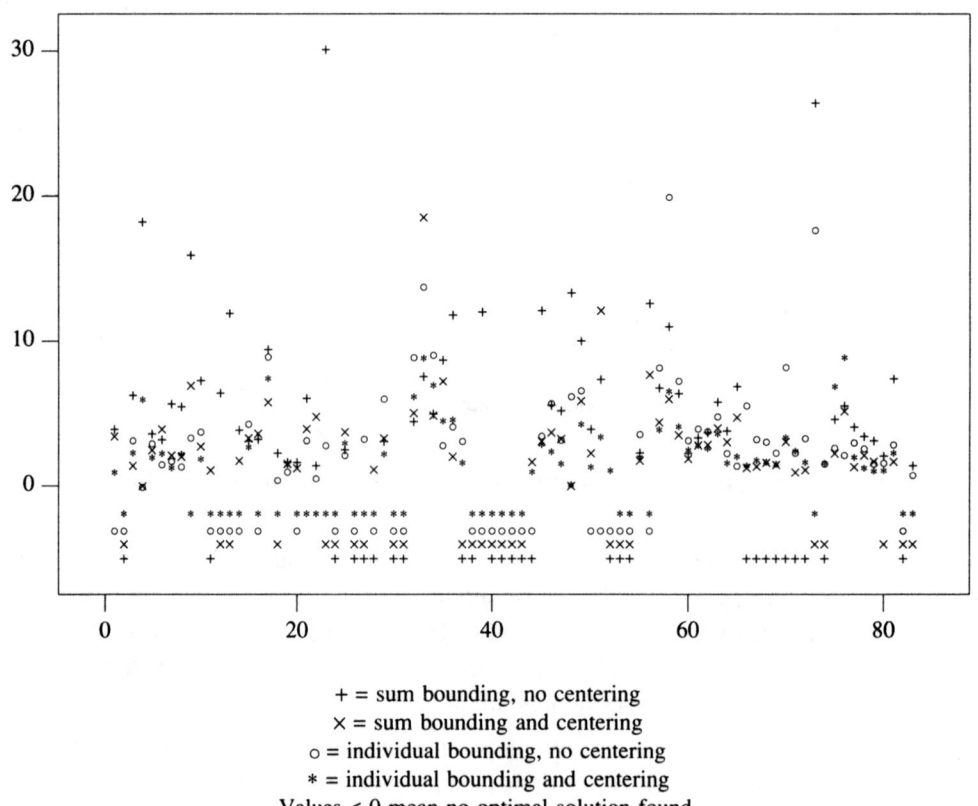

+ = sum bounding, no centering
× = sum bounding and centering
o = individual bounding, no centering
* = individual bounding and centering
Values < 0 mean no optimal solution found.

**Figure 2. Percentages of time for optimality heuristics.**

Figure 2 shows, for each run that yielded an optimal solution, the percentage of run time taken by the stopping tests. On most runs, the stopping tests accounted for less than ten percent of the run time; often they accounted for less than five percent.

Of course, with a different choice of $y^0$ or other algorithmic variations that reduced the number of iterations, the tests would probably take a greater fraction of the run time for some problems.

The two runs with the highest percentages in Figure 2 (problems BNL2 and E226 with sum-bounding and no centering) were the two in which it was helpful to reduce the component of $y$ corresponding to the sum-bounding constraint when rounding errors gave a negative component of $s$. Both of these problems required iterative refinement during computation of the optimal solution (2 iterations of (12) for BNL2, 14 for E226), and for both the Newton iteration (15) was slightly preferable to (12) (saving one iteration for BNL2 when trying an incorrect $B$ and saving one iteration for E226 during the successful verification of optimality).

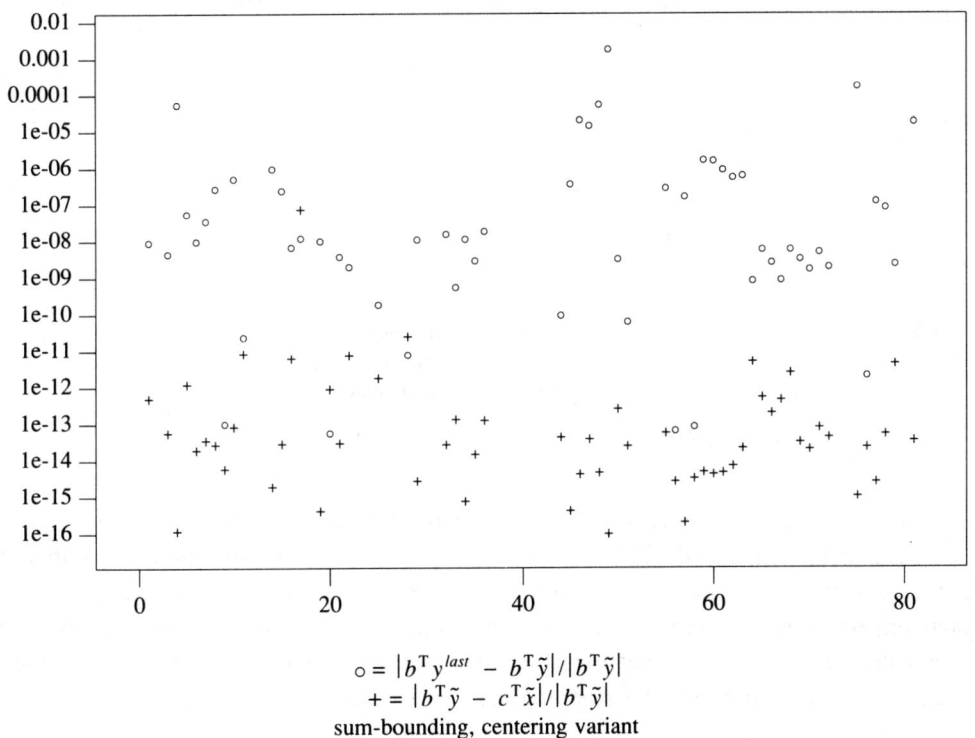

$\circ = |b^T y^{last} - b^T \tilde{y}|/|b^T \tilde{y}|$
$+ = |b^T \tilde{y} - c^T \tilde{x}|/|b^T \tilde{y}|$
sum-bounding, centering variant

**Figure 3. Objective gaps.**

Figure 3 shows how close to optimal the current objective values were when the stopping tests identified optimal solutions. This figure is for the sum-bounding, centering variant (the one with the most successes — see Table 1); such figures for the other variants are qualitatively similar to Figure 3. The values plotted are $|b^T y^{last} - b^T \tilde{y}|/|b^T \tilde{y}|$ and $|b^T \tilde{y} - c^T \tilde{x}|/|b^T \tilde{y}|$, where $b^T y^{last}$ is the dual objective value after the last step taken, and $b^T \tilde{y}$ and $c^T \tilde{x}$ are the approximations to the optimal dual and primal objective function values obtained by the tests. Of course, with exact arithmetic we would have $b^T \tilde{y} = c^T \tilde{x}$, so the plotting related to $|b^T \tilde{y} - c^T \tilde{x}|$ shows effects of the floating-point arithmetic we used.

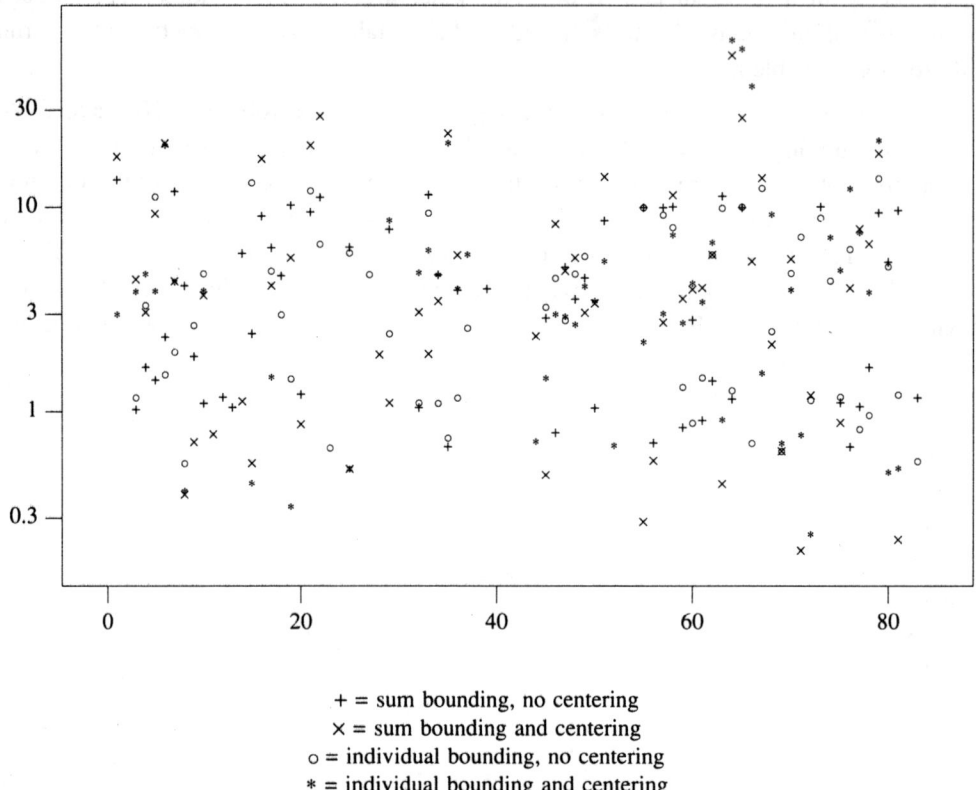

+ = sum bounding, no centering
× = sum bounding and centering
o = individual bounding, no centering
∗ = individual bounding and centering

**Figure 4.** $b^T(y^{last} - y^{last-1}) / b^T(\tilde{y} - y^{last})$.

Since some authors have suggested stopping tests based on the relative change in the objective (e.g., [1] and [21]), it is of some interest to see how good an indicator relative objective function changes were of closeness to optimality in our computational experiments. Figure 4 compares the objective improvement on the last step taken with the improvement yet possible after that step by showing the ratio of these values for the runs in which the stopping tests succeeded.

**7. Discussion.** Our stopping tests could be used with most interior-point algorithms for linear programming, but assessing their utility in connection with any particular algorithm requires numerical experiments. Figure 3 suggests that the tests may sometimes allow one to stop an interior-point algorithm slightly sooner than one otherwise might, but how often this makes much difference depends on both the algorithm and the problems solved.

One reason for my interest in this paper's topic is my desire, as a numerical analyst, to understand the proper role of calculations based on the normal equations in the context of interior-point algorithms. That is, I wish to understand when explicitly forming and factoring $AD^2A^T$ (henceforth called "using the normal equations") is reasonable. If one uses the normal equations in an interior-point algorithm applied to a degenerate problem and iterates long enough in finite-precision arithmetic, then numerical difficulties are virtually guaranteed to arise. However, if one can compute an optimal solution before these difficulties appear, then using the normal equations is perfectly reasonable.

Numerical difficulties prevented further progress in many of the runs in which the tests failed to deliver an optimal solution. Sometimes the computed $s = c - A^T y$ acquired negative components (a mathematical impossibility); other times we reached an iteration limit of 200. In both cases, more accurate search directions might have led to success. Our computational experience thus suggests that using the normal equations may often be reasonable, but that one should try to detect numerical loss of accuracy (perhaps by monitoring "pivots" — the diagonal elements whose square roots are computed — during Cholesky factorization) and should be prepared to switch to a numerically more robust alternative when the going gets tough. Such alternatives include computing a sparse Cholesky factorization (see, e.g., [12]), using LSQR [23, 22], or solving the augmented system [2]; these alternatives are usually slower than using the normal equations. Experimentation with such alternatives is one topic for future research.

Several other topics for future research suggest themselves. How much, for example, does preprocessing, i.e., automatic problem simplification, affect the success of the stopping tests? The computational results in [1, 17] suggest that preprocessing may significantly affect the performance of interior-point algorithms on some problems. What is the complexity of the iterations proposed in §4 and §5? How do pivoting techniques compare with our stopping tests combined with Megiddo's scheme [19] for computing an optimal basis from primal and dual optimal solutions?

**Acknowledgement.** I thank Margaret Wright and Bob Vanderbei for helpful comments on the manuscript.

**Appendix A. Centering Steps.** Consider first the primal affine algorithm (3), in which the search direction is the constrained steepest-descent direction after the linear change of variables $\bar{x} = D_{x^k}^{-1} x$ (which transforms the current iterate into the vector of ones). Let $D \equiv D_{x^k}$ denote the current scaling matrix, so that the transformed problem is that of minimizing $\bar{c}^T \bar{x}$ subject to $\bar{A}\bar{x} = b$ and $\bar{x} \geq 0$, where $\bar{A} = AD$ and $\bar{c} = Dc$. The affine search direction is then

(A1) $$\bar{p} = P_{\bar{A}} \bar{c},$$

in which $P_{\bar{A}} = I - \bar{A}^T (\bar{A}^T \bar{A})^\dagger \bar{A}$ denotes orthogonal projection onto the null space of $\bar{A}$. In terms of the transformed variables $\bar{x}$, we wish to consider the centering problem

(A2) $$\text{maximize } \psi(\bar{x}) \equiv \sum_{i=1}^{n} \log(\bar{x}_i)$$

subject to

$$\bar{A}\bar{x} = b,$$

(A3) $$\bar{c}^T \bar{x} \leq \bar{c}^T \bar{x}^{k+1} \text{ and } \bar{x} \geq 0,$$

where $\bar{c}^T \bar{x}^{k+1}$ denotes the linear objective value resulting from the current affine step. The projected steepest-descent direction for (A2) is

(A4) $$\bar{q} = P_{\bar{A}} \nabla \psi(\bar{x}^{k+1}).$$

If this direction makes a positive inner product with $\bar{c}$, then we must modify it to enforce (A3). In the current transformed scaling, the obvious modification is to project (A4) orthogonal to the affine search direction $P_{\bar{A}}\bar{c}$, which results in the search direction

(A5) $$\bar{q} - \left[\frac{\bar{q}^T \bar{c}}{\bar{p}^T \bar{c}}\right]\bar{p}.$$

In terms of the original variables, the primal affine centering direction, computed at the new iterate $x^{k+1}$ in the scaling that gave the step from $x^k$ to $x^{k+1}$, is $D$ times (A4) or (A5), i.e., is either

(A6) $$q = DP_{AD}D(x^{k+1})^{-1},$$

wherein $(x^{k+1})^{-1}$ denotes the component-wise reciprocal of the new iterate $x^{k+1}$, or

(A7) $$q - \left[\frac{q^T c}{p^T c}\right]p,$$

with $p = D\bar{p}$.

Now the dual problem (D) may be expressed in primal form in terms of the slack variables $s = b - A^T y$ (see [10] and [25]) as

(A8a) $$\text{minimize } (A^+ b)^T s - c^T A^+ b$$

subject to

(A8b) $$(I - A^T A^{+T})s = (I - A^T A^{+T})c$$

and

(A8c) $$s \geq 0,$$

where $A^+$ is any generalized inverse of $A$ (i.e., $AA^+A = A$). Since the rows of $A$ span the null space of $I - A^T A^{+T}$ and, similarly, the rows of $AD^{-1}$ span the null space of $(I - A^T A^{+T})D$, we have

(A9) $$P_{(I - A^T A^{+T})D} = D^{-1}A^T(AD^{-2}A^T)^\dagger AD^{-1}.$$

Using (A1–7) to express the centering direction for (A8), computing the relevant projections by (A9), and expressing the search direction in terms of the dual variables $y$, we arrive at the centering direction computation in (20).

**Appendix B. Problem Statistics and Result Tables.** Tables B1 and B2 give some statistics about our test problems. The "#" column gives each problem's *x*-axis coordinate in Figures 1–4. The "original" numbers of rows, columns and nonzeros reflect the problems as they are in the *netlib* collection — without added slack and surplus variables and with fixed columns. The "adjusted" numbers show the effects of adding slack and surplus variables, omitting fixed columns and any resulting empty rows, and omitting the objective row.

We define the *primal degeneracy* of a linear programming problem to be the affine dimension of (i.e., degrees of freedom in) the set of dual solutions. Similarly, we define the *dual degeneracy* of a linear programming problem to be the affine dimension of the set of primal solutions. The degeneracy estimates in Tables B1 and B2 are based on an estimate $\rho$ of the rank $\tilde{\rho}$ of $AD(\tilde{B})A^T$. Specifically, we estimate the primal degeneracy to be $m - \rho$ and the dual degeneracy to be $|B| - \rho$, where $|B|$ is the number of elements of $B$. The rank estimate involves deciding when a "pivot" is zero during the Cholesky factorization of $AD(\tilde{B})A^T$, and it is sometimes wrong — different runs sometimes give different rank estimates. It would be better to use a sparse condition estimator to compute $\rho$, but we leave that for future work. For problems where we obtained more than one degeneracy estimate, Tables B1 and B2 show the maximum values obtained.

Tables B3 and B4 show the numbers behind the sum-bounding, centering run in Figures 1 and 2, as well as several iteration counts for the stopping heuristics in this run. The "RC" column tells the reason for stopping:

"OPT" means optimal, with $B$ chosen by (6) and (7);
"OPR" means optimal, with $B$ chosen by (9) and (5) and $\delta^k \leq \tau_{rel} \cdot |b^T y^k|$;
"OPA" means optimal, with $B$ chosen by (9) and (5) and $\delta^k \leq \tau_{abs}$;
"SIN" means $s = c - A^T y$ acquired a negative component;
"ITL" means iteration limit reached.

The "time" column gives seconds of user time taken (on a 25MHz MIPS R3000 processor); the column labeled "%stop" shows the percentage of that time consumed by the stopping heuristics. Of course, the times and percentages are approximate; the times for repeated runs often vary by 3% or 4%. The "iters" column shows the number of dual-affine iterations. The columns labeled "CI" give the number of initial consistency-check iterations (8). Those labeled "OC" tell how many choices of $B$ were tested. Testing a choice of $B$ required at least one iteration of (11); the "DI" column gives the total number of such iterations. On the other hand, (12) sometimes required no iterations; the "PI" column tells how many times (12d–g) were needed. Sometimes (12d) required iterative refinement; the "PR" column gives the total number of such iterative refinement iterations.

| Problem | # | original | | | adjusted | | | degeneracy | |
|---|---|---|---|---|---|---|---|---|---|
| | | rows | cols | nonzeros | rows | cols | nonzeros | primal | dual |
| 25FV47 | 1 | 822 | 1571 | 11127 | 820 | 1876 | 10705 | 68 | 18 |
| 80BAU3B | 2 | 2263 | 9799 | 29063 | 2235 | 11516 | 22648 | ? | ? |
| ADLITTLE | 3 | 57 | 97 | 465 | 56 | 138 | 424 | 1 | 16 |
| AFIRO | 4 | 28 | 32 | 88 | 27 | 51 | 102 | 7 | 2 |
| AGG | 5 | 489 | 163 | 2541 | 488 | 615 | 2862 | 34 | 8 |
| AGG2 | 6 | 517 | 302 | 4515 | 516 | 758 | 4740 | 6 | 40 |
| AGG3 | 7 | 517 | 302 | 4531 | 516 | 758 | 4756 | 4 | 39 |
| BANDM | 8 | 306 | 472 | 2659 | 305 | 472 | 2494 | 11 | 0 |
| BEACONFD | 9 | 174 | 262 | 3476 | 173 | 295 | 3408 | 51 | 4 |
| BLEND | 10 | 75 | 83 | 521 | 74 | 114 | 522 | 6 | 2 |
| BNL1 | 11 | 644 | 1175 | 6129 | 632 | 1576 | 5522 | 31 | 230 |
| BNL2 | 12 | 2325 | 3489 | 16124 | 2280 | 4442 | 14952 | 415 | 504 |
| BOEING1 | 13 | 352 | 384 | 3865 | 348 | 723 | 3824 | 36 | 15 |
| BOEING2 | 14 | 167 | 143 | 1339 | 140 | 279 | 1332 | 36 | 13 |
| BORE3D | 15 | 234 | 315 | 1525 | 233 | 333 | 1446 | 103 | 0 |
| BRANDY | 16 | 221 | 249 | 2150 | 182 | 292 | 2191 | 34 | 5 |
| CAPRI | 17 | 272 | 353 | 1786 | 271 | 466 | 1864 | 7 | 34 |
| CYCLE | 18 | 1904 | 2857 | 21322 | 1886 | 3367 | 21230 | 1046 | 410 |
| CZPROB | 19 | 930 | 3523 | 14173 | 927 | 3331 | 10020 | 51 | 60 |
| D2Q06C | 20 | 2172 | 5167 | 35674 | 2171 | 5831 | 33081 | 248 | 32 |
| DEGEN2 | 21 | 445 | 534 | 4449 | 444 | 757 | 4201 | 209 | 0 |
| DEGEN3 | 22 | 1504 | 1818 | 26230 | 1503 | 2604 | 25432 | 777 | 0 |
| E226 | 23 | 224 | 282 | 2767 | 223 | 472 | 2768 | 28 | 5 |
| ETAMACRO | 24 | 401 | 688 | 2489 | 400 | 734 | 2188 | ? | ? |
| FFFFF800 | 25 | 525 | 854 | 6235 | 524 | 1028 | 6401 | 125 | 51 |
| FINNIS | 26 | 498 | 614 | 2714 | 497 | 1019 | 2542 | ? | ? |
| FORPLAN | 27 | 162 | 421 | 4916 | 135 | 463 | 4539 | 32 | 0 |
| GANGES | 28 | 1310 | 1681 | 7021 | 1309 | 1706 | 6937 | 109 | 147 |
| GFRD-PNC | 29 | 617 | 1092 | 3467 | 616 | 1160 | 2445 | 274 | 0 |
| GREENBEA | 30 | 2393 | 5405 | 31499 | 2389 | 5495 | 30908 | ? | ? |
| GREENBEB | 31 | 2393 | 5405 | 31499 | 2389 | 5483 | 30869 | ? | ? |
| GROW15 | 32 | 301 | 645 | 5665 | 300 | 645 | 5620 | 0 | 233 |
| GROW22 | 33 | 441 | 946 | 8318 | 440 | 946 | 8252 | 0 | 409 |
| GROW7 | 34 | 141 | 301 | 2633 | 140 | 301 | 2612 | 0 | 97 |
| ISRAEL | 35 | 175 | 142 | 2358 | 174 | 316 | 2443 | 0 | 19 |
| KB2 | 36 | 44 | 41 | 291 | 43 | 68 | 313 | 0 | 0 |
| LOTFI | 37 | 154 | 308 | 1086 | 153 | 366 | 1136 | 2 | 16 |
| NESM | 38 | 663 | 2923 | 13988 | 662 | 2930 | 13260 | ? | ? |
| PEROLD | 39 | 626 | 1376 | 6026 | 625 | 1442 | 5962 | 55 | 37 |
| PILOT | 40 | 1442 | 3652 | 43220 | 1440 | 4656 | 42299 | ? | ? |
| PILOT.JA | 41 | 941 | 1988 | 14706 | 924 | 1956 | 12100 | ? | ? |
| PILOT.WE | 42 | 723 | 2789 | 9218 | 722 | 2850 | 9001 | ? | ? |
| PILOT4 | 43 | 411 | 1000 | 5145 | 410 | 1093 | 5164 | ? | ? |
| PILOTNOV | 44 | 976 | 2172 | 13129 | 951 | 2242 | 12460 | 41 | 1203 |
| RECIPE | 45 | 92 | 180 | 752 | 87 | 178 | 652 | 16 | 58 |

**Table B1.** Problem Statistics (part 1)

| Problem | # | original | | | adjusted | | | degeneracy | |
|---|---|---|---|---|---|---|---|---|---|
| | | rows | cols | nonzeros | rows | cols | nonzeros | primal | dual |
| SC105 | 46 | 106 | 103 | 281 | 104 | 162 | 339 | 12 | 0 |
| SC205 | 47 | 206 | 203 | 552 | 204 | 316 | 664 | 13 | 0 |
| SC50A | 48 | 51 | 48 | 131 | 49 | 77 | 159 | 4 | 0 |
| SC50B | 49 | 51 | 48 | 119 | 48 | 76 | 146 | 0 | 0 |
| SCAGR25 | 50 | 472 | 500 | 2029 | 471 | 671 | 1725 | 44 | 0 |
| SCAGR7 | 51 | 130 | 140 | 553 | 129 | 185 | 465 | 0 | 0 |
| SCFXM1 | 52 | 331 | 457 | 2612 | 330 | 600 | 2732 | 34 | 12 |
| SCFXM2 | 53 | 661 | 914 | 5229 | 660 | 1200 | 5469 | ? | ? |
| SCFXM3 | 54 | 991 | 1371 | 7846 | 990 | 1800 | 8206 | ? | ? |
| SCORPION | 55 | 389 | 358 | 1708 | 388 | 466 | 1534 | 129 | 0 |
| SCRS8 | 56 | 491 | 1169 | 4029 | 490 | 1275 | 3288 | 189 | 16 |
| SCSD1 | 57 | 78 | 760 | 3148 | 77 | 760 | 2388 | 58 | 12 |
| SCSD6 | 58 | 148 | 1350 | 5666 | 147 | 1350 | 4316 | 64 | 30 |
| SCSD8 | 59 | 398 | 2750 | 11334 | 397 | 2750 | 8584 | 20 | 174 |
| SCTAP1 | 60 | 301 | 480 | 2052 | 300 | 660 | 1872 | 64 | 101 |
| SCTAP2 | 61 | 1091 | 1880 | 8124 | 1090 | 2500 | 7334 | 301 | 230 |
| SCTAP3 | 62 | 1481 | 2480 | 10734 | 1480 | 3340 | 9734 | 482 | 278 |
| SEBA | 63 | 516 | 1028 | 4874 | 515 | 1036 | 4360 | 76 | 0 |
| SHARE1B | 64 | 118 | 225 | 1182 | 117 | 253 | 1179 | 0 | 0 |
| SHARE2B | 65 | 97 | 79 | 730 | 96 | 162 | 777 | 10 | 6 |
| SHELL | 66 | 537 | 1775 | 4900 | 536 | 1527 | 3058 | 164 | 1 |
| SHIP04L | 67 | 403 | 2118 | 8450 | 360 | 2166 | 6380 | 84 | 1 |
| SHIP04S | 68 | 403 | 1458 | 5810 | 360 | 1506 | 4400 | 65 | 1 |
| SHIP08L | 69 | 779 | 4283 | 17085 | 712 | 4363 | 12882 | 272 | 0 |
| SHIP08S | 70 | 779 | 2387 | 9501 | 712 | 2467 | 7194 | 247 | 0 |
| SHIP12L | 71 | 1152 | 5427 | 21597 | 1042 | 5533 | 16276 | 318 | 1 |
| SHIP12S | 72 | 1152 | 2763 | 10941 | 1042 | 2869 | 8284 | 296 | 0 |
| SIERRA | 73 | 1228 | 2036 | 9252 | 1222 | 2715 | 7951 | 222 | 10 |
| STAIR | 74 | 357 | 467 | 3857 | 356 | 532 | 3813 | 1 | 1 |
| STANDATA | 75 | 360 | 1075 | 3038 | 358 | 1257 | 3172 | 261 | 12 |
| STANDMPS | 76 | 468 | 1075 | 3686 | 466 | 1257 | 3820 | 255 | 14 |
| STOCFOR1 | 77 | 118 | 111 | 474 | 117 | 165 | 501 | 10 | 0 |
| STOCFOR2 | 78 | 2158 | 2031 | 9492 | 2157 | 3045 | 9357 | 332 | 0 |
| STOCFOR3 | 79 | 16676 | 15695 | 74004 | 16675 | 23541 | 72721 | 3124 | 0 |
| TUFF | 80 | 334 | 587 | 4523 | 294 | 617 | 4550 | 137 | 51 |
| VTP.BASE | 81 | 199 | 203 | 914 | 197 | 327 | 943 | 95 | 0 |
| WOOD1P | 82 | 245 | 2594 | 70216 | 244 | 2595 | 70216 | ? | ? |
| WOODW | 83 | 1099 | 8405 | 37478 | 1098 | 8418 | 37487 | 538 | 706 |

**Table B2.** Problem Statistics (part 2)

| Problem | RC | time | %stop | iters | CI | OC | DI | PI | PR |
|---|---|---|---|---|---|---|---|---|---|
| 25FV47 | OPR | 103.3 | 3.4 | 48 | 0 | 1 | 1 | 0 | 0 |
| 80BAU3B | SIN | 517.1 | 0.0 | 129 | 0 | 0 | 0 | 0 | 0 |
| ADLITTLE | OPT | 0.7 | 1.4 | 24 | 0 | 1 | 1 | 0 | 0 |
| AFIRO | OPT | 0.1 | 0.0 | 12 | 0 | 1 | 1 | 0 | 0 |
| AGG | OPR | 34.1 | 2.5 | 62 | 0 | 1 | 1 | 0 | 0 |
| AGG2 | OPR | 38.5 | 3.9 | 41 | 0 | 1 | 1 | 0 | 0 |
| AGG3 | OPT | 39.6 | 2.1 | 43 | 0 | 1 | 1 | 0 | 0 |
| BANDM | OPR | 11.8 | 2.0 | 67 | 0 | 1 | 1 | 0 | 0 |
| BEACONFD | OPT | 11.6 | 6.9 | 63 | 3 | 2 | 2 | 2 | 0 |
| BLEND | OPT | 0.7 | 2.7 | 21 | 0 | 1 | 1 | 0 | 0 |
| BNL1 | OPT | 52.9 | 1.1 | 75 | 0 | 1 | 1 | 0 | 0 |
| BNL2 | SIN | 1004.2 | 0.4 | 109 | 0 | 1 | 1 | 2 | 0 |
| BOEING1 | SIN | 19.6 | 2.2 | 64 | 0 | 1 | 2 | 0 | 0 |
| BOEING2 | OPT | 4.6 | 1.8 | 44 | 0 | 1 | 1 | 0 | 0 |
| BORE3D | OPR | 4.8 | 3.3 | 39 | 0 | 1 | 1 | 0 | 0 |
| BRANDY | OPR | 6.1 | 3.6 | 39 | 0 | 1 | 1 | 0 | 0 |
| CAPRI | OPT | 9.2 | 5.8 | 38 | 0 | 2 | 4 | 0 | 0 |
| CYCLE | SIN | 619.8 | 1.3 | 94 | 0 | 4 | 8 | 0 | 0 |
| CZPROB | OPR | 80.1 | 1.5 | 82 | 0 | 2 | 3 | 0 | 0 |
| D2Q06C | OPT | 1988.1 | 1.3 | 87 | 0 | 2 | 2 | 2 | 0 |
| DEGEN2 | OPR | 32.2 | 4.0 | 35 | 0 | 1 | 1 | 0 | 0 |
| DEGEN3 | OPR | 674.8 | 4.8 | 47 | 0 | 4 | 7 | 0 | 0 |
| E226 | SIN | 16.4 | 5.4 | 85 | 4 | 1 | 1 | 2 | 0 |
| ETAMACRO | SIN | 64.6 | 6.7 | 112 | 0 | 6 | 12 | 0 | 0 |
| FFFFF800 | OPR | 75.9 | 3.7 | 82 | 0 | 2 | 3 | 0 | 0 |
| FINNIS | SIN | 19.3 | 0.0 | 67 | 0 | 0 | 0 | 0 | 0 |
| FORPLAN | SIN | 16.8 | 0.0 | 74 | 0 | 0 | 0 | 0 | 0 |
| GANGES | OPT | 115.8 | 1.1 | 80 | 0 | 1 | 1 | 0 | 0 |
| GFRD-PNC | OPT | 9.1 | 3.3 | 43 | 0 | 2 | 3 | 0 | 0 |
| GREENBEA | SIN | 621.7 | 0.0 | 111 | 0 | 0 | 0 | 0 | 0 |
| GREENBEB | SIN | 1021.2 | 0.9 | 191 | 2 | 0 | 0 | 0 | 0 |
| GROW15 | OPT | 12.1 | 5.0 | 39 | 0 | 2 | 3 | 0 | 0 |
| GROW22 | OPT | 25.2 | 18.5 | 48 | 0 | 4 | 6 | 7 | 13 |
| GROW7 | OPT | 5.3 | 4.9 | 40 | 0 | 2 | 3 | 0 | 0 |
| ISRAEL | OPR | 28.0 | 7.2 | 35 | 1 | 1 | 1 | 0 | 0 |
| KB2 | OPT | 0.5 | 2.0 | 25 | 0 | 1 | 1 | 0 | 0 |
| LOTFI | SIN | 3.4 | 0.0 | 37 | 0 | 0 | 0 | 0 | 0 |
| NESM | SIN | 143.5 | 0.0 | 102 | 0 | 0 | 0 | 0 | 0 |
| PEROLD | SIN | 131.7 | 10.1 | 79 | 0 | 5 | 10 | 0 | 0 |
| PILOT | SIN | 4206.5 | 0.0 | 118 | 0 | 0 | 0 | 0 | 0 |
| PILOT.JA | SIN | 410.2 | 5.2 | 96 | 0 | 3 | 6 | 0 | 0 |
| PILOT.WE | SIN | 106.5 | 0.0 | 103 | 0 | 0 | 0 | 0 | 0 |
| PILOT4 | SIN | 92.1 | 1.5 | 106 | 0 | 1 | 2 | 0 | 0 |
| PILOTNOV | OPT | 204.1 | 1.7 | 56 | 0 | 1 | 1 | 0 | 0 |
| RECIPE | OPT | 2.9 | 3.1 | 60 | 2 | 1 | 1 | 0 | 0 |

**Table B3: Solution Statistics (sum-bounding centering, part 1)**

| Problem | RC | time | %stop | iters | CI | OC | DI | PI | PR |
|---|---|---|---|---|---|---|---|---|---|
| SC105 | OPT | 0.5 | 3.7 | 19 | 0 | 1 | 1 | 0 | 0 |
| SC205 | OPT | 1.5 | 3.3 | 26 | 0 | 1 | 1 | 0 | 0 |
| SC50A | OPT | 0.2 | 0.0 | 15 | 0 | 1 | 1 | 0 | 0 |
| SC50B | OPT | 0.2 | 5.9 | 12 | 0 | 1 | 1 | 0 | 0 |
| SCAGR25 | OPT | 4.8 | 2.3 | 33 | 0 | 1 | 1 | 0 | 0 |
| SCAGR7 | OPR | 1.2 | 12.1 | 30 | 0 | 4 | 7 | 0 | 0 |
| SCFXM1 | SIN | 13.3 | 0.0 | 62 | 0 | 0 | 0 | 0 | 0 |
| SCFXM2 | SIN | 37.5 | 0.0 | 83 | 0 | 0 | 0 | 0 | 0 |
| SCFXM3 | SIN | 62.9 | 0.0 | 91 | 0 | 0 | 0 | 0 | 0 |
| SCORPION | OPR | 5.7 | 1.8 | 50 | 0 | 1 | 1 | 0 | 0 |
| SCRS8 | OPT | 25.8 | 7.7 | 65 | 4 | 2 | 2 | 2 | 0 |
| SCSD1 | OPT | 2.3 | 4.4 | 16 | 0 | 1 | 1 | 0 | 0 |
| SCSD6 | OPA | 8.9 | 6.0 | 32 | 0 | 2 | 3 | 0 | 0 |
| SCSD8 | OPT | 10.3 | 3.5 | 18 | 0 | 1 | 1 | 0 | 0 |
| SCTAP1 | OPT | 4.8 | 1.9 | 34 | 0 | 1 | 1 | 0 | 0 |
| SCTAP2 | OPT | 23.4 | 2.8 | 27 | 0 | 1 | 1 | 0 | 0 |
| SCTAP3 | OPT | 34.1 | 2.8 | 28 | 0 | 1 | 1 | 0 | 0 |
| SEBA | OPR | 302.0 | 4.0 | 45 | 0 | 1 | 1 | 0 | 0 |
| SHARE1B | OPR | 2.0 | 3.0 | 31 | 0 | 1 | 1 | 0 | 0 |
| SHARE2B | OPR | 1.1 | 4.7 | 24 | 0 | 1 | 1 | 0 | 0 |
| SHELL | OPT | 15.9 | 1.3 | 49 | 0 | 1 | 1 | 0 | 0 |
| SHIP04L | OPR | 26.4 | 1.4 | 52 | 0 | 1 | 1 | 0 | 0 |
| SHIP04S | OPR | 16.0 | 1.6 | 49 | 0 | 1 | 1 | 0 | 0 |
| SHIP08L | OPR | 101.1 | 1.5 | 98 | 0 | 2 | 3 | 0 | 0 |
| SHIP08S | OPR | 45.2 | 3.1 | 82 | 0 | 3 | 5 | 0 | 0 |
| SHIP12L | OPR | 105.3 | 0.9 | 79 | 0 | 1 | 1 | 0 | 0 |
| SHIP12S | OPR | 47.4 | 1.1 | 74 | 0 | 1 | 1 | 0 | 0 |
| SIERRA | SIN | 47.8 | 1.5 | 55 | 0 | 1 | 2 | 0 | 0 |
| STAIR | SIN | 51.3 | 6.4 | 59 | 1 | 2 | 4 | 0 | 0 |
| STANDATA | OPT | 7.1 | 2.3 | 27 | 0 | 1 | 1 | 0 | 0 |
| STANDMPS | OPT | 21.9 | 5.2 | 61 | 2 | 2 | 3 | 0 | 0 |
| STOCFOR1 | OPT | 0.8 | 1.3 | 19 | 0 | 1 | 1 | 0 | 0 |
| STOCFOR2 | OPT | 42.4 | 2.1 | 38 | 0 | 1 | 1 | 0 | 0 |
| STOCFOR3 | OPR | 690.3 | 1.7 | 71 | 0 | 1 | 1 | 0 | 0 |
| TUFF | SIN | 37.9 | 1.0 | 96 | 0 | 1 | 2 | 0 | 0 |
| VTP.BASE | OPT | 3.6 | 1.7 | 41 | 0 | 1 | 1 | 0 | 0 |
| WOOD1P | SIN | 275.6 | 1.2 | 76 | 0 | 1 | 2 | 0 | 0 |
| WOODW | SIN | 554.2 | 0.6 | 133 | 0 | 1 | 2 | 0 | 0 |

**Table B4: Solution Statistics (sum-bounding centering, part 2)**

## REFERENCES

[1] I. Adler, N. K. Karmarkar, M. G. C. Resende, and G. Veiga, "An Implementation of Karmarkar's Algorithm for Linear Programming," *Math. Programming* **44** #3 (1989), pp. 297–335.

[2] M. Arioli, I. S. Duff, and P. P. M. de Rijk, "On the Augmented System Approach to Sparse Linear Least-Squares Problems," *Numerische Math.* **55** #6 (1989), pp. 667–684.

[3] E. R. Barnes, S. Chopra, and D. L. Jensen, "A Polynomial Time Version of the Affine Scaling Algorithm," (1988), IBM T. J. Watson Research Center, Yorktown Heights, NY.

[4] A. R. Curtis and J. K. Reid, "On the Automatic Scaling of Matrices for Gaussian Elimination," *J. Inst. Maths. Applics.* **10** (1972), pp. 118–124.

[5] J. J. Dongarra and E. Grosse, "Distribution of Mathematical Software by Electronic Mail," *Communications of the ACM* **30** #5 (May 1987), pp. 403–407.

[6] S. I. Feldman, D. M. Gay, M. W. Maimone, and N. L. Schryer, "f2c — A Fortran-to-C Converter," Computing Science Technical Report No. ??? (1990), AT&T Bell Laboratories, Murray Hill, NJ.

[7] C. W. Fraser, "A Language for Writing Code Generators," *SIGPLAN Notices* **24** #7 (1989), pp. 238–245.

[8] D. M. Gay, "Electronic Mail Distribution of Linear Programming Test Problems," *COAL Newsletter* #13 (1985), pp. 10–12.

[9] D. M. Gay, "Pictures of Karmarkar's Linear Programming Algorithm," Computing Science Technical Report No. 136 (Jan. 1987), AT&T Bell Laboratories, Murray Hill, NJ 07974.

[10] D. M. Gay, "A Variant of Karmarkar's Linear Programming Algorithm for Problems in Standard Form," *Mathematical Programming* **37** #1 (1987), pp. 81–90.

[11] D. M. Gay, "Massive Memory Buys Little Speed for Complete, In-Core Sparse Cholesky Factorizations," Numerical Manuscript 88–04 (1988), AT&T Bell Laboratories, Murray Hill, NJ.

[12] A. George and J. W. Liu, *Computer Solution of Large Sparse Positive Definite Systems,* Prentice-Hall, Englewood Cliffs, NJ, 1981.

[13] N. Karmarkar, "A New Polynomial-time Algorithm for Linear Programming," *Combinatorica* **4** (1984), pp. 373–395.

[14] N. Karmarkar, private communication (1985).

[15] M. Kojima, "Determining Basic Variables of Optimal Solutions in Karmarkar's New LP Algorithm," *Algorithmica* **1** #4 (June 1986), pp. 499–515.

[16] K. O. Kortanek and Z. Jishan, "New Purification Algorithms for Linear Programming," *Naval Research Logistics* **35** (1988), pp. 571–583.

[17] I. J. Lustig, R. E. Marsten, and D. F. Shanno, "Computational Experience with a Primal-Dual Interior Point Method for Linear Programming," Tech. Report SOR 89-17 (Oct. 1989), Dept. of Civil Engin. and Operations Research, Princeton Univ., Princeton, NJ.

[18] R. E. Marsten, M. J. Saltzman, D. F. Shanno, G. S. Pierce, and J. F. Ballintijn, "Implementation of a Dual Affine Interior Point Algorithm for Linear Programming," Working Paper CMI-WPS-88-06 (Nov. 1988), Univ. of Arizona Center for the Management of Information, Tucson, AZ.

[19] N. Megiddo, "On Finding Primal- and Dual-Optimal Bases," RJ 6328 (61997) (July 1988), IBM Almaden Research Center, San Jose, CA.

[20] N. Megiddo and M. Shub, "Boundary Behavior of Interior Point Algorithms in Linear Programming," *Math. of Operations Research* **14** #1 (1989), pp. 97–146.

[21] C. L. Monma and A. J. Morton, "Computational Experience with a Dual Affine Variant of Karmarkar's Method for Linear Programming," *Operations Research Letters* **6** #6 (1987), pp. 261–267.

[22] C. C. Paige and M. A. Saunders, "Algorithm 583 LSQR: Sparse Linear Equations and Least Squares Problems," *ACM Trans. Math. Software* **8** (1982), pp. 195–209.

[23] C. C. Paige and M. A. Saunders, "LSQR: Sparse Linear Equations and Least Squares Problems," *ACM Trans. Math. Software* **8** (1982), pp. 43–71.

[24] K. Ponnambalam and A. Vannelli, "An Inexpensive Basis Recovery Procedure for Karmarkar's Dual Affine Method," manuscript (1989?), Faculty of Engineering, Univ. of Waterloo.

[25] L. D. Pyle, "The Generalized Inverse in Linear Programming. Basic Structure," *SIAM J. Appl. Math.* **22** (1972), pp. 335–355.

[26] R. A. Tapia and Y. Zhang, "A Fast Optimal Basis Identification Technique for Interior Point Linear Programming Methods," Report TR89-1 (Oct. 1989), Dept. of Mathematical Sciences, Rice Univ., Houston, TX.

[27] M. J. Todd, "Recent Developments and New Directions in Linear Programming," pp. 109–157 in *Mathematical Programming*, ed. M. Iri and K. Tanabe, Kluwer Academic Publishers (1989).

[28] M. J. Todd and B. P. Burrell, "An Extension of Karmarkar's Algorithm for Linear Programming Using Dual Variables," *Algorithmica* **1** #4 (1986), pp. 409–424.

[29] Y. Ye, "Recovering Optimal Basis in Karmarkar's Polynomial Algorithm for Linear Programming," manuscript (July 1987), Integerated Systems, Inc., Santa Clara, CA.

# CHAPTER 4

## Polyhedral Invariants and a Representative Theorem in Projective Geometry

Gilberto Calvillo Vives*
Guillermo López Mayo**

ABSTRACT. The improvement in the computational efficiency to solve linear programs achieved by N. Karmarkar using projective transformations of polyhedra motivated us to study polyhedra theory in the framework of projective geometry. In this paper we define the concepts of convexity and convex polyhedra in projective spaces. We show certain polyhedral invariants under projective transformations and present results similar to Minkowski and Weyl representation theorems.

**1. INTRODUCTION.** The ellipsoid method of L. Khachiyan [8] to solve linear programming in polynomial time is rather slow in practice compared to the simplex method which is not polynomial in a worse-case analysis [10], but it is so in the average case [2].

The search for a polynomial algorithm for linear programming faster than the ellipsoid method led N. Karmarkar to design, in 1984, his now famous algorithm. He uses projective transformations of polyhedra as the main tool of his algorithm. A logical question for a mathematician to ask is what does projective geometry has to say about polyhedra and linear programming?

As we were asking this question we were unable to find a definition of convexity in projective geometry and a proper notion of projective polyhedra. In this paper we introduce these concepts and prove they are projective invariants in the sense that images of convex sets under projective transformations are convex and a similar result holds for polyhedra. At the end, representation theorems similar to those of Minkowski and Weyl are presented.

Recently we have learned of a number of papers about projective convexity [12], [14], [6], [11] based in another definition of convex set in projective space. Ours is more general. Also, the results presented here seem to be new.

**2. CONVEX SETS IN THE N-DIMENSIONAL PROJECTIVE SPACE.** Among the several models for the n-dimensional projective space $P^n$, the one most suitable for our purposes is that in which the points of $P^n$ are the lines passing through the origin in $R^{n+1}$. For $z \epsilon R^{n+1}$, $[z]$ denotes the line through the origin and $z$.

---

 * Instituto Politécnico Nacional and Banco de México
 ** Universidad Autónoma de Puebla

Two points $A \equiv [x]$, $B \equiv [y]$, in $P^n$ define a line $l$ which corresponds to the plane in $R^{n+1}$ spanned by $x$ and $y$. The set of all lines through the origen contained in such a plane are the points of the projective line in $P^n$ joining $A$ and $B$. Clearly the points $A$ and $B$ divide $l$ into two sets.

$$(A,B)^+ = \{[z] \in P^{n+1} : z = \alpha x + \beta y, \ \alpha\beta > 0\}$$

$$(A,B)^- = \{[z] \in P^{n+1} : z = \alpha x + \beta y, \ \alpha\beta < 0\}$$

Called the open segments of $l$ defined by $A$ and $B$. When $A$ and $B$ are adjoint to the open segments, one obtains the closed segments $[A,B]^+$ and $[A,B]^-$.

A set $S$ contained in $P^n$ is said to be *convex* if given any pair of points $A$, $B$ in $S$, at least one of the two closed segments defined by $A$ and $B$ is contained in $S$.

## 3. SOME PROPERTIES OF PROJECTIVE CONVEX SETS.

Given four points $A$, $B$, $C$ and $D$ in a line $L$ we say that $A$, $B$ separates $C$, $D$ if $C \in (A,B)^+$ and $D \in (A,B)^-$. Clearly if $A$, $B$ separates $C$, $D$, then $C$, $D$ separates $A$, $B$.

LEMMA 1. Let $K$ be a subset of $P^n$ and let $A$, $B$, $C$ and $D$ be four points in a line, such that $A$, $B \in K$ and $C$, $D \in K^c$. If $A$, $B$ separates $C$, $D$, then neither $K$ nor $K^c$ are convex.

The proof follows immediately from the definition of convexity. Using Lemma 1 it is easy to prove the following:

THEOREM 1. A set $K$ in $P^n$ is a convex set if and only if its complement is convex.

PROOF. Assume that $K$ is convex and that $K^c$ is not. Let $A$, $B$ be two points in $K^c$. Since $K^c$ is not convex then there are points $C \in (A,B)$ and $D \in (B,A)$ which are in $K$. Lemma 1 then implies $K$ is not convex. So, $K^c$ must be convex. ◊

Although projective convexity is not preserved in general by taking intersections we have the following result:

THEOREM 2. If $K_1$ and $K_2$ are convex sets in $P^n$ and there is a hyperplane $H$ such that $K_1 \cap H = \phi$ and $K_2 \cap H = \phi$, then $K_1 \cap K_2$ is a convex set.

PROOF. If $K = K_1 \cap K_2$ is not convex then there exist points $A$, $B$ in $K$ such that neither $(A,B)^+$ nor $(A,B)^-$ are contained in $K$. Thus either $(A,B)^+ \subset K_1$ and $(A,B)^- \subset K_2$ or $(A,B)^+ \subset K_2$ and $(A,B)^- \subset K_1$. Thus, the line $L$ which passes through $A$ and $B$ is contained in $K_1 \cup K_2$. Since $L \cap H \neq \phi$, we have either $K_1 \cap H \neq \phi$ or $K_2 \cap H \neq \phi$, contradicting the hypothesis of the theorem. ◊

Theorem 2 also follows from the observation that the convex sets that fail to intersect a given hyperplane are precisely the affine convex sets which are contained in the affine chart obtained by removing such a hyperplane.

Just as affin transformations preserve convexity in $R^n$, projective transformations preserve convexity in $P^n$.

A *projective transformation* from $P^m$ to $P^n$ is one of the form

$$T([x]) = [Mx]$$

where $M$ is a matrix.

If $M$ is a square nonsingular matrix, then $T$ is said to be *regular*. It is straightforward to prove the following.

LEMMA 2. A projective transformation $T$ maps segments into segments. ◊

Notice that if $T$ is not regular the image of a segment can be a point. Notice also that if we restrict the projective transformation to an affine chart of $P^n$, then the lemma is no longer true.

THEOREM 3. *If $K$ is a convex set in $P^n$ and $T$ is a projective transformation from $P^m$ to $P^n$, then $T(K)$ is also a convex set.* ◊

## 4. CONVEX POLYHEDRA AND ITS INVARIANTS IN PROJECTIVE SPACES.

A polyhedron in $R^n$ which contains at least one vertex can be described as a convex combination of its vertices plus nonnegative linear combinations of its extreme rays. In projective geometry we use a similar notion to define the concept of a polyhedron though vertices and extreme rays become the same concept. A vector $s = (s_1, s_2, \ldots, s_k)$ where $s_1 \epsilon \{-1, 1\}$ is called a *combination of signs* of size $k$. The polyhedron in $P^n$ generated by the set $X = \{[x_1], \ldots, [x_k]\}$ and a combination of signs $s$, is the set

$$P(X, s) = \{[x] \epsilon P^n : x = \sum_{i=1}^k \alpha_i x_i,\ \alpha_i s_i \geq 0,\ i = 1, \ldots, k\}$$

Notice that $P(X, s) = P(X, -s)$, however any other change in $s$ may yield a different polyhedron. This is the main difference with euclidean geometry where there is exactly one polyhedron generated by a set of points, namely the so called convex hull of them. In projective geometry $k$ points can generate up to $2^{k-1}$ different polyhedra [14]. This is so, when the $k$ points are in general position. In such a case, the polyhedra are $k$-simplices that tessellate a $k$-plane of $P^n$.

It is easy to see that if $s^+$ denotes the set of signs where all $s_i = +1$ then

$$P(X, s) = P(\{[s_1 x_1],\ \ldots,\ [s_k x_k]\},\ s^+).$$

Thus any polyhedron can be represented by nonnegative linear combinations of a set of vectors in $R^{n+1}$. However, this does not imply a 1-1 correspondence between polyhedra in $P^n$ and polyhedral cones in $R^{n+1}$. Observe that a polyhedral cone in $R^{n+1}$ and the negative of it represent the same polyhedron in $P^n$.

It is immediate that polyhedra in $P^n$ are convex as shown by the following theorem.

THEOREM 4. *Every polyhedron $P = P(X, s^+)$ in $P^n$ is a convex set.*

PROOF. Pick $[y], [z] \epsilon P$. By definition

$$y = \sum_{i=1}^k \alpha_{1i} x_i, \qquad \alpha_{1i} s_i \geq 0$$

$$z = \sum_{i=1}^k \alpha_{2i} x_i, \qquad \alpha_{2i} s_i \geq 0$$

Hence, the segment

$$([y], [z]) = \{[x] : x = \sigma y + \pi z,\ \sigma, \pi > 0\}$$

is contained in $P$ since for all $[\omega] \epsilon ([y], [z])$ we have

$$\omega = \sigma y + \pi z$$
$$= \sigma(\sum_{i=1}^k \alpha_{1i} x_i) + \pi(\sum_{i=1}^k \alpha_{2i} x_i) =$$
$$= \sum_{i=1}^k (\sigma \alpha_{1i} + \pi \alpha_{2i}) x_i,$$

where $(\sigma \alpha_{1i} + \alpha_{2i}) s_i = \sigma \alpha_{1i} s_i + \pi \alpha_{2i} s_i > 0$ since each term is positive. ◊

From the above discussion it follows that there is a close relationship between projective polyhedra and polyhedral cones, and so, much of the theory of polyhedral cones translates directly to a theory of projective polyhedra. There are however some other results which belong completely to the realm of projective geometry and which we now present.

LEMMA 3. Let $P = P(X, s^+)$ be a polyhedron, $H$ a hyperplane that does not intersect $P$, and $Q$ a proper subset of $P$ which contains $X$. Then, there is a point $[y] \in P - Q$ and points $[z_1]$ and $[z_2]$ in $Q$ such that $[y] \in ([z_1, z_2])^+ \subset P$.

PROOF. Consider any point $[y]$ in $P - Q$. Then

$$y = \sum_{i=1}^{k} \alpha_i x_i, \quad \alpha_i \geq 0 \ i = 1, \ldots, k$$

Let $I$ be the set of indices such that $\alpha_i > 0$. So,

$$y = \sum_{i \in I}^{k} \alpha_i x_i, \quad \alpha_i > 0, \ \forall i \in I$$

If $[I] = 2$, then $\{[x_i] : i \in I\} = \{[z_1], [z_2]\}$ is the desired pair of points.

If $[I] > 2$, let $j \in I$ and write

$$y = q + \alpha_j x_j,$$

where

$$q = \sum_{i \in I - \{j\}} \alpha_i x_i, \quad \alpha_i \geq 0, \forall i \in I - \{j\}.$$

If $[q] \in Q$, we are done, since $([q], [x_j])^+ \subset P$ because for $\sigma > 0$,

$$\sigma q + \alpha_j x_j = \sum_{i \in I - \{j\}} \sigma \alpha_i x_i + \alpha_j x_j \text{ and } (\sigma \alpha_i) \geq 0, \ \alpha_j \geq 0$$

If $[q] \notin Q$ then by induction the proof is complete since $q$ is a linear combination of fewer terms than $y$. ◊

THEOREM 5. $P(X, s)$ is a minimal convex set that contains $X$ if there exists a hyperplane $H$ which does not intersect $P(X, s)$.

PROOF. Let $Q$ be a proper subset of $P$ which contains $X$. By lemma 3 there is a point $[y] \in P - Q$ and a segment $[[z_1], [z_2]] \subset P$ such that $[z_1], [z_2] \in Q$ and $y \in [[z_1], [z_2]]$. Consider the line $l$ that joins $[z_1]$ and $[z_2]$ and let $[r]$ be the intersection of $l$ and $H$. Since $[y]$ belongs to the segment $[[z_1], [z_2]]$ which does not intersect $H$, then $[r], [y]$ separate $[z_1]$ and $[z_2]$, but since $[z_1]$ and $[z_2]$ are in $Q$ and $[r]$ and $[y]$ are in $Q^c$ then by Lemma 1, $Q$ is not convex. ◊

THEOREM 6. A regular projective transformation $T$ of a polyhedron $P = P(X, s)$ in $P^n$ is a polyhedron in $P^n$ generated by the points $T[x_1], \ldots, T[x_k]$.

PROOF. Let $T([x]) = [Mx]$ be the transformation.

$$T(P) = \{[z] \in P^n : z = Mx, x \in P\}$$

$$= \{[z] \in P^n : z = M(\sum_{i=1}^{k} \alpha_i x_i) \quad \alpha_i s_i \geq 0 \}$$

$$= \{[z] \in P^n : z = \sum_{i=1}^{k} \alpha_i (Mx_i) \quad \alpha_i s_i \geq 0 \}$$

$$= P(Z, s)$$

where
$$Z = \{[Mx_1], \ldots, [Mx_k]\}$$
$$= \{T([x_1]), \ldots, T([x_k])\}. \quad \Diamond$$

Theorem 6 is important because it establishes polyhedrality as a projective invariant, which was our main question after reading Karmarkar's work.

**5.- REPRESENTATION THEOREMS.** Cornerstone theorems of classical polyhedra theory establish the equivalence between the representation of polyhedra by means of linear inequalities and as convex combinations of its vertices and non negative combinations of directions. Similar results hold for projective geometry:

THEOREM 7. Let $P(X, s^+)$ be a polyhedron which is a proper subset of $P^n$. Then, there exists a matrix $A$ such that
$$P(X, s^+) = \{[x] \epsilon P^n : Ax \leq 0\}.$$

This theorem is a generalization of Weyl's and can be proved essentially in the same way.

The following sketch of proof is inspired in the one given by Stoer and Witzgall ([13]), p. 57):

PROOF. We assume, without loss of generality, that $P = P(x, s^+)$ is full dimensional since otherwise it is enough to prove the theorem in the minimal subspace $\pi$ that contains $P$ and extend the relative hyperplanes to the whole space.

The proof is by induction on the dimension of the polyhedron. Clearly for dimension 0 and 1 the theorem is true. Assume that $P \subseteq P^n$.

If $P$ has no extreme points, then it has a point $x_0$ such that for any other point $x$ in $P$ the line $l_{x_0 x}$ is in $P$. Let $H$ be any hyperplane which does not contain $x_0$. $P \cap H$ is a polyhedron of dimension one less than $P$ generated by $\{l_{x_0 x_i} \cap H; i = 1, \ldots, k\}$. Thus by the induction hypothesis $P \cap H$ is bounded by a set of hyperplanes of $H$. Each one of those is then extended to a hyperplane of $P^n$ by taking the linear closure with $x_0$.

If $P$ has an extreme point, consider for each extreme point $x_i, i \epsilon I$, the set of lines through that point and the other points of $P$. Call $K_i$ the union of such lines. It can be proved that $P = \cap_i K_i$. Moreover the argument given above can be applied to each $K_i$ to prove that it can be represented as $\quad K_i = \{[x] \epsilon P^n : A^i x \leq 0\}$, where $A^i$ is a matrix.

Consequently
$$P = \{[x] \epsilon P^n : A^i x \leq 0 \quad i \epsilon I\}. \quad \Diamond$$

The converse of Weyl's theorem is Minkowski's theorem which can be stated in projective space as follows:

THEOREM 8. Any non-empty set $Q$ of the form $Q = \{[x] \epsilon P^n : a_i x \leq 0 \, i \epsilon I\}$, where $a_i \epsilon \mathbb{R}^{n+1}$, is a polyhedron.

The proof of this theorem is based on a proof of Jack Edmonds for Minkowski theorem in euclidean space.

PROOF. Assume first that $Q$ is a projective subspace. Then it is a polyhedron as we have proved already. If $Q$ is not a projective subspace, we proceed by induction on the dimension of $Q$. Assume without loss of generality that $Q$ is full dimensional. Let $H$ be the set of hyperplanes $< a_i > \equiv \{[x] \epsilon P^n : a_i x = 0\} \quad i = 1, \ldots, k$. Let $p$ be an interior point of $Q$, consider any line $l$ through $p$. Consider all points of $l$ which are not separated from $p$ by a pair of hyperplanes of $H$. If such a set is all of $l$ then the intersection of $l$ with all hyperplanes of $H$ is the same point $x_0$. In this case, the line which passes through $x_0$ and any point $q$ of $Q$ is in $Q$. Consider any hyperplane $< a_{k+1} >$ which does not contain $x_0$.

and therefore

$Q^1 \equiv \{[x]\epsilon\, P^n : a_i x = 0,\ i = 1, \ldots, k; a_{k+1} x = 0\}$ is one dimension lower than dimension of $Q$. Therefore, by induction, there is a set of points $\{x_1, \ldots, x_r\}$ such that

$$Q^1 = \{[x]\epsilon P^n : x = \sum_{i=1}^{r} \alpha_i x_i \quad \alpha_i s_i \geq 0 \quad i = 1, \ldots, k\}$$

It follows from this that

$$Q = \{[x]\epsilon P^n : x = \sum_{i=0}^{r} \alpha_i x_i \quad \alpha_i s_i \geq 0 \quad i = 1, \ldots, k\}$$

observe that $\alpha_0$ may have any sign.

The other case is when the line $l$ intercepts $Q$ in a segment $[p_0, p_1]$. Let $<a_{i_0}>$ and $<a_{i_1}>$ be the hyperplanes which contain $p_0$ and $p_1$. Consider the sets $Q^0 = \{[x] : a_i x \leq 0, i\epsilon I; a_{i_0} x = 0,\}$ and $Q^1 = \{[x] : a_i x \leq 0, i\epsilon I; a_{i_1} x = 0\}$. Both $Q^0$ and $Q^1$ are one dimension lower than $Q$ so $Q^0$ and $Q^1$ can be expressed as

$$Q^i = \{[x] : x = \sum_{j\epsilon I^i} \alpha_j x_j \quad \alpha_j s_j \geq 0\} \quad i = 0, 1$$

thus

$$p = \beta p_0 + \gamma p_1 \quad \beta\gamma > 0$$

and therefore,

$$p = \beta \sum_{j\epsilon I^0} \alpha_j x_j + \gamma \sum_{j\epsilon I'} \alpha_j x_j$$

$$= \sum_{j\epsilon I^0 \cup I^1} \sigma_j x_j$$

where $\sigma_j = \beta\alpha_j \quad j\epsilon I^0$ and $\sigma_j = \gamma\alpha_j \quad j\epsilon I^1$

Thus, we have proved that any point of $Q$ can be expressed as a signed linear combination of vertices of the arrangement of hyperplanes. ◊

## REFERENCES

[1] Baer, R., "Linear Algebra and Projective Geometry", Academic Press In., Publishers., New York, N. Y., 1952.

[2] Borwardt, K., "The average number of pivot steps required by the simplex method is polynomial", Zeitschrift für Operations Research 26, 157-177-1982.

[3] Edmonds, Jack, Verbal communication.

[4] Efímov, N.S., "Geometría Superior", Editorial Mir, Moscú 1978.

[5] Goldman, A.J., "Resolution and Separation Theorems for Polyhedral Convex Sets", Annals of Mathematics Studies, No. 38, Princeton University Press.

[6] De Groot, J. and De Vries, H., "Convex sets in projective space", Composition Math. 13 (1957), 113-118.

[7] Hatshorne, "Foundations of Projective Geometry", Cummings Publishing Co., 1967.

[8] Kachiyan, L. G. "Polynomial Algorithms in Linear Programming", Comput. Maths. Math. Phys. Vol. 20, 1980. Moscú.

[9] Karmarkar, n., "A new Polynomial-Time Algorithm for Linear Programming", AT&T Bell Laboratories, Murray Hill, New Jersey 07974, 1984.

[10] Klee V. and Minty. G.J. "How good is the simplex algorithm?", Inequalities, III (O. Shisha, ed.), Academic Press, 1972.

[11] Sinden, F. W., "Duality in Convex Programming and in Projective Space", J. Soc. Indust. Appl. Math. Vol. 11, No. 3, 1963, 535-552.

[12] Steinitz, E., "Bedingt Konvergente Reihen und Konvexe Systeme", I, II, III, J. Reine Angew Math., 1913, 1914, 1916.

[13] Stoer, J. & Witzgall C., "Convexity and Optimization in Finite Dimensions I", Springer Verlag, New York, 1970.

[14] Veblen, O. and Young, J.W., "Projective Geometry", II, Boston, 1918, p. 386.

# CHAPTER 5

## Towards a Theory of Optimization in Projective Space

Gilberto Calvillo-Vives*

**ABSTRACT.** Several researchers in the past have used tools borrowed from Projective Geometry to approach optimization problems. However, such applications of Projective Geometry have been done in Euclidean space, losing consequently part of the force of such concepts. In this note it is shown how the optimization problem can be defined in a completely projective setting and how some well-known problems fit into the formulation. Special attention is paid to the Linear Programming Problem.

**I. INTRODUCTION.** Projective geometry is not alien to optimization. Todd (1976), in the linear complementary problem; Davidon (1980), in Nonlinear Programming and Karmarkar (1984), in Linear Programming, used projective transformations to overcome some technical difficulties and make their algorithms more efficient. Fulkerson (1971) in his theory of Blocking and Antiblocking Polyhedra and Sewell (1987) in his new approach to maximum and minimum principles, used the concept of Polarity as the main tool of their research. Polarities are fundamental to establish concrete applications of duality between points and hyperplanes in Projective Geometry. However, all these applications of projective concepts were done in Euclidean Space. It is the purpose of this article to show that it is possible and may be convenient to develop a theory of Optimization in a completely projective setting. The main mathematical reason to put forward this idea is that the geometric duality between points and hyperplanes underlies all duality theory in optimization and such duality is perfect only in projective geometry. It is shown that several well-known problems belong to the realm of projective geometry, linear and Fractional Programming are redefined in this context; first order necessary conditions for optimality for unconstrained optimization are given and a gradient type algorithm is presented. Formal proofs are not given, but for most cases it is straightforward to establish the results. Some basic results are proved in other articles.

**II. PROJECTIVE CONVEX SETS AND POLYHEDRA.** The *Projective Space* $P^n$ is the space whose points are the lines through the origin (*homogeneous lines*) in $R^{n+1}$. A *k-dimensional subspace* of $P^n$ is the set of points that correspond to the set of homogeneous lines contained in a (k+1)-dimensional linear subspace of $R^{n+1}$. For $x$ in $R^{n+1}$, let $[x]$ denote the unique homogeneous line through $x$ in $R^{n+1}$. Similarly, for a finite subset X of $R^{n+1}$, [X] denotes the subspace of $R^{n+1}$ spanned by X. Thus $[x]$ is the point of $P^n$ represented by $x$; while [X] is the projective subspace of $P^n$ represented by X.

Given two points $[x]$ and $[y]$ in $P^n$ the unique line $[x,y]$ in $P^n$ defined by the two points can be thought as composed of two *segments* $[x,y]+$ and $[x,y]-$ defined by the nonnegative linear combinations of $x$ and $y$ and its complement, respectively. A subset K of $P^n$ is *convex* if for any two points $[x]$ and $[y]$ in K at least one of the segments of $[x,y]$ is in K.

---

* Instituto Politécnico Nacional de México and Banco de México.

It is immediate from the definition that complements of convex sets are *convex* while the intersection of two convex sets may not be convex. It was shown by Calvillo and López (1990) that images of convex sets under projective transformations are convex, i.e. convexity is a projective invariant.

For convex set A in $P^n$ its *type*, $\tau(A)$, is the maximum dimension of a projective subspace contained in A. It is convenient to define $\tau(\phi) = -1$. Euclidean convex sets are all homotopically equivalent to projective convex sets of type 0.

THEOREM 1. (Bracho and Calvillo). For any convex set A of $P^n$,

$$\tau(A) + \tau(P^n \setminus A) = n - 1$$

Proofs of this theorem are given in Bracho and Calvillo (1989) and Bracho and Calvillo (1990). In the first paper, it is shown how this theorem generalizes the classical separation theorems of euclidean convex sets.

Given a set of vectors $X = \{x_1, x_2, \ldots, x_k\}$ in $R^{n+1}$ and a set of signs $S = \{s_1, s_2, \ldots, s_k\}$, where a sign is 1 or -1, the *polyhedron* P(X,S) defined by X and S is

$$P(X,S) = \{[x] \epsilon P^n : x = \sum_{i=1}^{k} s_i \lambda_i x_i; \; s_i \lambda_i \geq 0\}$$

Notice that segments are special cases of polyhedra. Let $-S = \{-s_1, -s_2, \ldots, -s_k\}$, then $P(X,S) = P(X,-S)$. A linearly independent set X generates $2^{k-1}$ polyhedra, one for each sign set. When X is the cannonical basis of $R^{n+1}$, the polyhedra generated by X are a tessellation of $P^n$ into $2^n$ projective n-dimensional simplices. Each simplex corresponds to a pair of opposite orthants of $R^{n+1}$. More about Projective Polyhedra can be found in Calvillo and López (1990).

A less general definition of projective convexity was given by Steinitz (1913) and Veblen and Young (1918), who, among other things studied the tessellation of $P^n$ into simplices. Further work on that line was done by Dekker (1955) and Groot and de Vries (1957). Very recently, the author learned from Jack Edmonds about the work of Sinden (1963) who, using a definition of convexity equivalent to Steinitz', started a theory of optimization in projective space. In his work Sinden arrived at the conclusion that a more general definition of convexity was needed. Such a generalization may be achieved by our definition.

### III. ORIENTED MATROIDS AND MINTY'S LEMMA.

Fulkerson (1968), Rockafellar (1969) and Camion (1968) noticed that the fundamental properties of digraphs and digraphoids generalize naturally to euclidean matroids where orientation was taken into account, in particular Minty's lemma of painted arcs. Rockafellar suggested that it should be possible to develop a theory of oriented matroids. Bland (1974), Las Vergnas (1976) and Folkman and Lawrence (1976) developed independently the axiomatics of such a theory. Bland and Las Vergnas (1979) proved that Minty's lemma characterizes oriented matroids. It is interesting to notice that the theory of oriented representable matroids belongs to the realm of real projective spaces, since all its propositions are about the relationship between linear subspaces of $R^n$ (projective subspaces of $P^{n-1}$) and the coordinate subspaces (tessellation of $P^{n-1}$ into simplices). The following describes how oriented matroids arise in projective geometry when the proper translation from real vector spaces is performed:

The tessellation of $P^n$ into $2^n$ simplices described above corresponds to the tessellation of $R^{n+1}$ into $2^{n+1}$ orthants, identifying pairs of orthants which are symetric relative to the origen. A linear subspace and its orthogonal complement in $R^{n+1}$ become complementary projective subspaces of $P^n$, as in theorem 1. The tessellation of $P^n$ can be thought as a simplicial complex C whose faces form a partially ordered set under inclusion. The simple fact that two points in a projective line divide it into two segments serves to attach a sign vector to each face of C. For every projective subspace N, there are two sets C and $C^*$ of faces of C which are the minimal ones that intersect N and $N^\perp$ respectively. C and $C^*$ are the set of circuits and cocircuits of a matroid defined over the set of vertices X of C. Moreover, if O and $O^*$ are the signed sets associated with the faces of C and $C^*$ respectively, then (X,O), (X,$O^*$) is a dual pair of oriented matroids. Thus, Minty's lemma holds for (X,O) and (X,$O^*$) since it holds for any dual pair of oriented matroids. Figure 1 shows this projective representation of oriented matroids. Regions I, II, III, and IV are the 4 simplices that tesselate the proyective plane. Line N intersect 3 one dimensional faces given rise to a 3 circuit matroid. $N^{\perp}$ is the point in face I. These two oriented matroids come from the digraphs shown in figure 1.

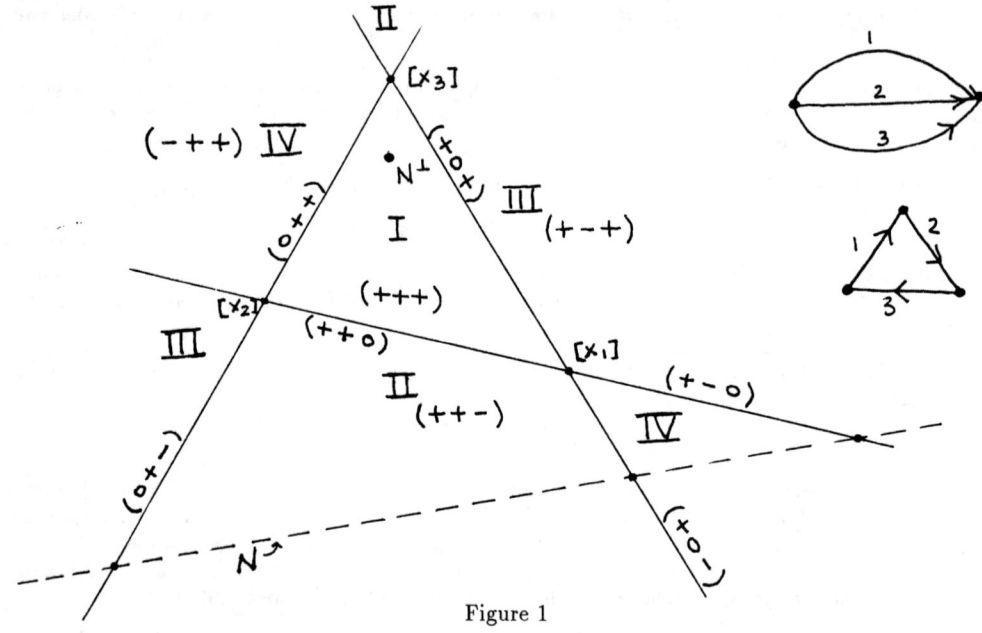

Figure 1

Another geometrical representation of euclidean oriented matroids is obtained by considering the subspace N as the host space and look at the intersection of N with C as an arrangement of hyperplanes in N. In this case, N itself can be viewed as a complex whose faces are the intersections of the faces of C with N. These faces inherited the signed vectors of their parents. Such set of signed vectors is an oriented matroid according to Edmonds and Mandel's definition (1981). Since the complex in N is generated by an arrangement of hyperplanes of N, it is possible to think of such oriented matroid as a description of the arrangement of hyperplanes by means of signed vectors.

The representation of general oriented matroids in geometrical terms was achieved by Folkman and Lawrence by means of what they call arrangements of pseudohemispheres. It happens that the description given above of how an euclidean oriented matroid can be represented as an arrangement of projective subspaces can be generalized to topological projective systems to characterize all oriented matroids. The equivalence between arrangements of pseudohemispheres and projective systems is established in chapter 4 of Mandel's thesis.

It is very interesting to notice that the elements of O* correspond precisely to the maximal faces of C which do not interesect N. This follows almost immediately from the orthogonality of O and O* as defined by Bland (1977).

**IV. THE OPTIMIZATION PROBLEM IN PROJECTIVE SPACE.** Let K be a convex set of $P^n$ and $F:K \rightarrow P$ be a continuous function. Since K is connected, its image under F, F(K), is either a proper segment of the projective line P or the whole line. In the first case the two boundary points of F(K) are called the *extrema* of F. *The optimization problem in $P^n$* is to find the extrema of a given function and their preimages, called the *extremizers*. Notice that the concepts of maximum and minimum are meaningless in this context since the projective line P does not admit a linear order, but only a cyclical one. When $K = \phi$ or $F(K) = P$ it is said that the problem has no solution. The first possibility corresponds to the classical notion of infeasibility, the second, to the one of unboundedness. When $K = P^n$ it is said that the problem is unconstrained.

**IV.1 Examples.**

**IV.1.1 The Eigenvalue Problem.** Let $K = P^n$ and $F([x]) = [(x^t Q x, x^t x)]$. Where Q is a $(n+1) \times (n+1)$ symetric matrix. Notice that F is well defined since $F([\alpha x]) = F([x])$. In this case the extrema of F are given by the eigenvectors of Q associated with the smallest and largest eigenvalues. Notice that $[(1,0)]$ is not the image of any point in $P^n$, so if it is removed from the projective line, one can get several well-known euclidean representations of the same eigenvalue problem. For example, one can write $F([x]) = [(x^tQx/x^tx, 1)]$ and then search for the maximum and minimum of the real-valued function $f(x) = x^tQx/x^tx$. In the same spirit one can rewrite the set K as $K = \{[x]\epsilon P^n : x^tx = 1\}$ and then the problem becomes:

"Find the extrema of the function $F([x])=[(x^tQx,1)]$ where $x^tx = 1$."

This again translates to the euclidean space as the well-known problem

Max (min) $x^tQx$ s.t. $x^tx = 1$.

**IV.1.2 Portafolio selection.** Let V be the Variance-Covariance matrix of yields of several stocks and let $r$ be the vector whose entries are the average yields per dollar of the given stocks. The decision variable is the vector $x$ of percentages to be invested in each stock. A model that aims at maximizing the yield while minimizing risk is, in Euclidean Space, the following:

$$\text{Max } (r^tx - \alpha) / (x^tVx)^{1/2}$$
$$\text{s.t. } 1^tx = 1$$
$$x \geq 0$$

Where $\alpha$ is a treshold value of yield and $1^t$ is the row vector of ones. In order to translate this problem to the projective geometry framework rewrite the objective function as

$$f(x) = (r^tx - \alpha 1^tx)/(x^tVx)^{1/2} = (r^t - \alpha 1^t)x/(x^tVx)^{1/2}$$

observe then that the first constraint can be removed since the objective function is now constant over each half line $\{\lambda x : \lambda \geq 0\}$. If

the function $f$ is extended to the negative orthant by making

$$f(x) = f(-x) \; \forall x \leq 0,$$

then $f$ is constant over each line $[x]$, $x \geq 0$. Thus the nonnegativity condition can be replaced by the set $K = \{[x] : x \geq 0\}$ which is a convex set in $P^n$. So, the problem in $P^n$ can be written as:

$$\text{Opt } F([x]) = [(f(x), 1)]$$
$$\text{s.t. } [x] \epsilon K$$

Notice that again $F([\lambda x]) = F([x]) \; \forall \lambda \epsilon \; R\text{-}\{0\}$, so F is a well defined function over $P^n$. As in the previous example, the formulation of the problem in projective form renders other equivalent formulations in Euclidean Space.

## V. LINEAR AND FRACTIONAL PROGRAMMING

Consider the linear programming problem in standard form:

(1)
$$\begin{aligned} \text{Min} \quad & cx \\ Ax &= b \\ x &\geq 0 \end{aligned}$$

This problem can be rewritten as:

(2) $\quad$ Min $x_0$

(2.1) $\quad x_0 - cx = 0$

(2.2) $\quad Ax - bx_{n+1} = 0$

(2.3) $\quad x_{n+1} = 1$

$\quad x \geq 0$

If one lets $\xi = (x_0, x, x_{n+1})$ and

(3)
$$\Lambda = \begin{pmatrix} 1 & -c & 0 \\ 0 & A & -b \end{pmatrix}$$

then the problem becomes

(4)
$$\begin{array}{rl} \text{Min} & x_0 \\ & \Lambda \xi = 0 \\ & x_{n+1} = 1 \\ & x \geq 0 \end{array}$$

The set $W = \{x \in R^{n+2} : \Lambda x = 0\}$ is a linear subspace of $R^{n+2}$. So, another way to write the LP problem is

(5)
$$\begin{array}{rl} \text{Min} & x_0 \\ & x_{n+1} = 1 \\ & x \geq 0 \\ & \xi = (x_0, x, x_{n+1}) \in W \end{array}$$

Let $y_0$, $y = (y_1, \ldots, y_m)$ and $y_{m+1}$ be the dual variables associated to (2.1), (2.2) and (2.3) respectively. Thus, the dual problem is

(6)
$$\begin{array}{rl} \text{Max} & y_{m+1} \\ & y_0 = 1 \\ & -y_0 c + yA \leq 0 \\ & -yb + y_{m+1} = 0 \end{array}$$

Let $\zeta = (z_0, z, z_{n+1}) = (y_0, y) \Lambda$. Then

$$z_0 = y_0 = 1$$

$$z = (y_0, y) \begin{pmatrix} -c \\ +A \end{pmatrix} = -y_0 c + yA \leq 0, \text{ and}$$

$$z_{n+1} = -yb = y_{m+1}$$

Since $\zeta \in W^\perp$, then the dual problem becomes

(7)
$$\begin{array}{rl} \text{Max} & z_{n+1} \\ & z_0 = 1 \\ & z \leq 0 \\ & \zeta = (z_0, z, z_{n+1}) \in W^\perp \end{array}$$

This formulation of dual pairs of LP problems is useful because both of them are formulated in the same space and orthogonality arises quite naturally. The purpose of doing this here is to motivate the formulation of the LP problem in projective space.

Let
$$W_+ = \{[\xi] \in P^{n+1} : \xi \in W, x \geq 0\}$$
and
$$W_+^\perp = \{[\zeta] \in P^{n+1} : \zeta \in W^\perp, z \leq 0\}$$

Consider the function

$$f : P^{n+1} \to P$$

defined as:

$$f([\xi]) = [(x_0, x_{n+1})]$$

The *linear programming problem in projective space*, is to optimize $f$ over $W_+$. The corresponding dual problem is to optimize $f$ over $W_+^\perp$. This is very similar to the formulation of Bland (1977), except that here the images of the objetive function are in the projective line instead of the real line, and no distinction is made between the roles of $x_0$ and $x_{n+1}$.

THEOREM 2. (Weak Duality). Let $[\xi] \in W_+$ and $[\zeta] \in W_+^\perp$. If $(x \geq 0$ and $z \leq 0)$ or $(x \leq 0$ and $z \geq 0)$ then $x_0 z_0 + x_{n+1} z_{n+1} \geq 0$. If $(x \geq 0$ and $z \geq 0)$ or $(x \leq 0$ and $z \leq 0)$ then $x_0 z_0 + x_{n+1} z_{n+1} \leq 0$.

Note that $\xi \cdot \zeta = 0 \Rightarrow x_0 z_0 + xz + x_{n+1} z_{n+1} = 0$. Taking into account all possibilities for the signs of $x$ and $z$ gives immediately the result.

THEOREM 3. (Strong Duality). If $W_+$ and $W_+^\perp$ are non-empty then there exist points $x^-$, $x^+$ in $W_+$ and $z^+$ and $z^-$ in $W_+^\perp$ such that

$$x_0^+ z_0^+ + x_{n+1}^+ z_{n+1}^+ = 0 \quad \text{and} \quad x_0^- z_0^- + x_{n+1}^- z_{n+1}^- = 0.$$

The traditional weak and strong duality of LP are special cases of theorems 3 and 4, with $x_{n+1}$ and $z_0$ equal to 1. A constructive proof of theorem 4 is obtained from a generalization of the simplex method. It is remarkable that in the projective linear programming problem the asimetry between the variable $x_0$, associated to the objective function, and the variable $x_{n+1}$, associated to the right hand side $b$, completely disappears.

EXAMPLE. Consider the following pair of linear programming problems in euclidean space.

Primal

$$\begin{aligned} \text{Min} \quad & x_1 + x_2 \\ & x_1 + 2x_2 = 6 \\ & x_1 \geq 0 \\ & x_2 \geq 0 \end{aligned}$$

Dual

$$\begin{aligned} \text{Max} \quad & 6y_1 \\ & y_1 \leq 1 \\ & 2y_1 \leq 1 \end{aligned}$$

The corresponding problems in projective space are

Primal

Opt. $[(x_0, x_3)]$
$[(x_0, x_1, x_2, x_3)] \epsilon W_+$

Dual

Opt. $[(z_0, z_3)]$
$[(z_0, z_1, z_2, z_3)] \epsilon W_+^\perp$

where

$$W_+ = \{[(x_0, x_1, x_2, x_3)] : \begin{array}{rrrrl} x_0 & -x_1 & -x_2 & & = 0 \\ & x_1 & +2x_2 & -6x_3 & = 0 \\ & x_1 & & & \geq 0 \\ & & x_2 & & \geq 0 \end{array} \}$$

and

$$W_+^\perp = \{[(z_0, z_1, z_2, z_3)] : (z_0, z_1, z_2, z_3) = (y_0, y_1) \begin{pmatrix} 1 & -1 & -1 & 0 \\ 0 & 1 & -2 & -6 \end{pmatrix}; z_1 \leq 0, z_2 \leq 0\}$$

The image of $W_+$ under the function $[(x_0, x_3)]$ is given by the equations (8)
$x_1 + x_2$
$$x_3 = 1/6 x_1 + 1/3 x_2 \qquad x_1, x_2 \geq 0.$$

The extrema are the images of $(1,0)$ and $(0,1)$ under (8):

$$(x_0^+, x_3^+) = (1, 1/6)$$

and

$$(x_0^-, x_3^-) = (1, 1/3).$$

The image of $W_+^\perp$ under $[(z_0, z_3)]$ is obtained by solving

$$\begin{aligned} z_0 &= y_0 \\ z_1 &= -y_0 + y_1 \leq 0 \\ z_2 &= -y_0 + 2y_1 \leq 0 \\ z_3 &= -6y_1 \end{aligned}$$

which is equivalent to

$$\begin{aligned} -z_0 - 1/6 z_3 &\leq 0 \\ -z_0 - 1/3 z_3 &\leq 0 \end{aligned}$$

or to

$$(1, 1/16) \cdot (z_0, z_3) \geq 0$$
$$(1, 1/3) \cdot (z_0, z_3) \geq 0.$$

The geometry of this situation is illustrated in figure 2.

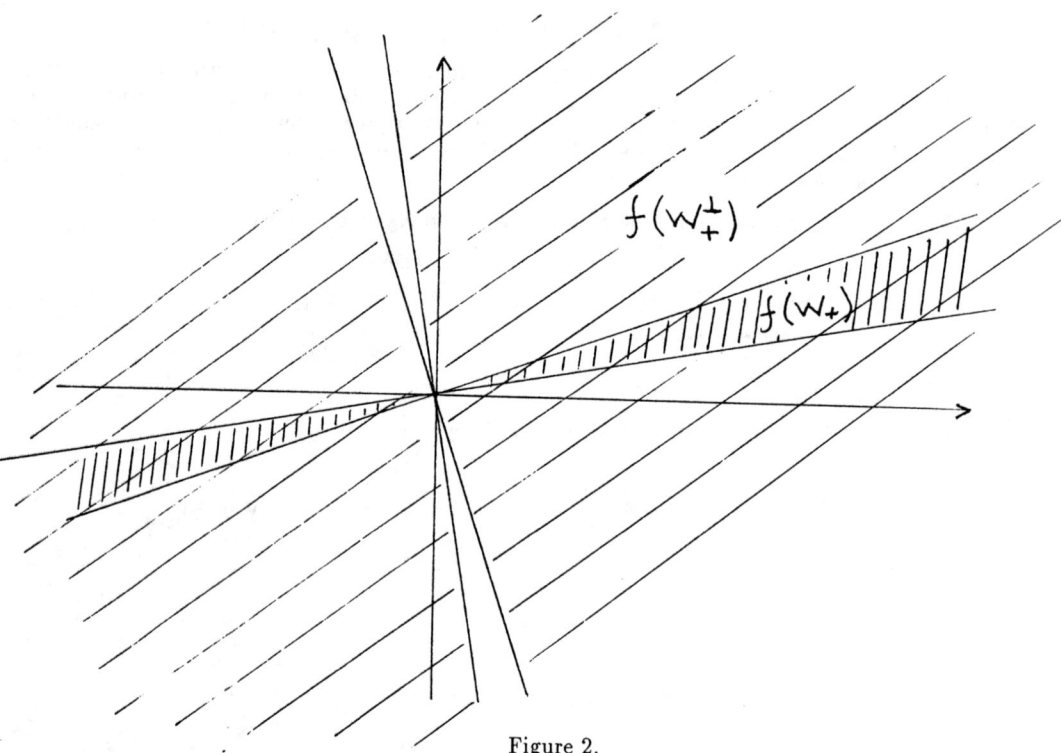

Figure 2.

Notice that problems in projective space are more general than their counterparts in euclidean space, since they deal with the two extrema of the images of $W_+$ and $W_+^\perp$ while the euclidean versions deal only with one of them.

The linear fractional programming problem in standard form is

(9)  $\qquad\qquad$ Min $cx/dx$, $Ax = b$, $x \geq 0$.

Since $cx/dx$ is constant over homogeneous lines, (9) is equivalent to

$\qquad\qquad$ Min $cx/dx$, $Ax - bx_{n+1} = 0$, $x \geq 0, x_{n+1} > 0$.

This motivates the formulation in projective space as:

(10) $\qquad\qquad$ Opt $[(cx, dx)]$

$\qquad\qquad [\xi] = [(x, x_{n+1})] \epsilon$ K

where $\qquad\qquad$ K $= \{[\xi] \epsilon$ P $^n : A^1 \xi = 0, \xi \geq 0\}$

and $\qquad\qquad A^1 = (A \mid -b)$

From this formulation it is obvious that linear and linear fractional programming in projective space are very closely related. Indeed, if one lets $x_0 = cx$ and $x_{n+2} = dx$, (10) becomes a projective linear programming problem. Thus euclidean linear and linear fractional programming can be considered as two particular cases of the same general projective linear programming problem. Equivalence of linear fractional programs with linear programming was established by Charnes and Cooper (1962). Bradley and Frey (1974) extended Charnes and Cooper results to non-linear homogeneous functions of degree one, giving as application the portafolio selection problem discussed above. In both cases, it seems that the projective approach renders simpler proofs.

Having defined Linear Programming in projective space, it is natural to ask about how Karmarkar method fits into this framework. The main technical problem to assure the validity of Karmarkar's algorithm is that the linearity of the objective function is lost when the problem is transformed projectively. Thus, a potential function has to be brought into play. The projective formulation given here is invariant under such transformations since the objective function itself is a projection from $P^{n+1}$ to $P$. Therefore, it is expected that Karmarkar's method can be generalized nicely to the completely projective sitting. This has been done at some extent by Anstreicher (1985). He claims that Karmarkar's method is fundamentally an algorithm for fractional programming, noticing that quotients of linear functions are invariant under projective transformations. One can say, in the same spirit, that Karmarkar's method is really a method for projective linear programming.

**VI. UNCONSTRAINED OPTIMIZATION.** Let $f : P^n \to P$ be a continous function. A point $x \epsilon \; P^n$ a *local optimum of f* if there exists a neighborhood N of $x$ such that $x$ is an extremizer of $f$ on N. Consider a function $g : R^{n+1} \to R^2$ that represents $f$. That is to say:

$$f([x]) = [g(x)] = [(g_1(x), g_2(x))]$$

The following theorem gives first order conditions for local optimality.

**THEOREM 4.** Let $g : R^{n+1} \to R^2$ differentiable and let $f([x]) = [g(x)]$. If $x^*$ is a local optimum of $f$, then

(11) $$g_1(x^*) \nabla g_2(x^*) - g_2(x^*) \nabla g_1(x^*) = 0$$

This is of course the analog of the condition that the gradient vanishes at an unconstrained local optimum in $R^n$.

In order to understand condition (11), consider a point $[x]$ which moves to the point $[x + d]$ where $d \epsilon \; R^{n+1}$. Using a first order aproximation in Taylor series one obtains that

$$g_1(x + d) \cong g_1(x) + \nabla g_1(x)d$$

and

$$g_2(x + d) \cong g_2(x) + \nabla g_2(x)d$$

Thus, in order to have a local optimum one needs that:

$$g_1(x) + \nabla g_1(x)d = \lambda g_1(x)$$

and

$$g_2(x) + \nabla g_2(x)d = \lambda g_2(x)$$

That is to say:

$$\nabla g_1(x)d = (\lambda - 1)g_1(x)$$

and

$$\nabla g_2(x)d = (\lambda - 1)g_2(x)$$

which immediately gives condition (11).

It is easily seen that if condition (11) does not hold, then the image of

(12) $$[x^{(1)}] = [x + \alpha(g_1(x^*) \nabla g_2(x^*) - g_2(x^*) \nabla g_1(x^*))]$$

can be forced to be in either side of $f([x])$ by a suitable choice of $\alpha$.

If one chooses consistently the sign of $\alpha$, (12) gives a gradient type algorithm.

# REFERENCES

Anstreicher, K.M. (1985). A Monotonic Projective Algorithm for Fractional Programming, Preprint, Yale School of Organization and Management. Yale University, New Haven Com.

Bland, R.G. (1974). Complementary Orthogonal Subspaces of $R^n$ and Orientability of Matroids, 'Doctoral Dissertation', Cornell University. May 1974.

Bland, R.G. (1977). A Combinatorial Abstraction of Linear Programming, *Journal of Combinatorial Theory*, Series B 23, 33-57.

Bland, R.G. and M. Las Vergnas (1979). Minty Colorings and Orientations of Matroids, *Annals of the New York Academy of Sciences*, Vol. 319, pp. 86-92.

Bracho, J. and Calvillo, G. (1989). Duality of Projective Convex Sets, *Publicaciones preliminares del Instituto de Matemáticas*, No. 185, *Universidad Nacional Autónoma de México*.

Bracho, J. and Calvillo, G. (1990). Homotopy Classification of Projective Convex Sets, Submitted for publication at Geometriae Dedicata.

Bradley, S. P. and S.C. Frey, Jr. (1974) Fractional Programming with Homegeneous Functions, *Oper. Res.* 22,350-357.

Calvillo, G. and G. López (1990) Polyhedral Invariants and a Representation Theorem in Projective Geometry, these proceedings.

Calvillo, G. and Sánchez, A. (1987). An Out of Killer Method for Linear Programming, *'Segundo Coloquio de Teoría de las Gráficas, Combinatoria y sus Aplicaciones'*. Xalapa, Veracruz, México.

Camion, P. (1968). Modules Unimodulaires, *J. Combinatorial Theory* 4 (1968), 301-362.

Charnes, A. and W.W. Cooper (1962). Programming with Linear Fractional Functions, *Nav. Res. Log. Quart.* 9, 181-186.

Davidon, W.C. (1980). Conic Approximations and Collinear Scallings for Optimizers, *SIAM J. Numer. Anal.*, Vol. 17 (2), pp. 268-281.

Dekker, D. (1955). Convex Regions in Projective n-Space, *The Amer. Math. Monthly*, 62, No. 6, 430-431.

De Groot, J. and H. de Vries (1957). Convex Sets in Projective Space, *Compos. Math.* 13, 113-118.

Folkman J. and J. Lawrence (1976). Oriented Matroids, *Journal of Combinatorial Theory*, Series B 25, 199-236.

Fulkerson, D.R. (1968). Networks, Frames, Blocking Systems,in 'Mathematics of the Decision Sciences', Lectures in Applied Mathematics, Vol. 11 (G.B. Dantzig and H.F. Veinott, Jr., Eds.) pp. 303-334. *American Mathematical Society*, Providence, R.I., 1968.

Fulkerson, D.R. (1971). Blocking and Antiblocking Pairs of Polyhedra, *Mathematical Programming* 1 (1), pp. 168-194.

Karmarkar, N. (1984). A New Polynomial-time Algorithm for Linear Programming, *Combinatorica* 4 (4), pp. 373-395.

Mandel, A. (1981). Topology of Oriented Matroids, Doctoral Dissertation. University of Waterloo, Canada.

Rockafellar, R.T. (1969). The Elementary Vectors of a Subspace of R, in 'Combinatorial Mathematics and its Applications', Proceedings of the Chapel Hill Conf. (R.C. Bose and T.A. Dowling, Eds.) pp. 104-127, *University of North Carolina Press*, Chapel Hill, 1969.

Sewell, M.J. (1987). 'Maximum and Minimum Principles; A unified approach, with applications', *Cambridge texts in Applied Mathematics, Cambridge University Press*.

Sinden, F. (1963). Duality in Convex Programming and in Projective Space, *SIAM Journal*. Vol. II, No. 3. pp. 535-552.

Steinitz, E. (1916). Bedingt Konvergente Reihen und Konvexe Systeme *J. für die reine angew.* Math 146, p. 34.

Todd, M. (1976). Extensions of Lemke's Algorithm for the Linear Complementarity Problem, *J. Opt. Theory and Applications*, Vol. 20 (4), pp. 397-416.

Veblen O. and J. W. Young (1918). 'Projective Geometry', Vol. 2, *Ginn*, Boston.

# CHAPTER 6

## Iterative Gradient-Newton Type Methods for Steady Shock Computations*

Bjorn Enquist†
Quyen Quang Huynh**

A class of modified Newton's methods are applied to difference approximations of the two-dimensional steady Burgers' equation and the transonic small disturbance equation. The solutions have sharp gradients which correspond to boundary layers and shock waves in fluid dynamics. The nonlinear terms in the differential equations are approximated by modern shock capturing schemes. The regularity of the coefficients is analyzed theoretically and its effect on the convergence on the Newton's method is studied numerically. Computational results from different types of gradient iterative methods and different types of preconditioners are presented. These methods are applied to the linear systems of the Newton iteration. The relative residuals in the Newton iterations are controlled such that a superlinear rate of convergence is preserved.

**1. Introduction.** We shall consider numerical solutions of nonlinear hyperbolic conservation laws. In two space dimensions the equation or system of equations have the form,

$$u_t + f(u)_x + g(u)_y = 0, \qquad (1.1)$$

with appropriate initial and boundary conditions. These equations are used as mathematical models in many applications. In gas dynamics, for example, the unknown vector valued function $u(x, y, t)$ has four components representing density, momentum (2 components) and energy.

---

\* Research supported by NSF-grant No. DMS88-11863, ONR-grant No. N00014-86-K-01691,AFOSR-grant No. AFOSR-87-0341 and NASA Consortium No. NCA2-372.
† Department of Mathematics, University of California, Los Angeles, CA 90024, U.S.A.
\*\* Department of Computer Sciences, Uppsala University, Sturegatan 4B 2TR, 75223 Uppsala, Sweden

Even with smooth initial values the solution of (1.1) generically develops discontinuities. These discontinuities or shocks cause both theoretical and computational difficulties.

Weak solution must be considered and uniqueness might be a problem. For scalar equations there are unique solutions if extra constraints (entropy conditions) are added [Smoller, 83]. The solution is e.g. given as the limit of vanishing viscosity solutions,

$$\lim_{\varepsilon \to 0} \int \int |u^\varepsilon(x,y,t) - u(x,y,t)|dxdy = 0,$$
$$u^\varepsilon_t + \delta(u^\varepsilon)_x + g(u^\varepsilon)_y = \varepsilon \Delta u^\varepsilon. \tag{1.2}$$

The numerical approximations of (1.1) must work well when $u$ is smooth but also at discontinuities of $u$. Traditionally numerical schemes mimicked the equation (1.2). The standard schemes contain such added artificial viscosity, [Rizzi, Engquist, 87]. During the last fifteen years new classes of so called high resolution schemes have been developed. See e.g. the survey [Colella, Woodward, 84]. These algorithms are based more directly on the properties of (1.1) at discontinuities. The result is often approximations with sharp shocks without numerical oscillations. These algorithms are nonlinear and change structure depending on the solution. This adaptivity is a source of difficulties for direct steady state computations. The purpose of this paper is to study a few questions in the coupling of high resolution schemes with modern algebraic methods for steady state computations.

A three point difference approximation of (1.1) has the form,

$$u^{n+1}_{i,j} = u^n_{i,j} - \frac{\Delta t}{\Delta x}(f(u^n_{i+1,j}, u^n_{i,j}) - f(u^n_{i,j}, u^n_{i-1,j}))$$
$$- \frac{\Delta t}{\Delta y}(g(u^n_{i,j+1}, u^n_{i,j}) - g(u^n_{i,j}, u^n_{i,j-1})), \quad t_n = n\Delta t. \tag{1.3}$$
$$u^n_{i,j} \sim u(x_i, y_j, t_n), \quad x_i = i\Delta x, y_j = j\Delta y.$$

The functions $f(\,,\,)$ and $g(\,,\,)$ are called numerical flux functions and are related to $f$ and $g$ in (1.1) via the consistency relation,

$$f(u,u) = f(u), \quad g(u,u) = g(u). \tag{1.4}$$

It is common to compute with (1.3) for large time in order to approximate the steady solution of (1.1).

We shall here consider direct approximations of the steady equation,

$$f(u)_x + g(u)_y = 0, \tag{1.5}$$

with boundary conditions. When direct approximations are feasible they are usually much faster than time evolution techniques. There is no time index $n$ in the equations and the following system of algebraic equations has to be solved

$$F(U) = 0, \tag{1.6}$$

where $F$ and $U$ are vectors with components,

$$F_{i,j} = \frac{1}{\Delta x}(f(u_{i+1,j}, u_{i,j}) - f(u_{i,j}, u_{i-1,j}))$$
$$+ \frac{1}{\Delta y}(g(u_{i,j+1}, u_{ij}) - g(u_{i,j}, u_{i,j-1})),$$
$$U = (u_{ij}).$$

Some of the equations in $F$ should also contain the boundary conditions.

We mentioned earlier the adaptive feature of the high resolution schemes. This means that $f(\,,\,)$ and $g(\,,\,)$ depends strongly on the solution and often not in a smooth way. Thus $F$ is not a smooth function of $U$ which causes trouble when solving (1.6).

In the following section we shall study the regularity of $F$ and show that very sharp shock resolution and $f, g \in C^1$ is mutually exclusive. We have to relax the sharpness of the discontinuities in the numerical solution in order to have an algebraic system which is suitable for numerical methods.

Some natural algebraic methods are outlined in section 3. The linear system resulting from a damped inexact Newton method is approximated by preconditioned gradient type methods.

In section 4 we shall apply these methods to two hyperbolic conservation laws in two space dimensions. The Burgers' equation

$$(\frac{1}{2}u^2)_x + (\frac{1}{2}u^2)_y = \varepsilon \Delta u \tag{1.7}$$

and the transonic small disturbance equation,

$$(K\phi_x - \frac{1}{2}(\gamma+1)\phi_x^2)_x + \phi_{yy} = 0 \tag{1.8}$$

are studied. In (1.7) we have a conservation law plus added viscosity. As in equation (1.2) we are interested in small values of $\varepsilon$.

**2. Regularity of Numerical Fluxes.** Consider Newton's method applied to the system (1.1): $F(U) = 0$. In the Kantorovich convergence theorem, the mapping $F$ is assumed to be continuously differentiable.

Let us consider a three point formula for the one dimensional Burgers' equation. The nonlinear term $f(u)_x = 0.5(u^2)_x$ is discretized as follows

$$[f(u)]_x = 1/\Delta x [f(u_i, u_{i+1}) - f(u_{i-1}, u_i)] \tag{2.1}$$

where $f(\,,\,)$ represents the numerical flux.

We shall now present various numerical fluxes for the Burgers' equation and compare their accuracy and smoothness properties. The shock speed is given by $u_S = 0.5(u_\ell + u_r)$, $f_+(u) = 0.5\max(u,0)^2$, $f_-(u) = 0.5\min(u,0)^2$.

*Godunov flux* (first order) [Godunov, 59]

$$f_G(u_\ell, u_r) = \max[f_+(u_\ell), f_-(u_r)]$$

If $u_S = 0$ i.e. $u_\ell = -u_r$, then $\partial f_G/\partial u_\ell = u_\ell$ or $0$

*Roe flux* (first order) [Roe, 85]

$$f_R(u_\ell, u_r) = 0.5[f(u_\ell) + f(u_r)] - 0.5|0.5(u_\ell + u_r)|(u_r - u_\ell)$$

If $u_S = 0$ then $\partial f_R/\partial u_\ell = u_\ell$ or $0$

*E-O flux* (first order) [Engquist, Osher 80]

$$f_{E-O}(u_\ell, u_r) = f_+(u_\ell) + f_-(u_r)$$

If $u_S = 0$ then $\partial f_{E-O}/\partial u_\ell = u_\ell$

*TVD flux* (second order, with van Leer limiter) [van Leer, 74]

$$\begin{aligned}f_{\text{TVD}}(u_{i-1}, u_i, u_{i+1}, u_{i+2}) =& f_{E-O}(u_i, u_{i+1}) = 0.5\Psi(R_{i+1}^-)\\ & (f_{E-O}(u_i, u_{i+1}) - f(u_i)) + 0.5\Psi(R_{i+1}^+)\\ & (f(u_{i+1}) - f_{E-O}(u_i, u_{i+1}))\end{aligned}$$

where

$$R_i^+ = (f(u_i) - f_{E-O}(u_{i-1}, u_i))/(f(u_{i+1}) - f_{E-O}(u_i, u_{i+1}))$$
$$R_i^- = (f_{E-O}(u_i, u_{i+1}) - f(u_i))/(f_{E-O}(u_{i-1}, u_i) - f(u_{i-1}))$$

$$\Psi_{VL}(R) = (|R| + R)/(1 + |R|)$$

Clearly the first derivatives of $f_G$ and $f_R$ exhibit jumps while $f_{E-O}$ is a $C^1$ function. The flux limiter $\Psi_{VL}(R)$ is differentiable.

It was shown in [Engquist, Osher, 80], that the E-O scheme admits a discrete representation of a steady shock with two interior states. Next, we shall show that for a three points scheme admitting a steady shock profile with at most one interior state, its numerical flux cannot be a $C^1$ function. Thus the sharpest possible steady discrete shock profile has two interior states for schemes with $C^1$ numerical fluxes. The Godunov and Roe schemes have discrete shock profiles with one interior state but their corresponding numerical fluxes are not $C^1$.

Consider a scalar hyperbolic conservation law

$$u_t + [f(u)]_x = 0 \tag{2.2}$$

with $f'' > 0$ and $f'(0) = 0$.

Eq. (2.2) is approximated by a three points scheme in conservation form,

$$u_i^{n+1} = u_i^n + \lambda[f(u_i^n, u_{i+1}^n) - f(u_{i-1}^n, u_i^n)], \tag{2.3}$$

where $\lambda = \Delta t/h$.

The regularity property of numerical flux is given in the following theorem.

THEOREM. *Let the scalar hyperbolic conservation law (2.2) be approximated by a consistent three points scheme in conservation form. Assume that this scheme admits the following discrete representation of a steady shock* $u_\ell > 0; f(u_\ell) = f(u_r)$, *and for any* $u_m, u_\ell \geq u_m \geq u_r$

$$u_i = u_\ell, \quad i \leq -1$$
$$u_0 = u_m,$$
$$u_i = -u_\ell, \quad i \geq 1$$

*Then the numerical flux* $f(u_\ell, u_r)$ *cannot be a* $C^1$ *function.*

PROOF. If (2.3) is applied to the point $i = -1$, then in terms of the states $u_1$ and $u_0$, we have

$$u_{-1}^{n+1} = u_{-1}^n + \lambda[f(u_{-1}^n, u_0^n) - f(u_1^n, u_1^n)] \tag{2.4}$$

Since there are multiple discrete representations of the same steady shock profile with one interior state $u_0$, $f(u_1, u_0)$ depends only on $u_1$ i.e.

$$\begin{aligned} f(u_0, u_r) - f(u_r), & \quad u_r < 0 \\ u_1 > u_0 > u_r & \end{aligned} \tag{2.5}$$

Hence from (2.4) and (2.5)

$$f(u,v) = f(u), \; u > 0, \; u > v > \underline{u} \text{ with } f(\underline{u}) = f(u) \tag{2.6}$$
$$f(u,v) = f(v), \; v < 0, \; \underline{v} > u > v \text{ with } f(\underline{v}) = f(u) \tag{2.7}$$

From (2.6) and (2.7), it follows that $f(u,v)$ cannot be a $C^1$ function since $\nabla f(\;,\;)$ is discontinuous at $(u,v) = (u, \underline{u})$ or equivalently $(u,v) = (\underline{v}, v)$.

**3. Algebraic Methods.** Consider the inexact Newton's method (IN) applied to (1.6),

$$U^{k+1} = U^k + S^k, \tag{3.1a}$$
$$J(U^k)S^k = -F(U^k) + r^k. \tag{3.1b}$$

The method is called inexact if there is an error ($r^k \neq 0$) in the solution of the linear system (3.1b). The error is controlled by a sequence $\{q_k\}$ such that $\|r^k\|/\|f(U^k)\| \leq q_k$. The Jacobian matrix of $F$ is denoted by $J$. Modifications in order to improve the global convergence properties can be done at the updating step (3.1a) as follows

$$U^{k+1} = U^k + \alpha_k S^k$$

The method is then called damped inexact Newton (DIN). The idea of a global method is to make sure that each step decreases the value of some norm of

$F : R^n \to R^n$. If we choose the $l_2$ norm $\|f(U)\|$, solving the system of nonlinear equations $F(U) = 0$ is equivalent to minimizing $g = 1/2 F(U)^T F(U)$.

Naturally one wants to choose a direction $S$ such that in this direction

$$g(U^k + \alpha_k S^k) < g(U^k) \quad \text{for some } 0 < \alpha_k \leq 1$$

It is easy to show that the vector $S^k$ is a descent direction $\nabla g(U^k)^T S < 0$, if $\|r^k\|$ is small enough.

Our damped inexact Newton method is based on an algorithm in [Dembo, Steihaug, 83]. We have incorporated a very simple backtrack technique instead of a more complicated quadratic or cubic backtrack which is described in [Dennis, Schnabel, 83]. Starting with $\alpha_k = 1$, $\alpha_k$ is reduced by a factor $\frac{1}{2}$ until a descent condition is satisfied.

From (3.1b) we see that at each Newton step, a large linear system of the form

$$AX = b \qquad (3.2)$$

needs to be solved. Several iterative gradient methods have been proposed recently to solve (3.2), where $A$ is nonsymmetric and possesses a positive definite symmetric part. In our examples, we have found the truncated GCR method called ORTHOMIN(i) to be particularly attractive in terms of computational effort and storage [Vinsome, 76]. This method, is a modification of the GCR method where only the last i direction vectors need to be saved. It is worth pointing out that the truncated version of the GMRES method proposed in [Saad, Schultz, 86], does require the positive definiteness of the symmetric part of $A$ although its full version does not. The GMRES algorithm requires less storage.

A survey and comparison of generalized gradient methods for nonsymmetric problems, is given by in [Saad, Schultz, 85]. We choose to adopt here the minimal residual (MR) method and the ORTHOMIN(1) method because of their simplicity. In particular, the MR method which is identical to ORTHOMIN(0) is a simple two-steps algorithm.

For symmetric problems, the convergence of iterative gradient methods can be accelerated by reducing the condition number of $A$. It is also well known that the rate of convergence depends on the clustering of eigenvalues into groups. However a similar theory does not exist in general for nonsymmetric problems. It is therefore necessary to conduct extensive numerical experiments for nonsymmetric matrices.

Preconditioning techniques transform the original matrix into a matrix with better properties. If $C$ is a preconditioning matrix, instead of solving $AX = b$, we solve $AC^{-1}CX = b$.

All the preconditioners discussed in this paper were first constructed for symmetric matrices with $C = LL^T$. We generalize them to nonsymmetric cases by choosing $C = LU$ such that $\text{diag}(U) = I$. We also make sure that $C(= LU)$ is symmetric when $A$ is symmetric.

The incomplete factorization method which was first proposed in [Dupont, Kendall, Rachford, 68] for self-adjoint elliptic difference equations will be described here for five points schemes approximating the linear advection-diffusion equation, (3.3), which can be seen as a linearization of the Burgers' equation:

$$\varepsilon \nabla^2 \phi - u\phi_x - v\phi_y = 0 \tag{3.3}$$

Eq. (3.3) is written in finite difference form as

$$(A\phi)_{ij} = s_{ij}\phi_{i,j-1} + w_{ij}\phi_{i-1,j} + c_{ij}\phi_{i,j} + e_{ij}\phi_{i+1,j} + n_{ij}\phi_{i,j+1} = 0.$$

The resultant matrix $A$ is sparse and nonsymmetric, and has five diagonals. It is possible to approximate $A$ in the form, $C = A + R$, where $C$ is the product $LU$ and $R$ is the defect matrix. $L$ and $U$ are defined to be respectively the lower and upper triangular matrices with no more than three entries per row,

$$(L\phi)_{ij} = v_{ij}\phi_{ij} + t_{ij}\phi_{i-1,j} + g_{ij}\phi_{i,j-1} \tag{3.4}$$

$$(U\phi)_{ij} = \phi_{ij} + f_{ij}\phi_{i,j+1} + k_{ij}\phi_{i+1,j} \tag{3.5}$$

The product $LU$ has seven diagonals

$$\begin{aligned}(LU\phi)_{ij} = &s_{ij}\phi_{i,j-1} + w_{ij}\phi_{i-1,j} + d_{ij}\phi_{i,j} + e_{ij}\phi_{i+1,j} \\ &+ n_{ij}\phi_{i,j+1} + y_{ij}\phi_{i-1,j+1} + z_{ij}\phi_{i+1,j-1} = 0.\end{aligned} \tag{3.6}$$

The new points $y_{ij}$ and $z_{ij}$ involved in the product $LU$ are those corresponding to $\phi_{i-1,j+1}$ and $\phi_{i+1,j-1}$ respectively.

We choose to equate the non-zero elements of $A$ which are off the main diagonal with the corresponding elements of $LU$. We shall make one assumption which is row-sum$(A) \geq 0$. If we impose, row-sum$(A) = $ row-sum$(C)$, we can solve uniquely for the five elements $v, t, g, f$ and $k$ in terms of $w, e, s, n$ and $c$. They are given recursively by the following formulae

$$\begin{aligned} &W - Preconditioner \quad ([\text{Wong, 78}]) \\ &t_{ij} = w_{ij}, \quad g_{ij} = s_{ij} \\ &v_{ij} = c_{ij} - g_{ij}f_{i,j-1} - t_{ij}k_{i-1,j} - t_{ij}f_{i-1,j} - g_{ij}k_{i,j-1} \\ &f_{ij} = n_{ij}/v_{ij}, \quad k_{ij} = e_{ij}/v_{ij} \end{aligned}$$

Here we adopt the convention that the elements $t, g, v, f$ and $k$ are set to zero if they cannot be computed by the above algorithm. It is easy to see that the row-sum of $R$ is zero by construction.

DKR-*Preconditioner* [Dupont, Kendall, Rachford, 68] is the same as above, except for the $v$ formula ($\alpha$ is an iteration parameter),

$$v_{ij} = (1 + \alpha h^2)c_{ij} - g_{ij}f_{i,j-1} - t_{ij}k_{i-1,j} - t_{ij}f_{i-1,j} - g_{ij}k_{i,j-1}.$$

We denote the preconditioning in [Meijerink, Van Der Vorst, 77] by the MV preconditioning. It forces equality of the elements of the preconditioning matrix $C$ and the matrix $A$ on the diagonals in positions defined by the non-zero diagonals of the matrix $A$. The MV preconditioner is constructed by the following formula,

MV-*Preconditioner* is the same as $W$-preconditioner, except for another formula for $v$

$$v_{ij} = c_{ij} - g_{ij}f_{i,j-1} - t_{ij}k_{i-1,j}.$$

All the preconditioning techniques described above can be applied to symmetric and nonsymmetric matrices. As an example of a preconditioned gradient method, we show the preconditioned ORTHOMIN(1) method.

$X^0$ given, let $r^0 = b - AX^0$
Solve $CZ^0 = r^0$ and set $p^0 = Z^0$
For $k = 0$ step 1 until convergence *do*
$\alpha_{k+1} = (r^k, Ap^k)/(Ap^k, Ap^k)$
$X^{k+1} = X^k + \alpha_{k+1}p^k$, $r^{k+1} = r^k - \alpha_{k+1}Ap^k$
Solve $CZ^{k+1} = r^{k+1}$
$p^{k+1} = Z^{k+1} + \beta_k p^k$, $\beta_k = -(AZ^{k+1}, Ap^k)/(Ap^k, Ap^k)$

**4. Two Nonlinear Examples.** The first example is the *Burgers's equation* (1.7) in the square $0 \leq x, y \leq 1$. We shall use central differences to discretize the second derivative terms and the shock capturing schemes to approximate the nonlinear terms. We obtain an equation $F(U) = 0$ where the unknowns $u_{ij}$ are ordered as the components of the vector $U$.

Let us first verify directly the regularity of the E-O differencing

$$[f(u)]_x \approx 1/\Delta x [\Delta_+ f_-(u_i) + \Delta_- f_+(u_i)].$$

The following estimates are valid for the Burgers' equation:

$$|f_+(a) - f_\pm(b) - f'_\pm(b)(a - b)| \leq 0.5 \, (a - b)^2 \quad \forall \, a, b \tag{4.1}$$
$$|f'_\pm(a) - f'_\pm(b)| \leq |a - b| \quad \forall \, a, b \tag{4.2}$$

Using (4.1) and (4.2), we can show that there exists constants $C_1$ and $C_2$ which depend on the grid size $h$ such that:

$$\|F(U) - F(V) - J(V)[U - V]\| \leq C_1 \|U - V\|^2 \quad \forall \, U, V \in R^n \tag{4.3}$$
$$\|J(U) - J(V)\| \leq C_2 \|U - V\| \quad \forall \, U, V \in R^n \tag{4.4}$$

Therefore the mapping $F$ is twice Frechet differentiable, which guarantees convergence of the Newton's method.

Let $U^*$ be the solution of (4.1). In order to ensure convergence, the initial guess $U^0$ ought to be chosen so that [Dennis, Schnabel, 83]

$$\begin{aligned} &\|U^0 - U^*\| \leq 1/2\alpha\beta, \quad \text{when} \\ &\|J(U^*)^{-1}\| \leq \alpha, \|J(U^0) - J(U^*)\| \leq \beta\|U^0 - U^*\|. \end{aligned} \quad (4.5)$$

The relation (4.5) tells us that the radius of convergence is inversely proportional to the product $\alpha\beta$. The size of $\alpha$ depends strongly on the structure of the solution. If there is a boundary layer solution (out Type A below) the eigenvalues of $J(U^*)$ are well bounded away from zero. In the one dimensional case $\sigma(J) \leq -C/\varepsilon$, [Kreiss, Kreiss, 86]. For shock solutions (our Type B below) the radius of convergence is much smaller. There are eigenvalues of $J$ of the order $e^{-1/\varepsilon}$.

The 2-D Burgers's equation (4.1) was solved on a square with two sets of boundary conditions (Figs. 1 and 2).

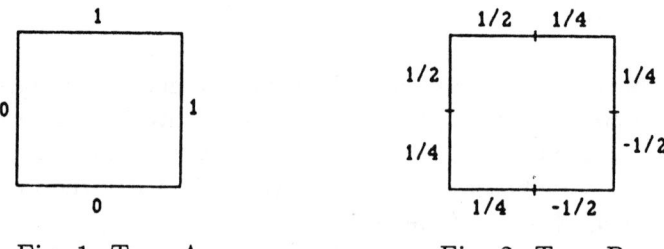

Fig. 1. Type A.            Fig. 2. Type B.

Note that in the case of type B boundary condition (Fig. 2), there is a jump in the middle of each side of the square. These jumps will indeed trigger switching mechanism of various upwind schemes.

The computations were done with a $31 \times 31$ grid if not otherwise noted. The iterative procedure stops when the norm of the residual $\|R\|$ is less than $10^{-5}$. At each Newton step, the linear system is solved iteratively by a minimal residual method (MR) and the convergence criteria are fixed at 20 MR iterations or inner residual $\|r\| < 10^{-5}$. In table 1 are listed the numbers of Newton iterations for the Godunov, Roe and E-O schemes. The number of iterations are the same for these three schemes which are identical for the type A boundary conditions.

With the type A boundary conditions there are no switchings involved because the boundary layers are at $x = 1$ and $y = 1$. The rapid convergence is indicated by the computational results. Note that the total number of MR iterations decreases as $\varepsilon$ becomes smaller. This is due to the fact that the Jacobian matrix reduces to a triangular matrix for $\varepsilon = 0$ and the LU factorization of the Jacobian matrix is exact.

In table 2 are reported the number of Newton iterations for the type B boundary conditions. The linear system at each Newton step is preconditioned by the DKR, W and MV techniques. the MV preconditioner proves to be the most robust while the inner iteration together with the DKR and W preconditioners fails to converge for $\varepsilon \leq 10^{-2}$. In table 3, as predicted the Newton iteration together with the E-O scheme converges for all values of $\varepsilon$ and the number of iterations does not vary greatly with respect to $\varepsilon$.

Table 4 reports the number of Newton iterations for three different grid sizes. We observe that the total work is still very modest with respect to the number of unknowns. The TVD scheme did not converge for $\varepsilon \leq 10^{-3}$. For $\varepsilon = 10^{-2}$, 12 Newton iterations were needed for convergence.

To illustrate the efficiency of different preconditioners on the eigenvalues of the matrix A in (3.2) we shall consider the linearization (3.3) of the Burgers' equation. A linear upwind scheme with the same character as our earlier algorithms for the nonlinear equations is used in the discretization. The velocities are given by $u = 1 - x$, $v = y$ and the boundary conditions are

$$\phi(0, y) = \phi(x, 0) = 0 \quad \phi(1, y) = \phi(x, 1) = 1 \tag{4.6}$$

We choose a matrix $A$ of order 225 and compute the eigenvalues of $A$ and $C^{-1}A$. The matrix $C$ is constructed by the three preconditioning algorithms described above. Figs. 3 to 6 show plots of eigenvalues in increasing order of magnitude for $\varepsilon = 1$. All eigenvalues computed are real and negative. In terms of the ratio $r = \lambda_{\max}/\lambda_{\min}$ for $\varepsilon = 1$, the W and DKR preconditioners perform better than the MV preconditioner as predicted from symmetric cases.

The coefficient matrix $A$ resulting from a certain ordering of the finite difference equations, becomes a triangular matrix as $\varepsilon$ tends to zero. Moreover if $A$ is a triangular matrix, the DKR, W and MV are all exact factorizations i.e. $A = C = LU$. Hence the matrix $C^{-1}A$ is "better conditioned" for smaller $\varepsilon$.

Our final examples is the *transonic small disturbance equation* (TSD) which is written in conservative form

$$[K\phi_x - 1/2(\gamma + 1)\phi_x^2]_x + \phi_{yy} = 0, \tag{4.7}$$

or

$$-[f(u)]_x + v_y = 0, \tag{4.8}$$

where

$$f(u) = 1/2(\gamma + 1)\phi_x^2 - K\phi_x,$$

1. Number of iterations , ( . ) = total number of MR iterations, type A boundary conditions , DKR ($\alpha = 1$) preconditioner

| $\epsilon$ | $10^{-1}$ | $10^{-2}$ | $10^{-3}$ | $10^{-4}$ | $10^{-5}$ | $10^{-6}$ |
|---|---|---|---|---|---|---|
| | 5 | 6 | 9 | 10 | 10 | 10 |
| | (95) | (79) | (57) | (37) | (27) | (16) |

2. Number of iterations , ( . ) = total number of MR iterations, type B boundary conditions , E-O scheme

\* no convergence

| $\epsilon$ | $10^{-1}$ | $10^{-2}$ | $10^{-3}$ | $10^{-4}$ | $10^{-5}$ | $10^{-6}$ |
|---|---|---|---|---|---|---|
| DKR $\alpha = 1$ | 5 (96) | \* | \* | \* | | |
| W | 6 (115) | \* | \* | \* | | |
| MV | 13 (225) | \* (142) | 9 (121) | 12 (147) | 15 (200) | 16 (214) |

3. Number of Newton iterations , ( . ) = total number of MR iterations, type B boundary conditions , MV preconditioner

\* no convergence

| $\epsilon$ | $10^{-1}$ | $10^{-2}$ | $10^{-3}$ | $10^{-4}$ | $10^{-5}$ | $10^{-6}$ |
|---|---|---|---|---|---|---|
| Godunov | 10 (182) | \* (146) | 10 (119) | 20 (217) | 30 (368) | \* |
| Roe | 12 (215) | 7 (140) | 14 (200) | \* | \* | \* |
| E-O | 13 (225) | \* (142) | 9 (121) | 12 (147) | 15 (200) | 16 (214) |

4. Number of Newton iterations , ( . ) = total number of MR iterations E-O schemes , type B boundary conditions , MV preconditioner

| $\epsilon$ | $10^{-1}$ | $10^{-2}$ | $10^{-3}$ | $10^{-4}$ | $10^{-5}$ | $10^{-6}$ |
|---|---|---|---|---|---|---|
| $h = 1/14$ | 4 (77) | 5 (63) | \* (54) | 9 (55) | 11 (54) | 11 (56) |
| $h = 1/30$ | 13 (225) | \* (142) | 9 (121) | 12 (147) | 15 (200) | 16 (214) |
| $h = 1/44$ | 10 (400) | 7 (280) | 9 (287) | 14 (353) | 17 (419) | 22 (567) |

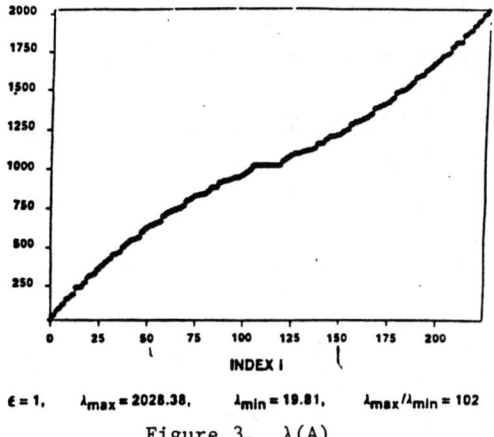

$\epsilon = 1,$ $\lambda_{max} = 2028.38,$ $\lambda_{min} = 19.81,$ $\lambda_{max}/\lambda_{min} = 102$

Figure 3. $\lambda(A)$

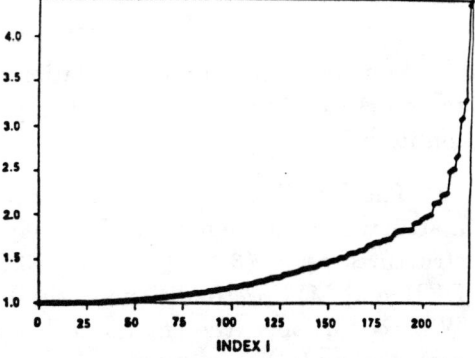

$\epsilon = 1,$ $\lambda_{max} = 4.43,$ $\lambda_{min} = 0.99,$ $\lambda_{max}/\lambda_{min} = 4.43$

Figure 4. $\lambda(C^{-1}A)$, w preconditioner

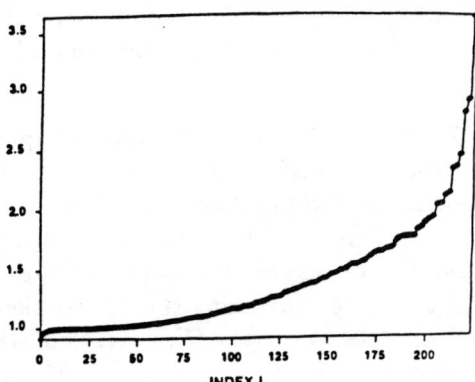

$\epsilon = 1,$ $\lambda_{max} = 3.64,$ $\lambda_{min} = 0.9,$ $\lambda_{max}/\lambda_{min} = 4$

Figure 5. $\lambda(C^{-1}A)$, DKR preconditioner

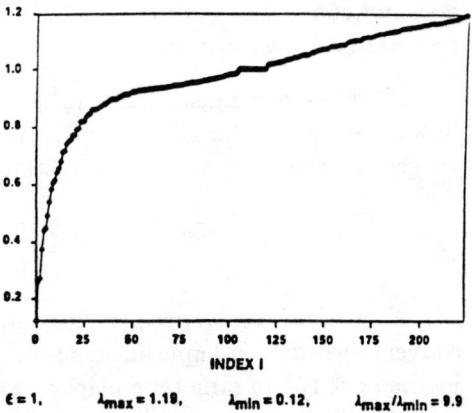

$\epsilon = 1,$ $\lambda_{max} = 1.19,$ $\lambda_{min} = 0.12,$ $\lambda_{max}/\lambda_{min} = 9.9$

Figure 6. $\lambda(C^{-1}A)$, MV preconditioner

$$u = \phi_x, \quad v = \phi_y, \quad K = (1 - (M_\infty)^2)/(\delta^{2/3} M_\infty)$$

Then the E-O approximation of Equation (5.2) gives us ($u_i = \Delta^x_- \phi_{ij}/h$)

$$-1/h[\Delta^x_+ f_-(u_i) + \Delta^x_- f_+(u_i)] + 1/h^2 \Delta^y_+ \Delta^y_- \phi_{ij} = 0 \qquad (4.9)$$

Let $f'(u) = (\gamma + 1)u - K$ and $\underline{u} = K/(\gamma+1)$.

Note that $\underline{u}$ is determined from $f'(\underline{u}) = 0$. Then

$$f'_+(u) = f'(u) \text{ if } u \geq \underline{u}; \quad f'_+(u) = 0 \text{ if } u < \underline{u}$$

and

$$f'_-(u) = f'(u) \text{ if } u \leq \underline{u}; \quad f'_-(u) = 0 \text{ if } u > \underline{u}$$

Similar to the Burgers' equation, we can show that the numerical fluxes are twice Frechet differentiable and this property guarantees convergence of the Newton iteration.

The Jacobian matrix $J$ with respect to the linearized E-O schemes possesses a structure of 6 diagonals. The matrices $L$, $U$ and $C$ have the same diagonal structures as in (3.4), (3.5) and (3.6) respectively. In the derivations of the DKR and MV preconditioners, the leftmost diagonal of $J$ is ignored. For the W-preconditioner, the entries of $L$ and $U$ are functions of all the elements of $J$ with rowsum($C$) = rowsum($J$).

For the TSD equation, ORTHOMIN(1) was used to solve the Newton equation. The computations were done for a parabolic arc airfoil with a thickness ratio $\delta = 6\%$. Unless mentioned explicitly otherwise, all computations were done on a grid $51 \times 30$. The iterative process stops when $||R||_2 < 10^{-5}$. The E-O scheme gave the smallest number of iterations and the displayed results are with this scheme. Godunov's and Roe's schemes did also give converging results. Table 5 shows performance of the Newton's method with respect to the W, DKR and MV preconditioners which are applied to the inner iterations. For $M_\infty = 0.895$ and $M_\infty = 0.916$, the ORTHOMIN(1) method with the W and DKR preconditioners does not converge. As in the case of the Burgers' equation, the MV preconditioner proves to be the most robust.

Next the forcing sequence $\{q^k\}$ will be invoked to control how accurately the Newton equation should be solved. In all the computational results presented here we use the initial $q^0 = 0.1$. Recall that $q^k$ is given by

$$q^{k+1} = c||r^k||/||F(\Phi^k)||, \quad 0 < c < 1.$$

Numerical results are presented in table 6 for different values of $c$. Clearly the convergence rate is superlinear as predicted by the theory. A well balanced adjustment of the parameter $c$ helps to minimize both the number of outer iterations and the total number of inner iterations. Convergence histories of the damped

# ITERATIVE GRADIENT-NEWTON TYPE METHODS

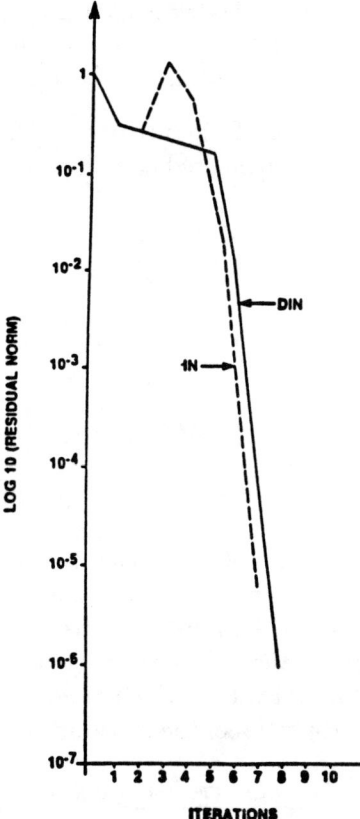

Figure 7.

Number of Newton iterations, E-O schemes

* no convergence

| M∞ | 0.839 subsonic | 0.872 transonic | 0.895 transonic | 0.916 transonic |
|---|---|---|---|---|
| W | 4 | 6 | * | * |
| DKR α = 1 | 4 | 6 | * | * |
| MV | 4 | 6 | 7 | 9 |

Table 5

Number of Newton iterations, ( . ) = total number of ORTHOMIN (1) iterations

| M∞ | 0.839 subsonic | 0.872 transonic | 0.895 transonic | 0.916 transonic |
|---|---|---|---|---|
| e = 1 | 5 (68) | 7 (110) | 11 (117) | 17 (237) |
| e = 0.5 | 4 (62) | 6 (154) | 7 (152) | 10 (336) |

Table 6

inexact and the inexact Newton's methods are shown in figure 7 for $M_\infty = 0.895$. Oscillations of the IN method are more pronounced as the Mach number goes up. The DIN method is more robust but both methods converged in most cases.

Finally, we shall compare the performance of the DIN method and an SLOR method. Our SLOR method is implemented as follows:

$$B(\phi^k)\delta\phi^k = -L(\phi^k) \qquad (4.10)$$
$$\phi^{k+1} = \phi^k + \delta\phi^k \qquad (4.11)$$

$$L(\phi) = [k - (\gamma+1)\phi_x]\phi_{xx} + \phi_{yy} = 0 \qquad (4.12)$$

$$s\delta\phi_{i,j-1} + c\delta\phi_{ij} + n\delta\phi_{i,j+1} = -\omega L(\phi_{ij}) - \omega w\delta\phi_{i-1,j} - z\delta\phi_{i-2,j} \qquad (4.13)$$
$$\phi_{ij}^{k+1} = \phi_{ij}^k + \delta\phi_{ij}^k \qquad (4.14)$$

In (4.13), the correction $\delta\phi_{ij}$ is solved on each successive vertical line $i$. The coefficients $s, c, n, w$ and $z$ are the entries of the Jacobian matrix $J$. $\omega$ is the relaxation parameter. Table 7 gives the number of iterations and computing time of the DIN method and SLOR method. The relaxation parameter $\omega$ in the SLOR method is assumed to have optimal values which are determined by numerical experiments. The convergence properties of the Newton's method is clearly superior.

Number of iterations , ( . ) = CPU time , E-O scheme

| $M_\infty$ | 0.839 subsonic | 0.872 transonic | 0.895 transonic | 0.916 transonic |
|---|---|---|---|---|
| SLOR | 142 (1:22.78) | 294 (2:40.23) | 408 (3:38.12) | 564 (4:58.26) |
| DIN | 5 (29.11) | 7 (34.86) | 12 (47.04) | 19 (1:02.60) |

Table 7

## REFERENCES

1. P. COLELLA and P. R. WOODWARD, *The numerical simulation of two-dimensional fluid flow with strong shocks*, J. Comp. Phys., 54 (1984), pp. 115-173.

2. R. S. DEMBO and T. STEIHAUG, *Truncated-Newton algorithms for large-scale unconstrained optimization*, Mathematical Programming, 26 (1983), pp. 190-212.

3. J. E. DENNIS, Jr. and R. B. SCNABEL, *Numerical Methods for Unconstrained Optimization and Nonlinear Equations* Prentice-Hall, New Jersey, 1983.

3. T. DUPOND, R. KENDALL, and H. H. RACHFORD, Jr., *An approximate factorization procedure for solving self-adjoint elliptic difference equations*, SIAM J. Numer. Anal., 5 (1968), pp. 559-573.

4. B. ENGQUIST and S. OSHER, *Stable and entropy satisfying approximations for transonic flow calculations*, Math. Comp., 34 (1980), pp. 45-75.

5. S. K. GODUNOV, *Finite difference method for numerical computation of discontinuous solutions of the equations of fluid dynamics*, Mat. Sbornik, 47 (1959), pp. 271-306. (In Russian.)

6. G. KREISS and H.-O. KREISS, *Convergence ot steady state of solutions of Burgers' equation*, in Advances in *Numerical and Applied Mathematics*, J. C. South, Jr. and M. Y. Hussaini, eds. (ICASE Report No. 86-18), 1986.

7. J. A. MEIJERINK and H. A. VAN DER VORST, *An iterative solution method for linear systems of which the coefficient matrix is a symmetric M-matrix*, Math. Comp., 31 (1977), pp. 148-162.

8. A. RIZZI and B. ENGQUIST, *Selected topics in the theory and practice of computational fluid dynamics*, J. Comp. Phys., 72 (1987), pp 1-69.

9. P.L. ROE, *Some contributions to the Modelling of Discontinuous Flows*, in Lectures in *Applied Mathematics*, Amer. Math. Soc., Providence, RI, 1985, Vol. 22.

10. Y. SAAD and M. H. SCHULTZ, *Conjugate gradient-like algorithms for solving nonsymmetric linear systems*, Math. Comp., 44 (1985), pp 417-424.

11. Y. SAAD and M. H. SCHULTZ, *A generalized minimal residual algorithm for solving nonsymmetric linear systems*, SIAM J. Sci. Statist. Comput., 7 (1986), pp 856-869.

12. J. SMOLLER, *Shock wave and reaction-diffusion equations*, Springer-Verlag, New York, 1983.

13. B. VAN LEER, *Towards the ultimate conservative difference scheme II. Monotonicity and conservation combined in a second order scheme*, J. Comput. Phys., 14 (1974), pp. 361-370.

14. P. K. W. VINSOME, *Orthomin, an iterative method for solving sparse sets of simultaneous linear equations*, in Proc. Fourth Symposium on *Reservoir Simulation*, Society of Petroleum Engineers of AIME, 1976, p. 149.

15. Y. S. WONG, *Iterative methods for problems in numerical analysis*, Ph.D. thesis, 1978, Oxford University, England.

# CHAPTER 7

## Applications of Adaptive Grid-Refinement Methods*

R. E. Ewing†
P. G. Jacobs‡
R. R. Parashkevov‡
J. Shen‡

**Abstract.** Highly localized phenomena can often dominate important physical processes. In large-scale simulation processes, attempts to implement local grid refinement can often destroy the efficiency of existing codes through complex data structures and associated solution algorithms. Patch refinement methods arising from domain decomposition techniques are described which are accurate and can be incorporated in existing simulation codes. Preconditioned iterative methods are presented that allow a wide variety of applications of these adaptive refinement techniques. Computational results are included and compared with theoretical convergence rates for local refinement of mixed finite element methods, incomplete factorization preconditioners, and local time-stepping applications.

**1. Introduction.** Adaptive grid refinement techniques are essential for resolving important local phenomena arising in many large-scale physical applications. The use of local refinement must not destroy the critical efficiency aspects of large-scale simulation codes. In this paper, we present patch refinement methods derived from domain decomposition techniques that can be easily incorporated into large existing codes. These methods maintain accuracy of the discretization across grid refinement interfaces. They also can take full advantage of the emerging supercomputers with parallel and vector computing capabilities.

There are two distinct classes of local grid refinement techniques – fixed and dynamic. For problems with fixed singularities or local structures, certain fixed local refinements have proven to be very effective.

---

*This research was supported in part by the Office of Naval Research Contract No. 0014–88–K–0370, and by funding from the Institute for Scientific Computation at the University of Wyoming through NSF Grant No. RII-8610680.

†Department of Mathematics, Chemical Engineering, and Petroleum Engineering, Univeristy of Wyoming, Laramie, Wyoming 82071.

‡Department of Mathematics, University of Wyoming, Laramie, Wyoming 82071.

Dynamic and adaptive grid refinement to follow moving phenomena is much more complex. Techniques which work well for fixed refinement can involve a data structure which is so complex that it can be very inefficient for dynamic applications. In this paper, we present methods that can be applied to both fixed and dynamic refinement problems in an efficient and accurate manner.

For ease of exposition, we consider a simple example problem to illustrate our local refinement techniques. The methods have been applied successfully to time-dependent, multiphase petroleum-related problems in a field-scale industrial simulator in [19]. As model physical problems for these extremely versatile techniques, we will consider the applications of fluid flow in porous media. The pressure $p$ of an incompressible fluid in a horizontal reservoir $\Omega \subset \mathbf{R}^2$ satisfies [15,19]

$$-\nabla \cdot \frac{k}{\mu} \nabla p = q \quad \text{in } \Omega. \tag{1.1}$$

We assume no-flow boundary conditions and ignore effects due to gravity. For the existence of $p$, we assume that the mean value of $q$ is zero, and for uniqueness, we fix $p$ at some point. We assume the fluid flow rates at injection and production wells to be specified via Dirac-delta functions. Extensions of these spatial refinement ideas to the simple parabolic time-dependent model for pressure transients are straightforward. Implementation for nonlinear time-dependent systems is discussed in [19].

In Section 2, we present preconditioned iterative techniques for local patch grid refinement which are derived from domain decomposition methods. We compare various methods to illustrate their properties, similarities, and differences. Computational results illustrating the potential of incomplete LU factorization as preconditioners are presented. In Section 3, mixed finite element methods are briefly discussed, along with versions that incorporate local grid refinement. Computational results are presented and compared with asymptotic error estimates developed in [24]. Local time-stepping applications are discussed in Section 4. The same domain decomposition techniques described in Section 2 are utilized to develop efficient preconditioned iterative methods to efficiently incorporate local time stepping in large, existing codes. Again, computational results are presented and compared with theoretical asymptotic convergence rates.

**2. Local Grid Refinement via Domain Decomposition Techniques.** In this section, we discuss efficient preconditioned conjugate gradient methods for treating local grid refinement problems. The methods are derived via techniques which were initially developed for domain decomposition applications. We will present two related methods and compare their properties. Each can be incorporated easily in existing large-scale codes. Application of these techniques in a three-dimensional, three-phase, black oil, industrial simulator appears in Ewing *et al.* [19].

For fixed local behavior, many authors have presented fully local techniques which allow arbitrary levels of refinement about any spatial point or set of points. A technique termed patch refinement [5,15,29,33,34] is an attractive alternative to truly local refinement. This method does not require as complex a data structure but does involve ideas of passing information from one grid to another. The idea of a local patch-refinement method is to pick a patch that includes most of the critical behavior requiring better resolution, and to use a special, possibly uniform, refinement within this patch. If a uniform or logically structured fine grid is utilized in the patch, very fast solvers, perhaps utilizing vector-based algorithms, can be applied locally in this region using boundary data from the original coarse grid. The local patch-refinement techniques have proven to be very effective [5,20,21] for obtaining local resolution around fixed singular points, such as wells, in a reservoir.

We have developed fast solution methods for the approximation of problems requiring mesh refinement. These techniques are related to various domain decomposition methods [4,6–10,12,15,29,34,36]. High accuracy throughout the computational region is obtained by incorporating local refinements around points with singular local behavior, such as wells in porous media flow. A composite grid is obtained by superimposing these local refinements on a quasi-uniform grid on the original domain. Previous techniques usually have no systematic way of dealing with such questions as interface interpolation, mass conservation, and degree of grid overlap. They also usually involve the solution of the coarse-grid problems with the regions corresponding to the refinement removed. This destroys the banded structure and ease of vectorization of the coarse-grid regions and often generates ill-conditioned matrix systems.

In the methods discussed below, the problem is formulated with a composite operator on the composite grid. The techniques are iterative procedures which drive the residual of this composite-grid operator to zero. Composite-grid operators for finite element discretization are common and relatively easy to describe and analyze. Examples of accurate finite difference based composite-grid operators for variable coefficient problems are presented in [20,21]. Complete error analysis for these difference stars appears in [25]. A new domain decomposition variant is presented to efficiently solve the resulting matrix equations [26]. This involves the development of a preconditioner. This preconditioner is novel in that the task of computing its inverse applied to a vector reduces to the solution of separate matrix systems for the local refinements and the matrix system for the quasi-uniform grid on the original domain. Note that this quasi-uniform grid overlaps the regions of local refinement, and that its corresponding matrix problem remains invariant when local refinements are dynamically added or removed. This local refinement technique can be incorporated in existing reservoir codes without extensive modification. The local problems on each of the subdomains are independent and can be solved separately and concurrently on different processors. Furthermore, if the nodes on the quasi-uniform grid are chosen in a regular pattern, highly vectorizable algorithms for the solution of the corresponding linear system can be developed.

We first describe our local refinement methods in terms of finite element approximations and bilinear forms. We then extend these concepts to a linear-algebraic formulation which includes finite difference discretizations. We feel that each formulation is valuable in understanding different aspects of the methods.

Multiplying (1.1) by an arbitrary (sufficiently regular) function $\phi$, integrating by parts and using the no-flow boundary conditions, we see that the solution $p$ satisfies

$$A(p, \phi) = (f, \phi), \tag{2.1}$$

where

$$A(u, v) = \int_\Omega \frac{k}{\mu} \nabla u \cdot \nabla v \, dx \tag{2.2}$$

and $(u, v)$ is the standard $L^2$ inner product of functions on $\Omega$.

We start with a regular grid on $\Omega$. After partitioning $\Omega$ as $\Omega = \Omega_1 \cup \Omega_2$, we construct a composite grid on $\Omega$ with local refinement on $\Omega_2$ and regular refinement on $\Omega_1 = \Omega \setminus \Omega_2$. The Galerkin approximation problem for (2.1) is then to find a function $P$ in a suitable finite-dimensional subspace $M_h(\Omega)$ (defined on the composite grid) of the Sobolev space $H^1(\Omega)$, e.g. the set of piecewise linear functions of the composite mesh, such that

$$A(P, \phi) = (f, \phi), \quad \text{for all} \ \ \phi \in M_h(\Omega). \tag{2.3}$$

Since the bilinear form $A(\cdot, \cdot)$ corresponds to a composite operator on the composite grid, (2.3) is, in general, difficult to solve efficiently for $P$. Instead, we use a preconditioned iterative method to obtain $P$. We must find a comparable form $B(\cdot, \cdot)$ such that, given $g$, the problem of finding $W \in M_h(\Omega)$ satisfying

$$B(W, \phi) = (g, \phi), \quad \text{for all} \ \ \phi \in M_h(\Omega), \tag{2.4}$$

is relatively easy.

As was described in [5], the problem of calculating the action of the inverse of the preconditioner essentially reduces to the solution of discrete mixed problems on the refined subgrids, and discrete Neumann problems on the original grid. Due to the regularity of the mesh geometry, such problems are generally easier to solve than the system resulting from the composite grid discretization.

We first split the bilinear form into parts $A(u, v) = A_1(u, v) + A_2(u, v)$, where

$$A_i(u, v) = \int_{\Omega_i} \frac{k}{\mu} \nabla u \cdot \nabla v \, dx \ . \tag{2.5}$$

Then, we decompose any $V \in M_h(\Omega)$ as follows: $V = V_p + V_r$ where $V_r$ equals $V$ on $\Omega_1$, $V_p \in M_h(\Omega_2) = \{\varphi \in M_h(\Omega) | \text{supp} \, \varphi \subset \Omega_2\}$ and $V_r \in M_h(\Omega)$ satisfies

$$A(V_r, \phi) = 0 \ \ \text{for all} \ \ \phi \in M_h(\Omega_2). \tag{2.6}$$

Then, as in [5], we see that for $V \in M_h(\Omega)$,

$$A(V, V) = A_1(V, V) + A_2(V_p, V_p) + A_2(V_r, V_r) \ . \tag{2.7}$$

The action of the inverse of (2.7) is not easy to obtain. However, by replacing $A(V, V)$ by

$$B(V, V) = A_1(V, V) + A_2(V_p, V_p) + A_2(V_c, V_c) \ , \tag{2.8}$$

where $V_c \in M_h^c(\Omega)$ ($M_h^c(\Omega)$ is analogous to $M_h(\Omega)$, but the original regular mesh is used, with $M_h^c(\Omega_2)$ being defined similarly to $M_h(\Omega_2)$) and satisfies $V_c = V$ in $\Omega_1$, and

$$A(V_c, \phi) = 0 \quad \text{for all} \quad \phi \in M_h^c(\Omega_2), \tag{2.9}$$

then the action of the inverse of (2.8) is relatively easy to obtain and the form $B(\cdot, \cdot)$ is comparable to the form $A(\cdot, \cdot)$ with comparability constants independent of the grid size $h$ [6–8]. As described in [5], the following algorithm suffices for solving

$$B(W, \phi) = (g, \phi) \quad \text{for all} \quad \phi \in M_h(\Omega), \tag{2.10}$$

given $g$.

Algorithm for Computing $W$ [5]:

1. Find $U_p$ by solving problems with mixed boundary conditions on the region $\Omega_2$.

2. Pass the local information to the right-hand side of the original problem and compute any solution $U_c$ of the coarse grid problem

$$A(U_c, \phi) = (g, \overline{\phi}) - A_2(U_p, \overline{\phi}) \quad \text{for all} \quad \phi \in M_h^c(\Omega) \tag{2.11}$$

where $\overline{\phi}$ is any function in $M_h(\Omega)$ which equals $\phi$ on $\Omega_1$.

3. Find $U_r$ on $\Omega_2$ by computing the discrete harmonic extension (2.6) with respect to the refinement subspaces.

4. Compute $\overline{U}$, the mean value of $U = U_p + U_r$. Set $W = U - \overline{U}$.

In the same way that the bilinear form $A$ from (2.3) generates a linear system upon a particular choice of finite dimensional space $M_h, (\Omega)$ we can obtain a similar discrete system via finite difference or collocation discretization of (1.1). We now move to a discrete linear-algebra formulation of the problem which includes aspects of each of these forms of discretizations. We do emphasize that the matrices arising from finite element or finite difference formulations may have fundamentally different properties which effect both the analysis of the convergence of the methods and their efficient implementation. We point out these differences later in this section.

We first consider a regular, quasi-uniform grid on $\Omega$, the computational domain. We consider the matrix $\mathbf{A}^c$, generated by a finite element or finite difference approximation of the flow equations using a coarse, quasi-uniform mesh. Let $\Omega_2 \subset \Omega$ be a region that contains some local phenomenon that may require better approximation. Let $\Omega_b$ be the grid points or cells in $\Omega \backslash \Omega_2$ that are adjacent to $\Omega_2$. This corresponds to the boundary grid for $\Omega_2$ since the communication links between these points or cells will be changed upon local refinement in $\Omega_2$. Let $\Omega_1 = \Omega \backslash (\Omega_2 \cup \Omega_b)$. Then $\Omega_1$, $\Omega_b$, and $\Omega_2$ produce a natural decomposition of $\Omega$ that is pictured in Figure 1a for finite element or point-centered finite difference methods and in Figure 1b for cell-centered finite difference discretizations. Let the solution vector $\mathbf{x}$ of the original coarse-grid problem be decomposed in the form $\mathbf{x} = (\mathbf{x}_1, \mathbf{x}_b, \mathbf{x}_2)^T$, where $\mathbf{x}_1, \mathbf{x}_b$, and $\mathbf{x}_2$ are the parts of the coarse-grid solution in $\Omega_1, \Omega_b$, and $\Omega_2$, respectively. The corresponding decomposition of the matrix $\mathbf{A}^c$ can be described in

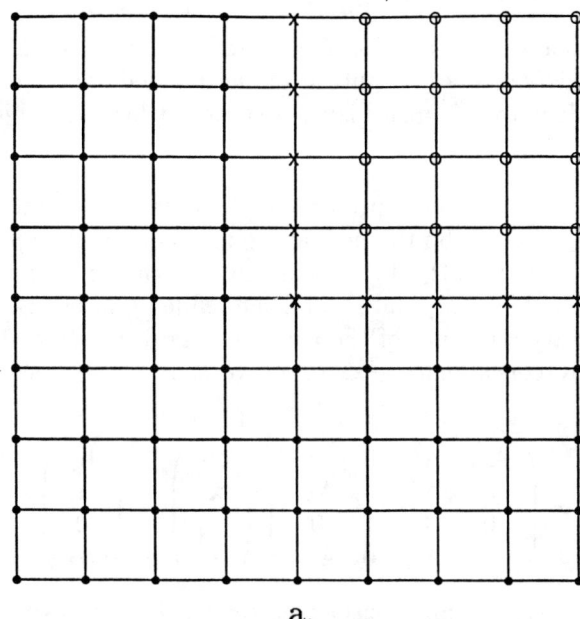

$o \in \Omega_2$
$x \in \Omega_b$
$\bullet \in \Omega_1$

*Figure 1. Grid decomposition: a) finite element or point centered finite difference, b) cell centered finite difference.*

$$\mathbf{A}^c \begin{pmatrix} \mathbf{x}_1 \\ \mathbf{x}_b \\ \mathbf{x}_2 \end{pmatrix} = \begin{pmatrix} \mathbf{A}_{11}^c & \mathbf{A}_{1b}^c & 0 \\ \mathbf{A}_{b1}^c & \mathbf{A}_{bb}^c & \mathbf{A}_{b2}^c \\ 0 & \mathbf{A}_{2b}^c & \mathbf{A}_{22}^c \end{pmatrix} \begin{pmatrix} \mathbf{x}_1 \\ \mathbf{x}_b \\ \mathbf{x}_2 \end{pmatrix}. \tag{2.12}$$

We assume that a code exists or can easily be written to solve (2.12) for a quasi-uniform grid; the code can take advantage of the banded structure of the matrix $\overline{\mathbf{A}}^c$, which is equivalent to $\mathbf{A}^c$, in utilizing a standard lexicographical ordering of the unknowns.

Next, we assume that due to the presence of some localized process, grid refinement is desired in $\Omega_2$. Let $\mathbf{x}_r$ be the new approximation on the refined grid in $\Omega_2$ and $\mathbf{A}_{rr}$ be the local matrix on $\Omega_2$. Let $\mathbf{A}_{br}$ and $\mathbf{A}_{rb}$ be the new connection matrices between the interface between $\Omega_1$ and $\Omega_2$ and the refined grid on $\Omega_2$. Then, in order to maintain the sparsity structure of the composite-grid matrix and a simple data structure obtained by concatenating $\mathbf{x}_r$ to $\mathbf{x}$, we can write the composite-matrix problem in the form

$$\tilde{\mathbf{A}}\tilde{\mathbf{x}} = \begin{pmatrix} \mathbf{A}_{11}^c & \mathbf{A}_{1b}^c & 0 & 0 \\ \mathbf{A}_{b1}^c & \mathbf{A}_{bb}^c & 0 & \mathbf{A}_{br} \\ 0 & 0 & \mathbf{I} & 0 \\ 0 & \mathbf{A}_{rb} & 0 & \mathbf{A}_{rr} \end{pmatrix} \begin{pmatrix} \mathbf{x}_1 \\ \mathbf{x}_b \\ \mathbf{x}_2 \\ \mathbf{x}_r \end{pmatrix} = \begin{pmatrix} \mathbf{b}_1 \\ \mathbf{b}_2 \\ 0 \\ \mathbf{b}_3 \end{pmatrix}. \tag{2.13}$$

The natural decomposition of unknowns generated by the composite grid is pictured in Figure 2a for finite element or point-centered finite difference discretizations and in Figure 2b for cell-centered finite differences. We note that the identity submatrix, $\mathbf{I}$, on the diagonal of (2.13) and the zeroes in the corresponding row, column, and right-hand side enforce the removal of $\mathbf{x}_2$ from the system without destroying the relationship of

$$\begin{pmatrix} \mathbf{A}_{11}^c & \mathbf{A}_{1b}^c \\ \mathbf{A}_{b1}^c & \mathbf{A}_{bb}^c \end{pmatrix} \tag{2.14}$$

to $\mathbf{A}^c$ and hence $\overline{\mathbf{A}}^c$.

A common way to iteratively solve Equation (2.13) is to use a block Gauss-Seidel (BGS) iterative method, where $\mathbf{A}_{rr}$ and the block in (2.14) are inverted sequentially and the information is passed between these blocks from previous iterates via $\mathbf{A}_{br}$ and $\mathbf{A}_{rb}$. This breaks apart the important information which connects $\Omega_1$ and $\Omega_2$ in the inversion process and results in very slow convergence properties for many applications.

We next modify the BGS procedure with an iterative procedure defined by Ewing [17] as an extension of the block preconditioner described in Bramble et al. [5] These methods have been analyzed extensively for block-centered finite difference methods by Ewing et al. [25–27]. The iterative procedure defined below is uniformly well-conditioned for finite elements but not for the cell-centered finite differences and may require a scaling via an acceleration parameter $\tau$ in this case.

**2.1 Iterative Procedures.** The method is described as follows. Given a previous iterate for $\mathbf{x}_b$, denoted $x_b^n$, we solve the local problem on $\Omega_2$ with Dirichlet

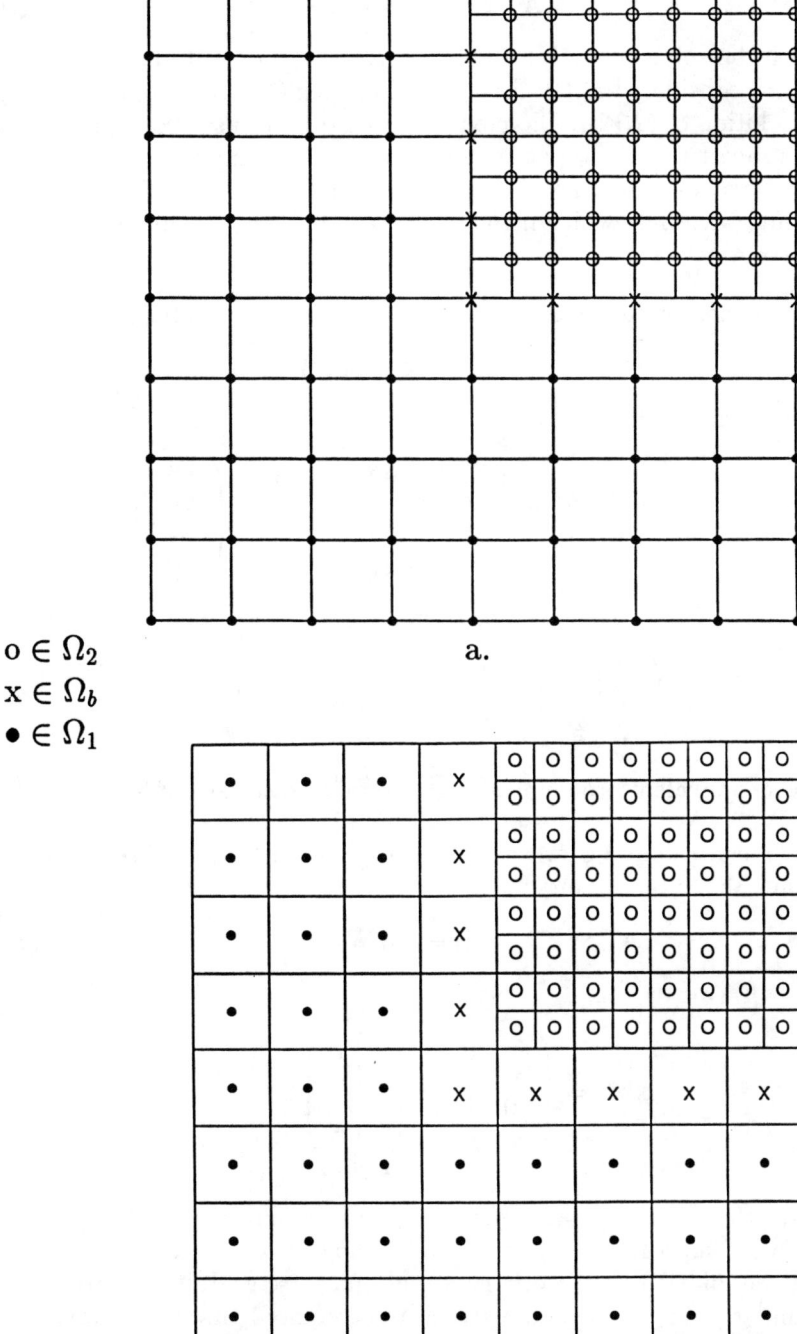

$o \in \Omega_2$
$x \in \Omega_b$
$\bullet \in \Omega_1$

*Figure 2. Composite grid: a) finite element or point centered finite difference, b) cell centered finite difference.*

boundary conditions on the interface between $\Omega_1$ and $\Omega_2$ (given by $\mathbf{A}_{rb}\mathbf{x}_b$):

$$\mathbf{x}_r^n = \mathbf{A}_{rr}^{-1}(\mathbf{b}_3 - \mathbf{A}_{rb}\mathbf{x}_b^n). \tag{2.15}$$

This problem can be solved exactly or approximately by some iterative technique. This step could be considered as the first part of a block Gauss-Seidel iterative procedure for the solution of (2.12). The next step would be to use the approximation for $\mathbf{x}_r^n$ and then invert the block (2.14) to obtain an approximation for $\mathbf{x}_1^n$ and $\mathbf{x}_b^n$. Since this block involves a complex region and may not be well-conditioned, we use an alternate solution method which involves a preconditioner, denoted by $\mathbf{B}$, for the composite matrix $\tilde{\mathbf{A}}$.

Using $\mathbf{B}$, we define, for each iterate $n$ and an iteration parameter $\tau$,

$$\tilde{\mathbf{x}}^{n+1} = \tilde{\mathbf{x}}^n + \tau \mathbf{B}^{-1}(\tilde{\mathbf{b}} - \tilde{\mathbf{A}}\tilde{\mathbf{x}}^n). \tag{2.16}$$

Let $Q$ be the residual vector given by

$$\tilde{\mathbf{b}} - \tilde{\mathbf{A}}\tilde{\mathbf{x}}^n = \begin{pmatrix} \mathbf{b}_1 - \mathbf{A}_{11}^c\mathbf{x}_1^n - \mathbf{A}_{1b}^c\mathbf{x}_b^n \\ \mathbf{b}_2 - \mathbf{A}_{b1}^c\mathbf{x}_1^n - \mathbf{A}_{bb}^c\mathbf{x}_b^n - \mathbf{A}_{br}\mathbf{x}_r^n \\ 0 \\ \mathbf{b}_3 - \mathbf{A}_{rb}\mathbf{x}_b^n - \mathbf{A}_{rr}\mathbf{x}_r^n \end{pmatrix} \equiv \begin{pmatrix} \mathbf{Q}_1^n \\ \mathbf{Q}_2^n \\ 0 \\ \mathbf{Q}_4^n \end{pmatrix}. \tag{2.17}$$

Next, we solve the original coarse grid problem

$$\mathbf{A}^c \begin{pmatrix} \mathbf{W}_1^{n+1} \\ \mathbf{W}_b^{n+1} \\ \mathbf{W}_2^{n+1} \end{pmatrix} = \begin{pmatrix} \mathbf{Q}_1^n \\ \mathbf{Q}_2^n - \mathbf{A}_{br}\mathbf{A}_{rr}^{-1}\mathbf{Q}_4^n \\ 0 \end{pmatrix} \tag{2.18}$$

(or its rearranged equivalent problem using $\overline{\mathbf{A}}^c$ to take advantage of banding of $\overline{\mathbf{A}}^c$) for $\mathbf{W}_1^{n+1}$ and $\mathbf{W}_b^{n+1}$. We have in essence inverted (2.14) in an efficient and vectorizable manner. Then, using $\mathbf{W}_b^{n+1}$, we complete the block Gauss-Seidel analogy on (2.13) and obtain $\mathbf{W}_r^{n+1}$ by solving

$$\mathbf{A}_{rr}\mathbf{W}_r^{n+1} = \mathbf{Q}_4 - \mathbf{A}_{rb}\mathbf{W}_b^{n+1}. \tag{2.19}$$

Finally, from (2.16), we set

$$\tilde{\mathbf{x}}^{n+1} = \begin{pmatrix} \mathbf{x}_1^n \\ \mathbf{x}_b^n \\ 0 \\ \mathbf{x}_r^n \end{pmatrix} + \tau \begin{pmatrix} \mathbf{W}_1^n \\ \mathbf{W}_b^n \\ 0 \\ \mathbf{W}_r^n \end{pmatrix}.$$

Since this algorithm only requires two separate solutions of mixed problems on the subregions (each subregion problem possibly being solved via a different parallel processor) and one solution on the original, uniform coarse grid, it is relatively easy to perform. Similarly, no complex data structure is required, and the algorithm can be implemented in existing large-scale codes without severely disrupting the solution process. Promising numerical results for the algorithm have appeared in [5,19,26].

As stated, the algorithm in its most general form involves two separate solutions on the subregions at each step. This iterative procedure is uniformly well-conditioned for finite element procedures or point-centered finite difference methods, but not for cell-centered finite differences [25-27]. For discretizations arising from cell-centered finite difference methods, a scaling of the iteration via the parameter $\tau$ in (2.16) may be necessary.

By considering the domain decomposition techniques presented by Bramble et al. [5-10] that led to this algorithm, we can see that if the subregion problems ((2.15) and its sequels with updated guesses for $x_b^n$) are solved exactly, then $\mathbf{Q}_4^n$ in (2.17) and (2.18) is identically zero. Preliminary computations indicate that if the subregion problem is solved iteratively with its own preconditioner, the full algorithm with two subregion solves will converge more slowly than the method

### TABLE 2.1
Convergence rates of the BEPS algorithm with inexact solving of the subproblems.

| $1/h_c$ | | 6 | | | | 12 | | | | 24 | | | |
|---|---|---|---|---|---|---|---|---|---|---|---|---|---|
| $h_f/h_c$ | $p_f$ | $p_c$ | q | iter | $p_f$ | $p_c$ | q | iter | $p_f$ | $p_c$ | q | iter | |
| | 8 | 5 | .002 | 2 | 17 | 11 | .002 | 2 | 35 | 23 | .001 | 2 | ★ |
| | 8 | 4 | .01 | 3 | 17 | 6 | .02 | 4 | 35 | 12 | .018 | 3 | |
| | 8 | 3 | .02 | 4 | 17 | 3 | .15 | 7 | 35 | 5 | .20 | 8 | |
| | 8 | 2 | .06 | 5 | 17 | 1 | .40 | 14 | 35 | 3 | .37 | 12 | |
| 1/3 | 8 | 1 | .19 | 8 | 9 | 11 | .27 | 9 | 18 | 23 | .23 | 8 | |
| | 6 | 5 | .13 | 6 | 5 | 11 | .45 | 15 | 9 | 23 | .52 | 18 | |
| | 4 | 5 | .28 | 10 | 5 | 5 | .47 | 17 | 9 | 12 | .54 | 19 | |
| | 4 | 3 | .30 | 10 | 5 | 3 | .50 | 18 | 9 | 3 | .59 | 22 | |
| | 2 | 5 | .43 | 15 | 5 | 1 | .57 | 23 | | | | | |
| | 14 | 5 | .003 | 2 | 29 | 11 | .002 | 2 | 59 | 23 | .002 | 2 | ★ |
| | 14 | 3 | .026 | 4 | 29 | 6 | .02 | 4 | 59 | 12 | .017 | 3 | |
| | 14 | 1 | .187 | 8 | 29 | 3 | .15 | 7 | 59 | 3 | .36 | 12 | |
| 1/5 | 7 | 5 | .33 | 11 | 29 | 1 | .39 | 14 | 30 | 23 | .26 | 9 | |
| | 4 | 5 | .46 | 17 | 15 | 11 | .30 | 10 | 15 | 23 | .53 | 19 | |
| | 7 | 3 | .33 | 12 | 7 | 11 | .57 | 22 | 15 | 12 | .55 | 21 | |
| | 4 | 3 | .48 | 17 | 7 | 5 | .57 | 22 | 15 | 3 | .61 | 25 | |
| | | | | | 7 | 1 | .64 | 27 | | | | | |
| | 20 | 5 | .00001 | 3 | 41 | 11 | .004 | 3 | 83 | 23 | .002 | 2 | ★ |
| | 10 | 5 | .35 | 12 | 41 | 6 | .02 | 4 | 83 | 12 | .016 | 3 | |
| | 10 | 3 | .36 | 12 | 41 | 3 | .14 | 7 | 83 | 3 | .35 | 12 | |
| 1/7 | 7 | 3 | .46 | 16 | 41 | 1 | .38 | 14 | 42 | 23 | .25 | 9 | |
| | 20 | 1 | .18 | 8 | 22 | 11 | .30 | 10 | 21 | 23 | .54 | 20 | |
| | | | | | 11 | 11 | .55 | 21 | 21 | 12 | .55 | 21 | |
| | | | | | 11 | 3 | .57 | 22 | 21 | 3 | .61 | 26 | |

★ These lines correspond to exact solving of both subproblems.

with direct solves unless the iterative method on the subdomain reduces the error sufficiently. Iterative solution of the unrefined region causes no difficulty with either version of the algorithm. This is an important consideration for the reservoir simulation applications when iterative solution of the unrefined problem is essential due to their size, since direct solution of the coarse-grid region problems is usually not possible.

To illustrate the above remarks, we present numerical results for the convergence rate of the PCG method, when we solve the subproblems, corresponding to the fine-grid and the coarse-grid subregions directly, but inexactly by incomplete block LDU decomposition. That is, we set to zero some of the off-diagonals of the inverse of the diagonal blocks in the exact block LDU decomposition [39]. In Table 2.1 we report the average reduction factor $q = (\Delta^{(iter)}/\Delta^{(0)})^{1/iter}$ and the number of iterations $iter$ that were necessary for the PCG method to reduce the norm of the residual to $10^{-12}$ as a function of the numbers $p_f$ and $p_c$ of the off-diagonals (counted from the main diagonal) that we keep in the incomplete LDU decomposition of the refined and coarse-grid matrices correspondingly. The presented results are for an elliptic problem with smooth coefficients in the unit square discretized by cell-centered finite differences.

**2.2 Preconditioned Iterative Methods.** Constructing an efficient iterative method for solving the system of linear equations (on the composite grid) is a crucial point. For a particular implementation, it is also important to take into account the data structure and the computer architecture to be used.

In many applied fields, there already exist large codes which do an excellent job of resolving most of the large length-scale phenomena. In order to improve the local resolution of these codes, it is necessary (in general) to use a local-refinement technique. This should be incorporated in these codes without destroying their data structure and efficiency.

Now, we discuss two preconditioned iterative methods based on domain decomposition ideas for solving the system on the composite grid: the FAC-method of McCormick and Thomas [29,33,34], and the BEPS-method of Bramble, Ewing, Pasciak, and Schatz [5]. These methods are also related to the two-level method of Bank and Dupont [3] as shown in [21].

It is perhaps easier to explain these methods in an algebraic way. Using a vector notation **y** for the unknown values on the composite grid $\omega$, we consider the problem

$$\mathbf{Ay} = \mathbf{b}.$$

We solve this problem using a preconditioned conjugate gradient method. Here, we present two methods for constructing a preconditioning matrix **B**.

We also introduce the matrix $\tilde{\mathbf{A}}$ corresponding to an approximation of our problem on the coarse grid $\tilde{\omega}$. We modify our earlier notation $\mathbf{A}^c$; we group $\Omega_b$ and $\Omega_1$ into a new $\Omega_1$ for the rest of our discussion. Hence, $\Omega = \Omega_1 \cup \Omega_2$ introduces a simpler decomposition than before. If **y** and $\tilde{\mathbf{y}}$ are partitioned into $\mathbf{y} = \begin{pmatrix} y_1 \\ y_2 \end{pmatrix}$ and

$\tilde{\mathbf{y}} = \begin{pmatrix} \tilde{\mathbf{y}}_1 \\ \tilde{\mathbf{y}}_2 \end{pmatrix}$, then this induces corresponding partitions for $\mathbf{A}$ and $\tilde{\mathbf{A}}$:

$$\mathbf{A} = \begin{pmatrix} \mathbf{A}_{11} & \mathbf{A}_{12} \\ \mathbf{A}_{21} & \mathbf{A}_{22} \end{pmatrix}, \quad \tilde{\mathbf{A}} = \begin{pmatrix} \tilde{\mathbf{A}}_{11} & \tilde{\mathbf{A}}_{12} \\ \tilde{\mathbf{A}}_{21} & \tilde{\mathbf{A}}_{22} \end{pmatrix}.$$

Both the FAC-preconditioner (McCormick and Thomas [33,34]) and the BEPS-preconditioner (Bramble, Ewing, Pasciak, and Schatz [5]) follow three common steps. We partition the points in the composite grid $\omega$ in the following way: the first group, $\omega_2$ contains all points in the refined subdomain $\Omega_2$ and the second group $\omega_1$ contains all coarse grid points in the remaining domain $\Omega_1$ (including the boundary interface). Then $\mathbf{y} = \begin{pmatrix} \mathbf{y}_1 \\ \mathbf{y}_2 \end{pmatrix}$, where $\mathbf{y}_2$ corresponds to the values at the grid points in $\omega_2$ and $\mathbf{y}_1$ to the values at the grid points in $\omega_1$. Similarly, $\tilde{\omega}_2$ are the coarse grid points in $\Omega_2$ and $\tilde{\omega}_1$ are the coarse grid points in $\Omega_1$.

Then, the common steps in the algorithm for solving $\mathbf{By} = \mathbf{b}$ are:

(i) solve the fine-grid problem in $\Omega_2$

$$\mathbf{A}_{22}\mathbf{y}_2^F = \mathbf{b}_2;$$

(ii) restrict the defect on the coarse grid

$$\tilde{\mathbf{d}} = \mathbf{P}^T \left( \mathbf{b} - \mathbf{A} \begin{pmatrix} 0 \\ \mathbf{y}_2^F \end{pmatrix} \right) = \mathbf{P}^T \begin{pmatrix} \mathbf{b}_1 - \mathbf{A}_{12}\mathbf{A}_{22}^{-1}\mathbf{b}_2 \\ 0 \end{pmatrix}$$

($\mathbf{P}^T$ is the restriction operator);

(iii) solve the coarse-grid correction

$$\tilde{\mathbf{A}}\tilde{\mathbf{c}} = \tilde{\mathbf{d}};$$

The FAC and BEPS methods differ in the last step:

(iv) (FAC) interpolate this correction over the fine grid in $\Omega_2$

$$\mathbf{c} = \mathbf{P}\tilde{\mathbf{c}}.$$

Then,

$$\mathbf{y} = \mathbf{B}^{-1}\mathbf{b} = \begin{pmatrix} 0 \\ \mathbf{y}_2^F \end{pmatrix} + \mathbf{c}.$$

(iv) (BEPS) find the harmonic component $\mathbf{y}_1^H$ on the fine grid

$$\mathbf{A}_{22}\mathbf{y}_2^H = -\mathbf{A}_{21}\tilde{\mathbf{c}}_1, \quad \tilde{\mathbf{c}} = \begin{pmatrix} \tilde{\mathbf{c}}_1 \\ \tilde{\mathbf{c}}_2 \end{pmatrix}.$$

Then

$$\mathbf{y} = \mathbf{B}^{-1}\mathbf{b} = \begin{pmatrix} \tilde{\mathbf{c}}_1 \\ \mathbf{y}_2^F + \mathbf{y}_2^H \end{pmatrix}.$$

An important feature of these methods is that both preconditioning matrices **B** are spectrally equivalent to the composite-grid matrix **A** with constants that do not depend on the mesh size. Therefore, the preconditioned iterative procedures based upon gradient-type methods are optimal.

These two methods have one common step—solving the coarse grid problem, which can often be done very efficiently by some fast solver or within the technology of the existing code maintaining a high level of vectorization and/or parallelization (see e.g. [19] where these techniques were implemented in an existing three-dimensional, multiphase, industrial reservoir simulation code). FAC and BEPS differ on one important issue: on the last step, BEPS solves one more problem on the fine grid (harmonic component), securing in this way the symmetry of the preconditioning matrix **B** (see [17,21,35]). Instead of that, FAC interpolates the coarse grid correction and adds it to the fine grid component. Thus, in general, in FAC we solve one fine-grid problem less, but we use interpolation. A restriction is that the interpolation operator **P** should be equal to the transposed restriction operation (from fine to coarse grid); moreover, the composite-grid matrix **A** and coarse-grid matrix **Ã** should satisfy the relation $\mathbf{A} = \mathbf{P}\mathbf{\tilde{A}}\mathbf{P}^T$ (so-called variational condition) [33,34]. This is automatically satisfied only for finite element problems. The FAC also produces a nonsymmetric preconditioning matrix **B** and generalizations of the standard CG-methods that do not take advantage of conjugacy should be used. However, as has been shown in [26], this matrix is symmetric in a certain subspace. Performing the iteration within this subspace will lead exactly to the last local problem of the BEPS preconditioner. In some cases, we can skip the first solve of the fine-grid problem, as can be seen from the following argument. Let us rewrite the BEPS preconditioner in the following form (see [26]):

$$\mathbf{B}_{BEPS} = \begin{pmatrix} \mathbf{\tilde{S}} & \mathbf{A}_{12} \\ 0 & \mathbf{A}_{22} \end{pmatrix} \begin{pmatrix} \mathbf{I} & 0 \\ \mathbf{A}_{22}^{-1}\mathbf{A}_{21} & \mathbf{I} \end{pmatrix},$$

where **S̃** is the corresponding Schur complement in the coarse grid matrix **Ã**. Suppose:

i.) We can solve exactly the system $\mathbf{A}_{22}\mathbf{x}_2 = \mathbf{b}_2$.

ii.) As an initial guess for the Generalized CG method we choose

$$\mathbf{x}^{(0)} = \begin{pmatrix} 0 \\ \mathbf{A}_{22}^{-1}\mathbf{b}_2 \end{pmatrix}.$$

Then the initial residual will be

$$\mathbf{r}^{(0)} = \mathbf{b} - \mathbf{A}\mathbf{x}^{(0)} = \begin{pmatrix} \mathbf{b}_1 \\ \mathbf{b}_2 \end{pmatrix} - \begin{pmatrix} * \\ \mathbf{b}_2 \end{pmatrix} = \begin{pmatrix} \mathbf{r}_1^{(0)} \\ 0 \end{pmatrix}.$$

The solution of the system $\mathbf{d}^{(0)} = \mathbf{B}_{BEPS}^{-1}\mathbf{r}^{(0)}$ is reduced to the following:

$$\begin{aligned}\mathbf{d}_1^{(0)} &= \tilde{\mathbf{S}}^{-1}\mathbf{r}_1^{(0)}, \\ \mathbf{d}_2^{(0)} &= -\mathbf{A}_{22}^{-1}\mathbf{A}_{21}\tilde{\mathbf{S}}^{-1}\mathbf{r}_1^{(0)}.\end{aligned} \qquad (2.20)$$

In other words, we can skip the first fine grid solving. Moreover,

$$\mathbf{A}\mathbf{d}^{(0)} = \mathbf{A}\mathbf{B}_{BEPS}^{-1}\mathbf{r}^{(0)} = \begin{pmatrix} \mathbf{A}_{11}\mathbf{d}_1^{(0)} + \mathbf{A}_{12}\mathbf{d}_2^{(0)} \\ 0 \end{pmatrix}. \qquad (2.21)$$

Since the new search vector in the GCG method is the current $\mathbf{d}^{(i)}\left(=\mathbf{B}^{-1}\mathbf{r}^{(i)}\right)$ plus a linear combination of the previous ones for all $i$ we have:

$$\mathbf{r}^{(i)} = \begin{pmatrix} \mathbf{r}_1^{(i)} \\ 0 \end{pmatrix}.$$

Therefore, on every step the solution of $\mathbf{B}^{-1}\mathbf{r}^{(i)}$ can be simplified exactly as in (2.20) in step 0. Moreover the calculation of the next search vector can also be simplified as in (2.21) in step 0. In this particular case, the computational costs of BEPS and FAC preconditioners are basically the same, but if the matrices $\mathbf{A}$ and $\tilde{\mathbf{A}}$ are symmetric, the BEPS algorithm will produce a symmetric preconditioner.

The BEPS [5] preconditioner was derived for finite element methods with local refinement and formulated in terms of bilinear forms with functions of continuous arguments. The FAC method has been designed for finite difference approximations satisfying the variational condition [33,34].

The cell-centered difference approximations are neither produced by a finite element approach nor satisfy, in general, the variational condition. In many cases, these schemes lead to systems of linear equations with nonsymmetric matrices. In order to incorporate these techniques into existing codes, the important problems are: to develop general procedures for deriving the finite difference approximation near the interface between a coarse and fine region, to investigate the error, to study the algebraic properties of the composite system, and to analyze methods for solving this system in both symmetric and nonsymmetric cases. For regular patch refinements, these problems have been studied by Ewing, Lazarov, and Vassilevski [25–27]; the general theoretical framework developed in [25,26] extends to local time stepping procedures [23], mixed finite element methods, and multilevel techniques [1,27]. Preliminary numerical results [26] show the efficiency of these methods for elliptic problems with fixed regular patch refinements.

## 3. Local Refinement for Mixed Finite Element Methods.

Mixed finite element methods have been used in a wide variety of applications when high accuracy is desired for both a function and its gradient. These methods are applied to both elliptic and parabolic partial differential equations, alone and coupled with other equations. Often the applications are of sufficiently large scale that important local phenomena are not well resolved on a global coarse grid and local grid refinement techniques are needed. Because of the large size of these problems, efficient solution of the algebraic equations resulting from the local refinement procedures is essential.

In this section, domain decomposition ideas are discussed for accurate and efficient techniques for combining local grid refinement and mixed finite element methods.

We consider mixed finite element methods for second-order elliptic equations with Dirichlet boundary conditions. Neumann boundary conditions do not change the results in any significant way. In its mixed form, the problem is to find $(\mathbf{u}, p) \equiv (u_1, u_2, p)$, such that

$$
\begin{aligned}
\text{(a)} \quad & \mathbf{u} + \alpha \nabla p = 0, \quad \text{in } \Omega, \\
\text{(b)} \quad & \text{div } \mathbf{u} = f, \quad \text{in } \Omega,
\end{aligned}
\tag{3.1}
$$

with the boundary condition

$$p = -g, \quad \text{on } \partial\Omega, \tag{3.2}$$

where $\partial\Omega$ indicates the boundary of $\Omega$, and $\alpha = \alpha(x)$ is a positive continuous function on $\Gamma = \Omega \cup \partial\Omega$. Here $\mathbf{u}$ and $p$ can be considered to be fluid velocity and pressure, respectively, in a problem describing fluid flow in porous media. (3.1a) is a representation of Darcy's law and (3.1b) is a conservation law.

For the sake of simplicity, we take the domain to be the square $(0,1) \times (0,1)$. The results presented in [24] and summarized here can be easily extended to more general domains and to 3–D elliptic and parabolic problems. We restrict our attention to the Raviart-Thomas [37] approximating spaces for a rectangular non-uniform partition of the domain, induced by local grid refinement. We consider rectangular elements for ease of extension to large, three-dimensional problems. Details of our 2-D codes appeared in [31].

[24] introduced the concept of "slave" nodes in the mixed method, constructed the corresponding spaces, and gave the finite element approximation to the weak saddle-point formulation of the boundary value problem. The constructed finite element spaces on the composite grid satisfy the Babuška-Brezzi condition [1,11,32]. In the particular case of rectangular elements with local refinement, this condition appears to be a simple consequence of the construction of the spaces if slave nodes are used on the refinement interface.

Error estimate for the finite element solution were also derived in [24]. The lowest-order Raviart-Thomas spaces, combined with numerical integration by a trapezoidal rule for the mass matrices, produce the 5-point finite difference approximation for the pressure equation (see [38]). For regular stars, near the fine- and coarse-grid interface, this approximation produces only an $O(h^{0.5})$ error estimate; the locally-refined mixed finite element approximation using slave nodes produced a similar error estimate. We were able to prove only $O(h^{r+0.5})$ convergence of the finite element solution to the true solution in an $H(\text{div}; \Omega)$-norm for the velocity and $L^2$ norm for the pressure, i.e., we have a loss of $h^{0.5}$ compared to the optimal error estimate for grids without refinement. This reduction of the convergence rate is due to the introduction of "slave" nodes along the interface for coarse/fine-grid domains.

Let $\mathbf{V} = H(\text{div}; \Omega)$ be the set of vector functions $\mathbf{v} = (v_1, v_2)$, $v_i \in L^2(\Omega)$, $i = 1, 2$, such that $\nabla \cdot \mathbf{v} \equiv \text{div } \mathbf{v} \in W$, with $W = L^2(\Omega)$.

For $\mathbf{u}, \mathbf{v} \in \mathbf{V}$ and $p, w \in W$, define the bilinear forms

$$\begin{aligned}\text{(a)} \quad \mathcal{A}(\mathbf{u}, \mathbf{v}) &= (\alpha^{-1} u_1, v_1) + (\alpha^{-1} u_2, v_2) \\ \text{(b)} \quad \mathcal{B}(\mathbf{u}, w) &= (\operatorname{div} \mathbf{u}, w)\end{aligned} \quad (3.3)$$

Then, the problem (3.1–3.2) is equivalent to solving the saddle-point problem given by

$$\begin{aligned}\mathcal{A}(\mathbf{u}, \mathbf{v}) - \mathcal{B}(\mathbf{v}, p) &= \langle g, \mathbf{v} \cdot \mathbf{n} \rangle, & \mathbf{v} \in \mathbf{V}, \\ \mathcal{B}(\mathbf{u}, w) &= (f, w), & w \in W,\end{aligned} \quad (3.4)$$

where $\langle \cdot, \cdot \rangle$ denotes the inner product in $L^2(\partial \Omega)$ and $\mathbf{n}$ is the outward unit normal vector to $\Omega$.

*Remark 3.1.* In many important practical problems, the system (3.4) (for the pressure and the velocity) is coupled with the equations for the concentrations of various components (see, for example, Douglas, Ewing, and Wheeler [14], Ewing, Russell, and Wheeler [30]). In this case, it is natural to combine the mixed method for the pressure with the standard finite element method for the concentrations [30], or with methods tailored to transport-dominated equations [30].

A description of a set of discrete spaces which use the concept of "slave nodes" to develop analogues of $\mathbf{V}$ and $\mathbf{W}$ for refined grids was presented in [24]. The construction of these spaces followed the ideas of Dryja and Widlund [14]. The new saddle-point problem is similar to (3.4) with $\mathbf{V}$ and $\mathbf{W}$ replaced by the composite spaces $V_h^r$ and $W_h^r$ (see [24] for details). The solution ideas of the two level approach described in the last section carry over naturally to this problem. Details of the use of various preconditioned iterative processes to solve saddle-point problems efficiently are given by Ewing *et. al.* [22]. We have performed extensive computations on the lowest-order Raviart-Thomas spaces in order to determine convergence rates under various refinement schemes. In this section, we will describe our results and make several observations about their relation to theoretical error estimates obtained elsewhere.

The lowest-order Raviart-Thomas spaces have discontinuous constants as pressure approximations. The $x$-component of the approximate velocity is a continuous linear function in $x$ tensored with a constant in the $y$ direction; the corresponding $y$-component, $U_y$, has the roles of the $x$ and $y$ variables reversed in the description of $U_x$.

For grids without local refinement, the error estimate in the same norms as those in the estimate (3.9) is $O(h)$ for the lowest-order Raviart-Thomas spaces. Along special lines through the cell centers and parallel to the coordinate axes, superconvergence results for the velocity components in $L^p(\Omega)$, $1 \leq p < \infty$, of $O(h^2)$ were obtained in [28]; [28] also contained a corresponding result for $L^\infty(\Omega)$ of $O(h^2 \ln h^{-1})$. Computational results illustrating these asymptotic convergence results were presented in [31]. The computational results given below were produced by codes which are very similar to those described in detail in [31]. Therefore, details will not be given here.

For testing purposes, we considered the problem (1.1) with $\Omega = (0,1) \times (0,1)$ and with $f$ given by point sources and sinks, $f = \delta(0,0) - \delta(1,1)$. If $\alpha \equiv 1$, we have analytic series solutions for both $p$ and $\mathbf{u}$. In this case, one can compare the computed approximation to the analytic solution for various mesh sizes and local grid refinements to obtain estimates of the asymptotic convergence rates for our computational methods. Due to the Dirac delta sources, $p$ has a singularity of order $\ln r$ and $\mathbf{u}$ has a singularity of order $1/r$, where $r$ is the distance away from the point source or sink. Since $1/r$ is not in $L^2(\Omega)$, in order to obtain non-trivial convergence rates for the velocity approximation, we delete small squares with sides of length 0.2 around the sources and compute the convergence in the rest of the domain to obtain "interior" estimates.

Computational results are presented in Tables 3.1 and 3.2 in the form of convergence rates of the form

$$\text{error} = Kh^\alpha \qquad (3.5)$$

for various levels of refinement (1 is no refinement; otherwise the given number is $h_c/h_f$). In Table 3.1, convergence rates in the $L^2$ norms are computed using quadrature points along the lines of superconvergence described in [28]. As expected, since $1/r$ is not square-integrable in the unit square, there is essentially no convergence for the velocity in the full region. Although there are currently no theoretical results yielding superconvergence in a regime of local refinement as described in this paper, we see from Table 3.1 that along the lines of superconvergence for the elements without refinement, we get approximately $O(h^{3/2})$ convergence for the $L^2$ norm of the velocity errors in the interior regions. This type of superconvergence will be discussed in more detail in a forthcoming paper. We also see superconvergence for the pressure errors at the cell centers of the form $O(h^2)$, approximately. Unlike many superconvergence results, the results from [28] are truly local, cell-by-cell results which do not require uniform gridding. We feel these local properties are also being reflected in the local grid refinement computations.

In order to reinforce the nature and location of the superconvergence computations, we also computed the $L^2$ norm of the velocity errors using $2 \times 2$ Gauss-quadrature rules where superconvergence is not present in the uniform-grid cases. As expected, the convergence rates dropped in the interior region to approximately $O(h)$. This is still higher than the $O(h^{1/2})$ results obtained in [24], but there are important differences. The estimate in [24] is for the $H(\text{div}; \Omega)$ norm for the velocity errors, while the computations are formally for the $L^2$ norm. However, in the interior region for this problem, $\text{div } \mathbf{u}$ and $\text{div } \mathbf{U}$ are both equal to zero and the $H(\text{div}; \Omega)$ norm reduces to the $L^2$ norm in this case. The approximation-theory result is $O(h)$ in the $L^2$ norm, so we hope to be able to improve the $O(h^{1/2})$ result to $O(h)$ in this case. This will also be the subject of a later report.

**·4. Local Time-Stepping Techniques.** In this section, we shall develop algorithms for parabolic problems which involve the use of local meshes in space and time. Thus, the schemes allow for different time step size in different parts of the

### TABLE 3.1
### Convergence Estimates Using $1 \times 2$ Gauss Points
$$\|p - P\| \leq K h_f^\alpha \qquad \|u - U\| \leq K h_f^\alpha$$

| Refinement $h_c/h_f$ | | Full Region $\|p - P\|$ | $\|u - U\|$ | Corners Removed $\|p - P\|$ | $\|u - U\|$ |
|---|---|---|---|---|---|
| 1 | $\alpha$ | 1.01 | 0.00 | 1.89 | 1.96 |
|   | $K$ | 0.01 | 0.01 | 0.01 | 0.52 |
| 2 | $\alpha$ | 1.00 | 0.02 | 1.93 | 1.60 |
|   | $K$ | 0.02 | 0.01 | 0.02 | 0.15 |
| 3 | $\alpha$ | 1.00 | 0.02 | 1.93 | 1.58 |
|   | $K$ | 0.02 | 0.01 | 0.01 | 0.13 |
| 4 | $\alpha$ | 1.00 | 0.02 | 1.93 | 1.58 |
|   | $K$ | 0.01 | 0.01 | 0.01 | 0.13 |
| 5 | $\alpha$ | 1.01 | 0.03 | 1.91 | 1.57 |
|   | $K$ | 0.01 | 0.01 | 0.01 | 0.12 |
| 6 | $\alpha$ | 1.01 | 0.03 | 1.90 | 1.56 |
|   | $K$ | 0.01 | 0.01 | 0.01 | 0.12 |

### TABLE 3.2
### Convergence Estimate Using $2 \times 2$ Gauss Points
$$\|u - U\| \leq K h_f^\alpha$$

| Refinement | | Full Region | Corners Removed |
|---|---|---|---|
| 2 | $\alpha$ | 0.01 | 0.98 |
|   | $K$ | 0.06 | 0.15 |
| 3 | $\alpha$ | 0.01 | 0.94 |
|   | $K$ | 0.06 | 0.12 |
| 4 | $\alpha$ | 0.01 | 0.92 |
|   | $K$ | 0.06 | 0.11 |
| 5 | $\alpha$ | 0.01 | 0.89 |
|   | $K$ | 0.06 | 0.10 |
| 6 | $\alpha$ | 0.01 | 0.89 |
|   | $K$ | 0.06 | 0.10 |
| 7 | $\alpha$ | 0.01 | 0.90 |
|   | $K$ | 0.06 | 0.10 |

domain. Such algorithms seem necessary for many applications which have solutions that behave nicely in most of the domain but vary much more rapidly in the remainder of the domain.

Since we want to study the basic properties of local time stepping algorithms, we only consider schemes based on Crank-Nicholson time approximation and finite element approximation in space and shall not consider many of the possible generalizations. This will allow us to focus on the fundamental underlying issues involved in multiple time scale formulation without the added complexity of the more general setting.

Let $\Omega$ be a domain in $d$ dimensional Euclidean space. We shall develop approximation schemes for the parabolic problem:

$$
\begin{aligned}
\frac{\partial u}{\partial t} + Lu &= f && \text{in } \Omega \times [0,T], \\
u(x,t) &= 0 && \text{on } \partial\Omega \times [0,T], \\
u(x,0) &= g(x) && \text{for } x \in \Omega,
\end{aligned}
\tag{4.1}
$$

where $L$ is given by

$$
Lu = -\sum_{i,j=1}^{d} \frac{\partial}{\partial x_i} a_{ij} \frac{\partial u}{\partial x_j} + au.
$$

We assume that the matrix $\{a_{ij}(x)\}$ is uniformly positive definite and $a(x) \geq 0$ in $\Omega$. We also consider the problem where the above boundary condition is replaced by

$$
\frac{\partial u(x,t)}{\partial n} = 0 \text{ on } \partial\Omega \times [0,T],
\tag{4.2}
$$

where $\partial/\partial n$ is the outward co-normal derivative on $\partial\Omega$.

For simplicity, we shall assume that the space domain $\Omega$ is partitioned into two subdomains

$$\Omega = \Omega_1 \cup \Omega_2$$

and that the solution is much smoother in the space-time domain $\Omega_1 \times [0,T]$. We shall develop approximation schemes which allow finer time steps in $\Omega_2 \times [0,T]$. These schemes lead to systems of equations for advancing one coarse time step which involve both the local time stepping $\Omega_2 \times [0,T]$ and a coarse time step on the remainder of the domain. A well conditioned iterative technique is described, for solving the system of equations for advancing one coarse time step. We apply domain decomposition in the space-time domain. This results in a well conditioned iteration (although nonsymmetric) which involves local time stepping in $\Omega_2 \times [0,T]$ and the solution of a standard coarse mesh time step with respect to a global coarse mesh (i.e. a coarse mesh on $\Omega$ without refinement). Moreover, this iteration only involves the coarse unknowns and thus the nodal values corresponding to the refinement section need not be stored.

Such an approach is important from a program development point of view since it enables the incorporation of these ideas into existing simulators without a complete redesign of the simulator. Note that code for the coarse grid solves already exists in the simulator. Thus, one need only add the necessary routines to do the fine grid local time stepping and provide an iteration procedure calling these routines. In [23], it was shown that the rate of convergence of the domain decompostion approach can be bounded independently of the mesh sizes in both space and time.

We use a global time step size $k_c \geq Ch_c^2$ which should adequately model the behavior on $\Omega_2 \times [0,T]$ and a local time step of size $k_f = k_c/m$ (for some integer $m \geq 1$) on $\Omega_1 \times [0,T]$. We derive our space-time discretization from the more accurate approximation which takes time steps of size $k_f$ throughout the entire space-time domain.

We use two nodal finite element approximation spaces. The first space is associated with a coarse mesh of quasi-uniform size triangles or quadrilaterals of size $h_c$ on $\Omega$ and will be denoted $M_h^c(\Omega)$. The second space $M_h(\Omega)$ is a superset of $M_h^c(\Omega)$ which comes from refining the grid in $\Omega_2$. The triangulation in $\Omega_2$ can be formed by using, for example, a uniform refinement in $\Omega_2$ and may introduce "slave nodes" on the internal boundary $\Gamma$ between $\Omega_1$ and $\Omega_2$. The spaces are then defined by using piecewise polynomial functions on the quadrilaterals or triangles. We also assume that the boundary $\Gamma$ between $\Omega_1$ and $\Omega_2$ coincides with the mesh lines of both triangulations.

We first define some operator notation. Let $L_h : M_h(\Omega) \mapsto M_h(\Omega)$ be the Galerkin operator defined by

$$(L_h u, \phi) = D(u, \phi) \text{ for all } \phi \in M_h(\Omega).$$

Here $(\cdot, \cdot)$ denotes the $L^2$ inner product and $D(\cdot, \cdot)$ denotes the generalized Dirichlet integral corresponding to $L$ given by

$$D(u,v) = \sum_{i,j=1}^{d} \int_\Omega a_{ij} \frac{\partial u}{\partial x_i} \frac{\partial v}{\partial x_j} dx + \int_\Omega auv dx.$$

Clearly, $L_h$ is symmetric and positive definite on $M_h(\Omega)$ with the $L^2$ inner product.

The Crank-Nicholson scheme with time step size $k_f$ is given by the sequence of functions $\{U^i, i = 1, \cdots, N\}$ defined by the recurrence

$$\begin{aligned}(I + \tfrac{k_f}{2}L_h)U^{i+1} &= (I - \tfrac{k_f}{2}L_h)U^i + k_f P_0 f(\cdot, t_{i+1/2}), \\ U^0 &= P_0 g.\end{aligned} \qquad (4.3)$$

Here, $t_{i+1/2} = (i + 1/2)k_f$ and $P_0$ denotes the $L^2$ projection operator onto $M_h(\Omega)$. $U^j$ approximates the solution $u$ of (4.1) at $t = t_j \equiv jk_f$.

We develop a procedure for advancing one global time step of size $k_c$ utilizing local time steps of size $k_f$ only in $\Omega_2 \times [0,T]$. The final time stepping algorithm will consist of repeatedly applying the procedure for one global time step $T/k_c$ times. To develop the equations for one global time step, we consider the process of marching $m$ steps in (4.3) from an initial vector $U^0$. The initial vector and forcing function $f$ are combined into a data vector $F$ given by

$$F = \begin{pmatrix} (I - \tfrac{k_f}{2}L_h)U^0 + k_f P_0 f(\cdot, t_{1/2}) \\ k_f P_0 f(\cdot, t_{3/2}) \\ \vdots \\ k_f P_0 f(\cdot, t_{m-1/2}) \end{pmatrix}$$

We consider the linear system for determining the function $U = (U^1, \ldots, U^m)^t$ given by

$$A_m U = \begin{pmatrix} (I + \frac{k_f}{2}L_h) & 0 & 0 & \cdots \\ -(I - \frac{k_f}{2}L_h) & (I + \frac{k_f}{2}L_h) & 0 & \cdots \\ \vdots & \vdots & \ddots & \vdots \\ 0 & \cdots & -(I - \frac{k_f}{2}L_h) & (I + \frac{k_f}{2}L_h) \end{pmatrix} \begin{pmatrix} U^1 \\ U^2 \\ \vdots \\ U^m \end{pmatrix} = F.$$

Note that $U \in M \equiv M_h(\Omega) \times \{1, 2, \ldots, m\}$. Let $(\cdot, \cdot)$ also denote the the $L^2$ inner product on $M$, i.e the sum of the $L^2$ inner product on the $m$ components. It easily follows that the symmetric part of $A_m$ is positive definite on $M$. This means that for any subspace $M_0$ of $M$, the "Galerkin system"

$$(A_m W, \phi) = (F, \phi) \text{ for all } \phi \in M_0, \tag{4.5}$$

has a unique solution $W \in M_0$. We shall develop the local time stepping schemes by choosing an appropriate subspace $M_0$ of $M$.

Domain decompostion algorithms, similar to the BEPS method described above, for solving the (4.6) were presented in [23]. Jacobs and Pasciak developed a code which implements these algorithms. The computational results presented below were obtained with this code.

Scheme (4.3) is a second order method in time. We shall formally preserve this order of accuracy in both $\Omega_1 \times [0, k_c]$ and $\Omega_2 \times [0, k_c]$. Our subspace $M_0$ will consist of functions which are linear in the second variable over $\Omega_1$, i.e. for $W \in M_0$,

$$W^i(x) = \frac{m-i}{m} U^0(x) + \frac{i}{m} W^m(x), \text{ for all } x \in \Omega_1.$$

Thus, the free nodes of $M_0$ consist of all nodes in $M$ in $\Omega_2 \times [0, k_c]$ but not on $\Gamma \times [0, k_c]$ and those in $M$ in $\overline{\Omega}_1$ at $t = k_c$.

The results in Table 4.1 were obtained using $d = 1, a_{11} = 1, a = 0, \Omega_2 = (0, .5)$ and $\Omega_1 = (.5, 1)$. The exact solution for the runs was such that for $(x, t) \in (0, 1) \times (0, 1)$

$$u(x, t) = \chi_{\Omega_1}(x) \exp\left(4 - \frac{1}{t} - \frac{1}{1-t}\right) \exp\left(8 - \frac{1}{x} - \frac{1}{.5 - x}\right) - .001 \sin(\pi x).$$

The mesh parameters were given by $h_c = \frac{1}{nx}, k_c = \frac{1}{nt}, h_f = \frac{h_c}{r}$, and $k_f = \frac{k_c}{m}$.

By chosing $M_h(\Omega)$ to be piecewise linears functions, the method is also formally second order in space. Thus, we expect to see a four fold decrease in the error ($L^2$ in space, $L^\infty$ in time) when $h_c$ and $k_c$ are halved ($r$ and $m$ held fixed). This is indeed what is observed for this test problem. We also see a four fold decrease in error by local refinement, i.e. doubling $r$ and $m$ but keeping $h_c$ and $k_c$ fixed.

**TABLE 4.1**
Test Problem Errors

| $r \times m$ | \multicolumn{4}{c}{$nx \times nt$} | | | |
|---|---|---|---|---|
| | 8×8 | 16×16 | 32×32 | 64×64 |
| 1×1 | .39001e-0 | .35845e-1 | .58682e-2 | .15115e-2 |
| 2×2 | .33992e-1 | .58684e-2 | .15115e-2 | .38035e-3 |
| 4×4 | .58688e-1 | .15115e-2 | .38034e-3 | .95237e-4 |
| 8×8 | .15115e-2 | .38031e-3 | .95229e-4 | .23817e-4 |
| 16×16 | .38020e-3 | .95197e-4 | .23808e-4 | .59527e-5 |
| 32×32 | .95145e-4 | .23795e-4 | .59493e-5 | .14873e-5 |
| 64×64 | .24043e-4 | .60127e-5 | .15034e-5 | |

## References

1. O. AXELSSON and P.S. VASSILEVSKI, *Algebraic multilevel preconditioning methods, I.*, Report 8811, Department of Mathematics, Catholic University, Nijmegen, 1988.

2. I. BABUŠKA, *Error bounds for the finite element method*, Numer. Math., 16 (1971), pp. 322–333.

3. R. BANK and T. DUPONT, *Analysis of a two-level scheme for solving finite element equations*, Report CNA-159, Center for Numerical Analysis, The University of Texas at Austin, 1980.

4. P.E. BJØRSTAD and O.B. WIDLUND, *Iterative methods for the solution of elliptic problems on regions partitioned into substructures*, SIAM J. Numer. Anal., 23 (1986), 1097–1120.

5. J.H. BRAMBLE, R.E. EWING, J.E. PASCIAK and A.H. SCHATZ, *A preconditioning technique for the efficient solution of problems with local grid refinement*, Comp. Meth. Appl. Mech. Eng., 67 (1988), 149–159.

6. J.H. BRAMBLE, J.E. PASCIAK, and A.H. SCHATZ, *An iterative method for elliptic problems on regions partitioned into substructures*, Math. Comp., 46 (1986), 361–370.

7. J.H. BRAMBLE, J.E. PASCIAK, and A.H. SCHATZ, *The construction of preconditioners for elliptic problems by substructuring, I*, Math. Comp., 47 (1986), 103–134.

8. J.H. BRAMBLE, J.E. PASCIAK, and A.H. SCHATZ, *The construction of preconditioners for elliptic problems by substructuring, II*, Math. Comp., 49 (1987), 1–16.

9. J.H. BRAMBLE, J.E. PASCIAK, and A.H. SCHATZ, *The construction of preconditioners for elliptic problems by substructuring, III*, Math. Comp., 51 (1988), 415–430.

10. J.H. BRAMBLE, J.E. PASCIAK, and A.H. SCHATZ, *The construction of preconditioners for elliptic problems by substructuring, IV,* Math. Comp., 53 (1989), 1–24.

11. F. BREZZI, *On the existence, uniqueness and approximation of saddle point problems arising from Lagrangian multipliers,* RAIRO Anal. Numer., 2 (1974), pp. 129–151.

12. B.L. BUZBEE and F.W. DORR, *The direct solution of the biharmonic equation on rectangular regions and the Poisson equation on irregular regions,* SIAM J. Numer. Anal., 11 (1974), 753–763.

13. J. DOUGLAS, Jr., R.E. EWING, and M.F. WHEELER, *The approximation of the pressure by a mixed method in the simulation of miscible displacement,* RAIRO Anal. Numer., 17 (1983), pp. 17–33.

14. M. DRYJA and O.B. WIDLUND, *On the optimality of an additive iterative refinement method,* Proceedings Copper Mountain Multigrid Conference, April 1989.

15. M.S. ESPEDAL and R.E. EWING, *Characteristic Petrov-Galerkin subdomain methods for two-phase immiscible flow,* Computer Methods in Applied Mechanics and Engineering, 64 (1987), 112–115.

16. R.E. EWING, M.S. ESPEDAL, T.F. RUSSELL, and O. SÆVAREID, *Reservoir simulation using mixed methods, a modified method of characteristics, and local grid refinement,* Proceedings of Joint IMA/SPE European Conference on The Mathematics of Oil Recovery, Robinson College, Cambridge University, July 25–27, 1988.

17. R.E. EWING, *Domain decomposition techniques for efficient adaptive local grid refinement,* Domain Decomposition Methods (T. F. CHAN, R. GLOWINSKI, J. PERIAUX, and O. WIDLUND, eds.), SIAM, Philadelphia, Pennsylvania, 1989, 192–206.

18. R.E. EWING, *Domain decomposition techniques and their efficient application on superconputers,* Special Issue on Supercomputing, Advances in Water Resources, (to appear).

19. R.E. EWING, B.A. BOYETT, D.K. BABU, and R.F. HEINEMANN, *Efficient use of locally refined grids for multiphase reservoir simulation,* SPE 18413, Proceedings Tenth SPE Symposium on Reservoir Simulation, Houston, Texas, February 6–8, 1989, 55–70.

20. R.E. EWING and R.D. LAZAROV, *Adaptive local grid refinement,* SPE No. 17806, Proceedings SPE Rocky Mountain Regional Meeting, Casper, Wyoming, May 11–13, 1988, 643–652.

21. R.E. EWING and R.D. LAZAROV, *Local refinement techniques in the finite element and finite difference methods,* ISC Report #1988–14; and Proceedings of Conference on Numerical Methods and Applications, Sofia, Bulgaria, August 22–27, 1988, 148–159.

22. R.E. EWING, R.D. LAZAROV, P. LU, and P.S. VASSILEVSKI, *Preconditioned conjugate gradient methods for mixed finite element systems*, Proceedings Conference on Preconditioned Conjugate Gradient Methods, Nijmegen, The Netherlands, June 15–17, 1989(to appear).

23. R.E. EWING, R.D. LAZAROV, J.E. PASCIAK, and P.S. VASSILEVSKI, *Finite element methods for parabolic problems with time steps variable in space*, (in preparation).

24. R.E. EWING, R.D. LAZAROV, T.F. RUSSELL, and P.S. VASSILEVSKI, *Local refinement via domain decomposition techniques for mixed finite element methods with rectangular Raviart-Thomas Elements*, Domain Decomposition Methods for Partial Differential Equations (T. F. CHAN, R. GLOWINSKI, J. PERIAUX, and O. W. WIDLUND, eds.), SIAM, Philadelphia, Pennsylvania, 1990, 98–114.

25. R.E. EWING, R.D. LAZAROV, and P.S. VASSILEVSKI, *Local refinement techniques for elliptic problems on cell-centered grids, I: Error analysis*, Math. Comp. (to appear).

26. R.E. EWING, R.D. LAZAROV, and P.S. VASSILEVSKI, *Local refinement techniques for elliptic problems on cell-centered grids, II: Two-grid iterative methods*, Math. Comp. (submitted).

27. R.E. EWING, R.D. LAZAROV, and P.S. VASSILEVSKI, *Local refinement techniques for elliptic problems on cell-centered grids, III: Algebraic multilevel BEPS preconditioners*, Numerische Mathematik (submitted).

28. R.E. EWING, R.D. LAZAROV, and J. WANG, *Superconvergence of the velocities along the Gaussian lines in the mixed finite element methods*, SIAM J. Numer. Anal., (to appear).

29. R.E. EWING, S. McCORMICK, and J. THOMAS, *The fast composite grid method for solving differential boundary-value problems*, Proceedings Fifth ASCE Specialty Conference, Laramie, Wyoming, 1984, 1453–1456.

30. R.E. EWING, T.F. RUSSELL, and M.F. WHEELER, *Convergence analysis of an approximation of miscible displacement in porous media by mixed finite elements and a modified method of characteristics*, Comput. Meth. Appl. Mech. Eng., 47 (1984), pp. 73–92.

31. R.E. EWING and M.F. WHEELER, *Computational aspects of mixed finite element methods*, Numerical Methods for Scientific Computing, (R.S. STEPLEMAN, ed.), North Holland Publishing Co., 1983, pp. 163–172.

32. R.S. FALK and J.E. OSBORN, *Error estimates for mixed methods*, RAIRO Anal. Numer., 14 (1980), pp. 249–277.

33. S. MCcORMICK, *Fast adaptive composite grid (FAC) methods: theory for the variational case*, Comp. Suppl., 5, (1984), 115–121.

34. S. McCORMICK and J. THOMAS, *The fast adaptive composite grid method for elliptic boundary value problems*, Math. Comp., 46 (1986), 439–456.

35. J.A. MEIJERINK and H.A. VAN DER VORST, *An iterative solution method for linear systems in which the coefficient matrix is a symmetric $M$-matrix*, Math. Comp., 31 (1977), 148–162.

36. J.E. PASCIAK, *Domain decomposition preconditioners for elliptic problems in two and three dimensions: First approach*, First International Symposium on Domain Decomposition Methods for Partial Differential Equations (R. GLOWINSKI, G. GOLUB, G. MEURANT, and J. PERIAUX, eds.), SIAM, Philadelphia, 1988.

37. P.A. RAVIAR and J.M. THOMAS, *A mixed finite element method for 2nd order elliptic problems*, Math. Aspects of the Finite Element Method, Lecture Notes in Math., 606, Springer-Verlag, 1977, pp. 292–315.

38. A. WEISER and M.F. WHEELER, *On convergence of block-centered finite differences for elliptic problems*, SIAM J. Numer. Anal., 25 (1988), pp. 351–375.

39. O. AXELSSON and B. POLMAN, *On approximate factorization methods for block-matrices suitable for vector and parallel processors*, Lin. Alg. Appl. 77(1986) 3–26.

# CHAPTER 8

## Approximate Inverse Preconditioning for Nonsymmetric Sparse Systems*

J. D. F. Cosgrove†
J. C. Díaz‡
C. G. Macedo, Jr.‡

**Abstract.** This article presents the use of approximate inverses based on minimization of the Frobenius norm as a preconditioning for a conjugate-gradient-type iterative method to solve sparse nonsymmetric linear systems. This work involves their use in connection with nested factorization techniques for highly structured sparse systems and their direct application for general sparse systems.

The approximate inverses in connection with the nested or block incomplete factorization are considered for highly structured systems, and a fully vectorizable formulation is presented. Some theoretical results concerning the properties of approximate tridiagonal inverses are also presented. Numerical results showing the advantages of this method for highly structured systems are presented.

The potential for application for the general sparse problems is great. General sparsity can be exploited in a straightforward fashion. The parallelism in the calculation of the approximate inverse preconditioner is also discussed. The uniqueness of the approximate inverse preconditioning for irreducible and nonsingular matrices can be established. Bounds on the condition number of the preconditioned system are introduced. Numerical criteria for determining what columns of the sparse approximate inverse require additional fill-in, and graph algorithms for the location of fill-in will be presented. Results are presented that illustrate the potential of the method.

### Introduction

Here, we present a method for the solution of large sparse systems of linear equations via the conjugate residual iterative method (**CRM**). The iterative method is preconditioned using some form of approximate inverses. Approximate inverses have been proposed by several authors either directly as preconditionings or in combination with a form of nested factorization [2,4,5,6,10,13,14,15]. Approximate inverses on the Frobenius norm have also been considered [6,10,13,14,15].

Here we will first discuss the use of the approximate inverses in conjunction with a nested factorization approach for equations arising from the discretization of partial differential equations. Then we will discuss the application of approximate inverse as preconditionings for general sparse matrices.

---

*Work partially supported by National Science Foundation Grants CDA 8820752 and RII-OK-8610676 (Task 10)
† Cameron University, Lawton, OK.
‡ Center for Parallel and Scientific Computing, University of Tulsa, Tulsa, OK.

Nested or block preconditioners take advantage of the natural block structure inherent in the linear systems that arise in the discretization of partial differential equations. In fact as the dimension of the PDE's increase so do the numbers of levels of nesting that can be exploited. It is hoped that by exploiting the structure we can construct a better preconditioning. Our approach relates to the nested block factorization of Appleyard and Cheshire [1], and it is also related to the **INV** and **MINV** preconditioners for symmetric systems introduced in [5,6]. In [14,15] a wider class of nested preconditioned is derived and their relative performance compared for certain transport dominated problems. The nested block preconditionings with the approximate inverses discussed here are similar to preconditionings considered in [13,14,15].

Consider the square system of equations $Ax = b$, the **CRM** iteration with right preconditioning is given below [3,12]. Let $x^0$ be a given vector; let the initial residual $r^0 = b - Ax$, and let $z^0$ be the solution of $Qz^0 = r^0$. For $k = 0, 1, \cdots$ perform the iterative steps:

$$\alpha_k = \frac{(r^k, Az^k)}{(Az^k, Az^k)}$$

$$x^{k+1} = x^k + \alpha_k z^k$$

$$r^{k+1} = r^k - \alpha_k Az^k$$

$$Qz^{k+1} = r^{k+1}$$

$$\beta_k = \frac{(Az^k, Az^{k+1})}{(Az^k, Az^k)}$$

$$z^{k+1} = z^{k+1} - \beta_k z^k$$

$$Az^{k+1} = Az^{k+1} - \beta_k Az^k$$

The matrix $Q$ is the preconditioning matrix. The method requires matrix-vector and vector-vector products, and the application of the preconditioning. The application of the preconditioning takes the form of the solution of a system with coefficient matrix $Q$. The focus on approximate inverses is to make this application step as efficient as possible.

For the matrices obtained using 5-, 7-, or 9-point discretization operators, the existing preconditionings usually require a block-recursive procedure which prevents vectorization. Preconditionings based on nested-incomplete-factorization require, at the innermost level, the solution of tridiagonal systems of equations [1] which have a recursive nature preventing full vectorization. Preconditioning schemes for symmetric problems using the calculation of approximate inverses of tridiagonal matrices were introduced in [4,5,13]. The approximate inverses introduced in [4,5] exploited known properties of the exact inverses of symmetric tridiagonal matrices. These form integral portions of **INV** and **MINV**. Preconditioning schemes using the Frobenius norm minimization for the determination of the approximate inverse of the inner most tridiagonal were presented in [13]. A symmetrization of these approximate inverses was proposed in [13] when the original matrix was symmetric.

Preconditioning schemes for nonsymmetric problems using the Frobenius norm minimization for the calculation of the approximate inverse were considered also in [14,15]. A formulation of this preconditioning that can be fully vectorized given appropriate data structures to represent the sparse matrices was first introduced in [10]. This formulation allows full vectorization of the calculation and application of the preconditioning when used in conjunction with **CRM**. Numerical experiments included here indicate that, for a class of nonsymmetric problems, application of the new preconditioning is up to 50 percent faster than other existing methods such as **ILU**. Calculation of the preconditioning is somewhat more expensive but the faster application and the reduced number of iterations more than compensate this expense. Some theoretical results concerning the properties of the approximate tridiagonal inverse are also discussed.

The comparison results obtained in [14,15] use the preconditioning as a left preconditioning. The results in [10] and those presented here use the preconditioning applied to the right. In fact the formulation of the **CRM** above implied a right preconditioning scheme.

The application of the approximate inverse alone as a preconditioner was first considered in [6]. The advantage is that the application of this preconditioning can be achieved via a matrix-vector multiply. In using an approximate inverse preconditioning $P$, the system $Ax=b$ is replaced by a new system of the form $By=b$ where $B=AP$ and $Py=x$. In fact, in the **CRM** algorithm above the $Qz^{k+1}=r^{k+1}$ is replaced by $z^{k+1}=Pr^{k+1}$. That is, a solution step is replaced by a matrix-vector multiply. This provides a significant advantage for concurrent implementation of the iterative method.

The rest of this article is presented in four major sections followed by a conclusion and an acknowledgement section. In the Approximate Inverse section, the Frobenius norm minimization process is defined and a condition for their uniqueness is derived. The next section describes various block preconditionings including **ILU, INV, AID,** and **AI3**. It also gives some representative comparative numerical results. The next section relates some interesting results for tridiagonal approximate inverses on the Frobenius norm. The last section describes the approximate inverses as a direct preconditioning for sparse systems. In that section criteria for fill-in are developed.

### Approximate Inverses

We now introduce the approximate inverses using minimization on the Frobenius norm. Recall the definition of the Frobenius norm of an $n \times n$ matrix $A$:

$$||A||_F^2 = \sum_{i=1}^{n} \sum_{j=1}^{n} \alpha_{ij}^2.$$

For a nonsingular $n \times n$ matrix $A$ the approximate inverse is defined as the $n \times n$ matrix $P$ that minimizes

$$||AP-I||_F^2.$$

In fact unless further restrictions are placed on $P$, we will get the full inverse of $A$. Thus, the above minimization is obtained among all such matrices $P$ that satisfy a given sparsity pattern. This pattern can be a tridiagonal matrix as considered in [10,13,14,15] or a pentadiagonal as considered in [6], or any given pattern. Given a sparsity pattern for the approximate inverse, the calculation of $P$ is performed making use of the fact that

$$||AP-I||_F^2 = \sum_{i=1}^{n} ||Ap_i - e_i||_2^2,$$

where $p_i$ is the $i^{th}$ column of $P$ and $e_i$ is the $i^{th}$ column of the identity matrix $I$. Thus,

$$min\,||AP-I||_F^2 = min \sum_{i=1}^{n} ||Ap_i - e_i||_2^2 = \sum_{i=1}^{n} min\,||Ap_i - e_i||_2^2.$$

This points out that the calculation of $P$ can be carried out independently as a collection of least squares subproblems for each column $p_i$.

The sparsity pattern of each column $p_i$ must be taken into account. Construct $A_i$ the coefficient matrix of each least squares subproblem, taking the columns of $A$ which correspond to the nonzero rows of $p_i$ and then deleting any rows which are made entirely of zeros. Let $k$ and $l$ be the number of rows and columns, respectively, of $A_i$. It can be shown that if $A$ is irreducible, then $k > l$. Further, it can also be shown that if $A$ is nonsingular, then $P$ is unique, as the columns of $A$ are linearly independent and thus those of $A_i$ are also. We summarize these results in the following.

**Theorem I** *Given a sparsity pattern for the approximate inverse $P$. Let $A$ be a square nonsingular irreducible matrix, let $k$ and $l$ be the number of rows and columns of the reduced matrix $A_i$. Then, $k > l$ and $P$ is unique.*

### Block preconditionings

We first examine the problems with block structure arising from the discretization of PDE's. Consider the system of equations $Ax=b$, where A has a block tridiagonal form given by

$$A = \begin{bmatrix} T_1 & U_1 & & & \\ L_2 & T_2 & U_2 & & \\ & \cdot & \cdot & \cdot & \\ & & L_{n-1} & T_{n-1} & U_{n-1} \\ & & & L_n & T_n \end{bmatrix}$$

Each $T_i$ is a tridiagonal matrix, and each $L_i$, ($i=1, \cdots, n$) or $U_i$, ($i=1, \cdots, n$) is a square matrix with 1, 2, or 3 diagonals depending on the choice of the discretization operator.

We first consider the matrix splitting $A = L + D + U$ with $D$ being the main diagonal, and $L$ ($U$) the lower(upper) triangular part of $A$. Note that the main diagonal $D$ is composed of diagonals of the tridiagonal matrices $T_i$, and that the lower (upper) diagonals of each $T_i$ are included in $L$ ($U$) along with the corresponding elements of each $L_i$ ($U_i$). The **ILU** preconditioning is defined as

$$Q = (L+DD)\,DD^{-1}\,(U+DD)$$

subject to $diag\,(Q - A) = 0$, where $DD$ is a diagonal matrix. This side condition implies

$$DD = D - L(DD)^{-1}U$$

which can be rewritten in its point formulation as

$$dd_i = d_i - \sum_{j=1}^{i-1} l_{jk}(dd_j)^{-1} u_{kj}$$

Note that because of the sparsity structure there are only a few elements in this summation.

The application of **ILU** as a preconditioning is done by solving the system of the form $Qx = y$ for a suitable vector $y$. Note the factored form of $Q$. The solution of this system is obtained as follows

$$(L+DD)z = y \quad \text{or} \quad z = DD^{-1}(y - Lz)$$
$$(DD^{-1}U + I)x = z \quad \text{or} \quad x = z - DD^{-1}(Ux)$$

For this method, only the elements of $DD^{-1}$ are required to be stored. In fact, the decomposition of $Q$ can be expressed in terms of $DD^{-1}$ as follows.

$$DD^{-1}Q = (DD^{-1}L + I)(DD^{-1}U + I).$$

The lower and an upper diagonal within each main triangular block $T_i$ result in a recursion in the calculation of each block of $dd_i$, and in their application. During the calculation of the preconditioning the contribution of each block to the following one can be calculated in vector mode, but the contribution of an element to the next one inside a given block is recursive. Similarly, during the forward elimination and the backward substitution processes involved, contributions from previous blocks are vectorizable but internal contributions inside a block are recursive.

Now we consider the splitting $A = L + T + U$ with $T$ containing the three main diagonals of $A$. $T$ contains each of tridiagonal $T_i$ subblocks of $A$. And $L$ ($U$) contains each of the $L_i$ ($U_i$) blocks of $A$. A block technique suggested in [4,5] defined the preconditioning $Q$ as:

$$S_1 = T_1$$
$$S_i = T_i - L_i S_{i-1}^{-1} U_{i-1} \qquad i=2, \cdots, n$$
$$Q = (L + S)\, S^{-1}\, (U + S)$$

Which can be expressed in terms of $S^{-1}$ alone as:

$$S^{-1}Q = (S^{-1}L + I)(S^{-1}U + I).$$

Eliminating the extra diagonals from the true inverse $S_i^{-1}$, it is replaced by a tridiagonal matrix $P_i$. The factorization then takes the form

$$PQ = (PL + I)(PU + I).$$

The method described in [4,5] requires that the matrix $A$ be symmetric and positive-definite although our experiments show that it is sufficient for the $T_i$ ($i = 1, \cdots, n$) to be symmetric and positive-definite for the method to work well. From the family of preconditionings presented in [4] two were selected for comparison: **INV** calculates an approximate inverse of the tridiagonal matrices involved using the outer product of formulation, while **MINV** incorporates zero-column sums to **INV**.

Having the block preconditioning, the solution of $Qx = y$ is obtained using

$$z = P(y - Lz)$$

and

$$x = z - P(Ux)$$

Note that only $P$ needs to be stored. However, this requires more storage space than for ILU of the size of two extra vectors. On the other hand, the recursion in the application step has been removed and each block of the solution can be obtained in vector mode using only vector-vector or vector operations. The drawback of this formulation arises in the calculation of the inverse of each tridiagonal block. The algorithm used is recursive. Our experiments have also shown potential sensitivity to overflow, necessitating the use of double precision.

In order to remove this difficulty approximate inverses in the Frobenius norm have been proposed [10,13,14,15]. We would not calculate the approximate inverse of $A$ but of each sublock $S_i$. Let us drop the subindex $i$ temporaly for the sake of simplicity. Let $m \times m$ be the dimension of the tridiagonal matrix $S$. The approximate inverse in the Frobenius norm is the $m \times m$ real matrix $P$ that minimizes

$$||SP - I||_F.$$

As we have observed above, this is equivalent to minimizing

$$||Sp_j - e_j||_2, \quad j = 1, \cdots, m$$

where $e_j$ is the $j^{th}$ column of the identity matrix, and $p_j$ is the $j^{th}$ column of $P$. Here we restrict $P$ to be tridiagonal. This keeps storage requirements low, reduces computational effort, and yields a preconditioning with similar sparsity pattern to the original matrix. Thus $P$ has the form:

$$P = \begin{bmatrix} y_1 & z_1 & & & \\ x_2 & y_2 & z_2 & & \\ & \cdot & \cdot & \cdot & \\ & & x_{m-1} & y_{m-1} & z_{m-1} \\ & & & x_m & y_m \end{bmatrix}$$

It is clear that each $p_j$ can be calculated separately solving a least squares problem. Setting up the normal equations for each of the least squares problems for each column $p_i$ we obtain two sets of 2x2 systems of equations representing the first and last columns of $P$, and $m - 2$ sets of 3x3 systems of equations for the inner columns of $P$.

All systems could be solved at once in parallel if desired. But the granularity is low and we chose to pipeline these calculations. A pipelined formulation for the $z$ values uses Cramer's rule on each of the 3x3 subsystems. Given these values the calculation of the $y$ values is pipelined by symbolically eliminating the unknown $x$ from two of the three equations for each subsystem. And finally, the $x$ values are obtained in a pipelined fashion using the $y$ and $z$ values previously calculated. All these steps can be coded as simple vectorizable loops provided the elements of the 3x3 matrices are stored in a convenient form.

The calculation of the preconditioning proceeds as follows.

$$S_1 = T_1,$$
$$P_1 \text{ minimizes } ||S_1 P_1 - I||_F,$$
$$S_i = T_i - d_x(L_i P_{i-1} U_{i-1}), \quad i = 2, \cdots, n$$
$$P_i \text{ minimizes } ||S_i P_i - I||_F, \quad i = 2, \cdots, n.$$

where $d_x(B)$ is the matrix composed of the 2x-1 main diagonals of some matrix B for $x=1,2$.

The solution of the system $Qx = y$ can be obtained using

$$z = P(y - Lz),$$

and

$$x = z - P(Ux),$$

which consists of matrix-vector products and is completely vectorizable. Note that our preconditioning has the same implicit form as the block preconditionings **INV** presented above.

$$PQ = (PL + I)(PU + I)$$

Note also that only the blocks $P_i$ need to be stored. This technique of nested-factorization with approximate inverses can be easily expanded to three dimensions as shown in [14,15].

The preconditioning presented in this section was compared with **INV**, **MINV**, and **ILU**, solving a series of problems. We illustrate the results here with a representative problem. The code was optimized for the Alliant FX/8 and the CRAY-2 vector computers, compiler versions 3.0.14 and cft3.0b, respectively. Since we were interested in speed-up due only to vectorization, concurrency was not allowed during the runs.

Four different versions of the approximate inverse were used as follows:
  AID: update only main diagonal of $S_i$ ($d_x = d_1$)
  MAID: AID with zero column-sum
  AI3: update all three diagonals of $S_i$ ($d_x = d_2$)
  MAI3: AI3 with zero column-sum

We used only pentadiagonal matrices that arise from the 5-point star discretization. The times presented in the tables were obtained using the **gprof** facility in the Alliant FX/8 and the **flow** facility in the CRAY-2. The sample problem was solved several times in a loop in order to provide the tracing routines with significant breakdowns per sub-routine. In the Alliant the sample problem was solved 10 times, and 200 times in the CRAY-2. The iteration process was stopped when the norm of the error was smaller than the right hand side of original system by a factor of $10^{-6}$.

The results (see Table I) show speed-ups of up to 5 for the CRAY-2 on the application of our preconditioning. In all cases the preconditioning performs better than **ILU**, with at least a 50 percent gain in the total time, when running on the CRAY-2. Also notice that our preconditioning needs fewer iterations than **INV**, **MINV**, or **ILU** for the method to converge, for non-symmetric matrices. This suggests that the preconditioning is doing a better job than the other three methods in clustering the eigenvalues of the preconditioned matrix. The lower speed-ups on the Alliant are due to the FX compiler using associative transformations capable of some vector optimization on recursive loops.

## Table I
### Preconditioner comparison times, 2500 variables, vector mode
### each solve performed 200 times on CRAY 2, 10 times on Alliant

Non-symmetric matrix (-1 -2 6 -1 -2)

| | CRAY-2 | | | | Alliant FX/8 | | | |
|---|---|---|---|---|---|---|---|---|
| Prec | Calc. | Appl. | Total | Iter. | Calc. | Appl. | Total | Iter. |
| INV | 1.14 | 10.37 | 22.52 | 62 | 0.72 | 10.40 | 25.43 | 62 |
| MINV | 5.69 | 10.47 | 27.41 | 61 | 2.43 | 10.45 | 26.59 | 61 |
| ILU | 0.46 | 39.54 | 49.97 | 52 | 0.21 | 13.12 | 25.01 | 52 |
| AID | 0.47 | 8.85 | 18.70 | 50 | 0.45 | 9.26 | 21.31 | 52 |
| MAID | 0.58 | 8.60 | 18.31 | 48 | 0.56 | 8.61 | 20.42 | 50 |
| AI3 | 0.51 | 8.56 | 18.13 | 49 | 0.52 | 9.00 | 21.49 | 53 |
| MAI3 | 0.50 | 8.32 | 17.43 | 49 | 0.55 | 9.21 | 21.55 | 53 |

Further results comparing the above methods and others that illustrate the power of the method for nonsymmetric problems are included in [11].

**Properties of the Tridiagonal Approximate Inverse.**

We are interested in the properties satisfied by the approximate inverses obtained from the minimization in the Frobenius norm. The case of tridiagonal M-matrices is of special interest. Since the inverse of an M-matrix is positive, we would like to know that the tridiagonal approximate inverse of a tridiagonal M-matrix is also positive. We have been able to show this for two specific cases.

**Theorem II** *Let S be an m x m tridiagonal matrix with main diagonal elements b and off diagonal elements $-a$, with b and a greater or equal zero. If $b \geq \sqrt{2}a$ then the elements of the approximate inverse will be positive.*

**Theorem III** *Let T be an m x m tridiagonal matrix, with main diagonal elements b, lower diagonal elements $-a$ and upper diagonal elements $-c$, with $a, b, c > 0$ and $b \geq a+c$. Further, require that $c = a + \varepsilon$ with $\varepsilon \geq 0$. Under these conditions S is a positive-definite, M-matrix, although not necessarily symmetric. Then, there is a function $\alpha(a)$ such that if $\varepsilon \leq \alpha(a)$, the approximate tridiagonal inverse is positive.*

Complete proofs of the Theorems II and III are included in [11]. Also from Theorem III, a restriction is imposed on the Cell Peclet number that can be used for certain n-dimensional problems.

The convergence rate of **CRM** is improved by preconditioning since the new coefficient matrix $AQ^{-1}$ (where $Q$ is the preconditioning matrix) has more clustered eigenvalues than $A$. We have under certain assumptions been able to describe this clustering.

**Theorem IV** *Let S be an m x m tridiagonal matrix, with main diagonal elements b, lower diagonal elements $-a$ and upper diagonal elements $-c$, with $a, b, c > 0$ and $b \geq a+c$. The eigenvalues of S satisfy $0 \leq \lambda \leq 2b$. If P is an m x m diagonal matrix, obtained by minimizing $||SP - I||_F$, then the eigenvalues of SP satisfy $0 \leq \lambda \leq 2$.*

**Proof:** Let $y_i$ ($i=1, \cdots, n$) be the ith element of the main diagonal of $P$. Minimization of $||SP - I||_F$ gives

$$y_1 = \frac{b}{b^2+a^2}$$
$$y_i = \frac{b}{b^2+a^2+c^2} \quad i=2, \cdots, n$$
$$y_n = \frac{b}{b^2+c^2}$$

$SP$ is also tridiagonal and applying Gerschgorin's theorem to its columns we find that its eigenvalues are in the union of circles $(c, r)$, with center $c$ and radius $r$, given by

$$\left[\frac{b^2}{b^2+a^2}, \frac{ab}{b^2+a^2}\right], \left[\frac{b^2}{b^2+a^2+c^2}, \frac{ab+bc}{b^2+a^2+c^2}\right], \left[\frac{b^2}{b^2+c^2}, \frac{cb}{b^2+c^2}\right]$$

All centers and radii can be easily shown to be less or equal 1. Therefore, the eigenvalues of *SP* satisfy $0 \leq \lambda \leq 2$. ■ [QED]

### Approximate Inverses as Sparse Preconditionings

In the remainder of this paper we will concentrate on discussing the approximate inverses as a direct preconditioning for the **CRM** algorithm. As noted in the introduction, when using an approximate inverse preconditioning $P$, the system $Ax=b$ is replaced by a new system of the form $By=b$ where $B=AP$ and $Py=x$. In fact, in the **CRM** algorithm the $Qz^{k+1}=r^{k+1}$ is replaced by $z^{k+1}=Pr^{k+1}$. That is, the application of the preconditioning takes the form of a direct matrix-vector multiply.

The application of the approximate inverse alone as a preconditioner was first considered in [6]. There the matrix of a five-point discretization was considered. Both the original matrix and the preconditioning matrix had the same five-star pattern. Speed up and efficiency results for this pattern are summarized in Table II. These numerical results indicate that the concurrent calculation of the columns of the preconditioning has excellent potential for this particular sparsity pattern. As Table II shows, the efficiency in terms of processor utilization for calculation of the preconditioning is near optimal. Certainly, this technique makes excellent use of the processors available during the calculation of the preconditioning.

In fact the sparsity pattern of the approximate inverse, $P$, does not have to be the same as in the the coefficient matrix $A$. This is especially important for general sparse matrices where the structure of the approximate inverse may need to be different and perhaps fuller than the original matrix. The first question that can be raised is what is the effect of altering the sparsity pattern of the approximate inverse on its performance as a preconditioning. The following Theorem shows that it may be beneficial to augment the sparsity pattern of $P$. By filling-in some positions in a column $p_i$ of the sparse approximate inverse $P$, the residual $||Ap_i-e_i||_2$ does not increase, and, in fact, it may be reduced. Thus, the augmented $P$ is a better approximation to the true inverse of $A$.

### Table II

| Number of Processes | Speedup in Preconditioning Calculation (OU Alliant) | Speedup in Preconditioning Calculation (ANL Alliant) | Speedup in Over all Calculation (OU Alliant) | Speedup in Over all Calculation (ANL Alliant) |
|---|---|---|---|---|
| 2 | 1.988 | 1.993 | 1.770 | 1.853 |
| 3 | 2.950 | 2.971 | 2.158 | 2.308 |
| 4 | 3.887 | 3.935 | 2.561 | 2.851 |
| 5 |  | 4.892 |  | 3.144 |
| 6 |  | 5.831 |  | 3.401 |
| 7 |  | 6.753 |  | 3.548 |
| 8 |  | 7.659 |  | 4.020 |

**Theorem V:** *Given the matrix $C \in R^{k \times l}$, $p$ is an $l$-vector, and $e$ is an $k$-vector with all entries 0 except one entry, which is set to 1. Let $C'$, $p'$ and $e'$ be such that*

$$C' = \begin{bmatrix} C & c \\ 0 & \delta c \end{bmatrix}, \quad p' = \begin{bmatrix} p \\ \delta p \end{bmatrix}, \quad \text{and} \quad e' = \begin{bmatrix} e \\ 0 \end{bmatrix},$$

where $c$ is an $k \times t$ matrix, $\delta c$ in an $h \times t$ matrix, and $\delta p$ is a $t$-vector. Then

$$\min_{p'} ||C'p' - e'||_2 \leq \min_p ||Cp - e||_2.$$

Moreover, the equality is achieved only when $c^T(Cp - e) = 0$.

This result tells us that filling-in the sparsity of the approximate inverse may result in a better preconditioning. It does not yet tell us which columns of $P$ need to be filled-in, nor does it tell us where to place that fill-in. We now provide criteria that answer those two questions.

The convergence rate of the conjugate gradient algorithm is dependent on the condition number of the matrix. The following Theorem provides an actual bound for the condition number of the preconditioned matrix in terms of the norm of the residual.

**Theorem VI** *Let $A$ be a nonsingular matrix, and $P$ be the approximate inverse obtained by minimizing $||AP - I||_F$ subject to a given sparsity pattern. Let $\varepsilon = ||AP - I||_F$, the optimal residual. If $\varepsilon < 1.$, then*

$$\kappa(AP) \leq \frac{1+\varepsilon}{1-\varepsilon}.$$

*Further, if for each column $p_j$ of $P$, $||Ap_j - e_j||_1 < 1.$, then, Richardson's iterative method with $AP$ as the iteration matrix converges.*

The second assertion of Theorem VI provides a numerical criteria for determining if a single row needs to receive fill-in. Proofs of the Theorems V and VI are presented in [9]. Now we turn to the question of where the new fill-in should be located.

Ideally, we would like the least squares subproblems to introduce new columns of $A$ which have entries in rows already involved by previous columns of $A$. This will allow potentially more freedom to adjust the values of the elements of $p_i$ to reduce the residual, for we will have the same equations involved previously with as many unknowns as possible. We want to do this while at the same time introducing as few new rows in the augmented matrix as possible. We know from Theorem I that $k > l$. Hence, we want to look for those columns that involve rows already considered but that introduce as few new rows as possible. To discover when this situation will occur we look at the representation of the least squares problem graphically.

Each of the least squares matrices must be represented by a bipartite graph. However, the discussion can be limited to the graph of square nonsymmetric matrices. Hence, the set of nodes is numbered from 1 to $n$, and it is sufficient to present the discussion in terms of directed graphs.

Let $A$ be an $n \times n$ sparse matrix, the adjacency graph of $A$, denoted by $G^A(X^A, E^A)$, is a graph in which the n vertices $X^A$ of $G^A$ are numbered from 1 to $n$, and the arc $(i, j) \varepsilon E^A$ if and only if $a_{ij} \neq 0$. A subgraph $G'(X', E')$ of $G(X, E)$ is a graph for which $X' \subset X$ and $E' = E \cap (X' \times X')$. The in-degree of a node $x$ in a graph $G(X, E)$, $indeg(x, G)$, is the number of elements in the set $\{y \varepsilon X \mid (y, x) \varepsilon E\}$.

The following construction of a subgraph of the original matrix $A$ will be central to the selection of fill-in candidates.

Let $G^A(X^A, E^A)$, be the graph of the matrix $A$, and be $G^P(X^P, E^P)$, the graph of the matrix $P$, where $P$ is the approximate inverse of $A$.

Construct the subgraph $G_i$ of $G^A$ and associate it with the $i^{th}$ column of $P$ by the following steps:

1. Let $CS_i = \{$nodes in $X$ which are adjacent in $G^P$ to node i$\}$.

2. Let $RS_i = \{$nodes in $X$ which are adjacent in $G^A$ to $j \varepsilon CS_i\}$.

3. Let $F = CS_i \cup RS_i$.

4. Then $G_i(X', E')$ is the subgraph of $G^A$, with $X' = F$.

Note that the $CS_i$ represents the columns involved in the least squares subproblem, and $RS_i$ represents the rows. Because of Theorem I, there are more elements in $RS_i$ than in $CS_i$. Thus, potentially the best elements to consider to help reduce the least squares residual are the ones that are in $RS_i$

but not in $CS_i$. These elements are classified according to the number of extra rows that they may introduce to the least squares matrix. Define

$$W_i(h) = \left\{ x \mid x \ \varepsilon \ RS_i - CS_i \quad \text{and} \quad (\text{indeg}(x,G^A) - \text{indeg}(x,G_i)) = h \right\}.$$

If a node $j \ \varepsilon \ W_i(h)$, then allowing a non zero in the $j^{th}$ position of the $i^{th}$ column of $P$ will introduce $h$ extra rows to the corresponding least squares submatrix. The cardinality of $RS_i$ is increased by $h$. Thus, it is clear that the most desirable nodes to consider introducing are those that belong to $W_i(h)$ for the lowest $h$ possible.

There are many possible renditions of algorithms to make this selection. These could yield different results according to the order in which the graph is sampled. Heuristically, the initial graph should ensure that the $i^{th}$ row is included in the determination of the $p_i$. This can be ensured, for instance, by initiating the graph with the sparsity of the transpose of the matrix or an appropriate subset of it. We close the presentation of this section with the following interesting result.

**Theorem VII** *Let $G^A$ be the graph of $A$, an $n^2 \times n^2$ matrix with the symmetric pattern of a five point operator on an $n \times n$ grid. Let $G^P$ be the graph of $P$, an $n^2 \times n^2$ matrix with a block diagonal pattern where each of the blocks is an $n \times n$ tridiagonal. Let the sparsity structure of $P'$ be given by its column sets defined by*

$$CS'_i = CS_i \cup W_i(0) \cup W_i(1).$$

*Then*

$$G^{P'} \supset G^A.$$

In fact the elements of $G^{P'}$ not included in $G^A$ are fill-in introduced around the boundary of the grid. A proof of the Theorem is found in [8]. Further results illustrating the techniques developed here are presented in [7]. Formulations for solving each of the least squares subproblems updating the $QR$ factorization without recomputing it are also presented in [9].

## Conclusions

We present two uses of the approximate inverses derived using minimization on the Frobenius norm. We have shown formulations of a preconditioning that allow high vectorization of the conjugate residual method by utilizing only vector-vector or matrix-vector products.

One preconditioning presented combines a nested-factorization technique with the calculation of an approximate inverse for a tridiagonal matrix using minimization on the Frobenius norm. Experimental results for the solution of symmetric and non-symmetric matrices show that this preconditioning compares favorably with other methods proposed, with speed-ups of up to 5 on the application step, and at least 2 for the total time when compared with the traditional **ILU** method, code running on a CRAY-2 supercomputer. Some important questions remain open. Most relevant is the need to establish the stability of the calculation of the nested algorithm.

The other technique discussed uses sparse approximate inverses of the whole matrix directly as preconditionings. Uniqueness of these sparse preconditionings is established for nonsingular matrices. Bounds on the condition number for the preconditioned systems are derived. Further, criteria for determine fill-in locations to improve the quality of the preconditionings are presented. The numerical results illustrate the significant potential for parallelism of this technique.

## Acknowledgements

The authors wish to thank M. Minkoff and A. Griewank of Argonne National Laboratory for the helpful discussions and suggestions. We also wish to thank Argonne National Laboratory for the use of the Alliant FX/8 in their Advanced Computer Research Facility, and NSF for supporting the use of the CRAY-2 supercomputer at NASA-Ames.

This paper was presented at the V IIMAS Workshop on Numerical Analysis held in Mérida, Yucatan, México, January 3-6, 1989.

# References

[1] J. R. Appleyard, I. M. Cheshire, Nested factorization, SPE 12264, **Reservoir Simulation Symposium**, Nov 15-18, 1983, San Francisco, CA, pp. 315:324.

[2] Benson, M. W., and P. O. Frederickson, Iterative Solution of Large Sparse Linear Systems Arising in Certain Multidimensional Approximation Problems, **Utilitas Math., 22**, 1982, pp. 127-140.

[3] R. Chandra, Conjugate Gradient Methods for Partial Differential Equations, Ph.D. Thesis, Yale University, New Haven, Conn., 1978.

[4] P. Concus, G. Golub, G. Meurant, Block preconditioning for the conjugate gradient method, **SIAM J. Sci. Stat. Comp. 6**, pp. 220-252 (1985).

[5] P. Concus, G. Meurant, On computing INV block preconditionings for the conjugate gradient method,**BIT 26**, 1986, pp. 493-504.

[6] Cosgrove, J. D. F., and J. C. Díaz, Fully Parallel Preconditionings for Sparse Systems of Equations, **Proceedings of the Second Workshop on Applied Computing, (ed. Uselton)**, Tulsa, Oklahoma, March 1988, pp. 29-34.

[7] Cosgrove, J. D. F., and J. C. Díaz, A Fill-in Strategy for Sparse Approximate Inverse Preconditionings, **Proceedings of the 1989 ACM Sth C. Regional Conference (ed. Dhall, Berghel, & George) Tulsa, Oklahoma, November 1989, pp. 121-126.**

[8] Cosgrove, J. D. F., and J. C. Díaz, Structural Properties of the Graph of Augmented Approximate Inverses, **Proceedings of the 1990 Symposium on Applied Computing, (ed. Berghel, Talburt, & Roach)**, Fayetteville, Arkansas, April 1990, pp. 131-136.

[9] Cosgrove, J. D. F., J. C. Díaz, and A. Griewank, Approximate Inverse Preconditionings for Sparse Linear Systems, To appear.

[10] J. C. Díaz, and C. G. Macedo, Jr., Vectorizable Block Preconditionings for Non-symmetric Systems of Equations, **Proceedings of the Second Workshop on Applied Computing, (ed. Uselton)**, Tulsa, Oklahoma, March 1988, pp. 22-28.

[11] J. C. Díaz, and C. G. Macedo, Jr., Fully Vectorizable Block Preconditionings with Approximate Inverses for Non-symmetric Systems of Equations, **I. J. Numer. Meth. Eng., 27**,, pp. 501-522, 1989.

[12] H. Elmann, Iterative Methods for Large, Sparse, Non-Symmetric Systems of Linear Equations, Ph.D. Thesis, Yale University, New Haven, Conn., 1982.

[13] Kolotilina, L. Y., and A. Y. Yeremin, On a Family of Two-Level Preconditionings of the Incomplete Block Factorization Type, **Sov. J. Numer. Anal. Math. Modeling, 1**, 1986, pp. 293-320.

[14] G. K. Leaf, M. Minkoff, J. C. Díaz, Preconditioned iterative methods for partial differential equations, **Proceedings of the 6th IMACS International Symposium on Computer Methods for Partial Differential Equations (eds. Vichnevetsky, Stepleman)**, June 23-26, 1987, Bethlehem, PA, pp. 551-555.

[15] G. K. Leaf, M. Minkoff, J. C. Díaz, Nested Block Factorization Preconditioners for Convective-Diffusion Problems in Three Dimensions, **Mathematics for Large Scale Computing**, (ed. Díaz), Lecture Notes in Pure and Applied Mathematics, Marcel Decker, Inc, 1989.

# CHAPTER 9

On Generalized Finite Difference Methods for
Approximating Solutions to Integral Equations

G. D. Allen*
P. Nelson, Jr.**

**Abstract.** This paper is concerned with proving convergence and super convergence for the approximation of solutions to certain types of integral equations that include for example the discrete ordinates approximation to the monoenergetic transport equation for slab geometry. We define and analyze generalized $n^th$ order finite difference schemes by means of a basic projection on Hilbert space. These schemes include all the currently applied characteristic methods in transport theory literature. We provide a convergence analysis of these methods and prove that super convergence phenomena are present for cell-edge and cell-average fluxes. The proofs of our main results are different and more general than standard methods, and in fact, apply even to the transport equation with non constant cross sections over individual slabs. Further super convergence results are developed for weighted cell-averages.

§1. **Introduction.**

Over the past several years, numerous papers have been published [1,3-12] on the finite difference approximations to discrete-ordinates slab transport equations. For example, Larsen and Nelson [7] have given convergence analyses for several of the standard linear finite difference schemes including the diamond difference, step characteristic, and linear moments methods. Neta and Victory [10] extended these results to a higher order generalizations of the diamond difference method. Finally, Victory and Ganguly [12] unified all these results, for both characteristic and continuous methods of arbitrary order. We further note parenthetically that Keller and Nelson [5] recently showed that a large abstractly defined class of methods, including all of those mentioned above, inherits from one-step finite-difference methods for first-order linear systems the property that consistency implies stability (and hence convergence).

The results of Victory and Ganguly [12] provide the starting point for the analysis of this paper. We develop a general method of projections that includes the above results

---

*Department of Mathematics, Texas A&M University, College Station, TX 77843.
**Department of Nuclear Engineering, Texas A&M University, College Station, TX 77843.

and demonstrate its applicability to a wide variety of integro-differential equations which when applied to the monoenergetic transport equation allow for example non constant cross section and scattering kernel over individual slabs. This is an improvement over previously reported results [7,10-12], in which the piecewise constancy of these data are essential to the methods of proof.

In §2 we consider the basic types of integro-differential equations to be considered in this paper. In §3 we develop the projection method. In §4 we give our main results on convergence and superconvergence of the approximations. In §5 we prove the superconvergence of the cell averages of the approximations, and related results for higher order moments.

## §2. Preliminaries.

We begin by considering the basic integral equation defining the monoenergetic transport equation for slab geometry which we subsequently generalize to the more general form of integral equations for which the results of this paper apply. The monoenergetic transport equation for slab geometry has the form

$$(2.1) \quad \mu \frac{\partial \psi}{\partial x}(x,\mu) + \sigma(x)\psi(x,\mu) = c(x)\sigma(x) \int_{-1}^{1} k(x,\mu',\mu)\psi(x,\mu')d\mu' + q(x,\mu)$$

where $\mu \in [-1,1]$ denotes the cosine of the angle between the positive $x$-axis and the particle direction, and where $x$ is the spatial variable over the slab $[0,a], a > 0$. As usual [7,12], $\psi(x,\mu)$ is the angular flux and the given functions $\sigma, c$, and $q$ are nonnegative and piecewise constant on $[0,a]$. The scattering kernel $k$ is nonnegative, and piecewise constant in $x$, symmetric in $\mu$ and $\mu'$, and satisfies

$$(2.2) \quad \int_{-1}^{1} k(x,\mu,\mu')d\mu \equiv 1.$$

The discrete ordinates approximation to (2.1) is given by

$$(2.3) \quad \mu_i \frac{d\psi_i^e}{dx}(x) + \sigma(x)\psi_i^e(x) = c(x)\sigma(x) \sum_{j=1}^{N} \omega_j k(x,\mu_j,\mu_i)\psi_j^e(x) + \tilde{q}_i(x)$$

where $\{\mu_i\}, i = 1,\ldots,N$ denotes the set of quadrature points with associated weights $\{\omega_i\}, i = 1,\ldots,N$. The function $\psi_i^e(x)$ are the discrete ordinates approximation to $\psi(x,\mu_i)$ and $q_i(x) = q(x,\mu_i)$. We use vacuum boundary conditions

$$\psi_i(0) = 0 \quad \text{if} \quad \mu_i > 0 \quad \psi_i(a) = 0 \quad \text{if} \quad \mu_i < 0.$$

An interface point $x \in [0,a]$ is a point where one or more of the functions $c, k,$ or $q$ is discontinuous. We assume that there are only a finite number of these. We define the

Banach function space $C_N[0,a]$ to be the space of $N$-vectors of continuous functions on $[0,a]$ with the supremum norm. That is, if

$$f = (f_1, f_2, \ldots, f_N)$$

then

$$\|f\| = \max_{1 \leq i \leq J} \max_{x \in [0,a]} |f_i(x)|.$$

In other words, we can write $C_N[0,a]$ as the direct sum of $N$ copies of $C[0,a]$:

$$C_N[0,a] = \oplus_{i=1}^{N} C[0,a].$$

Similarly, we define $C_{N,p}[0,a]$ to be the Banach function space of $N$-vectors of piecewise continuous functions with left and right hand limits at interface points.

The operators we shall need below are defined as

$$(Sf)_i(x) = c(x) \sum_{j=1}^{N} \omega_j k(x, \mu_j, \mu_i) f_j(x)$$

and

$$(Lf)_i(x) = \frac{1}{\mu_i} \int_{a_i}^{x} \exp\left\{-\frac{1}{\mu_i} \int_{x'}^{x} \sigma(s) ds\right\} \sigma(x') f_i(x') dx', \quad i = 1, \ldots, N$$

where

$$a_i = \begin{cases} 0, & \mu_i > 0 \\ a, & \mu_i < 0 \end{cases}.$$

It is known [7, §2] that $LS$ is a compact operator from $C_{N,p}[0,a]$ to $C_N[0,a]$. The equation (2.3) can be rewritten as

(2.4) $$\psi^e = L(S\psi^e + q)$$

$q$ is a vector function with components $q_i(x) = \tilde{q}_i(x)/\sigma(x)$. We assume that the spectral radius of $LS$ ($\rho(LS)$) is less than unity so that (2.4) has a solution, in fact a positive solution.

With this example in mind we consider the definition below of a more general class of matrix valued kernels for which our analysis will hold.

First, we require several spaces of functions. Let $J$ be a finite disjoint collection of open subsets of $[0,a]$. Define $C^r(J)$ to be the set of functions that are $r$-times continuously differentiable on $J$. For $f \in C^r(J)$, we define the norm of $f$ by

$$\|f\|_{C^r} = \sum_{\ell=0}^{r} \sup_{x \in J} |f^{(\ell)}(x)|.$$

If $r = 0$ we use $C(J)$ to denote all functions continuous on $J$. Similarly, we denote by $C_p^r(J)$ the space of all piecewise $r$-times continuously differentiable functions on $J$. The multidimensional versions of the two spaces above are denoted by $C_N^r(J)$ and $C_{N,p}^r(J)$. Thus, if $f \in C_N^r(J)$, then $f$ is the vector-valued function

$$f(x) = (f_1(x), f_2(x), \ldots, f_N(x))$$

with norm defined by

$$\|f\|_{C^r} = \max_{1 \le i \le N} \|f_i\|_{C^r}.$$

We also need the spaces $L_\infty(J)$ of all measurable functions that are essentially bounded on $J$. The norm of $f \in L_\infty(J)$ is defined by

$$\|f\|_\infty = \sup_{x \in J} |f(x)|.$$

Note that $\|f\|_\infty = \|f\|_C$ if $f \in C_r(J)$. Analogously define $L_{\infty,N}(J)$ to be the vector version of $L_\infty(J)$.

**Definition.** Let $0 = y_1 < y_2 < \cdots < y_s = a$ be a sequence of points, and define $J = [0, a] - \{y_k\}_{k=0}^{s}$. Define $A(N, r, J)$ to be the space of all $N \times N$ matrix-valued kernels $K(x, y)$ defined on $J \times J$ that satisfy the following conditions:

(1) $K(x,y) \in C^r(J \times J - \{(y,y) | y \in J\})$

(2) For $\ell = 0, 1, 2, \ldots, r$, the functions

$$\lim_{x \to y^\pm} \frac{\partial^\ell K}{\partial x^\ell}(x,y), \quad \lim_{x \to y^\pm} \frac{\partial^\ell K}{\partial y^\ell}(x,y)$$

$$\lim_{y \to x^\pm} \frac{\partial^\ell K}{\partial x^\ell}(x,y), \quad \lim_{y \to x^\pm} \frac{\partial^\ell K}{\partial y^\ell}(x,y)$$

are in $C^{r-\ell}(J)$.

Denote by $\|K\|_A$ the sum (over $\ell$ and matrix indices) of all the $C^r$ norms on $J \times J$ of all the functions (entries) in (1) and $C^{r-\ell}$ norms on $J$ of the functions in (2). We assume throughout that the operator norm of $K$ is taken over $L_{\infty,N}$, and denote it simply as $\|K\|$. It is easily seen that the kernels $L$ and $LS$ of the monoenergetic transport equation are in $A(N, \infty, J)$, where the endpoints of $J$ are the slab interfaces. Moreover, $\|L\|_A$ and $\|LS\|_A$ are finite for every finite $r$. We therefore will consider (2.4) for any operators $L$ and $LS$ with $L \in A(N, r, J)$ and $LS \in A(N, r, J)$.

The class of equations that we consider have the form

(2.5) $$\psi^e = K\psi^e + \tilde{K}q$$

where $K$ and $\tilde{K}$ are in $A(N, r, J)$ and $q \in C_N^r(J)$. We assume that the norms of $\|K\|$ in both the $L_2$ and $L_\infty$ metrics are less than 1. Let us remark that for the operator $LS$ of the transport equation above, it can be shown that if $\|S\|_\infty$ is sufficiently small then $\|LS\|_\infty < 1$. It is also possible to show that $\|L\|_1$, the norm on $L_1[0, a]$, is bounded less than one and hence $\|LS\|_1 < 1$. Thus, by interpolation $\|LS\|_2$, the norm on $L_2[0, a]$ is bounded less than one.

The class of finite difference approximations we consider are equivalent to the form

(2.6) $$\psi = KM\psi + \tilde{K}Mq.$$

Here $M$ will be a linear operator (or more specifically an orthogonal projection) on $L_2[0, a]$, the space of Lebesgue measurable functions $f$ with

$$\int_0^a |f(x)|^2 dx < \infty.$$

In fact, $M$ will be a finite dimensional approximation to the identity. The difference between the approximate and exact solution is

(2.7) $$\psi - \psi^e = KM(\psi - \psi^e) + K(M - I)\psi^e + \tilde{K}(M - I)q.$$

Assuming $\|KM\|_2 < 1$ we have

$$(\psi - \psi^e) = [I - KM]^{-1}[K(M - I)\psi^e + \tilde{K}(M - I)q].$$

The type of questions we consider are as follows: What does the order of approximation of $(M - I)\psi^e$ and $(M - I)q$ imply about the difference $\psi - \psi^e$? The answer, as we shall see, depends upon $M$ and upon which aspects of $\psi - \psi^e$, e.g. pointwise estimates or moments, are considered.

### §3. The Projection Method.

We consider partitions $\pi$ of $[0, a]$ defined by the mesh points

$$0 = x_{1/2} < x_{3/2} < \cdots < x_{H+1/2} = a.$$

Each interval $C_m = (x_{m-1/2}, x_{m+1/2}), m = 1, \ldots, H$ is termed a cell and $x_m$ denotes the midpoint of the cell $C_m$. We assume the interface points of the transport equation in §2 are included in mesh points of each partition. More generally with respect to $A(N, r, J)$ we assume the mesh points of each partition include the set of values $[0, a] - J$.

Define $h_m = x_{m+1/2} - x_{m-1/2}, m = 1, \ldots, H$ and $h = \max_{1 \le m \le H} h_m$. In what follows we will consider sequences of partitions $\{\pi_h\}$ for which $h \to 0$. We will always assume that the sequence is *quasi uniform*. This means there is a fixed constant $c$ so that for each partition $\pi_h$ in the sequence $h < ch_m$.

Given $\pi_h$, define the mapping

$$(3.1) \quad (\rho_{h,m}f)(x) = \left[\frac{2}{h_m}\right]^{1/2} f\left(\frac{h_m}{2}x + x_m\right) \qquad m=1,\ldots,H, \quad -1 \le x \le 1.$$

Clearly $\rho_{h,m}$ is a unitary transformation from $L_2(C_m)$ onto $L_2[-1,1]$. The inverse $\rho_{h,m}^{-1}$, also a unitary transformation from $L_2[-1,1]$ to $L_2(C_m)$, is given by

$$(3.2) \quad (\rho_{h,m}^{-1}f)(x) = \left[\frac{h_m}{2}\right]^{1/2} f\left(\frac{2}{h_m}(x - x_m)\right).$$

Let $\chi_{C_m}$ denote the indicator function of the closure, $\overline{C}_m$, of $C_m$; that is

$$\chi_{C_m}(x) = \begin{cases} 1 & x \in \overline{C}_m \\ 0 & x \in \overline{C}_m \end{cases}.$$

Let $M$ be a bounded linear operator on $L_2[-1,1]$; with norm denoted by $\|M\|_2$. For a given partition $\pi_h$, define the linear transformation $M_h$ from $L_2[0,a]$ to $L_2[0,a]$ by

$$(3.3) \quad M_h = \sum_{m=1}^{H} \chi_{C_m} \rho_{h,m}^{-1} M \rho_{h,m} \chi_{C_m}.$$

Although (3.3) is defined for bounded linear operators $M$ on $L_2[-1,1]$, the same formula can be used for operators $M$ on $C[-1,1]$, replacing $\rho_{h,m}$ by $(\tilde{\rho}_{h,m}f)(x) = f\left(\frac{h_m}{2}x + x_m\right)$ and $\tilde{\rho}_{h,m}^{-1}$ by $(\tilde{\rho}_{h,m}^{-1}f) = f\left(\frac{2}{h_m}(x - x_m)\right)$. However as we observe in the next result, $M_h$ will be a bounded operator from $C[-1,1]$ to $C(J)$, where $J = [0,a] - \{x_{1/2},\ldots,x_{H+1/2}\}$.

**Theorem 3.1.** Let $\pi_h$ is a partition of $[0,a]$ with $H$ subdivisions and cells $C_m, m = 1\ldots H$. (i) If $M$ is a bounded operator on $L_2[-1,1]$, then $M_h$ is a bounded operator on $L_2[0,a]$ and $\|M_h\|_2 = \|M\|_2$. Moreover $\dim M_h = H \dim M$. (Here, $\dim M$ means the dimension of the range of $M$.) In particular if $M$ is a projection then so also is $M_h$. (ii) If $M$ is a bounded operator on $C[-1,1]$ then $M_h$ is a bounded operator on $C(J)$, where $J = [0,a] - \{x_{1/2},\ldots,x_{H+1/2}\}$, and $\|M_h\|_{C(J)} = \|M\|_{C[-1,1]}$.

This result is elementary as well as standard in the literature on projection methods and its proof is omitted. In the following examples we illustrate that the standard operators used in many papers in transport theory are of the type $M_h$ for various $M$.
**Examples.** (1) Define

$$(3.4) \quad (M_0 f)(x) = \frac{1}{2}\int_{-1}^{1} f(u)du \chi_{[-1,1]}(x)$$

where $\chi_{[-1,1]}(x)$ is the indicator function of $[-1,1]$. $M_0$ is a bounded operator on $C[-1,1]$ and also a one dimensional projection on $L_2[-1,1]$. Then $M_h$ is the usual *step characteristic* operator on $[0,a]$. [7, p. 337].

(2) Defining

$$(M_D f)(x) = \frac{1}{2}(f(-1) + f(1))\chi_{[-1,1]}(x)$$

we see that $M_D$ is a bounded operator on $C[-1,1]$. Then $M_h$ is the classical *diamond-difference* operator [7, p. 337].

(3) Now define

$$(M_{LC} f)(x) = (M_0 f)(x) + \frac{3x}{2}\int_{-1}^{1} u f(u) du$$

where $M_0$ is defined above. Then $M_{LC}$ is a bounded operator on $C[-1,1]$ and a two dimensional projection on $L_2[-1,1]$. The corresponding $M_h$ is the familiar *linear characteristic* operator.

(4) Finally, let $p_0, p_1, \ldots, p_r$ be the first $r+1$ normalized Legendre functions on $[-1,1]$. Define

(3.5) $$(M_{r+1} f)(x) = \sum_{j=0}^{r} \left( \int_{-1}^{1} f(u) p_j(u) du \right) p_j(x).$$

Then $M_{r+1}$ is a bounded operator on $C[-1,1]$ and a $(r+1)$-dimensional projection on $L_2[-1,1]$. The corresponding $M_h$ ($= M_{r+1,h}$) is exactly the type of operator considered by Victory and Ganguly [12]. Note that the step characteristic and linear characteristic operators defined in examples (1) and (3) above are special cases of (3.5).

It is the operators $M_h$ based on some $M$ that will be used in the finite difference approximation (2.6). A necessary condition for convergence of the approximation is that $M_h$ should approximate the identity $I$ in the strong operator topology. Each of the operators in the examples do this. In our next result we give necessary and sufficient condition for this to hold.

**Theorem 3.2.** (i) Suppose that $M$ is a bounded operator on $L_2[-1,1]$. Then $\lim_{h \to 0} M_h = I$ in the strong operator topology if and only if $M\chi_{[-1,1]} = \chi_{[-1,1]}$.
(ii) Let $M$ be a bounded operator on $C[-1,1]$. Then

$$\lim_{h \to 0} \|M_h f - f\|_\infty = 0$$

for every $f \in C[0,a]$ if and only if $M\chi_{[-1,1]} = \chi_{[-1,1]}$.

**Proof.** First we prove the necessity in both (i) and (ii). The partition $\pi_h = \cup_{m=0}^{H} C_m$. Suppose that $M\chi_{[-1,1]} = f(x)$. Then

$$M_h \chi_{[0,a]} = \sum_{m=1}^{M} f\Big(\frac{2}{h_m}(x - x_m)\Big) \chi_{C_m}$$

So,

$$(M_h - I)\chi_{[0,a]} = \sum_{m=1}^{M} \Big[f\Big(\frac{2}{h_m}(x - x_m)\Big) - 1\Big] \chi_{C_m}.$$

In the $L_\infty$ norm it is easy to see that the norm is invariant in each cell, namely

$$\sup_{-1 \leq u \leq 1} |f(u) - 1|.$$

And in the $L_2$ norm we see

$$\int_{C_m} \Big|f\Big(\frac{2}{h_m}(x - x_m)\Big) - 1\Big|^2 dx = \int_{-1}^{1} |f(u) - 1|^2 h_m dx.$$

Thus

$$\|(M_h - I)\chi_{[0,1]}\|_2^2 = \int_{-1}^{1} |f(u) - 1|^2 dx$$

and the result follows.

To prove the sufficiency we assume $M\chi_{[-1,1]} = \chi_{[-1,1]}$. Suppose $f \in C[0,a]$. Then for each $\varepsilon > 0$ there exists a partition $\pi_h$ and a step function $s_h$ with $M_h s_h = s_h$ and $\|s_h - f\|_\infty < \varepsilon$. Then

$$\begin{aligned}
\|M_h f - f\|_\infty &= \|M_h(f - s_h) + s_h - f\|_\infty \\
&\leq \|M_h(f - s_h)\|_\infty + \|s_h - f\|_\infty \\
&\leq \|M_h\|_\infty \|f - s_h\|_\infty + \|s_h - f\|_\infty \\
&\leq \varepsilon(1 + \|M_h\|_\infty).
\end{aligned}$$

Now suppose $f \in L_2$. We know there is a continuous function $g \in C[0,a]$ such that $\|g - f\|_2 < \varepsilon/4$. If $h$ is small enough there is a step function $s_h$ such that $M_h s_h = s_h$ and $\|s_h - g\|_2 < \varepsilon/4$ Thus

$$\begin{aligned}
\|M_h f - f\|_2 &= \|M_h f - s_h + s_h - g + g - f\|_2 \\
&\leq \|M_h(f - s_h)\|_2 + \|s_h - g\|_2 + \|g - f\|_2 \\
&\leq \|M_h\|_2 \|f - s_h\|_2 + \varepsilon/2 \\
&\leq \|M_h\|_2 [\varepsilon 2/ + \varepsilon/2] = \varepsilon \|M_h\|_2 \leq \varepsilon \|M\|_2. \quad \blacksquare
\end{aligned}$$

As mentioned above, it is the operators $M_h$ that will be used in (2.6). What is desirable is to determine approximations that converge to the exact solution according to some power of $h$. For this we require approximations of the identity $M_h$ as follows:

**Definition 3.3.** Let $M$ be any bounded operator on $L_2[-1,1]$ (or $C[-1,1]$). Suppose that $r \geq 1$ is any integer. We say that $M$ is of *type r* if for each function $f \in C^r$, the function

$$(M_h - I)f = O(h^r)$$

(pointwise) as $h \to 0$ over quasi-uniform partitions. We say that $M$ is *type* 0 if for each $f \in C[0, a]$, the function

$$(M_h - I)f = o(1)$$

as $h \to 0$ over quasi-uniform partitions.

An immediate consequence of Theorem 3.2 (ii) shows the equivalence of type 0 operators and the condition $M_h \to I$. Formally we state.

**Corollary 3.4.** Suppose $M$ satisfies the conditions of Theorem 3.2 (ii). Then $M$ is type 0 if and only if $M\chi_{[-1,1]} = \chi_{[-1,1]}$.

For example, the diamond difference operator is of type 1. The linear characteristic operator $M_{LC}$ is of type 2. (See [7, p. 337]). The projection to the Legendre polynomials $p_0, p_1, \ldots, p_{r-1}$ is of type $r$.

The following result establishes the fact that operators of type $r$ must be invariant on the powers $x^i, j = 0, 1, \ldots, r-1$. That is, $Mx^j = x^j, j = 0, 1, \ldots, r-1$, and must therefore be of dimension at least $r$.

**Theorem 3.5.** Suppose $M$ is a bounded operator on $C[-1,1]$, and $r > 0$ is an integer. Then $M$ is of type $r$ if and only if $M(s^j)(x) = x^j, j = 0, 1, \ldots, r-1$.

**Proof.** By Theorem (3.2) we have $M\chi_{[-1,1]} = \chi_{[-1,1]}$. Assume, by induction, that the result holds up to $j - 1$. Then

$$M_h x^j = \sum_{m=1}^{H} \chi_{C_m} \rho_{h,m}^{-1} M \rho_{h,m} \chi_{C_m} x^j.$$

Suppose that $Ms^j(x) = f(x)$. Then, restricted to $C_m$,

$$[\tilde{\rho}_{h,m}^{-1} M \tilde{\rho}_{h,m} \chi_{C_m} s^j](x) = [\tilde{\rho}_{h,m}^{-1} M \left(x_m + \frac{h_m}{2} t\right)^j](x)$$

$$= [\tilde{\rho}_{h,m}^{-1} M [\sum_{k=0}^{j} x_m^k \left(\frac{h_m}{2}\right)^{j-k} \binom{j}{k} t^{j-k}]](x)$$

$$= \tilde{\rho}_{h,m}^{-1} [\sum_{k=1}^{j} x_m^k \left(\frac{h_m}{2}\right)^{j-k} \binom{j}{k} u^{j-k} + \left(\frac{h_m}{2}\right)^j f(u)](x)$$

$$= \sum_{k=0}^{j-1} x_m^k \left(\frac{h_m}{2}\right)^{j-k} \binom{j}{k} \left(\frac{x - x_m}{2h_m}\right)^{j-k}$$

$$+ \left(\frac{h_m}{2}\right)^j f\left(\frac{x - x_m}{2h_m}\right)$$

$$= x^j - (x - x_m)^j + \left(\frac{h_m}{2}\right)^j f\left(\frac{x - x_m}{2h_m}\right).$$

Thus, on $C_m$, we have

$$\tilde{\rho}_{h,m}^{-1} M \tilde{\rho}_{h,m} s^j(x) - x^j = -(x - x_m)^j + \left(\frac{h_m}{2}\right)^j f\left(\frac{2}{h_m}(x - x_m)\right) = O(h_m^r).$$

Since $(x - x_m)^j = O(h_m^j)$, it follows that $f\left(\frac{x-x_m}{2h_m}\right) = \left(\frac{x-x_m}{2h_m}\right)^j$ or what is the same, $f(x) = x^j$. This proves the result.

To prove the converse we suppose $Mx^j = x^j, j = 0, 1, \ldots, r - 1$. Suppose $f \in C^r[0, a]$. Then, by Taylor's formula

$$(\chi_m \tilde{\rho}_{h,m}^{-1} M \tilde{\rho}_{h,m} \chi_m f)(x) = \chi_m \tilde{\rho}_{h,m}^{-1} M f(x_m + \frac{h_m}{2} s)(x)$$

$$= \chi_m \tilde{\rho}_{h,m}^{-1} M \bigg( \sum_{j=0}^{r} \frac{f^{(j)}(x_m)}{(j!)} \Big(\frac{h_m}{2}\Big)^j s^j$$

$$+ \frac{f^r(y)}{r!} \Big(\frac{h_m}{2}\Big)^r s^r \bigg)(x)$$

$$= \chi_m \tilde{\rho}_{h,m}^{-1} \sum_{j=0}^{r} \frac{f^{(j)}(x_m)}{j!} \Big(\frac{h_m}{2}\Big)^j t^j(x)$$

$$+ \chi_m \tilde{\rho}_{h,m}^{-1} M(f^{(r)}(y) s^r) \Big(\frac{h_m}{2}\Big)^r \frac{1}{r!}$$

$$= \chi_m \sum_{j=0}^{r} \frac{f^{(j)}(x_m)}{j!} \Big(\frac{h_m}{2}\Big)^j \Big(\frac{2(x - x_m)}{h_m}\Big)^j$$

$$+ \chi_m \tilde{\rho}_{h,m}^{-1} M(f^{(r)}(y) s^r) \Big(\frac{h_m}{2}\Big)^r \frac{1}{r!}$$

$$= \chi_m \sum_{j=0}^{r} \frac{f^{(j)}(x_m)}{j!} (x - x_m)^j$$

$$+ \chi_m \tilde{\rho}_{h,m}^{-1} M(f^{(r)}(y) s^r) \Big(\frac{h_m}{2}\Big)^r \frac{1}{r!}.$$

Since $M$ is bounded on $C[-1, 1]$ and since $f^{(r)}(y) s^r$ is continuous $M(f^{(r)}(y) s^r) \in C[-1, 1]$. Thus

(3.6)
$$(\chi_m \tilde{\rho}_{h,m}^{-1} M \tilde{\rho}_{h,m} f \chi_m - f)(x) = (h_m)^r [M(f^{(r)}(y) s^r)(x) - f^r(y) x^r] \frac{1}{2^r r!} = O(h_m^r). \blacksquare$$

We now come to an elementary but important estimate that is needed later.

**Corollary 3.6.** Suppose $M$ is of type $r$, and $\pi_h$ is any partition of $[-1, 1]$. Then for any $f \in C^r(J_{\pi_h})$

$$\|(I - M_h) f\|_{C(J_{\pi_h})} \leq C \|f\|_{C^r(J_{\pi_h})} h^r.$$

**Remark.** This Corollary also follows as an application of Theorem 3, p. 115 in Bramble and Hilbert [2]. Since our proof below is simple and requires little machinery, we include it for completeness.

**Proof.** First note that the relaxation of the condition that $f \in C^r[-1, 1]$ to $f \in C^r(J_{\pi_h})$ is possible by the nature of the definition of $M_h$, and the fact that the estimates of $(I - M_h) f$ are made cell-by-cell. Using Taylor's formula and Theorem (3.5), we see that on the cell $C_m$ of $\pi_h$

$$|(I - M_h f)(x)| = |\tilde{C}h^r(M(f^{(r)}(y)x^r) - f^{(r)}(y)x^r)|$$
$$\leq \tilde{C}h^r(\|M\| \|f^r(y)\|_{C(C_m)} + \|f^{(r)}(y)\|_{C(C_m)})$$
$$\leq C\|f\|_{C^r(J_{\pi_h})}h^r. \blacksquare$$

A natural question that occurs is whether there are type $r$ operators for functions $f \in C^p$ where $p < r$. For example, are there operators $M$ bounded on $C[-1,1]$ for which

$$(M_h - I)f = O(h)$$

for all $f \in C[0,a]$? For finite dimensional operators $M$ and even operators with proper closed range, the answer to this question is no as we demonstrate below.

**Theorem 3.7.** Let $M$ be a bounded operator on $C[-1,1]$ with closed range $X = M(C[-1,1])$ properly contained in $C[-1,1]$. Then there is a sequence of partitions $\pi_h$ and a function $f \in C[0,1]$ for which $\|(M_h - I)f\|/h$ becomes arbitrarily large as $h \to 0$.

**Proof.** For an arbitrary closed subspace $Y$ of $C[-1,1]$, and $f \in C[-1,1]$ we define

$$dist(f, Y) = \inf_{g \in Y} \|f - g\|.$$

Since $X$ is a proper closed subspace of $C[-1,1]$ it follows that there is a function $\tilde{f} \in C[-1,1]$ with $\tilde{f}(-1) = \tilde{f}(1) = 0$, such that $dist(\tilde{f}, X) \geq 1$. Let $f(x) = \tilde{f}(2(x - 1/2))$, $x \in [0,1]$. Then $f \in C[0,1]$ and $f(0) = f(1) = 0$. Let $\pi_n$ denote the partition of $[0,1]$ into $4n^n$ equal subdivisions, denoted by $I_{n,1}, I_{n,2}, \ldots, I_{n,4n^n}$. With $h = \frac{1}{4}n^{-n}$, and $g(x) \in C[0,1]$

$$\tilde{\rho}_{h,3}g(x) = g\left(\frac{h}{2}x + \frac{5}{8}n^{-n}\right).$$

Define

$$f_n(x) = \begin{cases} 0, & \text{if } x \in I_{n,1} \\ a_2 + b_2 x, & \text{if } x \in I_{n,2} \\ \tilde{f}(\frac{2}{h}(x - \frac{5}{8}n^{-n})), & \text{if } x \in I_{n,3} \\ a_4 + b_4 x, & \text{if } x \in I_{n,4} \\ 0, & \text{otherwise} \end{cases}$$

where $a_2, b_2, a_4$, and $b_4$ are chosen so that $f_n(x)$ is continuous on $[0,1]$. Note that $\frac{5}{8}n^{-n}$ is the midpoint of $I_{n,3}$. Note also that for $n = 2, 3, \ldots$ the $f_n(x)$ have disjoint supports. Now define

$$g(x) = \sum_{n=1}^{\infty} \frac{1}{2^n} f_n(x).$$

Clearly, $g(x)$ is also continuous on $[0,1]$. Now, for the third cell $I_{n,3}$ we have

$$(\tilde{\rho}_{h,3} f_n)(x) = f_n\left(\frac{h}{2}x + \frac{5}{8}n^{-n}\right) \quad -1 \leq x \leq 1$$
$$= \tilde{f}\left(\frac{4}{h}\left\{\left[\frac{h}{2}x + \frac{5}{8}n^{-n}\right] - \frac{5}{8}n^{-n}\right\}\right)$$
$$= \tilde{f}(x).$$

On $I_{n,3}$, with $k(x) = Mf(x)$, we have

$$(M_h - I)g = \tilde{\rho}_{h,3}^{-1} M\tilde{f}(x) - f_n(x)$$
$$= \tilde{\rho}_{h,3}^{-1} k(x) - f_n(x)$$
$$= k\left(\frac{2}{h}\left(x - \frac{5}{8}n^{-n}\right)\right) - \tilde{f}\left(\frac{2}{h}\left(x - \frac{5}{8}n^{-n}\right)\right)\frac{1}{2^n}$$

and so

$$\|(M_h - I)g|_{I_{n,3}}\| \geq \frac{1}{2^n} \mathrm{dist}(\tilde{f}, X) \geq \frac{1}{2^n}.$$

But $2^{-n}/n^{-n} = \left(\frac{n}{2}\right)^n$ is arbitrarily large as $n \to \infty$. This is inconsistent with $(M_h - I)f = O(h)$.

**Remark.** This result shows that $M$ cannot be of type 1 over $C^0 (= C[-1,1])$. A similar argument applies to establish the incompatibility of the condition that $M$ be type $r$ over $C^p, p < r$.

## §4. Main Results.

We begin this section by recalling the problem at hand: The exact equation with solution $\psi^e(x)$ is

(4.1) $$\psi^e = K\psi^e + \tilde{K}q$$

where $K, \tilde{K} \in A(N, r, J)$, and $q \in C_N^r(J)$. Throughout this section and the next we make the following assumptions:

(1) $\|K\|_\infty \, (= \|K\|)$ and $\|K\|_2$ are bounded by a constant $\lambda < 1$.
(2) $M$ is an orthogonal projection on $L_{2,N}[-1,1]$ of type $r, r \geq 1$.
(3) $\{\pi_h\}$ is a sequence of quasi-uniform partition of $[0, a]$, with $J_{\pi_h} \subset J$ for each $\pi_h$.
(4) $N$ is fixed.

Then the finite difference approximations are given by

(4.2) $$\psi = KM_h\psi + \tilde{K}M_h q.$$

(We suppress here and through this section the dependency of $\psi$ on $\pi_h$.) Since $\|K\|_2 \leq \lambda$, it follows that $\|KM_h\|_2 < \lambda$. The quantity which we analyze is the difference

(4.3). $$\psi - \psi^e = KM_h(\psi - \psi^e) + K(M_h - I)\psi^e + \tilde{K}(M_h - I)q.$$

Since $M_h$ is a projection (on $L_n^2[0,a]$) by Theorem 3.1 a, we can use the customary notation, $M_h^\perp$, for $I - M_h$. Note that $M_h^\perp$ is also a projection. Then (4.3) becomes

(4.4) $$\psi - \psi^e = KM_h(\psi - \psi^e) - KM_h^\perp \psi^e - \tilde{K}M_h^\perp q.$$

Our first goal is to establish that $\psi^e$ is in an appropriate differentiabity class.

**Theorem 4.1.** Suppose $K$ and $\tilde{K}$ are in $A(N,r,J)$, with $\|K\|_A, \|\tilde{K}\|_A < \infty$, and suppose $q \in C_N^{r-1}(J)$. Then $\psi^e \in C_N^r(J)$. Moreover, if $M$ is a bounded operator on $C_N^r[-1,1]$, then $\psi \in C_N^r(J \cap J_{\pi_h})$.

The proof is a simple consequence of the following:

**Lemma 4.2.** Suppose $r \geq 1, K \in A(N,r,J_K)$ and $f \in C_N^{r-1}(J_f)$. Define

(4.5)
(i) $$g_1(x) = \int_0^a K(x,y)f(y)dy$$
(ii) $$g_2(y) = \int_0^a f(x)^T K(x,y)dx.$$

Then $g_i \in C_N^r(J_K \cap J_f), i = 1,2$. If $f \in L_{\infty,N}[0,a]$, then $g_i \in C_N(J \cap J_f)$. Moreover, for $r \geq 1$, and $i = 1,2$

(4.6) $$\|g_i\|_{C_N^r(J_K \cap J_f)} \leq C\|f\|_{C_N^{r-1}(J_K \cap J_f)}$$

where $C = \|K\|_A$. If $K \in A(N,0,J)$ and $f \in L_N^\infty[0,a]$ then $g \in C_N(J), i = 1,2$.

**Proof.** We establish (4.5) (i) and (4.6). The proof of (4.5) (ii) is similar. Suppose without loss in generality that $N = 1$. Write

(4.7) $$g_1(x) = \int_0^x K(x,y)f(y)dy + \int_x^a K(x,y)f(y)dy.$$

Since $K \in A(N,0,J) \subset A(N,r,J)$, it follows that each of the integrals in (4.7) is continuous on $J$. Suppose now that $r \geq 1$, and $x \in J \cap J_f$. Then

$$\frac{dg_1}{dx}(x) = \lim_{y \to x^-} K(x,y)f(x) + \int_0^x \frac{\partial K}{\partial x}(x,y)f(y)dy$$
$$+ \lim_{y \to x^+} K(x,y)f(x) + \int_x^a \frac{\partial K}{\partial x}(x,y)f(y)dy.$$

By hypothesis each of the functions $\lim_{y \to x^{\pm}} K(x,y)$ and $f(x)$ are in $C^{r-1}(J \cap J_f)$. We also know $\partial K/\partial x(x,y) \in A(1, r-1, J)$. Thus $g_1'(x) \in C(J \cap J_f)$, and so $g_1 \in C^1(J \cap J_f)$. The proof is complete if $r = 1$. If $r \geq 2$ we consider the function

$$\varphi_1(x) = \int_0^x \frac{\partial K}{\partial x}(x,y) f(y) dy.$$

Then for $x \in J \cap J_f$

$$\varphi_1'(x) = \lim_{y \to x^-} \frac{\partial K}{\partial x}(x,y) f(y) + \int_0^x \frac{\partial^2 K}{\partial x^2}(x,y) f(x) dy.$$

Since the function $\lim_{y \to x^-} \partial K/\partial x(x,y)$ is in $C^{r-2}(J)$, $f(x) \in C^{r-1}(J)$, and $\partial^2 K/\partial K^2(x,y) \in A(1, r-2, J)$ we conclude that $\varphi_1'(x) \in C(J \cap J_f)$. As a similar argument applies to the integral

$$\int_x^a \frac{\partial K}{\partial x}(x,y) f(y) dy.$$

We conclude that $g_1 \in C^2(J \cap J_f)$. Continuing in this way it easily follows that $\varphi \in C^r(J \cap J_f)$ for any finite $r = 0, 1, 2, \ldots$, and Lemma 4.2 is proved. ∎

**Proof of Theorem 4.1.** Since the norm of $K = \lambda < 1$ the solution of $\psi^e$ exists and is in $C_N(J)$. Now since $q \in C_N^1(J)$, and $K, \tilde{K} \in A(N, r, J)$ it follows that $K\psi^e \in C_N^1(J)$ and $\tilde{K}q \in C_N^r(J)$ by Lemma 4.1. Thus $\psi^e \in C_N^1(J)$. Now applying Lemma 4.1 recursively we conclude that $\psi^e \in C_N^r(J)$. The proof that $\psi \in C_N^r(J \cap J_\pi)$ is identical. ∎

In our applications below recall that $J_\pi \subset J$. Having now established that $\psi^e$ and $\psi$ are both in $C^r(J_{\pi_h})$ we now begin our study of estimating the order of the error of approximation $\psi - \psi^e$.

**Lemma 4.3.** Suppose $K \in A(N, r, J)$ (with $\|K\|_A < \infty$). Suppose that $M$ is also a bounded operator on $C_N^r[-1, 1]$, and $g \in C^r(J)$. Then for $x \in J_{\pi_h}$

$$KM_h^\perp g(x) = O(h^{r+1}).$$

Moreover, if $x \notin J_{\pi_h}$, that is $x$ is a mesh point of $\pi_h$, for all $h$, then

$$KM_h^\perp g(x) = O(h^{2r}).$$

**Proof.** We have

$$(KM_h^\perp)g(x) = \int_0^a K(x,y)(M_h^\perp g)(y) dy.$$

Suppose $x \in C_\ell \cap J_{\pi_h}$. Thus $x$ is an interior point of $C_\ell$. Then for the fixed partition $\pi_h$,

(4.8)
$$\int_0^a K(x,y)(M_h^\perp g)(y) dy = \sum_{\substack{m=1 \\ m \neq \ell}}^H \int_{C_m} K(x,y)(M_h^\perp g)(y) dy + \int_{C_\ell} K(x,y)(M_h^\perp g)(y) dy.$$

If $C_m \cap C_\ell = \phi$ we have that

$$\int_{C_m} K(x,y)(M_h^\perp g)(y)dy = \int_{C_m} M_h^\perp K(x,y)(M_h^\perp g)(y)dy.$$

Since $\|K\|_A < \infty$ we must have that

$$M_h^\perp K(x,y) = O(h^r)$$

uniformly in $x$. Thus

$$\int_{C_m} (M_h^\perp K)(x,y)(M_h^\perp g)(y)dy = \int_{C_m} O(h^{2r})dy =$$
$$= O(h^{2r+1}).$$

The estimate of the other integral, when $x \in C_\ell$ is

$$\int_{C_\ell} K(x,y)(M_h^\perp g)(y)dy = \int_{C_\ell} O(h^r)dy = O(h^{r+1}),$$

because $K$ is not necessarily $C^r$ across the line $y = x$. This proves the first part of the Lemma. The second part follows easily since the second integral in (4.8) is not present. ∎

With Lemma 4.3 the following result is simple.

**Theorem 4.4.** Suppose $K, \tilde{K} \in A(N, r, J), q \in C_N^r(J)$, and $M$ is a projection of type $r$ and a bounded operator from $C_N^r[-1, 1]$ to $C_N^r[-1, 1]$. Then

(4.9) $$(\psi - \psi^e)(x) = O(h^{r+1})$$

uniformly for $x \notin J_{\pi_h}$.

**Proof.** This follows directly from Lemma 4.3 and equation (4.3) when written as

(4.10) $$\psi - \psi^e = -(I - KM_h)^{-1}[KM_h^\perp \psi^e + \tilde{K}M_h^\perp q]$$

since $\psi^e$ and $q$ satisfy the hypotheses of Lemma 4.3. Thus $KM_h^\perp \psi^e$ and $\tilde{K}M_h^\perp q$ are uniformly $O(h^{r+1})$ and therefore so is $\psi - \psi^e$. ∎

Theorem 4.4 is consistent with the corresponding result for all characteristic methods for the monoenergetic transport equation. In particular see [12, Theorem 1, p. 86].

## §5. Cell Average and Projection Estimates.

In many applications of the transport equation one is not interested in the solution (i.e. the $\psi_i$ for the discrete ordinate approximation) *per se*, but rather the corresponding integral over some subinterval of $[0, a]$. For such calculations it is customary to use some quadrature rule in which the value of $\psi_i^e$ over each cell is approximated by its approximate cell-average. From this viewpoint it is important to know how well the $M_0\psi_i$ approximate the $M_0\psi_i^e$. This section is devoted to that question and related considerations.

We begin with (4.10). Expanding the operator $(I - KM_h)^{-1}$ in a Neumann series, (4.10) becomes

$$(5.1) \qquad \psi - \psi^e = -\sum_{j=0}^{\infty}(KM_h)^j[KM_h^\perp \psi^e + \tilde{K}M_h^\perp q].$$

Let $M_0$ the average operator defined in (3.4), and the corresponding step characteristic operator, denoted now by $M_{0,h}$. Note that since $M$ is a projection of type $r$ then by Theorem 3.5, $MM_0 = M_0M = M_0$. Our goal, to estimate

$$(5.2) \qquad M_{0,h}(\psi - \psi^e) = -\sum_{j=0}^{\infty} M_{0,h}(KM_h)^j[KM_h^\perp \psi^e + \tilde{K}M_h^\perp q]$$

is contained in the following:

**Theorem 5.1** Suppose that $K, \tilde{K} \in A(N, r, J), q \in C_N^r(J)$, $M$ is a projection of type $r$ and a bounded operator from $C_N^r[-1,1]$ to $C_N^r[-1,1]$ and $\psi^e, \psi$ are defined by (4.1) and (4.2) respectively. Then

$$(5.3) \qquad M_{0,h}(\psi - \psi^e) = O(h^{2r}).$$

The proof below establishes by induction the result for each term of the Neumann series $j = 1, 2, \ldots$ and then for the sum. The case $j = 0$ is a consequence of Lemma 4.3. We proceed in a sequence of lemmas.

**Lemma 5.2.** Suppose $K \in A(N, r, J_K)$ and $L \in A(N, r, J_L)$. Then

$$(5.4) \qquad M(x,y) = \int_I K(x,x')L(x',y)dx'$$

is in $A(N, r, J_k \cap J_L)$. Moreover $\|M\|_A \leq \|K\|_A\|L\|_A$.

**Proof.** Let $J = J_k \cap J_L$. Assume without loss of generality that $N = 1$. Assume that $(x,y) \in J \times J$. First note that $M(x,y)$ is continuous in each variable separately if $x \neq y$, and also both $\lim_{x \to y^{\pm}} M(x,y)$ and $\lim_{y \to x^{\pm}} M(x,y)$ are continuous on $J$. For example,

$$\lim_{x \to y^+} M(x,y) = \lim_{x \to y^+}\left[\int_0^y K(x,x')L(x',y)dx' + \int_y^a K(x,y')L(x',y)dx'\right]$$
$$= \int_0^y K(y,x')L(x',y)dx' + \int_y^a K(y,x')L(x',y)dx'$$

and is therefore continuous because both $K$ and $L$ are piecewise continuous in $y$ over the respective intervals of integration. Now write

$$M(x,y) = \int_0^y K(x,x')L(x',y)dx' + \int_y^a K(x,x')L(x',y)dx'.$$

Then, for $x \neq y, (x,y) \in J \times J$

$$\frac{\partial M}{\partial y}(x,y) = K(x,y) \lim_{x' \to y^-} L(x',y) + \int_0^y K(x,x')\frac{\partial L}{\partial y}(x',y)dx'$$
$$+ K(x,y) \lim_{x' \to y^+} L(x',y) + \int_0^y K(x,x')\frac{\partial L}{\partial y}(x',y)dx'$$

exists and is continuous on $J \times J$. Similarly, so is $\frac{\partial M}{\partial x}(x,y)$. Since

$$\lim_{x' \to y^\pm} L(x',y) \equiv \tilde{L}_\pm(y)$$

is in $C^r(J)$ and since $K \in A(1,r,J_k)$, the product is in $A(1,r,J)$. Also

(5.5) $\lim_{x \to y^+} \frac{\partial M}{\partial y}(x,y) = \lim_{x \to y^+} K(x,y)\tilde{L}_-(y) + \int_0^y K(y,x')\frac{\partial L}{\partial y}(x',y)dx'$
$$+ \lim_{x \to y^+} K(x,y)\tilde{L}_+(y) + \int_y^a K(y,x')\frac{\partial L}{\partial y}(x',y)dx'$$

is in $C^r(J)$ because $\tilde{L}_\pm(y), \lim_{x \to y^+} K(x,y)$, and the two integrals above are. The same conclusion obtains for $\lim_{x \to y^-} \frac{\partial M}{\partial y}(x,y)$. In fact, more is true. Since $K \in A(1,r,J)$ and $\frac{\partial L}{\partial y} \in A(1, r-1, J)$ we note that both integrals (5.5) are $C^{r-1}(J)$ by Lemma 4.2 taking $f(y) \equiv 1$. The lemma is now proved for $r = 1$. To complete proof the partial derivatives of $M$ of all orders up to $r$ must be computed. But this is a repetition of the above steps for $\partial M/\partial x, \partial M/\partial y$, etc. This completes the proof. ∎

**Corollary 5.3.** If $K_j$ is the $j^{\text{th}}$ iterated kernel of $K$, and if $K \in A(N,r,J)$ then $K_j \in A(N,r,J)$.

**Lemma 5.4.** Let $\pi_h$ be any partition of $[0,a]$ for which $J_{\pi_h} \subset J$. Suppose $K \in A(N,r,J)$. Let $K_j$ be the $j^{\text{th}}$ iterated kernel of $K$. Define

(5.6) $$g_j^{(m)}(y) = \frac{1}{h_m} \int_{C_m} K_j(x,y)dx$$

where $C_m$ is the $m^{\text{th}}$ cell of $\pi_h$ and $h_m$ is the length of $C_m$. Then there is a constant $C$ (independent of $m$ and $j$) such that

(5.7)  (i) $\quad \|g_j^{(m)}(y)\|_{C^r(C_m)} < C\lambda^j/h_m$
  (ii) $\quad \|g_j^{(m)}(y)\|_{C^r(J_{\pi_h}-C_m)} C\lambda^j.$

**Proof.** Assume without loss of generality that $N = 1$. From Corollary 5.3 and Lemma 4.2 it follows that $g_j^{(m)} \in C^r(J_{\pi_h})$. Hence the result holds up to an arbitrarily large, but fixed $j$, for some constant $\tilde{c}$. Assume below that $j > r + 2$

$$g_j^{(m)}(y) = \frac{1}{h_m} \int_{C_m} \int_0^a K_{j-1}(x,x')K(x',y)dx'dx$$
$$= \frac{1}{h_m} \int_{C_m} \int_0^y K_{j-1}(x,x')K(x',x)dx'dx$$
$$+ \frac{1}{h_m} \int_{C_m} \int_y^a K_{j-1}(x,x')K(x',y)dx'dx.$$

To estimate the derivatives of $g_i^{(m)}(y)$, we first suppose that $y \in J_{\pi_h} - C_m$. Then

$$\frac{d}{dy}g_j^{(m)}(y) = \frac{1}{h_m} \int_{C_m} K_{j-1}(x,y)dx \Big(\lim_{x' \to y^-} K(x',y)\Big)$$
$$+ \frac{1}{h_m} \int_{C_m} \int_0^y K_{j-1}(x,x')\frac{\partial K}{\partial y}(x',y)dx'dx$$
$$- \frac{1}{h_m} \int_{C_m} K_{j-1}(x,y)dx \Big(\lim_{x' \to y^+} K(x',y)\Big)$$
$$+ \frac{1}{h_m} \int_{C_m} \int_y^a K_{j-1}(x,x')\frac{\partial K}{\partial y}(x',y)dx'dx.$$

Now

$$\frac{1}{h_m} \int_{C_m} K_{j-1}(x,y)dx = \frac{1}{h_m} \int_{C_m} \int_0^a K_{j-2}(x,x')K(x',y)dx'dx.$$

Since $K(x',y) \in A(1,r,J)$ and hence $K(x',y) \in C^r(J)$ uniformily in $y$, it follows that

$$\left|\int_0^y K_{j-2}(x,x')K(x',y)dx'\right| \le \lambda^{j-2}\|K\|_{C^r(J\times J)}$$

whence

$$\left\|\frac{1}{h_m}\int_{C_m} K_{j-1}(x,y)dx\right\|_{C(J_{\pi_h}-C_m)} \le \lambda^{j-2}\|K\|_{C^r(J\times J)} = c_1\lambda^j.$$

Similarly, since $\partial K/\partial y \in A(1, r-1, J)$ it follows that

$$\left\|\int_0^a K_{j-1}(x,x')\frac{\partial K}{\partial y}(x',y)dx'\right\| < \lambda^{j-1}\left\|\frac{\partial K}{\partial y}(x',y)\right\|_{C^{r-1}(J\times J)} = c_2\lambda^j.$$

Similar estimates can be obtained for each of the derivatives of $g_j^{(m)}(y)$ upto order $r$.

Now suppose that $y \in C_m$. We simplify the original notation for $C_m$ by denoting $C_m = (x_m^-, x_m^+)$. Then we write $g_j^m(y)$ as follows:

$$g_j^{(m)}(y) = \frac{1}{h_m} \int_{x_m^-}^{y} K_j(x,y)dx + \frac{1}{h_m} \int_{y}^{x_m^+} K_j(x,y)dx$$

$$= \frac{1}{h_m} \int_{x_m^-}^{y} \int_0^a K_{j-1}(x,x')K(x',y)dy'dx$$

$$+ \frac{1}{h_m} \int_{y}^{x_m^+} \int_0^a K_{j-1}(x,x')K(x',y)dx'dx$$

$$= \frac{1}{h_m} \int_{x_m^-}^{y} \left\{ \int_0^y K_{j-1}(x,x')K(x',y)dx' + \int_y^a K(x,x')K(x',y)dx' \right\} dx$$

$$+ \frac{1}{h_m} \int_y^{x_m^+} \left\{ \int_0^y K_{j-1}(x,x')K(x',y)dx' \right.$$

$$\left. + \int_y^a K_{j-1}(x,x')K(x',y)dx' \right\} dx.$$

Each of the four integrals above represents a function of $y$. We analyze below just the first of them - denoted by $f(y)$; similar estimates hold for the others. First of all, since $K(x',y) \in A(1,r,J)$ we see that the integral

$$\left| \frac{1}{h_m} \int_{x_m}^{y} \int_0^y K_{j-1}(x,x')K(x',y)dx'dx \right|$$
$$\leq \lambda^{j-1} \|K\|_{C^r(J \times J)}$$

whence

$$\|f\|_{C(C_m)} \leq \tilde{c}_1 \lambda^j.$$

Now differentiate $f(y)$ to obtain

$$\frac{d}{dy}f(y) = \frac{1}{h_m} \int_0^y K_{j-1}(y,x')K(x',y)dx'$$

$$+ \frac{1}{h_m} \int_{x_m^-}^{y} \frac{\partial}{\partial y} \left\{ \int_0^y K_{j-1}(x,x')K(x',y)dx' \right\} dx.$$

Arguing as above we estimate the $C^1(J_{\pi_h})$ norm to be bounded by a constant times $\lambda^{j-2}/h_m$, or what is the same, a constant times $\lambda^j/h_m$. Since $K \in A(1,r,J)$ similar estimate holds for each of the higher order derivatives, thus proving the result. ∎

**Corollary 5.5.** *Given the hypotheses of Lemma 5.4, let $L \in A(N,r,J)$. Define*

$$M_j(x,y) = \int_0^a K_j(x,x')L(x',y)dx'$$

and

(5.8) $$g_j^{(m)}(y) = \frac{1}{h_m}\int_{C_m} M_j(x,y)dx.$$

Then the estimates (5.7) hold, with possibly a different constant $C$.

**Lemma 5.6.** Suppose $K, L \in A(N, r, J)$, with the spectral radius of $K = \lambda < 1$ and $\|L\| < \infty$. Suppose $\{\pi_h\}$ is a quasi-uniform sequence of partitions of $[0, a]$ with $J_{\pi_h} \subset J$. Suppose $g \in C_N(J)$. Then there is a constant $C$ independent of $j$ such that

(5.9) $$\|M_{0,h}K_j LM_h^\perp g\| \le C\lambda^j h^r \|g\|_{C_N(J)}, \quad j = 1, 2, \ldots.$$

If $g \in C_N^r(J)$ then

(5.10) $$\|M_{0,h}K_j LM_h^\perp g\| \le \tilde{c}\lambda^j h^{2r}, \quad j = 1, 2, \ldots.$$

where $\tilde{c} = \|M_h\|C$.

**Proof.** Suppose as usual that $N = 1$. Then

$$(M_{0,h}K_j LM_h^\perp g)(s) = \sum_m \frac{1}{h_m}\int_{C_m}\int_0^a K_j(x,x')(LM_h^\perp g)(x')dx'dx\chi_{C_m}(s).$$

Fixing $m$, we show that the bound (5.9) holds in each cell $C_m$. Define

$$f(s) = \frac{1}{h_m}\int_{C_m}\int_0^a K_j(x,x')\int_0^a L(x',y)M_h^\perp g(y)dy\,dx'\,dx$$

$$= \int_0^a \frac{1}{h_m}\int_{C_m}\int_0^a K_j(x,x')L(x',y)dx'\,dx\,M_h^\perp g(y)dy.$$

The kernel

$$M_j(x,y) = \int_0^a K_j(x,x')L(x',y)dx'$$

is in $A(N, r, J)$ and has norm bounded by $C\lambda^j$. So, by the self-adjointness of $M_h^\perp$,

$$f(x) = \sum_\ell \int_{C_\ell} \frac{1}{h_m}\int_{C_m} M_j(x,y)dx\,M_h^\perp g(y)dy$$

$$= \sum_\ell \int_{C_\ell} M_h^\perp\left(\frac{1}{h_m}\int_{C_m} M_j(x,y)dx\chi_{C_\ell}\right)M_h^\perp g(y)dy$$

$$= \sum_{\ell\ne m} \int_{C_\ell} M_h^\perp\left(\frac{1}{h_m}\int_{C_m} M_j(x,y)dx\chi_{C_m}\right)M_h^\perp g(y)dy$$

$$+ \int_{C_m} M_h^\perp\left(\frac{1}{h_m}\int_{C_m} M_j(x,y)dx\chi_{C_m}\right)M_h^\perp g(y)dy.$$

By Corollary 5.5 and Corollary 3.6, we have, that when $\ell \neq m, y \in C_\ell$.

$$(5.11) \qquad \left| M_h^\perp \left( \frac{1}{h_m} \int_{C_m} M_j(x,y) dx \right) \right| \leq C \lambda^j h^r.$$

Hence for any function $\tilde{g} \in C(J_{\pi_h})$

$$\int_{C_\ell} M_h^\perp \left( \frac{1}{h_m} \int_{C_m} M_j(x,y) dx \chi_{C_\ell}(y) dx \right) M_h^\perp \tilde{g}(y) dy \leq C h^{r+1} \|\tilde{g}\|_{C(J_{\pi_h})} \lambda^j$$

and therefore

$$(5.12) \qquad \sum_{\ell \neq m} \int_{C_\ell} M_h^\perp \left( \frac{1}{h_m} \int_{C_m} M_j(x,y) \chi_{C_\ell} dx \right) M_h^\perp \tilde{g}(y) dy \leq C h^r \|\tilde{g}\|_{C(J_{\pi_h})} \lambda^j.$$

In the case $y \in C_m$ we have (also by Corollary 5.5)

$$(5.13) \qquad \left| M_h^\perp \left( \frac{1}{h_m} \int_{C_m} M_j(x,y) dx \right) \right| \leq C h^{r-1} \lambda^j.$$

Thus

$$(5.14) \qquad \int_{C_m} M_h^\perp \left( \frac{1}{h_m} \int_{C_m} M_j(x,y) \chi_{C_m}(y) \right) M_h^\perp g(y) dy \leq C h^r \|\tilde{g}\|_{C(J_{\pi_h})} \lambda^j$$

which gives (5.9). With $\tilde{g} = M_h^\perp g, \psi \in C^r(J)$ we have $\|\tilde{g}\|_{C(J_{\pi_h})} = O(h^r)$, whence (5.10) follows. ∎

**Corollary 5.7.** *Suppose the hypotheses of Lemma 5.6 hold. If $g \in C^r(J_{\pi_h})$ then (5.9) and (5.10) hold.*

The estimates in (5.11) and (5.13) are pointwise. Therefore similar $L_2$ estimates of (5.12) and (5.14) are easily obtained. Thus (5.12) can be replaced by

$$(5.12)' \qquad \sum_{\ell \neq m} \int_{C_\ell} M_h^\perp \left( \frac{1}{h_m} \int_{C_m} M_j(x,y) \chi_{C_\ell} dx \right) M_h^\perp \tilde{g}(y) dy \leq \overline{C} h^r \|M_h^\perp \tilde{g}\|_2 \lambda^j$$

and (5.14) can be replaced by

$$(5.14)' \qquad \int_{C_m} M_h^\perp \left( \frac{1}{h_m} \int_{C_m} M_j(x,y) \chi_{C_m} dx \right) M_h^\perp g(y) dy \leq \overline{C} h^r \|M_h^\perp \tilde{g}\|_2 \lambda^j.$$

This remark, summarized below as Corollary 5.8, will be useful later.

**Corollary 5.8.** Suppose the hypotheses of Lemma 5.6 hold. If $g \in C_N(J_{\pi_h})$, then

$$\|M_{0,h}K_j LM_h^\perp g\| \le \overline{C}\lambda^j h^r \|M_h^\perp g\|_2 \tag{5.15}$$

**Lemma 5.9.** Suppose that $K, L \in A(N, r, J)$ with the spectral radius of $K = \lambda < 1$ (and $\|L\| < \infty$). Suppose that $g \in C_N^r(J)$ and $\{\pi_h\}$ is a quasi-uniform sequence of partitions of $[0, a]$ with $J_{\pi_h} \subset J$. Then there is a constant $C$, independent of $j$, such that

$$\|M_{0,h}(KM_h^\perp)^j LM_h^\perp g\| \le C^*(j+1)\lambda^j h^{2r}$$

**Proof.** We identify $K_j$ with $K^j$, for correspondence with the Neumann series. From Lemma 5.6 and Corollary 5.7 we have that both

$$\|M_{0,h}K^j LM_h^\perp g\| \le c_0 \lambda^j h^{2r}$$

where $c_0 = \tilde{c}$ in (5.10), and with $L = I$ (the identity)

$$\|M_{0,h}K^j M_h^\perp(LM_h^\perp g)\| \le \tilde{c}_1 \lambda^j h^r \|LM_h^\perp g\|$$
$$\le c_1 \lambda^j h^{2r}.$$

Therefore, since

$$M_{0,h}K^j LM_h^\perp g = M_{0,h}K^j M_h^\perp LM_h^\perp g + M_0 K^j M_h^\perp LM_h^\perp g$$

it follows that

$$\|M_{0,h}K^j M_h LM_h^\perp g\| \le (c_0 + c_1)\lambda^j h^{2r}$$

or what is the same

$$\|M_{0,h}K^{j-1}(KM_h)LM_h^\perp g\| \le (c_0 + c_1)\lambda^j h^{2r}.$$

Again, by Lemma 5.6 and Corollary 5.8, with $L = I$, it follows that

$$\|M_{0,h}K^{j-1}M_h^\perp(KM_h)LM_h^\perp g\| \le \tilde{c}_1 \lambda^{j-1} h^r \|(KM_h)LM_h^\perp g\|_2$$
$$\le c_2 \lambda^j h^{2r}.$$

Since

$$M_{0,h}K^{j-1}(KM_h)LM_h^\perp g = M_{0,h}K^{j-1}M_h(KM_h)LM_h^\perp g$$
$$+ M_{0,h}K^{j-1}M_h^\perp(KM_h)LM_h^\perp g$$

we have

$$\|M_{0,h}K^{j-2}(KM_h)^2 LM_h^\perp g\| \le (c_0 + c_1 + c_2)\lambda^j h^{2r}.$$

Now assume by induction that

$$\|M_{0,h}K^{j-\ell}(KM_h)^\ell LM_h^\perp g\| \leq \left(\sum_{m=0}^{\ell} c_m\right)\lambda^j h^{2r}.$$

Then, exactly similar reasoning as above allows us to conclude that

$$\|M_{0,h}K^{j-\ell-1}(KM_h)^{\ell+1} LM_h^\perp g\| \leq \left(\sum_{m=0}^{\ell+1} c_m\right)\lambda^j h^{2r}$$

and finally

$$\|M_{0,h}(KM_h)^j LM_h^\perp g\| \leq \left(\sum_{m=0}^{j} c_m\right)\lambda^j h^{2r}.$$

At each step of the calculation the constants $c_0, c_1, \ldots, c_j$ can be traced to the constant $c$ in (5.7), $\tilde{c}$ in (5.10) and $\bar{c}$ in (5.15) (which are uniformly bounded independent of $j$), the constant of approximation of $M_h^\perp g$ (which is also bounded), and the norm of $\|L\|$ (again bounded). Therefore

$$\sum_{m=0}^{j} c_m \leq (j-1)C^*$$

for some absolute constant $C^*$. This proves the result. ∎

**Proof of Theorem 5.1.** From Lemma 5.9 and equation (5.2) we have

$$\|M_{0,h}(KM_h)^j KM_h^\perp \psi^e\| \leq C_1^*(j+1)\lambda^j h^{2r}$$

and

$$\|M_{0,h}(KM_h)^j \tilde{K} M_h^\perp q\| \leq C_2^*(j+1)\lambda^j h^{2r}.$$

Thus the sum

$$\|M_{0,h}(\psi - \psi^e)\| \leq (C_1^* + C_2^*)\sum (j+1)\lambda^j h^{2r}$$
$$\leq C^* h^{2r}$$

that establishes (5.3). This proves the theorem. ∎

**Higher Order Moments**

Let $p \in C^r[-1,1]$, and $P$ the projection pertaining to $p$. Thus, for any $f \in L_2[-1,1]$

(5.16) $$(Pf)(x) = \left(\int_{-1}^{1} p(u)f(u)du\right)p(x)/\|p\|_2^2.$$

Let $P_h$ denote the projection defined by (3.3), with $P$ replacing $M$. It is clear, from Theorem 4.4 and from the boundedness of $p$ that

(5.17) $$P_h(\psi - \psi^e) = O(h^{r+1}).$$

Therefore, (4.9) cannot be generally improved for arbitrary one dimensional projections. However we can show (5.17) can be improved for particular choices of functions $p$, namely, for all polynomials $p$ of degree $< r$. (Recall, the operator $M$ is of type $r$). A generalization of Theorem 5.1, this result is stated be;ow.

**Theorem 5.10.** Let $K, \widetilde{K}, q, M, \psi$, and $\psi^e$ be the same as in Theorem 5.1 Suppose $p$ is a polynomial of degree $s < r$. Then, with $P$ defined by (5.16) and $P_h$ defined as above with respect to $P$,

(5.18) $$P_h(\psi - \psi^e) = O(h^{2r-s}).$$

**Proof of Theorem 5.10.** (Sketch) The steps of the proof are substantially the same as those of Theorem 5.1. First we replace (5.2) by

(5.19) $$P_h(\psi - \psi^e) = -\sum_{j=0}^{\infty} P_h(KM_h)^j [KM_h^\perp \psi^e + \widetilde{K}M_h^\perp q].$$

The estimates of the terms $P_h(KM_h)^j KM_h^\perp \psi^e$ and $P_h(KM_h)^j \widetilde{K}M_h^\perp q$ are based upon general pointwise estimates of $P_h K_j L M_h^\perp g$ for any $L \in A(N, r, J)$, $g \in C^r(J)$, and where $K_j (= K^j)$ is the $j^{th}$ iterated kernel of $K$.

Such estimates are to be analogous to (5.9) and (5.10). In fact we claim that

(5.20) $$\|P_h K_j L M_h^\perp g\| \leq \widetilde{C} \lambda^j h^{2r-s}.$$

where $\widetilde{C}$ is independent of $j$. To see this write (assuming $N = 1$, as usual)

$$(P_h K_j L M_h^\perp(g))(t) = \sum_{m=1}^{H} (\rho_{h,m}^{-1} p)(t) \chi_{C_m}(t) \int_{-1}^{1} p(u) \rho_{h,m}(K_j L M_h^\perp g)(u) du$$

$$= \sum_{m=1}^{N} (\rho_{h,m}^{-1} p)(t) \chi_{C_m}(t)$$

$$\int_{-1}^{1} p(u) \int_{0}^{a} M_j\left(x_m + \frac{h_m u}{2}, y\right) M_h^\perp(y) dy du$$

$$= \sum_{m=1}^{H} (\rho_{h,m}^{-1} p)(t) \chi_{C_m}(t)$$

$$\int_{0}^{a} \frac{1}{h_m} \int_{C_m} p\left(\frac{2(x - x_m)}{h_m}\right) M_j(x, y) dx M_h^\perp g(y) dy$$

$$= \sum_{m=1}^{H} (\rho_{h,m}^{-1} p)(t) \chi_{C_m}(t)$$

$$\int_{0}^{a} M_h^\perp \left[\frac{1}{h_m} \int_{C_m} p\left(\frac{2(x - x_m)}{h_m}\right) M_j(x, y) dx\right] M_h^\perp g(y) dy$$

where $M_j(x, y)$ is the kernel corresponding to $K_j L$. Note that the above expression is a sum of disjoint functions with support $C_m$ each bounded by $\|p\|_{C[-1,1]}$ times the factor

(5.21) $$\int_{0}^{a} M_h^\perp \left[\frac{1}{h_m} \int_{C_m} p\left(\frac{2(x - x_m)}{h_m}\right) M_j(x, y) dx\right] M_h^\perp g(y) dy.$$

Since $g \in C^r[0, a]$, we know that the term $M_h^\perp g(y) \leq c_1 h^r$ with a uniform constant over all cells $C_m$. Also, by Lemma 4.2 we know that

(5.22) $$\frac{1}{h_m} \int_{C_m} p\left(\frac{2(x - x_m)}{h_m}\right) M_j(x, y) dx$$

is in $C^r(J_h)$. From Corollary 3.6 we conclude that to estimate the first term in the integral (5.21) we need to estimate the derivatives (in $y$) of up to order $r$ of (5.22). To accomplish this we must carry out a calculation similar that in the proof of Lemma 5.4. Denote by $g_j^{(m)}(y)$ be the function given by (5.22). For this function the estimate (5.7)(ii) follows directly, and so we omit the proof. It is in the estimation analogous to (5.7)(i) that the difference occurs. So, assume $y \in C_m$. Then, (with $C_m = [x_m^-, x_m^+]$)

$$g_j^{(m)}(y) = \frac{1}{h_m} \int_{x_m^-}^{y} p\left(\frac{2(x - x_m)}{h_m}\right) M_j(x, y) dx$$

$$+ \frac{1}{h_m} \int_{y}^{x_m^+} p\left(\frac{2(x - x_m)}{h_m}\right) M_j(x, y) dx.$$

Denote the first of these functions by $f(y)$. Now write

$$f(y) = \frac{1}{h_m} \int_{x_m^-}^{y} p\left(\frac{2(x-x_m)}{h_m}\right) \int_0^a K_j(x,x')L(x',y)dx'dx$$

$$= \frac{1}{h_m} \int_{x_m^-}^{y} p\left(\frac{2(x-x_m)}{h_m}\right) \left[\int_0^y K_j(x,x')L(x',y)dx'\right.$$

$$\left. + \int_y^a K_j(x,x')L(x',y)dx\right]dx.$$

Since $p$ and $L$ are uniformly bounded and since $\|K_j\| \leq \lambda^j$, it follows that

$$\|f\| \leq \|p\|\|L\|_A \lambda^j$$

Denote by $f_1(y)$ the first of the above double integrals. Differentiate $f_1(y)$ to obtain

$$f_1'(y) = \frac{1}{h_m} p\left(\frac{2(y-x_m)}{h_m}\right) \int_0^y K_j(y,x')L(x',y)dx'$$

$$+ \frac{1}{h_m} \int_{x_m^-}^{y} p\left(\frac{2(x-x_m)}{h_m}\right) \left[\lim_{x'\to y^-}\left(K_j(x,x')L(x',y)\right)\right.$$

$$\left. + \int_0^y K_j(x,x')\frac{\partial}{\partial y}L(x,y)dx'\right]dx.$$

As should be apparent, the pointwise estimate of $f_1'(y)$ is

$$\|f_1'(y)\| \leq c_1 \frac{1}{h_m} \|p\| \|K\|_A \|L\|_A \lambda^j.$$

Each subsequent derivative of $f_1'(y)$ will increase the estimate by a constant factor times $1/h_m$, until the $(s+1)^{th}$ derivative, when, because $p$ is a polynomial of degree $s$, no further factors of $1/h_m$ will be contributed. Thus for $y \in C_m$

$$\|g_j^{(m)}\|_{C^r(C_m)} \leq \tilde{c}\lambda^j/h_m^s.$$

This sketch establishes (5.20). The rest of the proof follows that of Theorem 5.1. ∎

An immediate corollary is the result of Ganguly and Victory [9, Theorem 9 (b)] for the integral equation (2.1).

**Corollary 5.11.** Let $\{p_0, p_1, \ldots, p_s\}$, $s < r$, denote the normalized Legendre functions, and denote by $P_\ell$, and $P_{\ell,h}$ respectively the projections corresponding to $p_\ell$. Then for $\psi^e$ defined (2.3) we have

$$P_{\ell,h}(\psi - \psi^e) = O(h^{2r-\ell})$$

and consequently for

$$M_s = \sum_{\ell=0}^{s} P_\ell$$

we have

$$M_{s,h}(\psi - \psi^e) = O(h^{2r-s})$$

Let us note that the original proof of this result uses Padé approximants to the exponential function. Therefore the piecewise constancy of the functions $\sigma$, $c$, $q$, and $k$ in (2.3) over the slabs as described in §2 is essential and cannot be relaxed.

We conclude by mentioning that for the *initial value problems* (equivalently, in the transport-theoretic setting, for nonscattering systems), and for a generalized class of nodal methods, Hennart, et al. [4] have shown that the estimate (5.18) can be improved to $O(h^{2r})$.

## References

1. R.E. Alcouffe, E.W. Larsen, W.F. Miller, Jr. and B.R. Wienke, *Computational efficiency of numerical methods for the multigroup discrete-ordinates transport equation. The slab geometry case*, Nucl. Sci. Eng. 71 (1979) pp. 111-127.
2. J.H. Bramble and S. R. Hilbert, *Estimation of Linear Functinals on Sobolev Spaces with Application to Fourier Transforms and Spline Interpolation*, SIAM J. Numer. Anal., 7 (1970) pp. 112-124.
3. D.V. Gopinath, A. Natarajan, and V. Sundarazaman, *Improved interpolation schemes in anisotropic source-flux iteration techniques*, Nucl. Sci. Eng., 75 (1980) pp. 181-184.
4. J.P. Hennart, E. del Valle, F. Serrano, and J. Valdés, *Discrete- ordinates equations in slab geometry: A generalized nodal finite element formalism*, Int. Top Mtg, Advances in Reactor Physics, Mathematics and Computer, Paris, April 1987, CEC/OECD, pp. 1283-1288.
5. H.B. Keller and P. Nelson, *Closed linear one-cell functional spacial approximations: consistency implies convergence and stability*, Transport Theory and Statistical Physics 17 (1988) 191-208.
6. E.W. Larsen and W.F. Miller, Jr., *Convergence rates of spatial difference equations for the discrete-ordinates neutron transport equations in slab geometry*, Nucl. Sci. Eng., 73 (1980), pp. 76-83.
7. E.W. Larsen and P. Nelson, Jr., *Finite-difference approximation approximations and superconvergence for the discrete-ordinates equations in slab geometry*, SIAM J. Numerical Analysis, 19 (1982), pp. 334-348.
8. K.D. Lathrop, *Spatial differencing of the transport equation: positivity vs. accuracy*, J. Comp. Phys., 4 (1969), pp. 475-498.
9. S.M. Lee and R. Vaidyanathan, *Comparison of the order of approximation in several spatial difference schemes for the discrete-ordinates transport equation in one-dimensional plane geometry*, Nucl. Sci. Eng., 76 (1980), pp. 1-9.
10. B. Neta and H.D. Victory, Jr., *The convergence analysis for sixth-order methods for solving discrete ordinates slab transport equations*, Numer. Funct. Anal. and Optimiz., 5(1) (1982), pp. 85-126.

11. _____, *A new fourth-order finite-difference method for solving discrete-ordinates slab transport equations*, SIAM J. Numerical Analysis, 20 (1983), pp. 94-105.
12. H.D. Victory, Jr., and K. Ganguly, *On finite-difference methods for solving discrete-ordinates transport equations*, SIAM J. Numer. Anal. 23 (1986) pp. 78-108.

# CHAPTER 10

Discrete-Ordinates Equations in Slab Geometry:
A Mixed-Hybrid Nodol Version of the Diffusion
Synthetic Acceleration Method

E. Del Valle*
J. P. Hennart§
J. Valdés‡

**Abstract.** A mixed–hybrid version of the diffusion synthetic acceleration method is presented. In this version, the diffusion step is solved in its $P_1$ form which is the basis of the so called mixed formulation when the finite element method is used. Here moreover, the neutron current continuity conditions are relaxed by the use of Lagrange multipliers techniques leading finally to a mixed–hybrid version of the diffusion step and by extension of the diffusion synthetic acceleration method. Up to now, our numerical experiments have been limited to the quadratic and cubic continuous methods for the transport step and their mixed–hybrid equivalents for the diffusion step. The results are extremely satisfactory and unaccelerated cases which needed more than 4000 standard iterations can be solved in less than 20 accelerated iterations.

**1. Introduction.** When the standard iteration procedure, the so-called Source Iteration (SI) method is applied to the discrete-ordinates equations in slab geometry, it turns out to be very slow when $c = \sigma_s/\sigma_t \simeq 1$. In this case, the SI has to be replaced by the well-known Diffusion Synthetic Acceleration (DSA) method. This can be done by taking the first two moments of the original transport equation in its integrodifferential form to get a $P_1$ form of the diffusion equation, i.e. a system of first order equations. This form is the basis of a mixed formulation of the problem when the Finite Element Method (FEM) is used. A weak form is

---

* Departamento de Ingeniería Nuclear, Escuela Superior de Física y Matemáticas del IPN, Unidad Profesional "Adolfo López Mateos", 07738 México, D.F. (MEXICO)

§ Departamento de Métodos Matemáticos y Numéricos, Instituto de Investigaciones en Matemáticas Aplicadas y en Sistemas de la UNAM, Apartado Postal 20–726, 01000 México, D.F. (MEXICO)

‡ Gerencia de Tecnología Reglamentación y Servicios, Comisión Nacional de Seguridad Nuclear y Salvaguardias, 01030, México, D. F. (MEXICO). Present Address, Depto. de Energía Nuclear, Instituto de Investigaciones Eléctricas, Interior del Internado Palmira s/n, 62000 Cuernavaca, Morelos, (MEXICO)

first obtained after multiplication by test functions and integration by parts: as a result, the flux is no longer required to be continuous but the (normal) current must now be continuous. Here, we relax these current continuity conditions by using Lagrange multipliers techniques. This leads to the final Mixed–Hybrid (MH) version (Roberts and Thomas, (To Appear)) of the diffusion step and by extension to the Mixed–Hybrid version of the Diffusion Synthetic Acceleration method (MH–DSA). In the present work, we used the MH–DSA method in connection with the quadratic and cubic continuous methods for the transport step. In Section 2, we recall the steps required to obtain the diffusion synthetic acceleration system to be solved. In Sections 3 and 4, the numerical procedures to solve the neutron transport and diffusion steps are described, leading to the MH–DSA algorithm we have developed. The relationships between the methods described in Sections 3 and 4, and in particular their consistency, are shown in Section 5. The numerical results obtained with several model problems are extremely satisfactory and are presented in Section 6. Finally some conclusions are offered in Section 7.

**2. The DSA method.** The original transport step in its slab geometry discrete–ordinates form can be written as follows:

$$\mu_n \frac{d\psi_n^{(l+1/2)}}{dx} + \psi_n^{(l+1/2)} = c\phi_0^{(l)} + Q(x), \ n = 1,...N, \ \forall x \in (a,b), \qquad (2.1a)$$

subject to the boundary conditions:

$$\psi_n(a) = f_n \ \forall \ \mu_n > 0,$$
$$\psi_n(b) = g_n \ \forall \ \mu_n < 0, \qquad (2.1b)$$

where $c = \sigma_s/\sigma_t$ for a non–multiplicative and isotropic medium and $x$ is expressed in mean free paths.

When equations (2.1) are solved by the classical SI method, the superindex $(l)$ is an iteration index, $(l + 1/2)$ indicating a transport step with known second member. To obtain, $\phi_0^{(l+1)}$, for the next transport step, we calculate

$$\phi_0^{(l+1)}(x) = \int_{-1}^{+1} \frac{d\mu}{2} \psi^{(l+1/2)}(x,\mu), \qquad (2.2)$$

by means of a $N$ points Gauss-Legendre quadrature rule.

It is well known that, when $c \simeq 1$, the SI method is slow because its spectral radius, $\rho$, is very close to unity. Indeed $\rho = c \simeq 1$ in the infinite medium case. In this case, the DSA method (Alcouffe (1977) and Larsen (1982)) turns out to be a powerful acceleration algorithm. First of all, we derive the $P_1$–form of equation (2.1) by taking its first two angular moments to obtain:

$$\frac{d\phi_1^{(l+1)}}{dx} + (1-c)\phi_0^{(l+1)} = S_0, \qquad (2.3a)$$

$$\frac{d\phi_0^{(l+1)}}{dx} + 3\phi_1^{(l+1)} = -2\frac{d\phi_2^{(l+1/2)}}{dx}, \qquad (2.3b)$$

where

$$-2\frac{d\phi_2^{(l+1/2)}}{dx} = \frac{d\phi_0^{(l+1/2)}}{dx} + 3\phi_1^{(l+1/2)}, \qquad (2.3c)$$

and

$$\phi_n^{(l+1/2)}(x) = \int_{-1}^{+1} \frac{d\mu}{2} P_n(\mu)\psi^{(l+1/2)}(x,\mu). \qquad (2.3d)$$

Here $P_n$ is the normalized Legendre polynomial of degree $n$ over $[-1,+1]$ and $\phi_n^{(l+1)}(x)$ is obtained directly from the solution of equations (2.3). The system of DSA equations consists of equations (2.1) and (2.3).

It can be shown from a simple Fourier stability analysis (Larsen (1982)) that the spectral radius, $\rho$, of the DSA system for an infinite medium is:

$$\rho \leq 0.23c. \qquad (2.4)$$

At this point, is clear that the DSA method consists in two steps: a neutron transport step (Eqs. (2.1)) followed by a neutron diffusion step (Eqs. (2.3)).

**3. Numerical solution of the neutron transport step.** The transport step of the combined algorithm constitutes a system of coupled ordinary differential equations

$$\mu_n \frac{d\psi_n}{dx} + \sigma_t \psi_n = S_n, \quad n = 1, \ldots, N, \quad \forall\, x \in (a,b), \qquad (3.1)$$

subject to initial conditions given by (2.1b), where $x$ is now expressed in cms for the sake of simplicity.

For angular fluxes $\psi$ corresponding to a positive (resp. negative) $\mu$, an approximate solution $\psi_h$ is determined from left to right (resp. right to left) over each successive interval $\Omega_e = [x_{\ell e}, x_{re}], e = 1, \ldots, E$, partitioning $[a,b]$ (or generically $[x_\ell, x_r]$ when there is no risk of confusion). The source $S_n$ reads

$$S_n(x) := \sum_{m=1}^{N} w_m k_{nm}(x) + q_n(x), \qquad (3.2)$$

where the $w_m$ are the quadrature weights associated with the quadrature points $\mu_m \in (-1,+1)/\{0\}$ and $q_n$ is an independent neutron source.

In (3.1), $\sigma_t$ and $S_n$ (or rather $k_{nm}$) are piecewise functions of $x$, allowed to present jump discontinuites only at points belonging to the partition of $[a,b]$, but otherwise very smooth (i.e. piecewise $C^\infty$). Each transport step like (3.1) boils down to the solution of $N$ monodirectional monoenergetic transport equations of the form

$$\mu \frac{d\psi}{dx} + \sigma\psi = S, \qquad (3.3)$$

where $S$ is a known function of $x$. In the following, we shall assume that the particular angular direction $\mu$ considered is positive and that we solve (3.1) or rather (3.3) from left to right for the corresponding angular flux $\psi \equiv \psi(x,\mu)$.

Let $P_k(x)$ be the normalized Legendre polynomial of degree $k$ over $\hat{\Omega} \equiv [-1,+1]$ with the well-known properties

$$P_k(+1) = 1, \qquad (3.4a)$$
$$P_k(-1) = (-1)^k, \qquad (3.4b)$$

and

$$\int_{-1}^{+1} P_k(x) P_\ell(x)\, dx = \delta_{k\ell} N_k, \qquad (3.4c)$$

where

$$N_k = \int_{-1}^{+1} P_k^2(x)\, dx = 2/(2k+1). \qquad (3.4d)$$

Over $[x_\ell, x_r]$, the corresponding shifted Legendre polynomial $p_k(x)$ is given by

$$p_k(x) = P_k\left[(2x - x_\ell - x_r)/(x_r - x_\ell)\right]. \qquad (3.5)$$

Let us now define the cell moments $m_c^i(u)$ of a function $u(x)$ as

$$m_c^i(u) \equiv \int_{-1}^{+1} P_i(x) u(x)\, dx / N_i. \qquad (3.6)$$

Thanks to the normalization adopted, these expressions are valid over any interval $\Omega_e$ and not only over the reference interval $\hat{\Omega}$. For instance, over $\Omega_e$

$$m_c^i(u) \equiv \int_{\Omega_e} p_i(x) u(x)\, dx / \int_{\Omega_e} p_i^2(x)\, dx. \qquad (3.7)$$

Given any $\Omega_e$, going to the reference interval $\hat{\Omega}$ and then back to $\Omega_e$ is very appealing for the nodal finite element formalism used in this paper. For a general presentation of this formalism, see Hennart and Del Valle (In Preparation). In fact, the basis functions which shall be used later are quite conveniently developed over $\hat{\Omega}$ in terms of the $P_k$'s. Indeed, it is trivial to show that over $\hat{\Omega}$, (3.3) can be written as

$$\frac{d\psi}{d\xi} - \lambda\psi = S \tag{3.8}$$

where $\lambda = -\sigma h/2\mu$ and $S := Sh/2\mu$, $h$ being the length of $\Omega_e$, i.e. $h = x_r - x_\ell$, while $\xi = (2x - x_r - x_\ell)/h \in \hat{\Omega}$.

In the following, $\psi$ will be approximated over each cell $\Omega_e$, or equivalently over $\hat{\Omega}$, by some function $\psi_h$ of $x$, belonging to the following polynomial space

$$P_k \equiv \{1, x, \ldots, x^k\}. \tag{3.9}$$

In this paper, the quadratic ($k = 2$) and cubic ($k = 3$) cases are fully developed. This piecewise continuous behavior of $\psi_h \sim \psi$ justifies our reference to a *finite element formalism*. By *nodal*, we mean moreover that the basic parameters describing $\psi_h$ over each cell will be some of its edge values and of its cell moments, the edge values being in fact transverse moments in multidimensional situations.

In previous papers on space discretization for discrete ordinates transport equations in slab geometry, the schemes which were developed and analyzed were always referred to as *finite difference* schemes. They basically provide approximations to the cell edge (and possibly cell average) values of $\psi$, but they tend to ignore the possibility of some hidden continuous behavior within the cell. Here our approach is basically finite element oriented: the most conspicuous parameters are clearly cell edge and cell average approximations, but from them and behind them it is quite feasible to extract for instance to value of $\psi_h$ at any interior point of a given cell $\Omega_e$ or any moment which is not a basic degree of freedom.

With $P_k \equiv \{1, x, \ldots, x^k\}$, we shall assume that over each cell $\Omega_e = [x_\ell, x_r]$ the angular flux $\psi(x)$ in the direction $\mu > 0$ is approximated by $\psi_h(x)$ in $P_k$, defined in terms of $(k+1)$ degrees of freedom which will be $\psi_h(x_r) = \psi_r$, $\psi_h(x_\ell) = \psi_\ell$ as well as cell moments of $\psi_h$ up to order $k-2$, i.e. $\psi_c^i = m_c^i(\psi_h(x))$, $i = 0, \ldots, k-2$ if $k \geq 2$.

The determination of the corresponding basis functions can be performed once and for all over the reference cell $\hat{\Omega}$. Knowing the representation of $\psi_h$ over $\hat{\Omega}$, it is a simple matter to obtain $\psi_h$ over the real cell $\Omega_e$ by an inversible affine transformation: namely $x = [(x_r - x_\ell)\xi + x_r + x_\ell]/2$, $\forall \xi \in [-1, +1]$.

Over the reference cell $\hat{\Omega} = [-1, +1]$, $\psi_h$ is thus defined by a space $S$ and a set of degrees of freedom $D$ (actually linear functionals $L_i$ acting upon $S$) given by

$$\begin{aligned} S &= P_k \\ D &= \{\psi_\ell, \psi_r, \psi_c^i, \ i = 0, \ldots, k-2\}, \end{aligned} \tag{3.10}$$

with $\dim S = \text{card } D = k + 1 \equiv I$.

Over $\hat{\Omega}$, it is very convenient to work directly with the Legendre polynomials $P_\ell$ and to replace $\mathcal{P}_k$ by

$$\mathcal{P}_k = \{P_0, \ldots, P_k\}. \tag{3.11}$$

If we call $u_\ell, u_r$, and $u_c^i$, the basis functions corresponding to $\psi_\ell, \psi_r$, and $\psi_c^i$, so that

$$\psi_h = \psi_\ell u_\ell(x) + \psi_r u_r(x) + \sum_{i=0}^{k-2} \psi_c^i u_c^i(x), \tag{3.12}$$

it is easy to show (Hennart (1986)) that

$$u_\ell(x) = \frac{1}{2}(-1)^{k-1}[P_{k-1}(x) - P_k(x)],$$
$$u_r(x) = \frac{1}{2}\phantom{(-1)^{k-1}}[P_{k-1}(x) + P_k(x)],$$

and

$$u_c^i(x) = P_i(x) - P_{k-2+m(i)}(x), \quad i = 0, \ldots, k-2, \tag{3.13}$$

where $m(i) = 1$ or $2$ is such that $k - 2 + m(i)$ and $i$ have the same parity.

For $k = 2$, we have

$$u_\ell(x) = -\frac{1}{2}[P_1(x) - P_2(x)],$$
$$u_r(x) = +\frac{1}{2}[P_1(x) + P_2(x)],$$

and

$$u_c^0(x) = P_0(x) - P_2(x). \tag{3.14}$$

Clearly, $u_c^0(x_\ell) = u_c^0(x_r) = 0$ while $m_c^0(u_c^0(x)) = 1$. Moreover $u_\ell(x_\ell) = 1$, $u_\ell(x_r) = 0$, and $m_c^0(u_\ell(x)) = 0$, etc.

For $k = 3$, we have similarly

$$u_\ell(x) = +\frac{1}{2}[P_2(x) - P_3(x)],$$
$$u_r(x) = +\frac{1}{2}[P_2(x) + P_3(x)],$$
$$u_c^0(x) = P_0(x) - P_2(x),$$

and

$$u_c^1(x) = P_1(x) - P_3(x). \tag{3.15}$$

Over $[-1, +1]$ and with the Legendre polynomials, it is thus particularly simple to write down the basis functions of the generalized interpolation problem (3.10).

The basic degrees of freedom of $\psi_h$ are the cell moments $\psi_c^i$, $i = 0, \ldots, k - 2$ (if $k \geq 2$) plus $\psi_\ell$ and $\psi_r$. It is thus possible to calculate at least two more moments of $\psi_h$ from (3.13). Since we shall need them later on, their expressions will be given hereafter.

Let us take the simple example of $P_2$ ($k = 2$). We shall have $m_c^0(\psi_h) = \psi_c^0$, by construction,

and
$$m_c^1(\psi_h) = (\psi_r - \psi_\ell)/2,$$

$$m_c^2(\psi_h) = (\psi_r + \psi_\ell)/2 - \psi_c^0 . \tag{3.16}$$

Similarly for $k = 3$, we shall have

and
$$m_c^2(\psi_h) = (\psi_r + \psi_\ell)/2 - \psi_c^0,$$

$$m_c^3(\psi_h) = (\psi_r - \psi_\ell)/2 - \psi_c^1 . \tag{3.17}$$

With $\psi_h \in P_k$, the continuous moment methods (Hennart and Del Valle (In Preparation)) are based on the following ideas:

1. As we move through the slab from left to right, assuming $\mu > 0$, $\psi_h$ is continuous from cell to cell: that is $\psi_h(x = x_\ell) = \psi_\ell$ in each cell is taken to be $\psi_h(x = x_r) = \psi_r$ from the previous cell or the incoming flux at the left boundary (see (2.1b)) if we are in the first cell to the left.

2. Since one of the $k + 1$ parameters in $P_k$ is fixed by that continuity condition between successive cells, we still have $k$ parameters to determine. This is done by taking Legendre moments (of order 0 to $k - 1$) of the residual $\mu \frac{d\psi_h}{dx} + \sigma \psi_h - S \equiv L\psi_h - S$ and expressing that they are zero, namely

$$\int_{\Omega_e} (L\psi_h - S) p_i(x) \, dx = 0, \quad i = 0, \ldots, k - 1. \tag{3.18}$$

Going back to the reference interval $\hat{\Omega}$, these equations are

$$\frac{2\mu}{hN_i} \left\{ [\psi_r - (-1)^i \psi_\ell] - \int_{-1}^{+1} \psi_h \frac{dP_i}{d\xi} d\xi \right\} + \sigma \psi_c^i = m_c^i(S), \quad i = 0, \ldots, k - 1. \tag{3.19}$$

where $\sigma$ has been assumed to be constant over $\hat{\Omega}$ for the sake of simplicity and also to be able to make direct comparisons with well known schemes in the literature (where $\sigma$ is *practically always* assumed to be piecewise constant).

Let us write down explicitly some of these equations: for $i = 0$, with $\frac{dP_0}{d\xi} \equiv 0$ and $N_0 = 2$, we shall have

$$\frac{\mu}{h}(\psi_r - \psi_\ell) + \sigma\psi_c^0 = m_c^0(S), \qquad (3.20a)$$

while for $i = 1$, with $\frac{dP_1}{d\xi} \equiv P_0$ and $N_1 = 2/3$, (3.19) becomes

$$\frac{3\mu}{h}(\psi_r + \psi_\ell) - \frac{6\mu}{h}\psi_c^0 + \sigma\psi_c^1 = m_c^1(S). \qquad (3.20b)$$

Finally, for $i = 2$, with $\frac{dP_2}{d\xi} \equiv 3P_1$ and $N_2 = 2/5$, we have

$$\frac{5\mu}{h}(\psi_r - \psi_\ell) - \frac{10\mu}{h}\psi_c^1 + \sigma\psi_c^2 = m_c^2(S). \qquad (3.20c)$$

Now, we are going to consider the particular cases corresponding to $k = 2$ and $k = 3$.

In the $k = 2$ case, (3.12) becomes

$$\psi_h = \psi_\ell u_\ell(x) + \psi_r u_r(x) + \psi_c^0 u_c^0(x), \qquad (3.21a)$$

with $\psi_c^0 = m_c^0(\psi_h)$ and

$$\psi_c^1 \equiv m_c^1(\psi_h) = (\psi_r - \psi_\ell)/2, \qquad (3.21b)$$

from (3.16). Inserting this value of $\psi_c^1$ in (3.20b), we get a system of two equations for $\psi_r$ and $\psi_c^0$ knowing $\psi_\ell$, namely

$$\mu\psi_r + \sigma h \psi_c^0 = \mu\psi_\ell + h m_c^0(S), \qquad (3.21c)$$

and

$$\left(3\mu + \frac{\sigma h}{2}\right)\psi_r - 6\mu\psi_c^0 = \left(-3\mu + \frac{\sigma h}{2}\right)\psi_\ell + h m_c^1(S). \qquad (3.21d)$$

from (3.21c), and with $\epsilon \equiv \sigma h/\mu$

$$\psi_c^0 = -\frac{1}{\epsilon}(\psi_r - \psi_\ell) + \frac{m_c^0(S)}{\sigma},$$

so that (3.21d) becomes

$$\psi_r = \left\{ \left(1 - \frac{\epsilon}{2} + \frac{\epsilon^2}{12}\right) \bigg/ \left(1 + \frac{\epsilon}{2} + \frac{\epsilon^2}{12}\right) \right\} \psi_\ell \\ + \left[\frac{\epsilon}{\sigma} m_c^0(S) + \frac{\epsilon h}{6\mu} m_c^1(S)\right] \bigg/ \left(1 + \frac{\epsilon}{2} + \frac{\epsilon^2}{12}\right) , \quad (3.22)$$

which corresponds to the *Quadratic Continuous* (QC) scheme of Neta and Victory (1983) (see their equation (2.4)).

In the $k = 3$ case, (3.12) becomes

$$\psi_h = \psi_\ell u_\ell(x) + \psi_r u_r(x) + \psi_c^0 u_c^0(x) + \psi_c^1 u_c^1(x) , \quad (3.23a)$$

with $\psi_c^0 = m_c^0(\psi_h)$, $\psi_c^1 = m_c^1(\psi_h)$, and

$$\psi_c^2 \equiv m_c^2(\psi_h) = (\psi_\ell + \psi_r)/2 - \psi_c^0 \quad (3.23b)$$

from (3.17). Inserting this value of $\psi_c^2$ in (3.20c), we get a system of three equations for $\psi_r$, $\psi_c^0$, and $\psi_c^1$ knowing $\psi_\ell$, namely

$$\mu \psi_r + \sigma h \psi_c^0 = \mu \psi_\ell + h m_c^0(S) , \quad (3.23c)$$

$$3\mu \psi_r - 6\mu \psi_c^0 + \sigma h \psi_c^1 = -3\mu \psi_\ell + h m_c^1(S) , \quad (3.23d)$$

and

$$\left(5\mu + \frac{\sigma h}{2}\right) \psi_r - \sigma h \psi_c^0 - 10\mu \psi_c^1 = \left(5\mu - \frac{\sigma h}{2}\right) \psi_\ell + h m_c^2(S) . \quad (3.23e)$$

from (3.23c),

$$\psi_c^0 = -\frac{1}{\epsilon}(\psi_r - \psi_\ell) + \frac{m_c^0(S)}{\sigma} ,$$

which when used in (3.23d) leads us to

$$\psi_c^1 = -\frac{3}{\epsilon}\left(1 + \frac{2}{\epsilon}\right)\psi_r - \frac{3}{\epsilon}\left(1 - \frac{2}{\epsilon}\right)\psi_\ell + \frac{6}{\epsilon\sigma} m_c^0(S) + \frac{m_c^1(S)}{\sigma} .$$

When these two expressions are substituted in (3.23e), we finally get (after a little bit of algebra)

$$\psi_r = \left(1 + \frac{\epsilon}{2} + \frac{\epsilon^2}{10} + \frac{\epsilon^3}{120}\right)^{-1} \left\{\left(1 - \frac{\epsilon}{2} + \frac{\epsilon^2}{10} - \frac{\epsilon^3}{120}\right)\psi_\ell \right.$$
$$+ \left(\frac{\epsilon}{\sigma} + \frac{\epsilon^3}{60\sigma}\right) m_c^0(S)$$
$$+ \frac{\epsilon^2}{6\sigma} m_c^1(S)$$
$$\left. + \frac{\epsilon^3}{60\sigma} m_c^2(S)\right\}, \qquad (3.24)$$

which corresponds to the *Cubic Continuous* (CC) scheme of Neta and Victory (1982), as the reader can convince himself by comparing this equation to their equation (3.2a).

**4. Numerical solution of the neutron diffusion step.** The diffusion step of the DSA algorithm is formed by a coupled set of first order differential equations:

$$3\phi_1 + \frac{d\phi_0}{dx} = S_1^*, \qquad (4.1a)$$

$$\frac{d\phi_1}{dx} + (1 - c)\phi_0 = S_0, \qquad (4.1b)$$

which are deduced directly from Eqs. (2.3). The boundary conditions at $a$ and $b$ are of the following form (for instance)

$$\phi_0(a) + \alpha \phi_1(a) = 0, \qquad (4.2a)$$

in case of a vacuum boundary condition, or

$$\phi_1(b) = 0, \qquad (4.2b)$$

in case of a reflection boundary condition.

When the finite element method is used, the system (4.1) leads quite naturally to the mixed formulation. The strong form (4.1) is replaced by a weak form after multiplying Eq. (4.1a) by a test function $\psi_1$ and Eq. (4.1b) by a test function $\psi_0$. In more than one dimension, $\phi_1$ would become a vector and the scalar test function $\psi_1$ would be replaced by a vector test function. More details are given in Hennart and Del Valle (1990). After integration by parts and assuming that $\psi_1(b) = 0$, the final weak form reads:

Find $(\phi_1, \phi_0) \in U \times W$ such that

$$\int_a^b 3\phi_1\psi_1 dx - \alpha\phi_1(a)\psi_1(a) - \int_a^b \phi_0 \frac{d\psi_1}{dx} dx = \int_a^b \psi_1 S_1^* dx, \quad \forall \psi_1 \in U, \quad (4.3a)$$

$$-\int_a^b \frac{d\phi_1}{dx}\psi_0 dx - \int_a^b (1-c)\phi_0\psi_0 dx = -\int_a^b \psi_0 S_0 dx, \quad \forall \psi_0 \in W, \quad (4.3b)$$

where the second equation has been multiplied by $(-1)$ to make the overall system more symmetrical. Here $W \equiv L^2(\Omega)$ while $\phi_1$ must be looked for in $U \equiv H_0^1(\Omega)$ where

$$H_0^1 \equiv \{u | u \in L^2(\Omega), \frac{du}{dx} \in L^2(\Omega), u(b) = 0\}. \quad (4.4)$$

In more than one dimension, $H_0^1(\Omega)$ has to be replaced by a space known as $H_0(div; \Omega)$; for more details, see Hennart and Del Valle (1990).

$\phi_0$ only needs to be in $L^2(\Omega)$ and can in particular be discontinuous from one interval $\Omega_e$ to the following. However, $\phi_1 \in H_0^1$ must be continuous at all the interfaces. This condition may be relaxed and restored by the introduction of Lagrange multipliers leading to the MH formulation where $\phi_1$ is no longer in $H_0^1$. Consequently boundary terms appear at each $\Gamma_e = \hat{\Omega}_e - \Omega_e$ and the weak form (4.3) is replaced by:

Find $(\phi_1, \phi_0, \lambda) \in V \times W \times L$ such that

$$\int_a^b 3\phi_1\psi_1 dx - \sum_{e=1}^E \left\{ \int_{\Omega_e} \phi_0 \frac{d\psi_1}{dx} dx - [\lambda\psi_1]_{\Gamma_e} \right\} = \int_a^b \psi_1 S_1^* dx, \quad \forall \psi_1 \in V, \quad (4.5a)$$

$$-\sum_{e=1}^E \int_{\Omega_e} \frac{d\phi_1}{dx}\psi_0 dx - \int_a^b (1-c)\phi_0\psi_0 dx = -\int_a^b \psi_0 S_0 dx, \quad \forall \psi_0 \in W, \quad (4.5b)$$

and

$$\sum_{e=1}^E [\mu\psi_1]_{\Gamma_e} = 0, \quad \forall \mu \in L, \quad (4.5c)$$

where now

$$V \equiv \{u | u \in L^2(\Omega), \frac{du}{dx}\Big|_{\Omega_e} \in L^2(\Omega_e), e = 1, \ldots, E\},$$

and

$$L \equiv \{\lambda | \lambda \in \prod_{e=1}^{E} H^{1/2}(\Gamma_e)\},$$

$H^{1/2}(\Gamma_e)$ being the space of traces over $\Gamma_e$ of the functions in $H^1(\Omega_e)$ (Adams, (1975)).

A discrete mixed-hybrid formulation can be obtained by considering finite dimensional subspaces of $V, W$ and $L$, namely $V_h, W_h$ and $L_h$. This *discrete mixed-hybrid formulation* is given explicitly by

Find $(\phi_{1h}, \phi_{0h}, \lambda_h) \in V_h \times W_h \times L_h$ such that

$$\int_a^b 3\phi_{1h}\psi_{1h}dx - \sum_{e=1}^{E}\left\{\int_{\Omega_e}\phi_{0h}\frac{d\psi_{1h}}{dx}dx - [\lambda_h\psi_{1h}]_{\Gamma_e}\right\} = \int_a^b \psi_{1h}S_1^*dx, \quad \forall\, \psi_{1h} \in V_h, \tag{4.6a}$$

$$-\sum_{e=1}^{E}\int_{\Omega_e}\frac{d\phi_{1h}}{dx}\psi_{0h}dx - \int_a^b(1-c)\phi_{0h}\psi_{0h}dx = -\int_a^b \psi_{0h}S_0dx, \quad \forall\, \psi_{0h} \in W_h, \tag{4.6b}$$

and

$$\sum_{e=1}^{E}[\mu_h\psi_{1h}]_{\Gamma_e} = 0, \quad \forall\, \mu_h \in L_h, \tag{4.6c}$$

In fact $L_h \equiv L \equiv \mathbb{R}^{E+1}$, i.e. it consists of the vectors of reals defined at each mesh point. For $V_h$, the basis functions are given over each interval for a given index $k$ by their values at the left and right points, plus when $k > 2$ by the first $(k-2)$ moments over the interval. Namely

$$v_L(x) = \frac{1}{2}(-1)^{k-2}[P_{k-2}(x) - P_{k-1}(x)],$$
$$v_R(x) = \frac{1}{2}\quad\ [P_{k-2}(x) + P_{k-1}(x)], \tag{4.7a}$$

and

$$v_c^i(x) = P_i(x) - P_{k-2+m(i)}(x), \quad i = 0, \ldots, k-3, \tag{4.7b}$$

where $m(i) = 1$ or $2$ is such that $k - 2 + m(i)$ and $i$ have the same parity.

Locally, $\phi_{1h}$ reduces to a polynomial of degree $(k-1)$. For $W_h$, the basis functions are given over each interval by their first $(k-1)$ moments over each interval. Explicitly

$$w_c^i(x) = P_i(x), \quad i = 0, \ldots, k-2, \tag{4.8}$$

$\phi_{0h}$ is thus locally a polynomial of degree $(k-2)$. For the classical error analysis of the mixed or mixed–hybrid form to be valid (Roberts and Thomas, (To Appear)), we must have

$$divV_h \subset W_h, \tag{4.9}$$

the so called Babuška-Brezzi condition, which implies that $V_h$ and $W_h$ cannot be chosen independently.

The local basis functions for $\phi_1$ and $\phi_0$ for $k=2$ are given by

$$v_L(x) = \frac{1}{2}[P_0(x) - P_1(x)],$$
$$v_R(x) = \frac{1}{2}[P_0(x) + P_1(x)], \tag{4.10a}$$

and

$$w_c^0(x) = P_0(x). \tag{4.10b}$$

Similarly for $k=3$

$$v_L(x) = -\frac{1}{2}[P_1(x) - P_2(x)],$$
$$v_R(x) = +\frac{1}{2}[P_1(x) + P_2(x)],$$
$$v_c^0(x) = P_0(x) - P_2(x), \tag{4.11a}$$

and

$$w_c^0(x) = P_0(x),$$
$$w_c^1(x) = P_1(x). \tag{4.11b}$$

In this case the algebraic system to be solved is the following:

$$A\phi_1 + B\phi_0 + C\lambda = S_1^* \tag{4.12a}$$
$$B^T\phi_1 + D\phi_0 = S_0 \tag{4.12b}$$
$$C^T\phi_1 = 0 \tag{4.12c}$$

where now $\phi_1$ violates conformity (i.e. $\phi_1 \notin H_0^1(\Omega)$), a fact which is compensated by introducing additional equations involving Lagrange multipliers.

If one eliminates $\phi_1$ in terms of $\phi_0$ and $\lambda$ and then $\phi_0$ in terms of $\lambda$ namely

$$\phi_1 = A^{-1}[S_1^* - B\phi_0 - C\lambda] \tag{4.13}$$
$$\phi_0 = [D - B^T A^{-1} B]^{-1}[S_0 + B^T A^{-1} C\lambda - B^T A^{-1} S_1^*] \tag{4.14}$$

then the resulting algebraic system for $\lambda$ is given by

$$G\lambda = S. \tag{4.15}$$

where $G$ is a matrix with known coefficients and $S$ is a source term. From a practical point of view, one of the main advantages of the algebraic system involved in equation (4.15) is that the corresponding matrix $G$ is always tridiagonal independently of $k \geq 2$, since the interelement node is only coupled to its two neighbors. This nice feature is quite interesting in the determination of the $\lambda$ parameters because the computational effort is reduced and the time saving is increased. Besides, once the $\lambda$ parameters have been determined, the computational effort to calculate the neutron flux moments $(\phi_0)$ is very small because it can be done cell by cell since the matrix $[D - B^T A^{-1} B]^{-1}$ is block–diagonal, with blocks of order $(k-1)$. Finally, $\phi_1$ can be determined in a similar way using the values of $\lambda$ and $\phi_0$, and the related matrix $A$ is also block–diagonal, with blocks of order $k$. In other words, the solution of the diffusion step appearing in the combined DSA algorithm is particularly inexpensive, thanks to the MH formulation we have adopted.

**5. Consistency between transport and diffusion steps.** It is important to point out that the numerical procedures introduced in Sections 3 and 4 are consistent in the sense that both of them have the same type of interpolation parameters for the angular and scalar neutron flux, namely the edge values and each of the moments considered per cell. The Lagrange multipliers $\lambda$ give information on the variable dual to the one they relax: the $\lambda$ parameters are thus in fact the cell edge values of the neutron flux. Taking into account the $(k-1)$ moments of the neutron flux in a given cell plus these edge values, it is possible to build up an approximation, say $\tilde{\phi}_{0c}$ that interpolates over each cell the neutron flux at the edges ($\lambda_\ell$ and $\lambda_r$) and the $(k-1)$ neutron flux moments within the cell ($\phi_{0c}^i$, $i = 0, \ldots, k-2$). What is really important is that the basis functions related to this approximation are precisely the ones used for the transport step. Thus on each cell $[x_\ell, x_r]$, we have

$$\tilde{\phi}_{0h}(x) = \lambda_\ell u_\ell(x) + \lambda_r u_r(x) + \sum_{i=0}^{k-2} \phi_{0c}^i u_c^i(x), \quad k = 2, 3. \tag{5.1}$$

If one compares this equation with equation (3.12) previously divided by 2 and integrated for $\mu$ over $[-1 + 1]$, both of them are clearly consistent.

**6. Numerical results.** Using the methods we have described in this paper we developed a computer program called DSA–TRANS which obtains the numerical solution of the discrete ordinates neutron transport equation in slab geometry using a nodal diffusion synthetic acceleration method. The way we tested DSA–TRANS consisted essentially in the reproduction of one–dimensional seven model problems widely used in the literature (Reed (1971)) where the first six are in the $S_2$ approximation while the last one in the $S_8$ one. All the problems have vacuum boundary condition at both edges. The model problems M1 to M6, have characteristics which are summarised in Table I while the configuration of the model problem M7 is shown in Fig. 1. All of them were first solved by using the computer program TNOD1 (Serrano (1985)) employing the quadratic continuous nodal method but based on a classical source iteration method. For the M7 problem the mesh size was of 2cm.

| Problem | $c$ | $\sigma_t$ | $\sigma_s$ | $E$ | Thick |
|---|---|---|---|---|---|
| M1 | 1 | 1 | 1 | 30 | 30 |
| M2 | 1 | 2 | 2 | 30 | 30 |
| M3 | 1 | 1 | 1 | 60 | 60 |
| M4 | 1 | 1 | 1 | 30 | 60 |
| M5 | 2/3 | 3 | 2 | 30 | 30 |
| M6 | 1/2 | 4 | 2 | 30 | 30 |

Table I. Characteristics for the M1 to M6 sample problems.

| REFLECTOR | CORE | REFLECTOR |
|---|---|---|
| $\sigma_t = 1cm^{-1}$ | $\sigma_t = 1cm^{-1}$ | $\sigma_t = 1cm^{-1}$ |
| $c = 1.0$ | $c = 0.9$ | $c = 1.0$ |
| $Q = 0.0$ | $Q = 1.0 n/cm^3$ | $Q = 0.0$ |
| $500cm$ | $20cm$ | $500cm$ |

Figure 1. Slab configuration for M7 sample problem.

The seven problems were also solved using DSA–TRANS with the Quadratic Continuous as well as the Cubic Continuous nodal methods. The results in both cases are exhibited in Table II where the number of inner iterations and the error (in infinite norm) between two successive iterations are given.

| Model   | TNOD1/QC | | DSA-TRANS/QC | | DSA-TRANS/CC | |
|---------|------|-----------|-----|------------|-----|------------|
| Problem | It.  | Error     | It. | Error      | It. | Error      |
| M1      | 2792 | 1.000E-04 | 16  | 9.166E-05  | 19  | 8.860E-05  |
| M2      | 4000*| 1.884E-02 | 17  | 3.334E-05  | 18  | 8.860E-05  |
| M3      | 4000*| 3.772E-02 | 17  | 7.839E-05  | 20  | 8.981E-05  |
| M4      | 4000*| 3.771E-02 | 17  | 6.726E-05  | 19  | 8.512E-05  |
| M5      | 15   | 7.324E-05 | 8   | 5.722E-05  | 6   | 6.496E-05  |
| M6      | 13   | 6.244E-05 | 8   | 3.090E-05  | 6   | 5.163E-05  |
| M7      | **   | ———       | 13  | 8.485E-05  | 11  | 8.753E-05  |

Table II. Comparison of convergence characteristics between the SI and MH–DSA methods.

\* In these cases the convergence criteria was not satisfied and the numerical results provided correspond to iteration 4,000.

\*\* Computer time exceeded, results not reported.

**7. Conclusions.** As the numerical results show the number of inner iterations is drastically diminished in those cases when the parameter c is equal to one. The main advantages of the MH–DSA method proposed in this paper are the following:

1.-The representation of the neutron scalar and angular flux is done in an essentially consistent way in both the transport step and the diffusion step.

2.-The average angular and scalar fluxes are directly unknowns of the problem: they do not have to be calculated a posteriori, which is quite convenient for the estimation of the reaction rates.

3.-The discrete equations satisfy neutron balance over each cell, for the transport as well as the diffusion step. In the diffusion step, both the flux and the current are continuous at the interfaces.

4.-The postprocessing operation yielding a quadratic (resp. cubic) $\tilde{\phi}_0$ from a constant (resp. linear) $\phi_0$ can be done a posteriori at no cost, while the actual computation relies on a lower order approximation ($\phi_0$ and not $\tilde{\phi}_0$) with simpler algebraic equations.

5.-The final algebraic systems for the Lagrange multipliers are always tridiagonal, independently of the scheme considered. The corresponding $\phi_0$ and $\phi_1$ are obtained cell by cell.

At the time of this writing, we are currently working on extensions of the method to X–Y geometry (García (In Preparation)) as well as to the use of spaces including exponentials in $1D$, which lead to the entire class of characteristic schemes (Hennart and Del Valle (In Preparation)).

# REFERENCES

ADAMS, R.A. (1975), *Sobolev Spaces*, Academic Press, New York.

ALCOUFFE, R.E. (1977), "Diffusion synthetic acceleration methods for the diamond-differenced discrete-ordinates equations", *Nucl. Sci. Engng.* **64**, 344–355.

GARCÍA, L.M., In Preparation, *Aplicación del Método MH–DSA en la Solución de las Ecuaciones de Transporte de Neutrones en Geometría X–Y*, Master Thesis, ESFM–IPN.

HENNART, J.P. and DEL VALLE, E., In Preparation, "A generalized nodal finite element formalism for discrete-ordinates equations in slab geometry. Part I: Theory in the continuous moment case.

HENNART, J.P. (1986), "A general family of nodal schemes", *SIAM J. Sci. Stat. Comput.* **7**, 264–287.

HENNART, J.P. and DEL VALLE, E. (1990), "Fast elliptic solvers using mixed-hybrid nodal finite elements", Preprint.

LARSEN, E.W. (1982), "Unconditionally stable diffusion-synthetic acceleration methods for the slab geometry discrete-ordinates equations. Part I: Theory", *Nucl. Sci. Engng.* **82**, 47–63.

NETA, B. and VICTORY, H.D., JR. (1983), "A new fourth-order finite-difference method for solving discrete-ordinates slab transport equations", *SIAM J. Numer. Anal.* **20**, 94–105.

NETA, B. and VICTORY, H.D., JR. (1982), "The convergence analysis for sixth-order methods for solving discrete-ordinates slab transport equations", *Numer. Funct. Anal. and Optimiz.* **5**, 85–126.

REED, W.H. (1971), "New difference schemes for the neutron transport equation", *Nucl. Sci. Engng.* **46**, 309–314.

ROBERTS, J.E. and THOMAS, J.M., To Appear, "Mixed and Hybrid Methods", To Appear in *Handbook of Numerical Analysis*, Vol. II, P.G. Ciarlet and J.L. Lions, Eds., North Holland.

SERRANO, F. (1985), *Solución Numérica de las Ecuaciones de Transporte de Neutrones en Ordenadas Discretas y Geometría Plana*, Master Thesis, ESFM–IPN.

# CHAPTER 11

## Upwind Finite Element Approximations of the Advection-Diffusion Problem

Gonzalo Alduncin*
Jorge Carrera*

**Abstract.** From the symmetrization of the advection-diffusion problem a heuristic approach to upwind techniques is identified. In this manner the Petrov-Galerkin method and the direct upwind method due to Tabata are studied in an abstract framework of internal approximations. Numerical experimentation is performed utilizing finite element techniques in order to exemplify the high accuracy of the direct upwind scheme.

**1. Introduction.** In this paper upwind finite element numerical schemes are studied in the simulation of advection-diffusion phenomena in several dimensions. The corresponding mathematical models are non potential because of the non symmetric advective term and, consequently, the generation of numerical schemes requieres *ad hoc* techniques. As is well known, the standard approximations degenerate as advection dominates diffusion, being the upwind techniques one of the most efficient ways of handling such an inconvenience.

The purpose of this work is to recognize a heuristic for upwinding in the process of symmetrization of the problem and then, via abstract upwind internal approximations, to establish systematically finite element numerical schemes. The specific techniques that will be considered here are the Petrov-Galerkin's and the direct upwinding introduced by Tabata.

As a model problem we will consider the following one:

$$\begin{aligned} -\varepsilon \, \text{div}\{K\nabla u\} + \mathbf{w} \cdot \nabla u &= f && \text{in } \Omega, \\ u &= \bar{u}, && \text{on } \partial\Omega_1, \\ \varepsilon \, K\nabla u \cdot \mathbf{n} &= \varepsilon \, g, && \text{on } \partial\Omega_2. \end{aligned} \qquad (1)$$

---

*Instituto de Geofísica, UNAM, 04510-Mexico, D.F., Mexico.

Here $\Omega \subset \mathbb{R}^n$, $n \geq 1$, with boundary $\partial\Omega = \partial\Omega_1 \cup \partial\Omega_2$, $\partial\Omega_1 \cap \partial\Omega_2 = \emptyset$, $\varepsilon > 0$ is a parameter that weights the diffusion term in constrast with the advection term, $\mathbf{w}$ is the given velocity field, $f$ is the interior data, and $\bar{u}$ and $g$ are respectively the Dirichlet and Neumann boundary conditions. We shall assume that div $\mathbf{w} = 0$ in $\Omega$ and $\mathbf{w} \cdot \mathbf{n} \geq 0$ on $\partial\Omega_2$.

As the parameter $\varepsilon$ tends to zero the behavior of the governing equation $(1)_1$ becomes of hyperbolic type, and its typical characteristics may appear, as boundary layers and interior discontinuities. This is precisely the reason why standard numerical schemes may degenerate. As we shall see, one of the most effective methods for reproducing such a hyperbolic behavior is the one proposed by Tabata [1] and developed by Bristeau et al [2]. Some numerical experiments are performed to illustrate this technique with a third order scheme.

**2. Primal variational formulation of the model problem.** In this section, we present the *primal variational formulation* of the problem, from which we will derive numerical schemes on the basis of abstract internal approximations and upwind techniques.

Proceeding in a direct form, after integration and application of the corresponding Green's formula, the primal variational formulation of model problem (1) turns out to be:

Find $u \in K_{\bar{u}}$ such that

$$\varepsilon \int_\Omega K\nabla u \cdot \nabla v_0 \, d\Omega + \int_\Omega \mathbf{w} \cdot \nabla u \, v_0 \, d\Omega = \int_\Omega f \, v_0 \, d\Omega + \varepsilon \int_{\partial\Omega_2} g \, \gamma v_0 \, d\partial\Omega,$$

$$\forall \, v_0 \in K_0. \quad (2)$$

Here $K_{\bar{u}}$ is the set of admissible fields, which satisfy the prescribed Dirichlet boundary condition, $\bar{u}$, on $\partial\Omega_1$:

$$K_{\bar{u}} = \{v \in H^1(\Omega): \gamma v = \bar{u} \text{ on } \partial\Omega_1\}, \quad (3)$$

where $H^1(\Omega)$ is the usual Hilbert Sobolev space and $\gamma \in \mathcal{L}(H^1(\Omega), H^{1/2}(\partial\Omega))$ the corresponding trace operator. Similarly, $K_0$ is defined as the linear subspace of $H^1(\Omega)$ of functions with zero trace on $\partial\Omega_1$.

REMARK 2.1. From the advection term property,

$$\int_\Omega \mathbf{w} \cdot \nabla v_0 \, v_0 \, d\Omega \geq 0, \quad \forall \, v_0 \in K_0, \quad (4)$$

the classical results on existence and uniqueness of solutions for elliptic variational problems hold in this case (cf., e.g., [3]); i.e., there exists a unique solution to problem (2). Notice that due to the non potentiality of the problem, there is no characterization in terms of an optimization problem, which is not essential for deriving numerical schemes. □

Another type of variational formulations of problem (1), even with nonlinear conditions and constraints in the interior and on the boundary, can be constructed by applying subdifferential ideas [4]. (Cf. [5] for the case of control problems.)

**3. Standard internal approximations.** In the context of abstract variational approximations [6-8], an *internal approximation* of problem (2) consists in defining a family of finite dimensional subspaces $V_h$, $h > 0$, of $H^1(\Omega)$, with dimension $m_h \longrightarrow \infty$ as $h \longrightarrow 0$, and a corresponding family $K_{\bar{u}_h} \subset V_h$ of convex subsets that approximates $K_{\bar{u}}$. In general $K_{\bar{u}_h}$ is not a subset of $K_{\bar{u}}$. In this manner, the natural discrete problem associated to (2) is defined by:

Find $u_h \in K_{\bar{u}_h}$ such that

$$\varepsilon \int_\Omega K\nabla u_h \cdot \nabla v_{h0} \, d\Omega + \int_\Omega w \cdot \nabla u_h \, v_{h0} \, d\Omega$$
$$= \int_\Omega f \, v_{h0} \, d\Omega + \varepsilon \int_{\partial\Omega_2} g \, \gamma v_{h0} \, d\partial\Omega, \quad \forall \, v_{h0} \in K_{0_h}, \quad (5)$$

and is called its *abstract internal approximation*.

Analogously, because of property (4), problem (5) possesses a unique solution and, under the sufficient conditions of consistency and convergence for the family $\{K_{\bar{u}_h}\}_{h>0}$ (cf. [7, 8]) problem (5) defines a sequence converging in $H^1(\Omega)$ to the solution of (2); i.e.,

$$\lim_{h \longrightarrow 0} \|u_h - u\|_{H^1(\Omega)} = 0. \quad (6)$$

As is well known, even convergence result (6) holds, the presence of the advection term conditions the capability of this standard discrete scheme to approximate pointwisely the exact solution of the problem as $\varepsilon \longrightarrow 0$. In order to overcome such a deficiency, upwind techniques, among others, have proved to be very effective.

**4. The heuristic of upwinding: Symmetrization.** We next want to determine how to handle the advection term in order to obtain adequate numerical behavior like in the pure diffusion case. Thus, the following question arises: is there a characterization of the advection-diffusion problem as a pure diffusion problem?

One answer to the preceeding question can be obtained looking at a diffusion problem with a formal elliptic operator of the form:

$$-\text{div}\{\lambda K \nabla u\} \equiv -\lambda \text{ div}\{K \nabla u\} - K^T \nabla \lambda \cdot \nabla u. \tag{7}$$

Comparing this with the original differential operator

$$-\varepsilon \text{ div}\{K \nabla u\} + w \cdot \nabla u, \tag{8}$$

it is clear that the following result holds:

THEOREM 4.1. The advection-diffusion problem (1) is characterized by the pure diffusion problem:

$$\begin{aligned} -\varepsilon \text{ div}\{\lambda K \nabla u\} &= \lambda f, & \text{in } \Omega, \\ u &= \bar{u}, & \text{on } \partial\Omega_1, \\ \varepsilon \lambda K \nabla u \cdot n &= \varepsilon \lambda g, & \text{on } \partial\Omega_2, \end{aligned} \tag{9}$$

whenever the function $\lambda$ satisfies the differential equation

$$\varepsilon K^T \nabla \lambda + \lambda w = 0, \quad \text{in } \Omega. \tag{10}$$

That is, under condition (10), any solution of problem (1) is also a solution of problem (9), and viceversa. □

In the case of constant velocity field, $w$, homogeneous equation (10) has the general solution

$$\lambda(x) = C \exp(-K^{-T} w \cdot x), \quad x \in \Omega. \tag{11}$$

Therefore, once a solution of (10) has been determined, we can utilize problem (9) to obtain the solution of the original problem. The corresponding primal variational formulation of such a symmetric equivalent problem is (cf. (2)):

Find $u \in K_{\bar{u}}$ such that

$$\varepsilon \int_\Omega \lambda K \nabla u \cdot \nabla v_0 \, d\Omega = \int_\Omega \lambda f \, v_0 \, d\Omega + \varepsilon \int_{\partial\Omega_2} \lambda g \, \gamma v_0 \, d\partial\Omega, \quad \forall v_0 \in K_0, \tag{12}$$

with abstract internal approximation (cf. (5))

Find $u_h \in K_{\bar{u}_h}$ such that

$$\varepsilon \int_\Omega \lambda K \nabla u_h \cdot \nabla v_{h0} \, d\Omega = \int_\Omega \lambda f \, v_{h0} \, d\Omega + \varepsilon \int_{\partial\Omega_2} \lambda g \, \gamma v_{h0} \, d\partial\Omega,$$

$$\forall v_{h0} \in K_{0_h}. \tag{13}$$

This completes a process of symmetrization of the original non symmetric problem, in accordance with Theorem 4.1. Thus, we shall refer to problem (13) as the *abstract symmetric internal approximation* of problem (2).

Now we are in a position to interpret the above symmetrization process as an upwind technique. Indeed, first we observe that the function $\lambda$ appears in variational problem (12) as an integrating factor and, in the case of constant flow, it is a function exponentially increasing along the upstream direction. In this sense we can regard internal approximation (13) as an upwind numerical scheme.

A second interpretation arises by observing that, under condition (10),

$$\varepsilon\, \lambda K \nabla u \cdot \nabla v_0 = \varepsilon\, K \nabla u \cdot \nabla (\lambda v_0) + \mathbf{w} \cdot \nabla u\, (\lambda v_0), \qquad (14)$$

and, consequently, primal symmetric variational formulation (12) can be rewritten in the form:

$$\varepsilon \int_\Omega K \nabla u \cdot \nabla (\lambda v_0)\, d\Omega + \int_\Omega \mathbf{w} \cdot \nabla u\, (\lambda v_0)\, d\Omega =$$

$$\int_\Omega f\, (\lambda v_0)\, d\Omega + \varepsilon \int_{\partial \Omega_2} g\, \gamma(\lambda v_0)\, d\partial\Omega, \qquad \forall\, v_0 \in K_0. \qquad (15)$$

Similarly, abstract internal approximation (13) takes the form:

$$\varepsilon \int_\Omega K \nabla u_h \cdot \nabla (\lambda v_{h0})\, d\Omega + \int_\Omega \mathbf{w} \cdot \nabla u_h\, (\lambda v_{h0})\, d\Omega =$$

$$\int_\Omega f\, (\lambda v_{h0})\, d\Omega + \varepsilon \int_{\partial \Omega_2} g\, \gamma(\lambda v_{h0})\, d\partial\Omega, \qquad \forall\, v_{h0} \in K_{0_h}. \qquad (16)$$

That is, variational problems (12) and (15) (resp. (13) and (16)) are equivalent to each other under condition (10) for $\lambda$. Thus, the integrating factor $\lambda$ has in (15) and (16) the interpretation of a weighting function factor in "upstream" direction.

Therefore, we have arrived at the heuristic to upwinding that establishes the orientation of the present work. In the next section, we shall present Petrov-Galerkin techniques for upwind approximations. Notice that (15) (resp. (16)) is precisely a formulation of Petrov-Galerkin type.

The motivation of the above ideas was obtained from the article of Barrett and Morton [9], in which simmetrization of one dimensional problems is discussed (cf. also [10]).

**5. Upwind Petrov-Galerkin numerical schemes.** In general, an *abstract internal Petrov-Galerkin approximation* to problem (2) can be defined as follows: Let $V_h$ and $W_h$, $h > 0$, be two families of finite dimensional subspaces of $H^1(\Omega)$, both with dimension $m_h \longrightarrow \infty$ as $h \longrightarrow 0$, and let $K_{\underline{u}_h} \subset V_h$ and $C_{0_h} \subset W_h$ be corresponding families of convex subsets that approximate $K_{\underline{u}}$ and $K_0$, respectively. Then, in analogy to (5) we associate to (2) the discrete problem

Find $u_h \in K_{\bar{u}_h}$ such that

$$\varepsilon \int_\Omega K\nabla u_h \cdot \nabla v_{h0} \, d\Omega + \int_\Omega w \cdot \nabla u_h \, v_{h0} \, d\Omega$$

$$= \int_\Omega f \, v_{h0} \, d\Omega + \varepsilon \int_{\partial\Omega_2} g \, \gamma v_{h0} \, d\partial\Omega, \quad \forall \, v_{h0} \in C_{0_h}. \quad (17)$$

Motivated by interpretation (16) of discrete problem (13) and definition (17), we extend Theorem 4.1 to the following

*THEOREM 5.1.* Let $\lambda$ be any fixed sufficiently regular scalar function defined in $\Omega$. Then the advection-diffusion problem (1) is characterized by the problem:

$$-\varepsilon \, \text{div}\{\lambda K \nabla u\} + w^\lambda \cdot \nabla u = \lambda f, \quad \text{in } \Omega,$$

$$u = \bar{u}, \quad \text{on } \partial\Omega_1, \quad (18)$$

$$\varepsilon \, \lambda K \nabla u \cdot n = \varepsilon \, \lambda g, \quad \text{on } \partial\Omega_2,$$

where the velocity vector is given by

$$w^\lambda = \varepsilon \, K^T \nabla \lambda + \lambda w. \quad (19)$$

That is, any solution of problem (1) is also a solution of problem (18), and viceversa.

*PROOF.* Multiplying the advection-diffusion operator of (1) by $\lambda$, it follows that

$$-\lambda \, \varepsilon \, \text{div}\{K\nabla u\} + \lambda \, w \cdot \nabla u = -\varepsilon \, \text{div}\{\lambda K \nabla u\} + \{\varepsilon \, K^T \nabla \lambda + \lambda w\} \cdot \nabla u, \quad (20)$$

from which the validity of the assertion is evident. □

Hence, on the basis of characterization (18), a variational formulation of problem (1), equivalent to (2), is:

Find $u \in K_{\bar{u}}$ such that

$$\varepsilon \int_\Omega \lambda K \nabla u \cdot \nabla v_0 \, d\Omega + \int_\Omega w^\lambda \cdot \nabla u \, v_0 \, d\Omega = \int_\Omega \lambda f \, v_0 \, d\Omega + \varepsilon \int_{\partial\Omega_2} \lambda g \, \gamma v_0 \, d\partial\Omega,$$

$$\forall \, v_0 \in K_0. \quad (21)$$

*REMARK 5.1.* Notice that variational formulation (21) reduces to formulation (12) in case the modified velocity vector $w^\lambda$ is zero. This is precisely condition (10) imposed in Theorem 4.1. □

Furthermore, introduction of definition (19) into (21) leads us to the alternative form

$$\varepsilon \int_{\Omega} K\nabla u \cdot \nabla(\lambda v_0) \, d\Omega + \int_{\Omega} \mathbf{w} \cdot \nabla u \, (\lambda v_0) \, d\Omega$$

$$= \int_{\Omega} f \, (\lambda v_0) \, d\Omega + \varepsilon \int_{\partial\Omega_2} g \, \gamma(\lambda v_0) \, d\partial\Omega, \quad \forall \, v_0 \in K_0, \qquad (22)$$

which is similar to the form of (17). Thus, it is natural to refer to formulation (22) or, equivalently, (21) as the $\lambda$ *Petrov-Galerkin variational formulation* of problem (1).

Hence, we can associate to the $\lambda$ Petrov-Galerkin variational formulation (22) the discrete problem:

Find $u_h \in K_{u_h}$ such that

$$\varepsilon \int_{\Omega} K\nabla u_h \cdot \nabla(\lambda v)_{h0} \, d\Omega + \int_{\Omega} \mathbf{w} \cdot \nabla u_h \, (\lambda v)_{h0} \, d\Omega$$

$$= \int_{\Omega} f \, (\lambda v)_{h0} \, d\Omega + \varepsilon \int_{\partial\Omega_2} g \, \gamma(\lambda v)_{h0} \, d\partial\Omega, \quad \forall \, (\lambda v)_{h0} \in C_{0_h}, \qquad (23)$$

where the notation of problem (17) is in force. This is a specialized abstract internal Petrov-Galerkin approximation of our original problem.

According to the heuristic presented in the previous section, discrete problem (23), associated to (22), corresponds to an upwind numerical scheme whenever the parameter $\lambda$ is defined as an upstream weighting factor. For other approaches to the $\lambda$ Petrov-Galerkin formulations see [11-13].

**6. Direct upwind approximation of the advection term.** Analyzing the structure of the matrices of the discrete schemes (5), (13) and (23), we can observe the following. In the standard scheme (5) the advection skewsymmetric term dominates the diffusion symmetric term as $\varepsilon$ decreases, which explains the presence of the well known numerical spurious oscillations (see next section). Clearly this undesired behavior is eliminated via symmetrization, as in the case of numerical scheme (13), but the presence of exponential factors introduces numerical complications. The $\lambda$ Petrov-Galerkin scheme (23), a generalization of the symmetrization technique, offers an alternative to preclude these difficulties [13]. However the computational cost of these formulations is high, since it affects both the advection and diffusion terms, as well as the forced term.

Another approach for generating upwind numerical schemes has been proposed by Tabata [1] and Bristeau et al [2] (cf. also [8] and [14]), which essentially consists in approximating the advection term by means of an upwind interpolation of the directional derivative $|\mathbf{w}| \partial u/\partial w = \mathbf{w} \cdot \nabla u$ along streamlines.

Consider the standard abstract internal approximation of problem (2), discussed in Section 3:

Find $u_h \in K_{u_h^-}$ such that

$$\varepsilon \int_\Omega K\nabla u_h \cdot \nabla v_{h0}\, d\Omega + \int_\Omega \mathbf{w}\cdot\nabla u_h\, v_{h0}\, d\Omega$$
$$= \int_\Omega f\, v_{h0}\, d\Omega + \varepsilon \int_{\partial\Omega_2} g\, \gamma v_{h0}\, d\partial\Omega, \qquad \forall\, v_{h0} \in K_{0_h}. \qquad (24)$$

The purpose of the direct method is, unlike the previous ones, to modify only and in a direct form the advection term. In this manner the diffusion symmetric term and forced term remain unaltered, which is of great advantage. In general, the advective term of (24) is approximated in the following sense:

$$\int_\Omega \mathbf{w}\cdot\nabla u_h\, v_{h0}\, d\Omega \approx \left( |\mathbf{w}| \frac{\partial_h u_h}{\partial \mathbf{w}},\, v_{h_0} \right)_h, \qquad (25)$$

where $(\cdot,\cdot)_h$ denotes an approximate $L^2$-inner product, as a quadrature scheme, and $\partial_h u_h/\partial \mathbf{w}$ is an upwind interpolation of $\partial u/\partial \mathbf{w}$ along streamlines. Upwind interpolations of first order were worked out by Tabata [1] and of second order by Bristeau et al [2, 8]. In the next section we shall apply the method with an upwind interpolation of third order.

**7. A numerical example.** In order to exemplify and compare the standard and direct upwind numerical schemes, we consider the following particular case of model problem (1):

$$-\varepsilon \Delta u + \mathbf{w}\cdot\nabla u = 1, \qquad \text{in } \Omega \subset \mathbb{R}^2,$$
$$u = 0, \qquad \text{on } \partial\Omega, \qquad (26)$$

where $\mathbf{w} = (1, 0)$, and the geometry of the domain $\bar{\Omega}$ is that of Fig. 1. Notice that in this case the corresponding local Peclet number is given by

$$Pe_h = \frac{h}{\varepsilon}. \qquad (27)$$

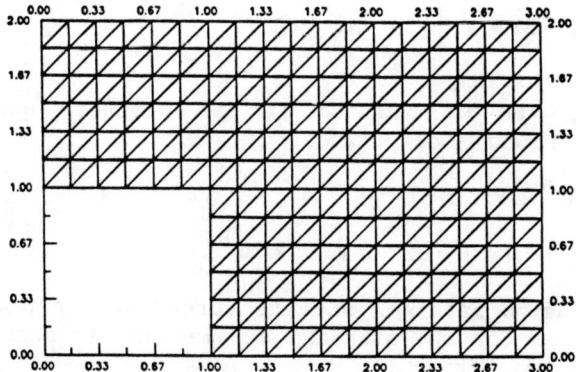

**Fig. 1.** The geometry of the domain and its mesh.

REMARK 7.1. This example has been numerically simulated by several authors in the experimentation of upwind techniques (cf., e.g., [8] and [14]). In the limit, as the parameter $\varepsilon$ tends to zero, the exact solution to (26) becomes :

$$u(x,y) = x, \qquad (x,y) \in (0,3) \times (1,2),$$
$$u(x,y) = x - 1, \qquad (x,y) \in (1,3) \times (0,1). \tag{28}$$

Observe that this solution is discontinuous along the internal line $(1,3) \times \{1\}$, and presents the boundary layer phenomenon along the boundary $\{3\} \times (0,2)$. Because of such characteristics, the particular flow direction $\mathbf{w} = (1,0)$ becomes a critical one, a fact that makes its choice interesting. □

First we proceed to deduce the $\mathbb{R}^{m_h}$ characterizations of discrete problems (5) and (24), (25) in terms of a general basis of $V_h$, say,

$$V_h = \text{span}\{\varphi_1, \varphi_2, \ldots, \varphi_{m_h}\}. \tag{29}$$

Thus, expressing $u_h$ and $v_{h0}$ in terms of their coordinate vectors $\underline{\alpha}$ and $\underline{\beta}_0 \in \mathbb{R}^{m_h}$, we obtain for the standard discrete problem (5)

Find $\underline{\alpha} \in K^{m_h}$ such that:

$$\underline{\beta}_0 \cdot \{\varepsilon \, A\underline{\alpha} + B\underline{\alpha}\} = \underline{\beta}_0 \cdot \mathbf{b}, \qquad \forall \, \underline{\beta}_0 \in K_0^{m_h}, \tag{30}$$

where $K^{m_h}$ (resp. $K_0^{m_h}$) is the $\mathbb{R}^{m_h}$ version of $K_{u_h}$ (resp. $K_{0_h}$) and, for $i, j = 1, \ldots, m_h$,

$$A_{ij} = \int_\Omega K \nabla \varphi_j \cdot \nabla \varphi_i \, d\Omega, \tag{31}$$

$$B_{ij} = \int_\Omega \mathbf{w} \cdot \nabla \varphi_j \, \varphi_i \, d\Omega \tag{32}$$

and

$$b_i = \int_\Omega f \, \varphi_i \, d\Omega + \varepsilon \int_{\partial \Omega_2} g \, \gamma \varphi_i \, d\partial\Omega. \tag{33}$$

Similarly, the $\mathbb{R}^{m_h}$ version of the discrete upwind numerical scheme (24), (25) turns out to be:

Find $\underline{\alpha} \in K^{m_h}$ such that

$$\underline{\beta}_0 \cdot \{\varepsilon \, A\underline{\alpha} + \tilde{B}\underline{\alpha}\} = \underline{\beta}_0 \cdot \mathbf{b}, \qquad \forall \, \underline{\beta}_0 \in K_0^{m_h}, \tag{34}$$

where the diffusion matrix $A$ and the forced vector $\mathbf{b}$ are defined by (31) and (33) -as for the standard scheme- and in accordance to (25) the matrix $\tilde{B}$ is given by

$$\tilde{B}_{ij} = \left( \frac{\partial_h \phi_j}{\partial w}, \phi_i \right)_h. \tag{35}$$

This is an upwind approximation of the advective matrix (32) along streamlines.

These are the concrete forms of the abstract discrete problems (5) and (24), (25), which become finite element numerical schemes once the basis functions $\{\varphi_1, \varphi_2, \ldots, \varphi_{m_h}\}$ are identified with global interpolating finite element functions [15, 16].

Here, we utilize lagrangian triangular finite elements of type (1) [15], with a uniform partition of the domain (see Fig. 1),

$$\bar{\Omega} = \bigcup_{\alpha=1}^{N} E_\alpha. \tag{36}$$

Consequently, the finite dimensional subspace $V_h$, (29), is characterized by

$$V_h = \{v_h \in C^0(\bar{\Omega}): v_h|_{E_\alpha} \in P_{\leq 1}(E_\alpha), \alpha = 1,\ldots,N\}, \tag{37}$$

where $P_{\leq 1}(E_\alpha)$ is the space of polynomials of degree $\leq 1$ over the real triangular finite element $E_\alpha$.

In this case, the proper approximate $L^2$-inner product for defining (35) is given by Simpson's quadrature scheme,

$$\tilde{B}_{ij} = \frac{1}{3} \sum_{\alpha=1}^{N} \operatorname{meas}(E_\alpha) \sum_{r=1}^{3} \frac{\partial_h \phi_j}{\partial w}(a_r^\alpha) \phi_i(a_r^\alpha), \tag{38}$$

where $a_r^\alpha$, $1 \leq r \leq 3$, denote the vertices of the triangular element $E_\alpha$.

The upwind interpolated directional derivative $\partial_h/\partial w$, at a given node, can in general be defined as the value of the derivative at such a node of a polynomial, which coincides upstreamly at specific points with the field at hand. For instance, in the case of an interpolation of third order, the use of a polynomial of third degree requieres the choice of three additional points, located along the streamline passing through the node, to determine the corresponding coefficients. For a technique to choose the additional upwind interpolation points see Glowinski [8].

We present some numerical results in Figs. 2 -5, where we have utilized an upstream approximation $\partial_h \cdot/\partial w$ of the directional derivative operator $\partial \cdot/\partial w$, third order accurate. That is, for a function v, $\partial_h v/\partial w$ is defined at a vertex $a_\alpha^r$ by the corresponding value of the polynomial of third order that interpolates v, at $a_\alpha^r$ and at three points upstreamly located on the streamline. For first and second order approximations see Tabata [1] and Glowinski [8], respectively.

As mentioned beforehand, the standard scheme (30) ((5)) degenerates as local Peclet number (27) increases. In Figs. 2 and 3 this phenomenon is shown. In particular, from values of $\varepsilon = 0.8$ ($Pe_h = 0.2083$) and higher, accurate results are obtained. However, in the interval $\varepsilon = 0.5$ ($Pe_h = 0.3333$) to $\varepsilon = 0.1$ ($Pe_h = 1.6666$) the behavior of the numerical solution is not so accurate. Furthermore, from $\varepsilon = 0.08$ ($Pe_h = 2.0833$) spurious oscillations become apparent and the standard scheme is no longer useful (see Fig. 3).

On the other hand, as is well known, the upwind numerical schemes of Petrov-Galerkin type have shown some improvement for treating advection dominated phenomena. However, in the case of exponential weighting factors as suggested by the symmetrization process, the matrices turn out to be numerically ill-conditioned. Other proposed schemes that replace the exponential factors by Padé approximations or

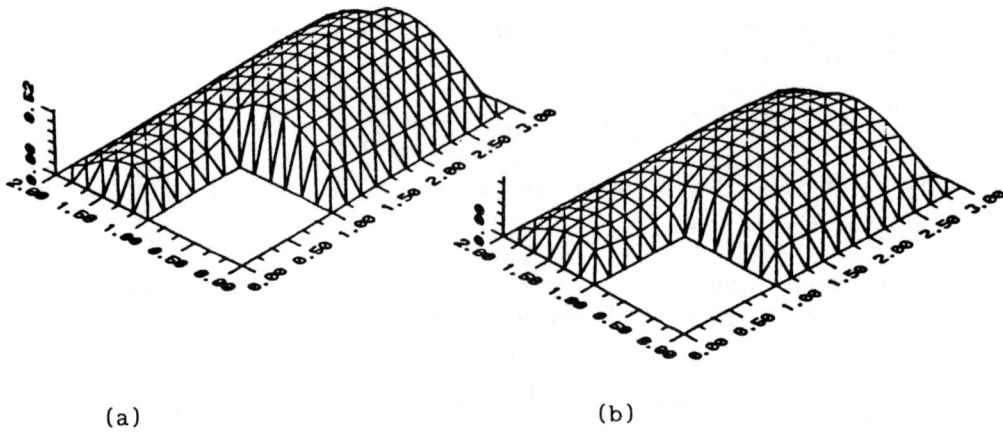

(a)          (b)

**Fig. 2.** Standard (a) and upwind (b) approximations for $\varepsilon = 0.8$ ($Pe_h = 0.2083$).

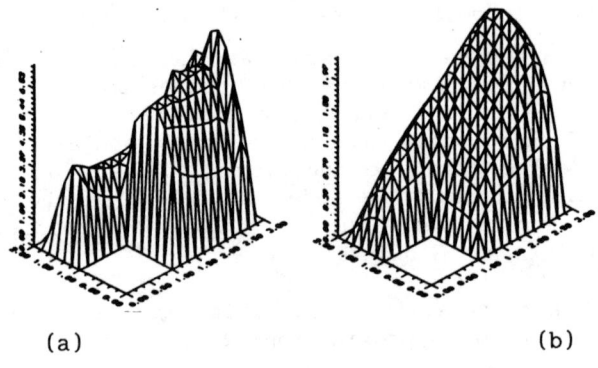

(a)          (b)

**Fig. 3.** Standard (a) and upwind (b) approximations for $\varepsilon = 0.05$ ($Pe_h = 3.3333$).

some other rational and polynomial approximations, have the general inconvenience of changing all the original numerical scheme, leading to a rigidization of the system (cf., e.g., [9], [13]).

In contrast, the use of the direct upwind approximation scheme (34) ((24), (25)) renders very accurate results, even in case of problems with nonlinearities, as well as interior and boundary constraints [5].

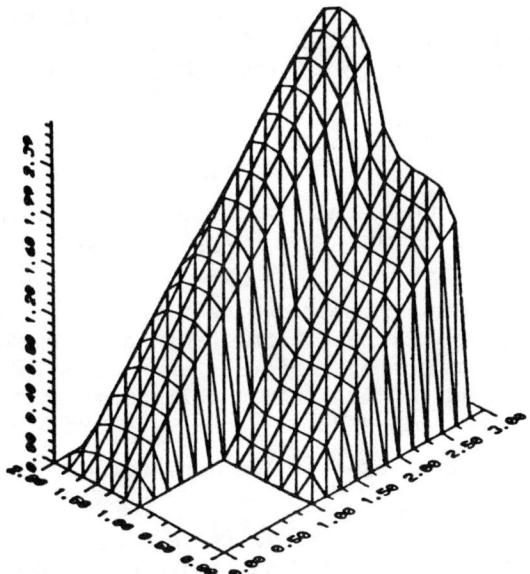

**Fig. 4.** Upwind approximation for $\varepsilon = 0.005$ ($Pe_h = 33.3333$).

In Figs. 4 and 5 where $\varepsilon$ has been taken as equal to 0.005 ($Pe_h = 33.3333$) and as small as 0.000001 ($Pe_h = 166,666.6667$), one can see the high accuracy of the method. It is worth mentioning that the computational effort is similar to the one of the standard scheme.

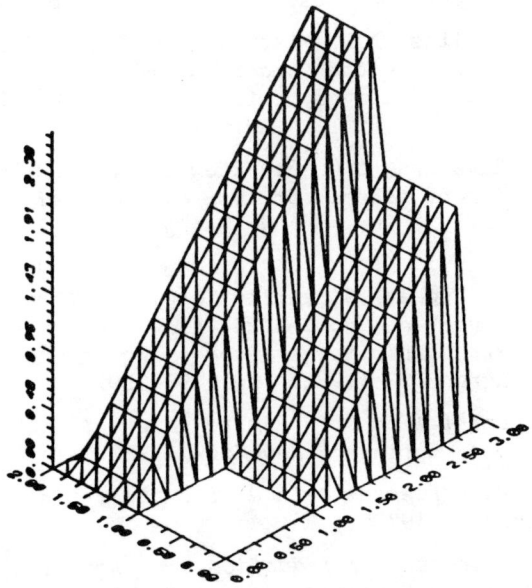

**Fig. 5.** Upwind approximation for $\varepsilon = 0.000001$ ($Pe_h = 166,666.67$).

## REFERENCES

[1] M. Tabata, *A finite element approximation corresponding to the upwind differencing*, Memoirs Num. Math., 1 (1977), pp. 47-63.

[2] M.O. Bristeau, R. Glowinski, J. Periaux, P. Perrier and O. Pironneau, *On the numerical solution of non linear problems in fluid dynamics by least squares and finite element methods*, Comput. Meth. Appl. Mech. Engrg., 17/18 (1979), pp. 619-657.

[3] R.E. Showalter, *Hilbert Space Methods for Partial Differential Equations*, Pitman, London, 1977.

[4] G. Alduncin, *Subdifferential and variational formulations of boundary value problems*, Comput. Meths. Appl. Mech. Engrg., 72 (1989), pp. 173-186.

[5] G. Alduncin and J. Carrera, *A control advection-diffusion problem and upwind finite element approximations*, in Applied Mechanics: Proceedings of the International Conferece on Applied Mechanics, Z. Zhemin, ed., International Academic Publishers and Pergamon Press, Oxford, 1989, pp. 1619-1624.

[6] R. Temam, *Analyse numérique*, Presses Universitaires de France, Paris, 1970.

[7] R. Glowinski, J.L. Lions and R. Trémolièrs, *Numerical Analysis of Variational Inequalities*, North-Holland, Amsterdam, 1981.

[8] R. Glowinski, *Numerical Methods for Nonlinear Variational Problems*, Springer, New York, 1984.

[9] J.W. Barrett and K.W. Morton, *Optimal finite element solutions to diffusion-convection problems in one dimension*, Int. J. Num. Meths. Engrg., 15 (1980), 1457-1474.

[10] T.J.R. Hughes and J.D. Atkinson, *A variational basis for "upwind" finite elements*, in Variational Methods in Mechanics of Solids, S. Nemat-Nasser, ed., Pergamon Press, Oxford, 1980, pp. 387-391.

[11] I. Christie, D.F. Griffiths, A.R. Mitchell and O.C. Zienkiewicz, *Finite element methods for second order differential equations with significant first derivatives*, Int. J. Num. Meths. Engrg. 10 (1976), pp. 1389-1396.

[12] J.C. Heinrich, P.S. Huyakorn, O.C. Zienkiewicz and A.R. Mitchell, *An "upwind" finite element scheme for two- dimensional convective transport equation*, Int. J. Num. Meths. Engrg., 11 (1977), pp. 131- 143.

[13] A.N. Brooks and T.J.R. Hughes, *Streamline upwind/Petrov- Galerkin formulations for convection dominated flows with particular emphasis on the incompressible Navier-Stokes equations*, Comput. Meths. Appl. Mech. Engrg., 32 (1982), pp. 199-259.

[14] F. Thomasset, *Implementation of Finite Element Methods for Navier-Stokes Equations*, Springer, New York, 1981.

[15] P.G. Ciarlet, *The Finite Element Method for Elliptic Problems*, North-Holland, Amsterdam, 1978.

[16] P.A. Raviart and J.M. Thomas, *Introduction à l' Analyse Numérique des Équations aux Dérivées Partielles*, Masson, Paris, 1983.

# CHAPTER 12

Fast Elliptic Solvers Using Mixed-Hybrid Nodol Finite Elements

J. P. Hennart*
E. Del Valle§

**Abstract.** After giving a general description of nodal methods within the framework of the finite element method (FEM), we introduce a mixed–hybrid implementation which is particularly attractive from the computational point of view.

**1. General Introduction to Nodal Methods** "Nodal" methods appeared in the 1960's in relation with the numerical solution of the neutron diffusion equations, which are the basis of the three–dimensional simulators currently used in numerical reactor calculation. The first nodal simulators exemplified by the code FLARE (Delp et al. (1964)) are based on three stages: the first one is the cell homogenization by which equivalent diffusion parameters are determined for each fuel or control cell. This first stage is typical of the early fine mesh finite difference calculations, which were quite expensive in computer time and memory: a typical light water reactor (LWR) may contain as many as maybe 160 assemblies and 225 fuel and control cells per assembly, meaning that 36,000 mesh points are needed per energy group and per axial mesh plane for the core alone. Nodal simulators are based on a second stage homogenization, whereby fuel and control cells are grouped together into an assembly (or in fact the quarter part of it) with equivalent diffusion parameters: these relatively large homogeneous regions are the "nodes" leading to the third stage, i.e. a coarse mesh calculation. As "node" is normally used in the finite element context for a particular point of an element to which degrees of freedom (values and/or derivatives) are attached, it is clear that the selection of such a word is not particularly relevant. In the following, we shall therefore prefer to use "cell" instead of "node".

---

* Departamento de Métodos Matemáticos y Numéricos, Instituto de Investigaciones en Matemáticas Aplicadas y en Sistemas de la UNAM, Apartado Postal 20–726, 01000 México, D.F. (MEXICO)

§ Departamento de Ingeniería Nuclear, Escuela Superior de Física y Matemáticas del IPN, Unidad Profesional "Adolfo López Mateos", 07738 México, D.F. (MEXICO)

The first nodal simulators (Gupta (1981)) following FLARE involved parameters adjustable to actual operating data or the results of more accurate calculations in less dimensions. In a sense, they were not "consistent" as the approximation to the solution of the basic equations would not tend to it in the limit of small meshes (when the size $h$ of the cells goes to zero). After 1975, "consistent" nodal schemes were developed and they are currently used in modern nuclear reactor calculation (Lawrence (1986)).

Over each large homogeneous cell, a set of parameters is chosen, in general cell and/or edge moments, which directly provide some average information about the solution of the basic equations. After the two stage homogenization, the third stage coarse mesh calculation is sometimes followed by a fourth stage "dehomogenization", whereby fine details of the solution are looked for within a particular large cell. The first two homogenization stages are covered by a recent review paper (Smith (1986)).

These first comments may give the impression that nodal methods have a very restricted range of applications. Fortunately, this is not true and there are many more fields of applications where the "coarse mesh" calculations are standard. This is the case in particular when the domain over which the given equations are to be solved and the corresponding coefficients in the equations are not well known. For such situations, the finite difference method is normally used over regular meshes, assuming that over each cell average coefficients are available. Typical nodal situations occur in fluid flow in porous media (underground hydrology, oil reservoir simulation, etc.) as well as in problems related to diffusion and transport of contaminants (in the underground or in the atmosphere). For such situations, the domain of interest $\Omega$ is in principle approximated by the union of fairly large rectangular homogeneous cells $\Omega_e, e = 1, \ldots, E$. In typical geophysical applications, these cells may have one or two kilometers in each direction. If over some part of the coarse mesh, more accuracy is needed because the solution is expected to present more variation on that part, local mesh refinements have been developed (Ewing (1986)) which avoid solving the problem over a completely fine grid.

In the engineering literature, there are two basic variations around the nodal method: the "coarse–mesh" methods developed by Langenbuch et al. (1977a and b) and the "nodal expansion" methods proposed by Finnemann et al. (1977). In the coarse–mesh methods, over each cell there is an explicit representation of the approximation in terms of its values at the centers of gravity of the cell and of its edges. In the nodal expansion method (NEM), no such explicit representation exists and only average cell and edge values appear in the final equations. In a recent paper (Hennart et al. (1988)), we have shown that such a distinction is in fact quite artificial. Given any distribution of cell and edge moments, it is possible to find a space of polynomials (or even "quasipolynomials" (Hennart (1988)) ) interpolating these moments. To be complete and in relation with the "quasipolynomials" we just mentioned , we should also briefly speak of the "analytical" nodal methods which after "transverse integration" of the original partial differential equations (PDE's) over all the variables minus one, solve the resulting equations in an "analytical" way, using fundamental as well as particular solutions for the corresponding one-dimensional differential equations (Shober et al. (1977), Greenman et al. (1979), and Azmy and Dorning (1983)): these methods will not be considered here as they have been analyzed in a recent paper (Hennart (1988)). Finally, in Lawrence and Dorning (1980), analytical nodal methods are also developed in an integral rather than differential way by the introduction of Green's functions.

**2. A Brief Description of the Simplest Nodal Methods** In static as well as dynamic studies, the basic equation is the one-group diffusion equation

$$-\nabla \cdot D\nabla \phi + \Sigma \phi = S, \quad \forall \mathbf{r} \epsilon \Omega, \tag{1a}$$

where in standard notations $D \equiv D(\mathbf{r})$ and $\Sigma \equiv \Sigma(\mathbf{r})$ are the diffusion coefficient and the removal cross-section in the energy group considered, while $S \equiv S(\mathbf{r})$ is some source term. $D$ and $\Sigma$ are piecewise constant, *i.e.* constant per cell. The neutron flux $\phi$ is moreover subject to boundary conditions on $\Gamma = \overline{\Omega} - \Omega$ which we shall choose as follows

$$\phi = 0, \quad \forall \mathbf{r} \epsilon \Gamma_1,$$

and

$$\frac{\partial \phi}{\partial n} = 0, \quad \forall \mathbf{r} \epsilon \Gamma_2,$$

$$\Gamma = \Gamma_1 \cup \Gamma_2, \quad \Gamma_1 \cap \Gamma_2 = \phi. \tag{1b}$$

In Hennart (1985b), the original Langenbuch–Maurer–Werner (LMW) (Langenbuch et al. (1977a and b)) coarse-mesh methods were analyzed and then extended in Hennart (1985a) and (1986) to a general family of nodal schemes, which "climb correctly in order". The LMW nodal schemes actually exhibit $L^2$ convergence of only $O(h^2)$, where $h$ is the maximum diameter of the cells $\Omega_e, e = 1, \ldots, E$, independently of the complexity of the scheme considered. This is due to the following combination of facts: first of all, the approximating polynomial spaces over each $\Omega_e$ only contain $P_1$, the space of polynomials of maximum degree one in all the variables, but never $P_k, k > 1$; moreover, only zeroth order moment continuity of the global approximation is ensured between adjacent cells. Numerical experiments presented in Fédon–Magnaud et al. (1983) confirm these theoretical rates of $O(h^2)$. The schemes presented in Hennart (1985a) and (1986) were designed to exhibit $L^2$ convergences of $O(h^{k+2})$ where $k \in \mathbb{N}$ is the index attached to a particular member of the family. This result should of course be taken with caution, as for the type of problems we are considering, the coefficients in the PDE's are typically constant per cell. In this case, it is well known that singularities arise which limit the asymptotic rates of convergence to $O(h^{2\alpha})$ where $0 < \alpha < 1$ (see Hennart and Mund (1977) and the references it contains). However, far from asymptotic, extensive numerical experiments presented in the same reference have shown that the convergence rates of

$O(h^{k+2})$ coming from the smooth part of the solution, can still be observed for $k = 0$ and 1, but not for $k > 1$. Consequently, there is still a strong interest for considering $O(h^3)$ nodal schemes, since the general philosophy of nodal methods is to remain far from asymptotic with cells as large as possible. Moreover, as we shall see later, the $k = 1$ case for the family of nodal elements developed in Hennart (1985a) and (1986) is particularly attractive from the computational point of view, within the mixed–hybrid implementation we are going to propose in Section 3.

At this point, it is appropriate to briefly describe the NEM as it was proposed by Finnemann et al. (1977). First of all, the neutron equation (1a) in energy group $g$ is written in the so called $P_1$ form, i.e. as a system of first order PDE's like in the mixed formulation for second order PDE's:

$$\mathbf{J} = -D\nabla\phi, \tag{2a}$$

$$\nabla \cdot \mathbf{J} + \Sigma\phi = S. \tag{2b}$$

The first step in the NEM consists in expressing the (neutron) balance over each particular cell $\Omega_e$ namely

$$\int_{\Omega_e} (\nabla \cdot \mathbf{J} + \Sigma\phi)\, d\mathbf{r} = \int_{\Omega_e} S\, d\mathbf{r}, \quad e = 1,\ldots,E,$$

or equivalently

$$\int_{\Gamma_e} \mathbf{J}(\mathbf{r}_e) \cdot \mathbf{1}_e\, d r_e + \Sigma\phi_C V_e = S_C V_e, \quad e = 1,\ldots,E,$$

where $\phi_C$ is the cell averaged flux while $S_C$ is the cell averaged source term in the group considered while $\Gamma_e = \overline{\Omega}_e - \Omega_e$. Over rectangular cells of horizontal (resp. vertical) dimension $h_e$ (resp. $k_e$), this basic balance equation becomes

$$(J_{x_R} - J_{x_L})/h_e + (J_{y_T} - J_{y_B})/k_e + \Sigma\phi_C = S_C, \quad e = 1,\ldots,E, \tag{3}$$

where we used the fact that $V_e = h_e \cdot k_e$. In this equation, $J_{x_L}$ (resp. $J_{x_R}$) is the net horizontal current through the left (resp. right) edge of the cell. In the following, we shall use $J_L$ and $J_R$ (or $J_H$ for the two horizontal components) and similarly $J_B$ (for bottom) and $J_T$ (for top) (or $J_V$ for the two vertical components).

The second step in the NEM consists in integrating transversally the equations (2). If $y_B$ and $y_T$ are the ordinates of the bottom and top edges of a given cell, we get after integration in the vertical direction

$$\frac{d}{dx}J_H(x) + \Sigma\phi_H(x) = S_H(x) - \frac{1}{k_e}L_V(x), \tag{4a}$$

and

$$J_H(x) = -D\frac{d}{dx}\phi_H(x), \tag{4b}$$

where the horizontal fluxes and currents are defined as follows

$$\phi_H(x) = \frac{1}{k_e}\int_{y_B}^{y_T} \phi(x,y)\, dy,$$

$$J_H(x) = \frac{1}{k_e}\int_{y_B}^{y_T} J_x(x,y)\, dy,$$

while $L_V(x)$ is a transversal leakage given by

$$L_V(x) = \frac{1}{k_e}\int_{y_B}^{y_T}\left[\frac{\partial}{\partial y}[-D\frac{\partial \phi}{\partial y}(x,y)]\right]dy = [J_y(x,y_T) - J_y(x,y_B)]/k_e.$$

With (4b), (4a) becomes

$$-\frac{d}{dx}D\frac{d}{dx}\phi_H(x) + \Sigma\phi_H(x) = S_H(x) - \frac{1}{k_e}L_V(x), \tag{5}$$

that is a one–dimensional diffusion equation for a flux averaged in the vertical direction. A transverse integration from $x_L$ to $x_R$ would yield a similar equation for a flux averaged in the horizontal direction. Clearly

$$\phi_C = \frac{1}{h_e}\int_{x_L}^{x_R}\phi_H(x)\, dx = \frac{1}{k_e}\int_{y_B}^{y_T}\phi_V(y)\, dy,$$

$$\frac{1}{h_e}\int_{x_L}^{x_R}L_V(x)\, dx = J_T - J_B,$$

and similarly

$$\frac{1}{k_e}\int_{y_B}^{y_T}L_H(y)\, dy = J_R - J_L.$$

In the simplest case, the transverse integrated fluxes $\phi_H(x)$ and $\phi_V(y)$ are expanded into quadratic polynomials interpolating $\phi_C$, $\phi_L$, and $\phi_R$ for $\phi_H(x)$ and similarly $\phi_C$, $\phi_B$, and $\phi_T$ for $\phi_V(y)$. For instance, over the reference interval $[-1,+1]$, we have

$$\phi_H(x) = \phi_C u_C(x) + \phi_L u_L(x) + \phi_R u_R(x), \tag{6}$$

where

$$\phi_H(x_L) = \phi_L \text{ and } \phi_H(x_R) = \phi_R.$$

At this point, it is convenient to introduce some notation. Let $P_i$ be the normalized Legendre polynomial of degree $i$ over $[-1, +1]$ which satisfies

$$P_i(+1) = 1,$$
$$P_i(-1) = (-1)^i,$$
$$\int_{-1}^{+1} P_i(x) P_j(x)\, dx = N_i \delta_{ij},$$

with $N_i = 2/(2i+1)$. With this notation, it is easy to see that

$$u_C(x) = P_0(x) - P_2(x), \qquad (7a)$$
$$u_L(x) = [P_2(x) - P_1(x)]/2, \qquad (7b)$$
$$u_R(x) = [P_2(x) + P_1(x)]/2. \qquad (7c)$$

We thus have in each direction ($H$ or $V$) three unknowns, for instance in the horizontal direction, $\phi_C$, $\phi_L$, and $\phi_R$. $\phi_C$ is common to both directions and satisfies the balance equation (3). If we use the same average fluxes $\phi_E$ over each edge of the cell on both sides, i.e. within the cell and its neighbor cell, we shall automatically have mean flux continuity. Using (6) in (4b), we shall moreover impose the mean continuity of the net currents through the edges of each cell. These equations are the basic equations of the NEM in its simplest form, which can be solved in a direct or iterative way.

There exist of course supposedly "high-order" versions of the NEM, which try to cope with the following problem: in the original NEM, the transverse leakages are essentially "flat", as they are represented by a mean $J_E$ on each edge. This, as it was shown in Appendix C of Hennart et al. (1988), practically reduces the scheme to second order. A pragmatic (but not easy to formalize) solution to that problem has been in more recent versions of the NEM to use what is called a "quadratic transverse leakage fit" whereby the flat leakage of a given cell is coupled with the flat leakages of its two neighbor cells in any direction to produce a quadratic approximation of the transverse leakage. This seems to work very well, but is hardly analysable.

In summary, the NEM results from the combination of three basic ingredients, which are 1/Cell balance equations, 2/Transverse integrated one-dimensional equations, and finally 3/Quadratic leakage fits.

Most of these features are in fact easily incorporated in the mixed-hybrid implementation we shall propose in the next section with the difference that this particular implementation lends itself very well to truly "high-order" versions.

### 3. An Approach Based on Mixed–Hybrid Nodal Finite Elements

There is a close relationship between the NEM as presented in the preceding section and the approach based on the use of nodal finite elements in a mixed–hybrid formulation. By "nodal" finite elements, we mean finite elements where the basic unknowns or "functionals" of $\phi_h$, the approximation of $\phi$, are in fact cell and edge moments, instead of values and/or derivatives of $\phi_h$ at specified points as with the classical finite element method (FEM). As with the NEM, the basic equation is written as a system of first order PDE'S, Eqs. (2), where the first equation is usually written as

$$\frac{1}{D}\mathbf{J} + \nabla\phi = \mathbf{0}, \tag{2a}$$

because in highly heterogeneous media where $D$ varies from cell to cell it is known that an harmonic instead of an arithmetic average of $D$ is much better (Bensoussan et al. (1978)). This "strong" form of the basic equations is usually replaced by a "weak" form after multiplying the vectorial equation (2a) by a test vector $\mathbf{K}$ and the scalar equation (2b) by a test scalar $\psi$. The resulting equations are integrated over $\Omega$ and in particular the term containing $\nabla\phi$ in the first equation is integrated by parts: The resulting surface integral vanishes if the boundary condition $\phi = 0$ on $\Gamma_1$ is taken into account and if moreover the boundary condition $\frac{\partial\phi}{\partial n} = 0$ on $\Gamma_2$ is imposed to the normal component of $\mathbf{K}$.

The final "weak" form reads

Find $(\mathbf{J}, \phi) \in (U \times V) \times W$ such that

$$\int_\Omega \frac{\mathbf{J} \cdot \mathbf{K}}{D} d\mathbf{r} - \int_\Omega \phi(\nabla \cdot \mathbf{K}) d\mathbf{r} = 0, \quad \forall \mathbf{K} \in U \times V, \tag{8a}$$

$$-\int_\Omega (\nabla \cdot \mathbf{J})\psi d\mathbf{r} - \int_\Omega \Sigma\phi\psi d\mathbf{r} = -\int_\Omega S\psi d\mathbf{r}, \quad \forall \psi \in W, \tag{8b}$$

where the sign has been changed in the second equation to make the overall result more symmetrical. Here $W \equiv L^2(\Omega)$, while $\mathbf{J}$ must be looked for in a subspace $U \times V$ of a space known as $H(div; \Omega)$:

$$H(div; \Omega) \equiv \{\mathbf{J} | \mathbf{J} \in (L^2(\Omega))^2, \quad \nabla \cdot \mathbf{J} \in L^2(\Omega)\}, \tag{9}$$

so that all the expressions appearing in (8) make sense. The members of $U \times V$ have moreover a null normal component on $\Gamma_2$: this subspace of $H(div, \Omega)$ is usually noted $H_0(div; \Omega)$.

As $\phi$ needs only to be in $L^2(\Omega)$, it can in particular be discontinuous at the interfaces but $\mathbf{J}$ now needs to be in $H(div; \Omega)$ which implies (Raviart and Thomas (1977)) that its normal component to any interface (for instance between two distinct neighbor elements) must be continuous. This "conformity" condition may be relaxed by hybridization, leading to the so-called *mixed-hybrid formulation* where $\mathbf{J}$ is no longer in $H(div; \Omega)$. As a consequence boundary terms appear on each $\Gamma_e = \overline{\Omega}_e - \Omega_e$ and a supplementary equation is needed. Namely the weak form (8) is replaced by

Find $(\mathbf{J}, \phi, \lambda) \in (U \times V) \times W \times L$ such that

$$\int_\Omega \frac{\mathbf{J} \cdot \mathbf{K}}{D} dr - \sum_{e=1}^{E} \left\{ \int_{\Omega_e} \phi (\nabla \cdot \mathbf{K}) dr - \int_{\Gamma_e} \lambda \mathbf{K} \cdot d\mathbf{s} \right\} = 0, \quad \forall \mathbf{K} \in U \times V, \quad (10a)$$

$$-\sum_{e=1}^{E} \int_{\Omega_e} (\nabla \cdot \mathbf{J}) \psi dr - \int_\Omega \Sigma \phi \psi dr = -\int_\Omega S \psi dr, \quad \forall \psi \in W, \quad (10b)$$

$$\sum_{e=1}^{E} \int_{\Gamma_e} \mu \mathbf{J} \cdot d\mathbf{s} = 0, \quad \forall \mu \in L, \quad (10c)$$

where

$$U \times V \equiv \{\mathbf{J} | \mathbf{J} \in (L^2(\Omega))^2, \quad \mathbf{J}|_{\Omega_e} \in H(div; \Omega_e), \quad e = 1, \ldots, E\}, \quad (11a)$$

$$W \equiv \{\phi \in L^2(\Omega)\}, \quad (11b)$$

and

$$L \equiv \{\lambda \in \prod_{e=1}^{E} H^{\frac{1}{2}}(\Gamma_e)\}, \quad (11c)$$

$H^{\frac{1}{2}}(\Gamma_e)$ being the space of the traces over $\Gamma_e$ of functions in $H^1(\Omega_e)$.

These equations are discretized by considering finite dimensional subspaces of $(U \times V), W$, and $L$, namely $(U_h \times V_h), W_h$, and $L_h$. This discrete mixed–hybrid formulation reads

Find $(\mathbf{J}_h, \phi_h, \lambda_h) \in (U_h \times V_h) \times W_h \times L_h$ such that

$$\int_\Omega \frac{\mathbf{J}_h \cdot \mathbf{K}_h}{D} dr - \sum_{e=1}^{E} \left\{ \int_{\Omega_e} \phi_h (\nabla \cdot \mathbf{K}_h) dr - \int_{\Gamma_e} \lambda_h \mathbf{K}_h \cdot d\mathbf{s} \right\} = 0,$$
$$\forall \mathbf{K}_h \in U_h \times V_h, \quad (12a)$$

$$-\sum_{e=1}^{E} \int_{\Omega_e} (\nabla \cdot \mathbf{J}_h) \psi_h dr - \int_\Omega \Sigma \phi_h \psi_h dr = -\int_\Omega S \psi_h dr, \quad \forall \psi_h \in W_h, \quad (12b)$$

and

$$\sum_{e=1}^{E} \int_{\Gamma_e} \mu_h \mathbf{J}_h \cdot d\mathbf{s} = 0, \quad \forall \mu_h \in L_h. \quad (12c)$$

Since $\mathbf{J}_h$ must now be in $H(div; \Omega_e)$ for each element $\Omega_e$, and no longer in $H_0(div; \Omega)$, it can be defined *element by element*. At each interface, the continuity of the normal component is imposed by Eq. (12c) so that finally the $\mathbf{J}_h$ obtained will be identical to the one which would have been obtained by a similar discretization of Eqs. (8). Since $\lambda_h$ is a variable dual to $\mathbf{J}_h \cdot \mathbf{1}_e$, it will provide a valuable information about the trace of $\phi_h$ on each $\Gamma_e$, a fact of which we can take advantage by a postprocessing operation proposed by Arnold and Brezzi (1985).

It is this discrete mixed-hybrid formulation which we have implemented on the computer using different choices of nodal finite elements. These possible choices are fully documented in Hennart and Del Valle (In Preparation), with numerical results given in Del Valle et al. (In Preparation).

We want to close this section by giving a brief description of the lowest order mixed-hybrid discretization to compare it in particular to the NEM of the preceding section. The classical finite dimensional subspaces of $H(div;\Omega)$ and $L^2(\Omega)$ for such a discretization are the Raviart-Thomas spaces in $2D$ (Raviart and Thomas (1977)) and the Nédélec spaces in $3D$ (Nédélec (1980)), or in general the Raviart-Thomas-Nédélec (RTN-$k$) spaces, where $k$ is any nonnegative integer. For each RTN-$k$ space, $\mathbf{J}_h = (J_{hx}, J_{hy}) \in U_h \times V_h$ and $\phi_h \in W_h$, when restricted to the reference cell $[-1,+1] \times [-1,+1]$, are defined by a set of degrees of freedom $D$ (the cell and/or edge moments) and a space of polynomials $S$, with card $D = \dim S$.

Let us introduce the spaces of polynomials of degree $i$ in $x$ and $j$ in $y$

$$\mathcal{Q}_{i,j}(x,y) \equiv \{x^a y^b | 0 \le a \le i, 0 \le b \le j\}, \quad \forall\, i,j \in \mathbb{N}, \qquad (13)$$

with in particular $\mathcal{Q}_i(x,y) \equiv \mathcal{Q}_{i,i}(x,y)$.

For $k = 0$, we have

$$U_h \begin{cases} D \equiv & \{J_L, J_R\} \\ S \equiv & \mathcal{Q}_{1,0} \end{cases}$$

$$V_h \begin{cases} D \equiv & \{J_B, J_T\} \\ S \equiv & \mathcal{Q}_{0,1} \end{cases}$$

$$W_h \begin{cases} D \equiv & \{\phi_C\} \\ S \equiv & \mathcal{Q}_0 \end{cases}$$

Finally, we discretize $L$ into $L_h$

$$L_h \begin{cases} D \equiv \lambda_E;\ E \equiv L, R, B, \text{and } T \\ S \equiv P_0(E);\ E = L, R, B, \text{and } T, \end{cases}$$

where $P_0$ is a constant (a polynomial of degree 0) on each edge. By picking up $\psi_h$ constant on $\Omega_e$, $e = 1, \ldots, E$, and zero elsewhere, the equations (12b) are strictly equivalent to equations (3). The resulting $\mathbf{J}_h$ has normal components continuous through the edges. The $\lambda_E$'s correspond in fact to the mean values of $\phi_h$ along the edges which may be combined (Arnold and Brezzi (1985)) with $\phi_C$ to provide a fully 2D function $\tilde{\phi}_h$ in $\mathcal{Q}_{2,0} \cup \mathcal{Q}_{0,2}$ interpolating $\phi_C$ as well as $\phi_L$, $\phi_R$, $\phi_B$, and $\phi_T$.

This function is in the mean continuous through the edges. On $\Gamma_1$, it will be zero in the mean, while $\mathbf{J}_h \cdot \mathbf{1}_e$ is zero on $\Gamma_2$. This postprocessed solution $\{\mathbf{J}_h, \tilde{\phi}_h\}$ is thus formally equivalent to the solution of the simplest NEM. For a demonstration in a more general situation, the reader is referred to Hennart (1985b).

Before leaving this section, it is interesting to point out that the simplest NEM and the RTN–0 scheme both satisfy neutron balance on each cell $\Omega_e$, as well as (mean) flux and current continuity through the edges, and the boundary conditions on $\Gamma_1$ (in the mean) and $\Gamma_2$. In other words, they follow the physics of the problem as closely as possible, which explains a posteriori their success.

**4. A Higher Order Mixed–Hybrid Nodal Scheme: RTN–1** In this Section, we describe the second scheme of the RTN–k family, namely the RTN–1 scheme, which has been implemented in the computer code MIXQUIC (Del Valle and Hennart (1988, 1989)) developed to solve the group diffusion equations in 2D geometries when the domain $\Omega$ is of the union of rectangles type. This geometry is typical of the cores of Pressurized Water Reactors (PWR) or of Boiling Water Reactors (BWR). The RTN–1 scheme provides (if the solution is smooth enough) a convergence of $O(h^3)$: indeed it ensures the continuity of two moments of $\tilde{\phi}_h$ through the edges and moreover $P_2$ (in fact $P_3$) is included in the approximation space. In the different "higher–order" NEM's which have been proposed only one moment of $\tilde{\phi}_h$ was continuous through the cell edges, which implies that only a convergence of $O(h^2)$ could be achieved (see Appendix C of Hennart et al. (1988)).

In the RTN–1 scheme, we allow higher order moments of the unknown function (say u, i.e. $\phi$ or $\mathbf{J}$) to be used. Over the left, right, bottom, and top edges of the reference cell, the following moments are defined

$$m_L^i(u) = \int_{-1}^{+1} P_i(y) u(-1, y)\, dy / N_i,$$

$$m_R^i(u) = \int_{-1}^{+1} P_i(y) u(+1, y)\, dy / N_i,$$

$$m_B^i(u) = \int_{-1}^{+1} P_i(x) u(x, -1)\, dx / N_i,$$

and

$$m_T^i(u) = \int_{-1}^{+1} P_i(x) u(x, +1)\, dx / N_i, \tag{14}$$

where the normalized Legendre polynomial of degree $i$, $P_i$, has been defined previously, as well as the normalization factor $N_i$. Moreover, over the cell, cell moments are defined as follows

$$m_C^{ij}(u) = \int_{-1}^{+1} \int_{-1}^{+1} P_{ij}(x, y) u(x, y)\, dx dy / N_i \cdot N_j, \tag{15}$$

where $P_{ij}(x,y) \equiv P_i(x)P_j(y)$.

In Section 2, for instance, we used with this notation

$$\phi_C = m_C^{00}(\phi_h(x,y)),$$
$$\lambda_E = m_E^0(\phi_h(x,y)), \; E = L, \; R, \; B, \; \text{or} \; T,$$

and

$$J_E = m_E^0(\mathbf{J}_h(x,y) \cdot \mathbf{1}_e), \; E = L, \; R, \; B, \; \text{or} \; T.$$

Here with RTN-1, $U_h, V_h, W_h$, and $L_h$ are described as follows

$$U_h \begin{cases} D \equiv & \{m_H^i(J_{hx}), m_C^{0i}(J_{hx}), \; i = 0,1; \; H = L \text{ and } R\}, \\ S \equiv & \mathcal{Q}_{21}, \end{cases}$$

$$V_h \begin{cases} D \equiv & \{m_V^i(J_{hy}), m_C^{i0}(J_{hy}), \; i = 0,1; \; V = B \text{ and } T\}, \\ S \equiv & \mathcal{Q}_{12}, \end{cases}$$

$$W_h \begin{cases} D \equiv & \{m_C^{ij}(\phi_h), \; i,j = 0,1\}, \\ S \equiv & \mathcal{Q}_1, \end{cases}$$

$$L_h \begin{cases} D \equiv & m_E^i(\lambda_h), \; i = 0,1; \; E = L, \; R, \; B, \text{ and } T, \\ S \equiv & P_1(E), \; E = L, \; R, \; B, \text{ and } T. \end{cases}$$

The resulting algebraic system to be solved is the following

$$AJ + B\phi + C\lambda = 0, \tag{16a}$$
$$B^t J + D\phi = S, \tag{16b}$$
$$C^t J = 0. \tag{16c}$$

In these equations, $J$, $\phi$, and $\lambda$ are vectors of unknowns. Per cell, there are twelve such unknowns for $J$ and four for $\phi$. On each edge, there are moreover two unknowns for $\lambda$.

If one eliminates $J$ in terms of $\phi$ and $\lambda$ and then $\phi$ in terms of $\lambda$ namely

$$J = -A^{-1}[B\phi + C\lambda], \tag{17a}$$
$$\phi = [D - B^t A^{-1} B]^{-1}[S + B^t A^{-1} C\lambda], \tag{17b}$$

the resulting algebraic system for $\lambda$ becomes

$$\left\{ -C^t A^{-1} C - C^t A^{-1} B [D - B^t A^{-1} B]^{-1} B^t A^{-1} C \right\} \lambda =$$
$$C^t A^{-1} B [D - B^t A^{-1} B]^{-1} S. \tag{18}$$

If we partition $\lambda$ into

$$\begin{pmatrix} m_H^0 \\ m_H^1 \\ m_V^0 \\ m_V^1 \end{pmatrix},$$

this induces the corresponding partition for the matrix on the lefthandside of (18)

$$\begin{pmatrix} M_{HH}^{00} & 0 & M_{HV}^{00} & 0 \\ 0 & M_{HH}^{11} & 0 & M_{HV}^{11} \\ M_{VH}^{00} & 0 & M_{VV}^{00} & 0 \\ 0 & M_{VH}^{11} & 0 & M_{VV}^{11} \end{pmatrix}.$$

One can easily see that this system is highly ADI–sable and it was in fact by using this technique that we solved eq. (18) for the unknown $\lambda$. In other words, after guessing the vertical moments, we solve for the horizontal ones and viceversa. All the systems involved are strictly (symmetric) tridiagonal ones as there is no coupling between the (horizontal or vertical) moments of order 0 and 1, a feature which is truly remarkable, and which by the way would be lost when we climb up to the next scheme (RTN–2) as couplings appear between the zeroth and second order moments. In computational costs, this means that the RTN–1 scheme is roughly twice as expensive as the RTN–0 one since the system (18) in $\lambda$ is the only coupled system. For the other variables, since they are defined cell by cell, it is sufficient to solve small algebraic systems, which are themselves quite decoupled and could ideally be solved in parallel: for $\phi$, these systems have order four but are simply diagonal; for $J$, they have order twelve but are actually block-diagonal with four blocks each of order three, as the horizontal moments are not coupled to the vertical ones while within one kind of moments there is also decoupling between moments of different orders. In static diffusion studies, the one–group diffusion equation (1.a) is only part of an overall iteration strategy to determine the fundamental eigenvalue $k$ of an eigenvector problem where the corresponding eigenvector is the fundamental flux of the reactor positive almost everywhere while $k$ is also known as the effective multiplication factor of the reactor: as $k$ is basically a ratio of reaction rates, an estimation of it is known as soon as $m_C^{00}(\phi_h)$ is available. The complete iteration procedure may thus consist in converging on $\lambda$ by iteratively solving (18), followed by an estimation of $k$ repeated until convergence. At the end of this complex iteration, $J$ is eventually determined. It is this iteration procedure that we have implemented in the computer code MIXQUIC (Del Valle and Hennart (1989)). With dynamic memory allocation techniques and for domains of the union of rectangles type (not simply rectangular), the most complicated configurations for the full reactor core of the Laguna Verde nuclear power plant never needed more than twenty thousand words in central memory and were perfectly run on a minicomputer.

# REFERENCES

ARNOLD, D.N. and BREZZI, F., (1985), "Mixed and nonconforming finite element methods: implementation, postprocessing and error estimates", $M^2AN$ **19**, 7–32.

AZMY, Y. Y. and DORNING, J.J. (1983), "A nodal integral approach to the numerical solution of partial differential equations", in Advances in Reactor Computations, American Nuclear Society, Salt Lake City, Utah, pp. 893–909.

BENSOUSSAN, A., LIONS, J.L. and PAPANICOLAU, G. (1978), *Asymptotic Analysis of Periodic Structures* North–Holland, Amsterdam.

DELP, D.L. et al., (1964), "FLARE, a three–dimensional boiling water reactor simulator", GEAP-4598.

DEL VALLE, E. and HENNART, J.P. (1988), "Desarrollo de programas de cómputo y su aplicación en el campo de la ingeniería nuclear", Comunicaciones Técnicas AM **91**, 16 pp., IIMAS-UNAM.

DEL VALLE, E. and HENNART, J.P. (1989), "MIXQUIC–Manual del Usuario", Internal Report.

DEL VALLE, E., HENNART, J.P., and MEADE, D. "A fast mixed hybrid nodal solver for the multidimensional static neutron diffusion equations. Part II. Numerical results", In Preparation.

EWING, R.E. (1986), "Efficient adaptive procedures for fluid flow applications", Comput. Meth. Appl. Mech. Eng. **55**, 89–103.

FEDON–MAGNAUD, C., HENNART, J.P., and LAUTARD, J.J. (1983), "On the relationship between some nodal schemes and the finite element method in static diffusion calculations", in Advances in Reactor Computations, American Nuclear Society Salt Lake City, Utah, pp. 987–1000.

FINNEMANN, H., BENNEWITZ, F., and WAGNER, M.R. (1977),"Interface current techniques for multidimensional reactor calculations", Atomkernergie **30**, 123–128.

GREENMAN, G., SMITH, K., and HENRY, A.F. (1979), "Recent advances in an analytical nodal method for static and transient reactor analysis", in Computational Methods in Nuclear Engineering, American Nuclear Society, Williamsburg, Virginia, p. 3.49–3.71.

GUPTA, N.K. (1981), "Nodal methods for three–dimensional simulators", Progress in Nuclear Energy **7**, 127–149.

HENNART, J.P. (1985a), "A general finite element framework for nodal schemes", in MAFELAP V, The Mathematics of Finite Elements and Applications, J.R. Whiteman, Ed., Academic Press, London, pp. 309–316.

HENNART, J.P. (1985b), "Nodal Schemes, mixed hybrid finite elements and block–centered finite differences", Rapports de Recherche **386**, 58 pp., INRIA.

HENNART, J.P. (1986), "A general family of nodal schemes", SIAM Journal on Scientific and Statistical Computing **7**, 264–287.

HENNART, J.P. (1988), "On the numerical analysis of analytical nodal methods", Numerical Methods for Partial Differential Equations **4**, 233–254.

HENNART, J.P. and DEL VALLE, E. (In Preparation), "A fast mixed–hybrid nodal solver for the multidimensional static neutron diffusion equations. Part I. Theory".

HENNART, J.P., JAFFRE, J., and ROBERTS, J.E. (1988), " A constructive method for deriving finite elements of nodal type", Numerische Mathematik **53**, 701-738.

HENNART, J.P. and MUND, E.H.. (1977), "Singularities in the finite element approximation of two–dimensional diffusion problems", Nuclear Science and Engineering **62**, 52–68.

LANGENBUCH, S., MAURER, W., and WERNER, W. (1977a), "Coarse–mesh flux–expansion method for the analysis of space–time effects large light water cores", Nuclear Science and Engineering **63**, 437–456.

LANGENBUCH, S., MAURER, W., and WERNER, W. (1977b), "High order schemes for neutron kinetics calculations, based on a local polynomial approximation", Nuclear Science and Engineering **64**, 508-516.

LAWRENCE, R.D. (1986), "Progress in nodal methods for the solution of the neutron diffusion and transport equations", Progress in Nuclear Energy **17**, 271–301.

LAWRENCE, R.D. and DORNING, J.J. (1980), "A nodal Green's function method for multidimensional neutron diffusion calculations", Nuclear Science and Engineering **76**, 218–231.

NEDELEC, J.C. (1980), "Mixed finite elements in $\mathbb{R}^3$", Numer. Math. **35**, 315–341.

RAVIART, P.A. and THOMAS, J.M. (1977), "A mixed finite element method for 2nd order elliptic equations", Lecture Notes in Mathematics **606**, 292–315, Springer–Verlag, Berlin.

SHOBER, R.A., SIMS, R.N., and HENRY, A.F. (1977), "Two nodal methods for solving time-dependent group diffusion equations", Nuclear Science and Engineering **64**, 582–592.

SMITH, K.S. (1986), "Assembly homogenization techniques for light water reactor analysis", Progress in Nuclear Energy **17**, 303–335.

# CHAPTER 13

## A New Discrete Functional For Grid Generation

P. Barrera S.*
J. L. Castellanos N.**
R. B. Ojeda C.*
A. Perez D.**

## ABSTRACT

In this work is presented, a new discrete functional for computing smooth convex grids in bidimensional regions. The practical solution of the grid problem is then solved by means of large scale unconstrained optimization methods and results concerning both topics: grid generation and large scale optimization are also given.

## 1. INTRODUCTION

This paper deals with the problem of generating a smooth convex grid over an irregular bidimensional region. One way to obtain that grid is constructing a functional that measures some geometrical properties that imply smoothing and convexity, and then one has to optimize them, *i.e.*, optimize the functional. Here we propose a new discrete functional that takes into account the conditions required and also some large scale optimization methods to minimize it. Furthermore, grid generation is then focused as a test problem generator for large scale optimization, since the finer the grid, the grater the number of variables of the functional.

The work is presented like this. In section number 2 the problem formulated in terms of a mapping from a coordinate system to another. Section 3 describes

---

*Facultad de Ciencias, U N A M, México.
**ICIMAF, Academia de Ciencias, Cuba.

some of the most useful continuos functionals to solve the grid problem. Section 4 explains the most used discrete ones, and ends with our own proposal, showing some examples of grids we have obtained, and which make our results encouraging. Section 5 is concerned with obtaining the optimal grid. Here are made some comments on our functional characteristics, that depend on the value of a parameter to be selected. Then we describe the optimization methods tested, making the corresponding notes on their behaviour for this specific problem. At the end, there are our conclusions on the subjects treated.

## 2. PROBLEM FORMULATION

The aim of this work is to present a new method to automatically generate a convex curvilinear grid on a bidimensional region. Grid generation is the numerical calculation of curvilinear coordinate systems, *i.e.*, the discrete analogue of constructing curvilinear coordinate systems whenever the boundary is conformed by coordinate lines. It is an excellent tool for solving partial differential equations on an arbitrary shaped region by finite difference methods.

Inheret in grid generation techniques is a mapping of some canonical domain such as a square or rectangle to the physical domain. The image of a mesh on the canonical domain will be a grid on the physical one. When the grid boundary coincides with the boundary of the physical domain, the system generated is called a boundary fitted coordinate system.

In general the problem may be stated as follows:

Given a region $\Omega \subset \Re^2$, we want to find a smooth mapping $\bar{c}$ from $\Omega$ to the "unit box" $B$, such that the boundary of $\Omega$ goes into the boundary of $B$, or inversely a smooth mapping $\bar{d}$ from $B$ to $\Omega$ such that the boundary of $B$ goes into the boundary of $\Omega$ (See Fig. 1).

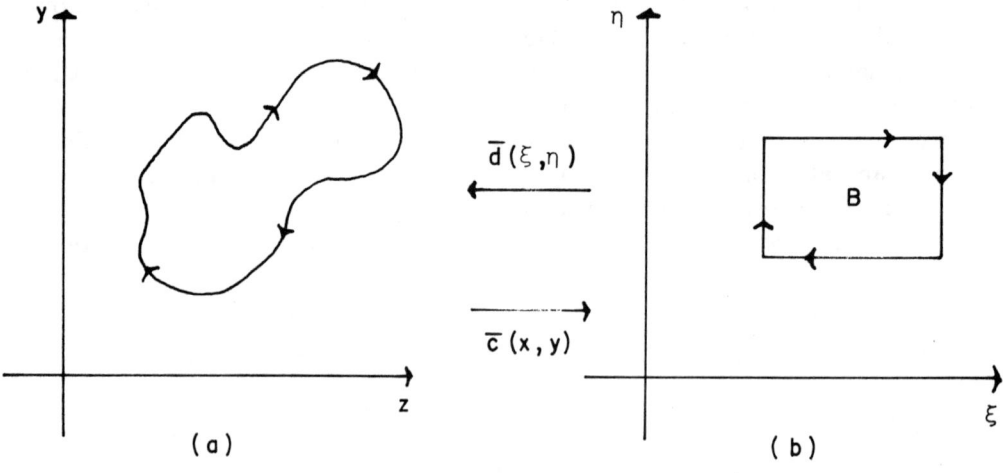

Fig. 1

Problem I: Find $\bar{d}: B \to \Omega$ such that $\bar{d}(\partial B) = \partial \Omega$
Problem II: Find $\bar{c}: \Omega \to B$ such that $\bar{c}(\partial \Omega) = \partial B$
Clasical example:

$$\Omega = \{(x,y) \mid r_1 \leq x^2 + y^2 \leq r_2\}$$
$$B = \{(\xi, \eta) \mid 0 \leq \xi, \eta \leq 1\}$$

Let $\theta(\xi) = 2\pi\xi : r(n) = r_1 + (r_2 - r_1)\eta$, then

$$x(\xi, \eta) = r(\eta) \cos \theta(\xi); \quad y(\xi, \eta) = r(\eta) \sin \theta(\xi);$$
$$\eta(x, y) = \frac{\sqrt{x^2 + y^2 - r_1}}{r_2 - r_1}; \quad \xi(x, y) = \frac{1}{2\pi} \text{ang} \tan\left(\frac{y}{x}\right)$$

In grid generation the object is to obtain numerically adequate curvilinear grids for arbitrary regions, Fig. 2, what is already available for regions as simple as those shown in Fig. 1

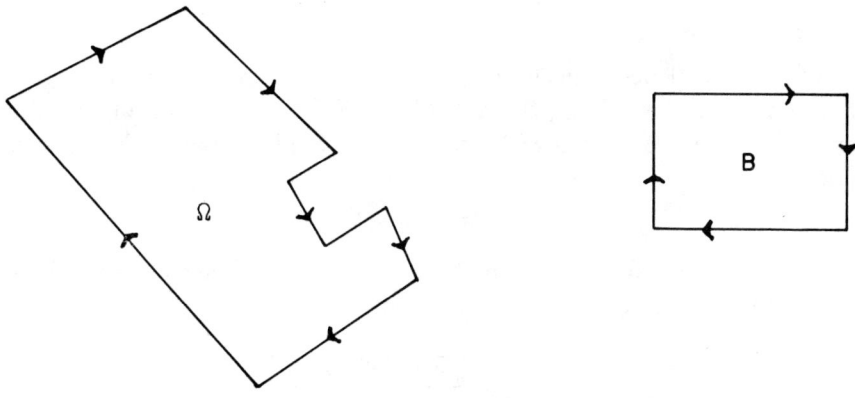

Fig. 2

## 3. DESCRIPTION OF METHODS

There is a great deal of interest in development of computational grid generation thechniques [13], [14], [15], [16]. In order that $\bar{d}(\xi, \eta)$ be well defined, it is necessary to impose some conditions wich determine $\bar{d}$ univocaly.

It has been found useful to associate some functionals that give us a measure of the geometrical properties of the grid. The most used are:

**i)** The smoothness functional

$$I_s(\bar{c}) = \iint_\Omega \left( \|\nabla \xi\|^2 + \|\nabla \eta\|^2 \right) dx\, dy$$

**ii)** The weighted-area functional

$$I_a(\bar{c}) = \iint_\Omega w\, J\, dx\, dy$$

where $J = \xi_x \eta_y - \xi_y \eta_x$ is the Jacobian of $\bar{c}$ and $w(x,y) \geq 0$ is a chosen function.

**iii)** The orthogonality functionals

$$I_o(\bar{c}) = \iint_\Omega (\nabla \xi \cdot \nabla \eta)^2\, dx\, dy$$
$$I'_o(\bar{c}) = \iint_\Omega (\nabla \xi \cdot \nabla \eta)^2\, J^3\, dx\, dy$$

Winslow [17] used the functional $I_s$ to build grids. He obtained the grid that minimizes $I_s$ by solving the discretized Euler's equations of the smoothness functional. Later, Brackbill & Salzman [3] extended Winslow's ideas using the functional

$$I = I_s + \lambda_a I_a + \lambda'_0 I'_0$$

Where $\lambda_a$, $\lambda'_0$ are positive numbers to be chosen. They changed the variables from $(x,y)$ to $(\xi, \eta)$, and find that $I_s$ transform into:

$$I_s(d) = \iint_B \frac{\|\nabla x\|^2 + \|\nabla y\|^2}{J}\, d\xi\, d\eta$$

where

$$J = x_\xi\, y_\eta - x_\eta\, y_\xi.$$

The idea behind these methods is to control: the smoothness, the area of the cells and the orthogonality of the grid by properly choosing $w(x,y)$, $\lambda_a$, $\lambda'_0 \geq 0$. Their approach to compute the grid is similar to Winslow's in that the Euler's equations are solved numerically.

Roache & Steinberg [10, 11] found out that Winslow's methods may produce non-convex grids, even for simple regions, so in 1986 they proposed [10,13] a new set of functionals for controling the geometric properties of the grid:

**i)** Smoothness functional:

$$\tilde{I}_s = \iint_B \left( \|\nabla x\|^2 + \|\nabla y\|^2 \right) d\xi\, d\eta$$

**ii)** Area functional:

$$\tilde{I}_a = \iint_B J^2 \, d\xi \, d\eta; \qquad J = x_\xi y_\eta - x_\eta y_\xi$$

**iii)** Orthogonality functional:

$$\tilde{I}_0 = \iint_B \left( x_\xi x_\eta + y_\xi y_\eta \right)^2 d\xi \, d\eta$$

Computationally, they proceeded still in the same way, by solving the discretized Euler's equations. Castillo [4] proposed a more direct approach to control the geometric properties of the grid. He considered the discrete analogue to the smoothness and area functionals. He directly discretized the functionals of Roache & Steinberg without using Euler's equations. Barrera and Castillo [1] stated the properties of this method and computed optimal grids using Shanno and Phua's conjugate gradient method [12].

Ivanienko and Sharajahian [7] presented a method for generating smooth convex grids. They constructed a discrete functional directly form $I_s(d)$. In order to compute the optimal grid they needed an initial convex grid and, afterwards, they solve Euler's equations making use of a Newton-type method to find a critical point. The need of an initial convex grid is the main drawback of their method.

In this paper, having in mind Ivanenko's and Castillo's ideas, we propose a new functional that is cheaper than the previous ones, and allows us to generate smooth convex grids, making use of the most recent techniques for large scale optimization to solve the minimization problem.

## 4. DISCRETE FUNCTIONALS

A region $\Omega \subset \Re^2$ is polygonal if its boundary $\partial \Omega$ is the union of simple closed polygons. A grid on a polygonal region $\Omega$ is a subdivision of $\Omega$ into quadrilaterals, the vertices of the quadrilaterals are called the points of the grid and the quadrilaterals are called the cells of the grid.

More precisely a $m \times n$ grid $G_{m,n}$ of $\Omega$ is an ordered set of $mn$ points:

$$G_{m,n} = \{ P_{i,j} \in \Re^2 \mid i = 1, 2, ..., m; j = 1, 2, ..., n \}$$

such that the boundary $\partial G_{m,n}$ is given by

$$\partial G_{m,n} = \{ P_{1,1}, P_{1,2}, ..., P_{1,m}, P_{2,m}, ..., P_{n,n}, P_{n,m-1}, ..., P_{n-1,1}, ..., P_{2,1} \}$$

where $\partial G_{m,n} \subset \partial \Omega$, (See Fig. 3,4).

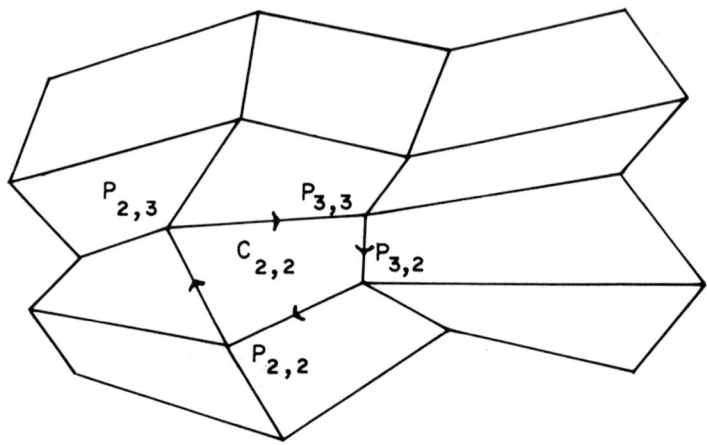

**Fig. 3** an $4 \times 5$ grid $G_{4,5}$

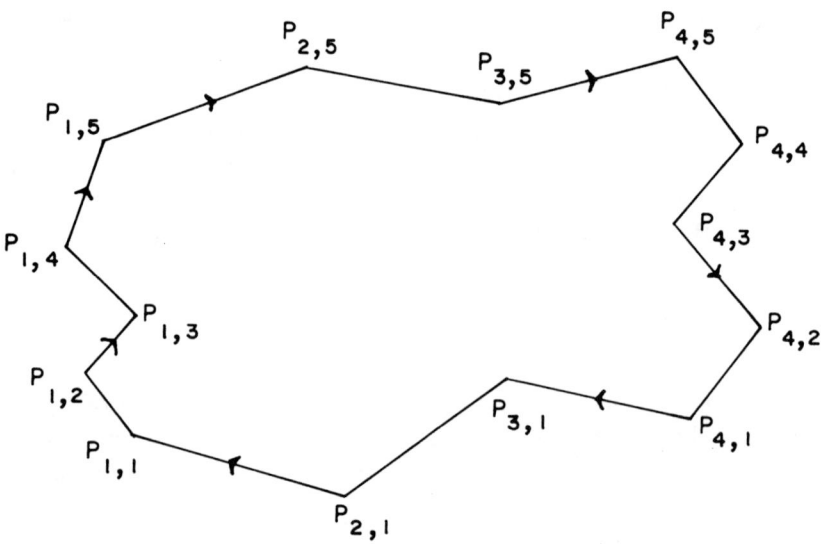

**Fig. 4** $\partial G_{4,5}$

In the following discussion we will assume that the grids $G_{m,n}$ of $\Omega$, have the same fixed boundary, *i.e.* we want to choose an optimal grid in the set of all grids of $\Omega$ that have the same boundary.

Let $c_{i,j}$ denote the oriented cell whose vertices are $P_{i,j}$, $P_{i,j+1}$, $P_{i+1,j+1}$, $P_{i+1,j}$ and let $\alpha_{i,j}$ denote its area, i.e. $\alpha_{i,j}$ = area ( $P_{i,j}$, $P_{i,j+1}$, $P_{i+1,j+1}$, $P_{i+1,j}$ ). We will consider the points $P_{i,j}$ as column vectors, i.e.

$$P_{i,j} = (x_{i,j}, y_{i,j})^t$$

Castillo [4] proposed to control the lengths of sides of the cells by mean of the functional:
$$f_s(G_{m,n}) = S_H(G_{m,n}) + S_V(G_{m,n})$$
where $S_H$ is the sum of the squares of the lengths of the "horizontal sides" and is given by:

$$S_H(G_{m,n}) = \sum_{i=1}^{m-1} \sum_{j=2}^{n-1} \|P_{i,j} - P_{i+1,j}\|^2$$

and $S_V$ is the sum of the squares of the lengths of the "vertical sides" and is given by:

$$S_V(G_{m,n}) = \sum_{i=2}^{m-1} \sum_{j=1}^{n-1} \|P_{i,j} - P_{i,j+1}\|^2,$$

and control the size of the area $\alpha_{i,j}$ of the cells with the functional:

$$f_a(G_{m,n}) = \sum_{i=1}^{m-1} \sum_{j=1}^{n-1} \alpha_{i,j}^2,$$

Barrera and Castillo [1] succeeded in constructing smooth grids using the functional:
$$f_\sigma(G_{m,n}) = \sigma f_s(G_{m,n}) + (1 - \sigma) f_a(G_{m,n})$$
where $0 \leq \sigma \leq 1$, is chosen to avoid the folding of the optimal grids.

They proved that for $m, n$ suficiently larges, the functional $f_\sigma$ has a unique minimum for $0 < \sigma \leq 1$ if the region $\Omega$ has an $m \times n$ grid in wich the areas of the cells are nearly equals [1]. The optimal grid was computed using Shanno & Phua's conjugate gradient method [12]. Barrera and Pérez found out that Castillo's functional might fail to produce convex grids on some regions like Habana Bay. Motivated by Ivanienko & Sharajahian's work [7], we propose to control the convexity of the cells by means of the following device:

Consider, in the $c_{i,j}$ cell, the triangles:

$$\triangle_{i,j}^{(1)} = \text{triangle } (P_{i,j}, P_{i,j+1}, P_{i+1,j+1});$$
$$\triangle_{i,j}^{(2)} = \text{triangle } (P_{i,j+1}, P_{i+1,j+1}, P_{i+1,j});$$
$$\triangle_{i,j}^{(3)} = \text{triangle } (P_{i+1,j+1}, P_{i+1,j}, P_{i,j});$$
$$\triangle_{i,j}^{(4)} = \text{triangle } (P_{i+1,j}, P_{i,j}, P_{i,j+1});$$

Let $\alpha_{i,j}^{(k)} =$ area $\left(\triangle_{i,j}^{(k)}\right)$, then it is easy to prove that if all $\alpha_{i,j}^{(k)}$ have the same sign then the grid is convex. Therefore, we try to obtain a convex grid by means of the functional.

$$\tilde{f}_a = \sum_{i=1}^{m-1} \sum_{j=1}^{n-1} \left( \sum_{k=1}^{4} \left(\alpha_{i,j}^{(k)}\right)^2 \right),$$

in order to prove that this functional achieves its goal, let us rewrite the functional in a different way.

Let us enumerate the triangles $\triangle_{i,j}^{(k)}$ by means of the function

$$\ell = \psi(i,j,k) = 4\left[(n-1)(i-1) + (j-1)\right] + k$$

so, the total number of triangles is

$$N = 4(n-1)(m-1)$$

Let

$$F_a^t = (f_1, f_2, ..., f_N);$$

where

$$f_\ell = \alpha_{i,j}^{(k)}$$

then

$$\tilde{f}_a = \|F_a\|^2; \qquad \nabla \tilde{f}_a = 2\tilde{J}_a^t F_a$$

and

$$\frac{1}{2}\nabla^2 \tilde{f}_a = \tilde{J}_a^t \tilde{J}_a + \sum_{\ell=1}^{N} f_\ell \nabla^2 f_\ell,$$

where $\tilde{J}_a$ is the Jacobian of $F_a$.

**LEMMA:**

Let $\alpha =$ area $(P, Q, R)$, then

$$2\frac{\partial \alpha}{\partial P} = J_2(Q-R); \quad 2\frac{\partial \alpha}{\partial Q} = J_2(R-P); \quad 2\frac{\partial \alpha}{\partial R} = J_2(P-Q)$$

and

$$\nabla^2 \alpha = \begin{pmatrix} 0 & J_2 & -J_2 \\ -J_2 & 0 & J_2 \\ J_2 & -J_2 & 0 \end{pmatrix}$$

where

$$2\frac{\partial \alpha}{\partial P} = \begin{pmatrix} \frac{\partial \alpha}{\partial x_p} \\ \frac{\partial \alpha}{\partial y_p} \end{pmatrix}; \quad 2\frac{\partial \alpha}{\partial Q} = \begin{pmatrix} \frac{\partial \alpha}{\partial x_q} \\ \frac{\partial \alpha}{\partial y_q} \end{pmatrix}; \quad 2\frac{\partial \alpha}{\partial R} = \begin{pmatrix} \frac{\partial \alpha}{\partial x_r} \\ \frac{\partial \alpha}{\partial y_r} \end{pmatrix}$$

and

$$J_2 = \begin{pmatrix} 0 & 1 \\ -1 & 0 \end{pmatrix}$$

**PROOF:** straight forward computation

## THEOREM:

Suppose that there exist a grid $\hat{G}$ such that all its triangles $\triangle_{i,j}^{(k)}$, $i = 1, 2, ..., m$; $j = 1, 2, ..., n$; $k = 1, 2, 3, 4$; have the same area, then $\hat{G}$ is a local minimum of $\tilde{f}_a$.

## PROOF

In order to show that the gradient is zero consider an interior point $P$ of the grid and let $A, B, C, D, E, F, G$ and $H$ be its neighbours as shown in fig. 5. In the same figure all the triangles that have $P$ as a vertice are shown and enumerated, using the previous lemma it can be verified that

$$\frac{\partial \tilde{f}}{\partial P} = g_1 + g_2$$

where

$$g_1 = J_2\,[\,\alpha_1(H - B) + \alpha_2(B - D) + \alpha_3(D - F) + \alpha_4(F - H)\,]$$
$$g_2 = J_2\,[\,\alpha_5(H - A) + \alpha_6(A - B) + \alpha_7(B - C) + \alpha_8(C - D) +$$
$$+ \alpha_9(D - E) + \alpha_{10}(E - F) + \alpha_{11}(F - G) + \alpha_{12}(G - H)\,]$$

with $\alpha_i$ = area of $i^{\text{th}}$ triangle. Clearly, $g_1 = g_2 = 0$ if $\alpha_1 = \alpha_2 = ... = \alpha_{12}$ given that $\nabla \tilde{f}_a = 0$

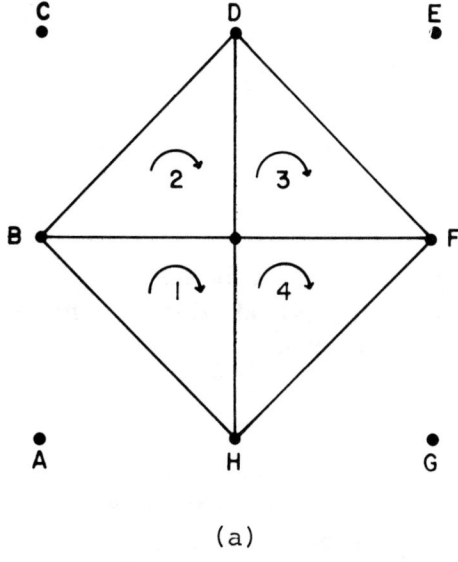
(a)

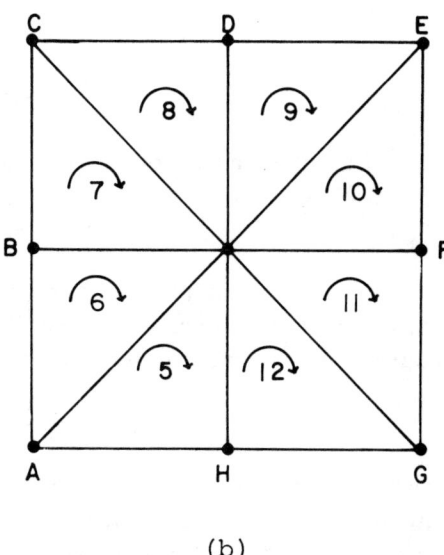
(b)

Fig. 5

We are now going to prove that $\nabla^2 \tilde{f}_a \geq 0$, let

$$\hat{H} = \sum_{\ell=1}^{N} f_\ell \nabla^2 f_\ell$$

where

$$f_\ell = \alpha_{i,j}^{(k)}; \qquad \ell = \psi(i,j,k)$$

The block in $\hat{H}$ corresponding to the points $P$ and $Q$ is null if $P$ and $Q$ do not belong to some $\Delta_{i,j}^{(k)}$ simultaneously.

The block in $\hat{H}$ corresponding to the points $P$ and $Q$, when they belong to some $\Delta_{i,j}^{(k)}$ is given by:

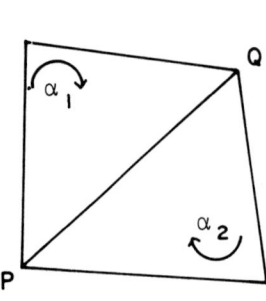 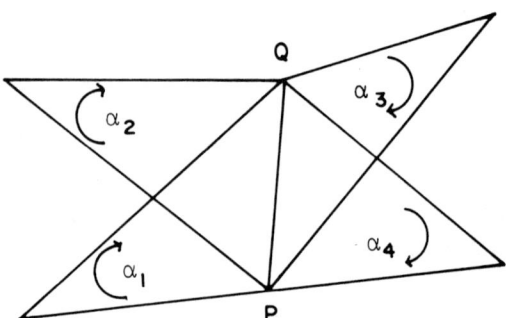

a) $J_2(\alpha_2 - \alpha_1)$ 	b) $J_2(\alpha_3 + \alpha_4 - \alpha_1 - \alpha_2)$

In each case the contribution is zero if the areas of the triangles are equal so that we have

$$\nabla^2 \tilde{f}_a = 2\tilde{J}_a^t \tilde{J}_a \geq 0 \qquad \blacksquare$$

This theorem justifies our claim, namely, that $\tilde{f}_a$ can be useful to build convex grids wherever they exist. In order that the optimal grids not only be convex but also smooth, we have been using the functional.

$$\tilde{f}_\sigma = \sigma f_s + (1-\sigma)\tilde{f}_a$$

with $0 < \sigma \leq 1$. It should be noticed that the selection of parameter $\sigma$ generates different functionals for each value, so, with $\sigma$ near 1 we are giving more importance to smoothnes and otherwise, with $\sigma$ near 0, we are choosing the convexity of the cells instead of smoothnes. So, to choose this parameter it is crucial to have these observations in mind, and we are now getting experience in this problem in order to adaptively choose a value of $\sigma$.

This functional has been implemented in a system that automatically generates grids. It has worked quite well, much better than others before, and we have got convex smooth grids in not very simple regions, as can be seen in the following pictures:

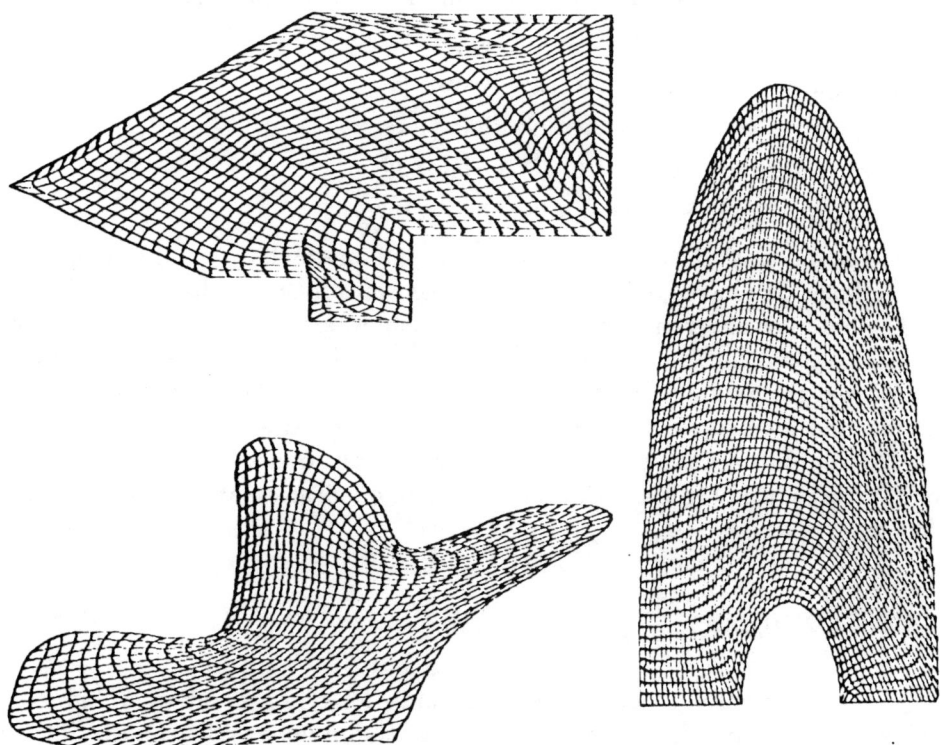

## 5. OPTIMIZATION ASPECTS AND COMPUTATIONAL RESULTS

To obtain the optimal grid, which means a smooth convex grid, it is necessary to apply a minimization algorithm to solve the problem

$$\min_{G_{m,n}} \tilde{f}_\sigma(G_{m,n})$$

First of all, it is worthwhile to have some knowledge about the functional that is going to be minimized. From comments below one can have an idea of how much difficult to solve will the problem be:

a)   for $\sigma = 1$ the functional is quadratic and it is relatively easy to compute the optimal grid.

b)   for $\sigma \neq 1$ the functional is quartic and may have several local minima, the optimum grid may be very hard to compute, if the boundary of the region is quite irregular, because usually the function is "flat" in a neighbourhood of a minimum, specially when $\sigma = 0$.

**c)** the number of variables is very large even for regular size grids. An $m \times n$ grid has $2(m-1)(n-2)$ variables.

Thus, a method for large scale unconstrained optimization is needed, that has minimum memory requirements to enable handling large grids.

Such method should also have nice convergence-rate properties.

In current testing the most promising methods seem to be:

**a)**  Conjugate – Gradient of Fletcher-Reeves [6]

**b)**  Conjugate – Gradient of Shanno-Phua [12]

**c)**  Limited memory BFGS of Liu-Nocedal [8,9]

**d)**  Truncated-Newton of Dembo-Steighaug [5]

At the meeting in Mérida, we reported some preliminary results comparing those methods. We have got some recent implementations, and have made lots of test. The final picture from a fair comparison is now emerging. At this point, we only want to comment that in the preliminary tests, the code of Shanno and Phua [12], with a slight modification that we introduced, has worked very well for most problems, and his main difficulty has been with quadratics ($\sigma = 1$). L-BFGS of Liu & Nocedal has always been quite reasonable and robust. Truncated Newton with line-search has been most times the best, on quadratics, and had never failed in that case.

## 6. CONCLUSIONS

We have presented a new discrete functional for grid generation on bidimensional regions, which seems to be cheaper than the previous ones, and is able to produce smooth convex grids.

Besides, the problem of our concern has been solved using tools from large scale optimization and we have implemented a system that, given the coordinate points of the boundary, generates optimal grids. Now, we are testing both aspects of the problem: the ability of our functional to build in "good" grids and, the best methods for solving the optimization problem. As was mentioned, these tests are not still concluded and final results will be published elsewhere.

## ACKNOWLEDGEMENTS

We want to thank: Facultad de Ciencias and IIMAS of UNAM for their support; to the unknown referee for his valuable comments on the paper that help us to improve it; to R. Pico and R. Morales for their programming help; to J. Nocedal of Northwestern Univ. for making available his implementation and the paper on the L-BFGS method; to D. Moreno for his suggestions; to L. Parra and M. Rosas their careful typing.

# REFERENCES

[1] P. Barrera-Sánchez and J. E. Castillo. *A large scale optimization problem arising from numerical grid generation* Tech. Rep. Dept. of Math. and Statistics Univ. of New Mexico, (1986).

[2] P. Barrera-Sánchez, L. Castellanos, R. Ojeda, *A comparison of some Optimization Methods for grid generation* to appear summer 1990.

[3] J. V. Brackbill and J. S. Saltzman, *Adaptive zoning for singular problems in two dimensional* J. Comp. Phys. 46, 1982, 342–368.

[4] J. E. Castillo, *Mathematical aspects of variational grid generation I*, Presented at the Int. Conference Numerical Grid Generation in Computational Fluid Dynamics Landshut, W. Germany, July 1986.

[5] R. S. Dembo, and T. Steihaug, *Truncated Newton algorithms for large scale unconstrained optimization.* Math. Programming, 26 (1983) 190–212.

[6] R. Fletcher and C. M. Reeves, *Function minimization by conjugate gradients*, Computer Journal 7 (1964); 149–154.

[7] S. A. Ivanienko and Sharajahian, *An algorithm for constructing currilinear grids of convex quadrilaterals*, Dokl. Akad. Nauk SSSR, 295, 2, 280–283, 1987,

[8] D. C. Liu and J. Nocedal, *On the limited memory BFGS Method for large scale optimization.* Technical Report NAM 03, Northwestern University, Dic. 1988.

[9] J. Nocedal, *Uptating Quasi-Newton Matrices with Limited Storage*, Math. Comp. 35, (1980), 773–782.

[10] P. J. Roache and Steinberg, *A new approach to grid generation using a variational formulation*, 7th Computational Fluid Dynamics Conference, Cincinnati, Ohio, July 1985.

[11] P. T. Roache, S. Steinberg and W. M. Moeny, *Interactive electric field calculations for lasers*, ARAA paper No. 84-1655, AIAA 17th Fluid Dynamics, Plasma Dynamics and Laser Conference, Snowman, Colorado, June 1984.

[12] D. F. Shanno and K. H. Phua. *A Variable Method Subroutine for unconstrained nonlinear minimization*, M. I. S. (1978) Technical Report No. 28 Univ. of Arizona. Tucson.

[13] S. Steinberg and P. J. Roache, *Variational grid generation*, Num. Meth. for Partial Diff. Eqns. 2, 1986, 71–96.

[14] J. F. Thompson, Z. V. Warsi and C. E. Mastin, *Boundary Fitted coordinate systems for numerical solution of partial differential equations*, A Review J. Comp. Phy. 47, 1982.

[15] J. F. Thompson (Ed.), Numerical Grid Generation, North Holland, 1982.

[16] J. F. Thompson, Z. V. Warsi and C. E. Mastin, Numerical Grid Generation: Foundations and Applications, North Holland, 1985.

[17] A. M. Winslow, *Numerical Solution of the quasilinear Poisson equation in a nonuniform triangle mesh*, J. Comp. Phys. Vol. 1, No. 2 (1967).

# CHAPTER 14

# Computing Eigenvalues of Large Matrices, Some Lanczos Algorithms and a Shift and Invert Strategy

Jane K. Cullum*
Ralph A. Willoughby*

### Abstract

We summarize our recent research on Lanczos algorithms with no reorthogonalization for the computation of eigenvalues and eigenvectors of large scale real symmetric and nonsymmetric eigenvalue problems and present a novel shift and invert strategy which is applicable to both our real symmetric and our nonsymmetric Lanczos algorithms. The summary consists of two parts. First we attempt to clarify the distinctions between our Lanczos algorithms with no reorthogonalization and the Lanczos algorithms which use selective reorthogonalization of the Lanczos vectors. Then we summarize the similarities and differences between our Lanczos algorithms for real symmetric problems and our Lanczos algorithms for nonsymmetric problems. We then show that for both types of our algorithms, that their look-around capabilities can be used very easily and profitably to construct an automatic shift strategy for shift and invert versions of these algorithms. Our proposed strategy provides one possible solution to the difficult question of locating those eigenvalues of a given nonsymmetric matrix which are in a user-specified box in the complex plane. Numerical examples of the use of this strategy in the nonsymmetric and the real symmetric procedures are included.

**1 Introduction.** We consider a specific family of algorithms for the solution of real symmetric and of nonsymmetric eigenvalue and eigenvector problems. Specifically, we are given a matrix $A$ or matrices $A$ and $B$ and we want to compute eigenvalues $\lambda$ and eigen-

---

*Mathematical Sciences Department, IBM Research Division, T.J. Watson Research Center, Yorktown Heights, NY 10598.

vectors $x \neq 0$ satisfying

$$Ax = \lambda x \quad \text{or} \quad Ax = \lambda Bx. \tag{1.1}$$

In some cases only a few eigenvalues are required, in other cases many may be needed. In all of our discussions $n$ will denote the order of the $A$ (and $B$) matrix (matrices).

Large scale eigenvalue and eigenvector problems are central to the solution of many different kinds of scientific and engineering problems. Such problems arise for example in physics, chemistry, structural analysis, medicine, and biology. In many cases the 'large' matrices being considered are not as 'large' as the user wants to make them. This is especially true when the matrix is generated by some appropriate discretization of a complicated system of differential equations. Increases in matrix size can result from either a finer discretization of the model, or from the use of a more complex model, and typically result in more realistic approximations.

When designing or selecting an algorithm for solving a particular eigenvalue problem the user has several things to consider. Principal considerations are (1) the amount of computer storage available; (2) the speed of the computer available; (3) the number of arithmetic operations required and (4) the associated data movement required by the algorithm being used. The user must also consider how much of the eigenvalue spectrum is desired and where the desired portion sits in the overall spectrum of the given problem. As computers continue to get faster and have larger memories some of these considerations may become less important. However, past history indicates that that need not happen. As computers have become more powerful, we have seen the expectations of scientists and engineers increase proportionately and watched them generate more complex or detailed models which again taxed the capabilities of the computer they were using.

The accepted algorithms for small to medium size matrices, which are found for example in the EISPACK library [1976, 1977], are direct methods which explicitly modify the given matrix. Real symmetric matrices are explicitly transformed into real symmetric tridiagonal matrices. Nonsymmetric matrices are transformed into Hessenberg matrices. (A matrix is lower Hessenberg if all the entries below the first subdiagonal are zero.) EISPACK utilizes orthogonal or unitary transformations to obtain the real symmetric tridiagonal or the nonsymmetric Hessenberg forms. Such transformations are numerically stable, since they do not amplify the sizes of the quantities they are transforming. The resulting canonical problem is then solved using an algorithm which has been optimized with respect to operation counts and numerical stability.

The EISPACK procedures require $O(n^2)$ storage since the matrix is explicitly transformed into a simpler form, and take $O(n^3)$ arithmetic operations. Limitations in storage and in computer speed have led to the development of 'iterative' algorithms for large scale problems. See for example Cullum and Willoughby [1985, 1986, 1989], Parlett and Scott [1979], Parlett [1980] and Saad [1980 - 1985]. The basic iterative algorithms do not modify the original matrix or matrices. Such algorithms see what the given matrix $A$ is by looking at what $A$ does to various vectors $x$. Some algorithms for nonsymmetric problems also look at what $A^T$ does to vectors $y$.

If the required matrix-vector multiplies can be generated in $O(n)$ storage and $O(n)$ operations, and if the iterative algorithm converges in $O(m)$ steps, then the storage requirements for the entire procedure are reduced to $O(n)$ and the arithmetic operation counts are reduced to $O(mn)$.

Iterative methods based upon Lanczos recursions have been devised for both large-scale real symmetric and nonsymmetric eigenvalue problems. 'Recursive' is perhaps a better term for the types of algorithms we discuss in this paper. The earlier simultaneous iteration algorithms required that the desired eigenvalues be dominant in magnitude in the spectrum of the matrix being used in the eigenvalue computations. In such a method the rate of convergence is proportional to the ratios of the magnitudes of the eigenvalues being computed to the largest of the magnitudes of the eigenvalues which are not being computed. If nondominant eigenvalues are to be computed, it is necessary to work with a transformed problem where the desired eigenvalues are made magnitude dominant.

Lanczos procedures do not require magnitude dominance, although it is still true that such dominance speeds convergence. The convergence of the eigenvalue approximations in a Lanczos procedure is controlled by a combination of the effects of the eigenvalue gap structure, the magnitude dominance or nondominance of the eigenvalues to be computed, and the relative location of these eigenvalues in the overall spectrum of the matrix being used in the eigenvalue procedure. Well separated, magnitude dominant eigenvalues which are on the extremes of the spectrum will typically converge first. However, extreme and/or dominant eigenvalues that are not well-separated need not converge first. This type of behavior is observed numerically in both the real symmetric and the nonsymmetric Lanczos procedures.

A variety of Lanczos procedures exist in the literature, especially for real symmetric problems. In Section 2 we focus on real symmetric procedures and consider some of the similarities and differences between these algorithms. In Section 3 we then look at the differences between our Lanczos algorithms with no reorthogonalization for real symmetric problems and for nonsymmetric problems. In Section 4 we introduce our proposed shift and invert strategy which is applicable to both our real symmetric and our nonsymmetric procedures, and in Section 5 we give several numerical examples illustrating its use.

## 2 Comparison of Lanczos Algorithms.

In this Section we summarize some of the differences between Lanczos algorithms which incorporate reorthogonalization and our Lanczos algorithms which do not use any reorthogonalization. These two types of procedures behave very differently and there has been some confusion in the literature and in the community with regard to the behavior of the Lanczos algorithms which do not use reorthogonalization.

Lanczos procedures for computing eigenvalues and eigenvectors of real symmetric matrices are based upon one or more variants of the basic single-vector Lanczos recursion for tridiagonalizing a real symmetric matrix $A$. In the next Section we will discuss a nonsymmetric variant but in this Section we confine our remarks to the real symmetric case. Given a starting vector $v_1$ which is typically generated randomly, the Lanczos recursion implements a Gram-Schmidt orthogonalization of the matrix-vector products $Av_i$ corresponding to the Lanczos vectors $v_i$ generated by the recursion. Specifically we have for $i = 1, 2, \ldots, m$

$$\beta_{i+1} v_{i+1} = A v_i - \alpha_i v_i - \beta_i v_{i-1} \equiv r_i \tag{2.1}$$

where $\alpha_i \equiv v_i^T (A v_i - \beta_i v_{i-1})$ and $\beta_{i+1} \equiv \| r_i \|$. The vector $\alpha_i v_i$ is the projection of the vector $A v_i$ onto the Lanczos vector $v_i$ and $\beta_i v_{i-1}$ is the projection of $A v_i$ onto the Lanczos vector $v_{i-1}$. The resulting vectors $V_m = \{v_1, v_2, \ldots, v_m\}$ are orthogonal for some $m \leq n$. Thus, in the real symmetric case, the Lanczos recursion maps the given matrix $A$ into a family of real symmetric tridiagonal matrices $T_j$, $1 \leq j \leq m$ where

$$T_j \equiv \begin{bmatrix} \alpha_1 & \beta_2 & & \\ \beta_2 & \alpha_2 & \ddots & \\ & \ddots & \ddots & \beta_j \\ & & \beta_j & \alpha_j \end{bmatrix}. \tag{2.2}$$

In matrix form we have that

$$AV_m = V_m T_m + \beta_{m+1} v_{m+1} e_m^T + E_m \quad (2.3)$$

where $E_m$ represents the errors incurred from the use of finite precision computer arithmetic.

A Lanczos procedure for computing eigenvalues of a matrix $A$ uses a variant of the Lanczos recursion to transform the given matrix into a family of tridiagonal matrices of varying sizes. The eigenvalue and eigenvector computations for $A$ are replaced by eigenvalue and eigenvector computations for one or more of the associated simpler, tridiagonal, Lanczos matrices. The procedure computes such eigenvalues and then selects a subset of them as approximations to eigenvalues of the original matrix $A$.

For any such eigenvalue $\mu$ we can obtain a corresponding eigenvector approximation for the original matrix by computing a corresponding eigenvector $u$ of an appropriately chosen Lanczos matrix $T(\mu)$ and forming the Ritz vector $z = V(\mu)u$. In the discussion below we will explain why our methods allow the Lanczos matrices used in the Ritz vector computation to differ from the one used in the eigenvalue computations.

When $A$ is a real symmetric matrix and the arithmetic is exact, we have that the Lanczos matrices are orthogonal projections of $A$ onto the Krylov subspaces $K_m \equiv \{v_1, Av_1, \ldots, A^{m-1}v_1\}$, and there are theorems which relate the eigenvalues of these projection matrices to the eigenvalues of the original matrix, and the corresponding Ritz vectors $z = V_m u$ for $T_m u = \mu u$ to eigenvectors of the original matrix.

In practice however, the arithmetic is not exact and losses in orthogonality occur. The Lanczos vectors are not orthogonal and may not even remain linearly independent. In such a situation the associated Lanczos matrices are not orthogonal projections of $A$ onto the Krylov subspaces. If the recursion is used directly, ignoring these losses in orthogonality and simply accepting all of the eigenvalues of the Lanczos matrices as being representative of the eigenvalues of the original matrix, then incorrect approximations are obtained.

These losses in orthogonality occur because of the finite precision arithmetic. However, these losses are not caused by a simple accumulation of roundoff errors. The principal losses occur whenever one of the eigenvalues of a Lanczos matrix becomes an accurate approximation to an eigenvalue of the original matrix, and they occur along the direction of the converged Ritz vector associated with that eigenvalue. For more details see for example, Cullum and Willoughby [1985] for a discussion of these results of Paige [1971].

In the literature we find basically three different proposals for 'correcting' for these losses in orthogonality. The first proposal was given by Lanczos and consists of reorthogonalizing each Lanczos vector as it is generated against all previously-generated Lanczos vectors. The second proposal was to selectively reorthogonalize the Lanczos vectors against all Ritz vectors which have converged up to the point in time at which the current Lanczos vector is being generated. In practice, for this approach, the objective is to maintain a level of orthogonality between the Lanczos vectors that guarantees that they remain 'sufficiently independent'. The third proposal simply accepts the losses in orthogonality, no reorthogonalization is performed, and the eigenvalues of the associated Lanczos matrices are sorted into 'good' eigenvalues, those which are approximations to eigenvalues of the original matrix, and 'spurious' eigenvalues, those which have appeared because of the losses in orthogonality of the Lanczos vectors.

Total reorthogonalization of the Lanczos vectors has more recently been replaced by Horst Simon's proposal [1984] to perform such reorthogonalizations periodically. Here we focus on the other two approaches, selective reorthogonalization and no reorthogonalization.

Selective reorthogonalization has been developed by Parlett and Scott [1979], Parlett [1980]. An earlier and cruder version was used in Cullum and Donath [1974] in a Block

Lanczos Algorithm. In the Parlett and Scott algorithm, losses in orthogonality of the Lanczos vectors are monitored as the recursion proceeds and when significant losses in orthogonality are perceived by the quantities tracking such losses, the recursion pauses and eigenvalue and eigenvector computations are performed on the current Lanczos matrix. Ritz vectors corresponding to those converged eigenvalues causing the losses in orthogonality are then computed and used to reorthogonalize the Lanczos vectors currently being generated. There are variations on this idea which allow for various levels of reorthogonalization but the basic idea is to monitor losses in orthogonality, to focus on converged Ritz vectors, and to use them to reorthogonalize the Lanczos vectors when required. For some problems such reorthogonalizations occur infrequently; for other problems they occur very frequently. For a discussion of this procedure see Parlett [1980].

Our Lanczos procedures with no reorthogonalization sort the eigenvalues of each Lanczos matrix considered into 'good' and 'spurious' eigenvalues. This identification test is described in detail in Cullum and Willoughby [1985] and discussed briefly in Section 2.4 below. No reorthogonalizations are performed. We list some of the key differences between a Lanczos procedure based upon selective reorthogonalization and our Lanczos procedures with no reorthogonalization and make a few comments about each of these differences.

### 2.1 Eigenvalue and Ritz Vector Computations.

In our Lanczos procedures with no reorthogonalization the computation of the eigenvalue approximations does not use or require the computation of any Ritz vectors. Thus, the storage requirements for the eigenvalue computations are minimal. For example, for our real symmetric Lanczos procedure, only 6.5 vectors of size $\max(n,m)$ are required for the Lanczos procedure plus whatever is required to generate any matrix-vector multiply, $Ax$. Here $n$ is the order of $A$ and $m$ is the size Lanczos matrix considered.

In contrast, Lanczos procedures based upon selective reorthogonalization intersperse eigenvalue computations with Ritz vector computations, thereby requiring additional storage for any Ritz vectors which are generated for use in the reorthogonalizations and for the Lanczos vectors which must be used to compute converged Ritz vectors as they converge. Each of these vectors is of size $n$, the order of the original matrix.

In our Lanczos procedures with no reorthogonalization the Ritz vectors are computed after the desired eigenvalues have been computed. Our separate eigenvector program allows the user to specify a subset of the computed eigenvalues as those eigenvalues for which an eigenvector approximation is required. We see from Equation (2.3) that the corresponding Ritz vector residual will be small only if the last component of the corresponding eigenvector of the Lanczos matrix is small. Using Sturm sequencing and inverse iteration on a sequence of Lanczos matrices, our eigenvector program determines such an appropriate Lanczos matrix for each eigenvalue supplied. Typically these matrices are not the same size as the size of the Lanczos matrix used to compute the eigenvalues and typically they are of different sizes for different eigenvalues.

It is also important to emphasize that in a Lanczos procedure with no reorthogonalization the Lanczos matrix can be generated first and then the user can do the tridiagonal eigenvalue computations on that or on any smaller size Lanczos matrix, and in addition, for real symmetric problems, on any subinterval desired. Once a Lanczos matrix is generated for a given problem it can be saved, and it and any of its submatrices can be reused at any later date. Moreover, if the matrix and the last two Lanczos vectors are saved, then that matrix can be extended to a larger Lanczos matrix if the user decides that he/she needs portions of the spectrum that were not converged in the spectrum of the original Lanczos matrix.

**2.2 Computer Storage Required.** Neither type of algorithm has a mechanism for determining a priori how much computer storage will be needed. However, the amount of storage required by our Lanczos eigenvalue procedures which do not use reorthogonalization is a simple linear function of the size of the Lanczos matrix being computed whereas the amount of storage required by a procedure which uses selective reorthogonalization grows in unpredictable chunks of size $n$ as the size of the Lanczos matrix is increased. In a selective reorthogonalization procedure, the number of Ritz vectors which must be computed and then used in the subsequent reorthogonalizations depends upon the eigenvalue distribution of the original matrix and upon which part of the spectrum is required. These vectors, along with the associated Lanczos vectors, can however be stored on secondary storage and then called back as needed. The cost of doing that is a function of the particular input/output facilities available on the computer being used.

**2.3 Eigenvalue Approximations.** In Lanczos procedures with selective reorthogonalization, every eigenvalue of the associated Lanczos matrices is taken to be an approximation to an eigenvalue of the original matrix. However, for Lanczos procedures which do not use reorthogonalization this is not valid. In fact for such a procedure there is no simple reason why the eigenvalues of the Lanczos matrices should continue to be related to the eigenvalues of the original matrix after orthogonality is lost. Paige [1976] demonstrated however, at least for the real symmetric case, a certain stability to the recursion. He proved that although the orthogonality of the Lanczos vectors is lost as eigenvalue approximations converge, the errors in the Lanczos recursion are 'small' and proportional to the product of the norm of $A$ and the machine $\epsilon$. He also gave numerical examples indicating that although orthogonality was lost, the desired eigenvalues appeared in the spectrum of the Lanczos matrices if those matrices were made large enough. Cullum and Willoughby [1985] used the arguments in Paige [1976] to demonstrate that the well-known equivalence, in exact arithmetic, of the Lanczos tridiagonalization of a real symmetric matrix to the conjugate gradient method for solving $Ax = b$, can be extended to finite precision arithmetic. They then used that equivalence to map the Lanczos vectors and scalars into conjugate gradient quantities which, under certain assumptions, were shown to 'converge'. This 'convergence' was used to demonstrate that if certain error terms do not grow too rapidly, then for large enough $m$ each distinct eigenvalue of $A$ should be a near-zero of an appropriately-scaled characteristic polynomial of the Lanczos matrix $T_m$.

Details of this argument are given in Cullum and Willoughby [1985]. The arguments use a combination of results and ideas from optimization theory and linear algebra. Nominally the arguments apply only to real symmetric positive definite matrices. However, the Lanczos recursions are shift invariant, so in fact the arguments extend to any real symmetric matrix, giving a plausible explanation for the behavior of real symmetric Lanczos procedures even when no reorthogonalizations are incorporated. These arguments also clarify the observation that once a good approximation to an eigenvalue of the original matrix appears in the spectrum of a Lanczos matrix, it will also be in the spectrum of any larger Lanczos matrices. Thus, we can expect that for large enough $m$, every eigenvalue we want to compute will appear in the spectrum of the Lanczos matrices, along with other eigenvalues which are not related to the original matrix, but are 'spurious', caused by the losses in orthogonality.

**2.4 Identification of Spurious Eigenvalues in the Spectrum of the Lanczos Matrices.** As discussed briefly above, the Lanczos procedures with no reorthogonalization are based upon the premise that if a large enough Lanczos matrix is generated, then all of the distinct eigenvalues of the original matrix will appear in the spectrum of that Lanczos matrix. However the Lanczos spectrum will also contain other 'spurious' eigenvalues generated by the losses of orthogonality. It is necessary to identify such eigenvalues. As demonstrated in Cullum and Willoughby [1983] it is not sufficient to use the computed error estimates to make this identification.

Two types of tests have been proposed for sorting the eigenvalues of any given Lanczos matrix into the 'good' eigenvalues which are retained as approximations to the eigenvalues of the original matrix, and the 'spurious' eigenvalues, which are caused by the losses in the orthogonality of the Lanczos vectors and are therefore discarded. The first test which was suggested is a fuzzy test in the sense that the user specifies a convergence tolerance and an increment $K$, and the procedure checks to see which eigenvalues of the two Lanczos matrices $T_m$ and $T_{m+K}$ are the same to within this convergence tolerance. Eigenvalues which agree to this tolerance are accepted as 'good' eigenvalues, the other eigenvalues are discarded. See for example van Kats and van der Vorst [1976, 1977] and Parlett and Reid [1981].

Our identification test is quite different. For a given Lanczos matrix $T_m$ it uses the associated submatrix which we denote by $T(2, m)$ obtained by deleting the first row and column of $T_m$. Furthermore, it does not identify the 'good' eigenvalues directly. Instead it identifies the 'spurious' eigenvalues directly as those simple eigenvalues which are eigenvalues of both $T_m$ and of $T(2, m)$ to within the precision of the associated eigenvalue computations. The remaining eigenvalues not classified as spurious are kept as the good eigenvalues. Thus, it obtains eigenvalues approximations which are accurate to varying degrees of approximation. This ability to 'see' eigenvalues before they appear to the desired convergence accuracy is used in the shift and invert strategy which we propose in Section 4. A justification for this identification test is given in Section 3.4.

For real symmetric matrices, the cost of this identification test is very cheap, a few multiplications per eigenvalue, and this test is incorporated directly into the bisection subroutine which is used to compute the eigenvalues of the Lanczos matrices. By running the bisection recursions backwards up through the Lanczos matrix it is possible to determine with a few extra multiplications both an eigenvalue of $T_m$ and whether or not that eigenvalue is multiple or an eigenvalue of the submatrix $T(2, m)$. In the nonsymmetric procedures this identification is done using inverse iteration on the Lanczos matrices. More will be said about the nonsymmetric case in Section 3.

**2.5 Residual Estimates of Convergence.** The Lanczos procedures with reorthogonalization and the procedures without reorthogonalization both rely on estimates of the residual norms in solving Eqns. (1.1) to estimate the accuracy of the computed eigenvalue approximations.

In both cases the error term $E_m$ in Eqn. (2.3) will be nonzero. For real symmetric matrices, Paige [1976, 1980] has proven that if there is no reorthogonalization, then $E_m$ is proportional to the product of the size of the machine $\epsilon$, and the norm of the original matrix. Thus, it is small. For any Ritz vector $z = V_m u / \| V_m u \|$, where $u$ is an eigenvector of $T_m$ corresponding to an eigenvalue $\mu$ of $T_m$, the corresponding residual vector $Az - \mu z$ satisfies the following inequality.

$$\| Az - \mu z \| \leq [| \beta_{m+1} \| u(m) | + | E_m |] / \| V_m u \|. \qquad (2.4)$$

Paige [1971] has also proven that $\| V_m u \|$ is 'not small'. This is stated imprecisely but sufficiently for this discussion. Thus, the size of this bound on the residual is controlled by the size of the last component of the corresponding eigenvector of the Lanczos matrix which is used to generate the Ritz vector. For any real symmetric matrix the size of the residual is an upper bound on the error in the eigenvalue approximation $\mu$. This is not however true for nonsymmetric matrices. For nonsymmetric matrices you must incorporate the condition number of the particular eigenvalue being considered. However, even in the nonsymmetric case, Lanczos procedures and other iterative procedures rely upon similar estimates of convergence since typically these are the only computable quantities.

**2.6 Orthogonality of the Ritz Vectors Generated.** It is not a priori evident that the Ritz vectors which a real symmetric Lanczos procedure with no reorthogonalization produces for different eigenvalues will be orthogonal because the corresponding Lanczos vectors are not orthogonal, whereas the Ritz vectors obtained from a procedure based on selective reorthogonalization are nearly orthogonal because the associated Lanczos vectors are nearly orthogonal. In fact, in a Lanczos procedure which does not use reorthogonalization the Lanczos vectors may not even be linearly independent. However, the real symmetric eigenvector procedure which we use produces Ritz vectors which are orthogonal.

Consider the following argument which rests upon the assumption that the principal losses in orthogonality occur along converged Ritz vectors and that such losses occur only if some eigenvector of a Lanczos matrix has a small last component. In our real symmetric procedure, for each 'converged' eigenvalue approximation $\mu$ obtained from some Lanczos matrix $T_M$, we use Sturm sequencing to determine a size $J(\mu)$ at which $\mu$ first appears in the spectrum of $T_j$ as 'converged' to some tolerance. Then starting at the size matrix obtained from the Sturm sequencing, we recursively determine a larger Lanczos matrix $T_{K(\mu)}$ for which the corresponding eigenvector $u$ of $T_K$ corresponding to $\mu$ has a sufficiently small last component $|u(K)|$. We then compute the corresponding Ritz vector by $z = V_{K(\mu)}u$. If $\eta$ is another 'converged' eigenvalue, and $K(\eta)$ is the corresponding size Lanczos matrix for $\eta$, then we define the corresponding Ritz vector by $x = V_{K(\eta)}w$ where $T_{K(\eta)}w = \eta w$.

Paige [1971] proved that losses of orthogonality of the Lanczos vectors occur along the directions of converged Ritz vectors when convergence occurs, as measured by the size of the last components of eigenvectors of Lanczos matrices. Numerical experiments indicate that these correspond to the principal losses of orthogonality in the Lanczos vectors. Therefore, assuming without loss of generality that $K(\eta) > K(\mu)$, we have that $x$ must still be orthogonal to $V_{K(\mu)}$ so that $x^T z = x^T V_{K(\mu)} u = 0$.

**2.7 Observed Convergence Behavior.** For both types of Lanczos procedures, those with selective reorthogonalization and those with no reorthogonalization, and within any localized portion of the spectrum of the original matrix, it is observed numerically that the convergence is dominated by the gaps between the eigenvalues in that region and that the computed error estimates can be used to determine whether or not all of the eigenvalues in such a region have converged. Non-convergence is reflected in poor error estimates. For the real symmetric Lanczos procedures with no reorthogonalization, this information can then be used to identify subintervals which encompass the eigenvalues that have not yet converged. The eigenvalue computations can then be redone on a larger Lanczos matrix but just on those subintervals.

**2.8 Repetitions of Eigenvalues in the Spectrum of the Lanczos Matrices.** If the Lanczos vectors are sufficiently orthogonal, then there should not be any repetition of eigenvalues in the spectra of the associated Lanczos matrices. However, in procedures with no reorthogonalization, multiple copies of eigenvalues of the original matrix can and do appear in the spectra of the associated Lanczos matrices even if the original matrix has only simple eigenvalues, and in spite of the fact that if all of the subdiagonal entries of an associated Lanczos matrices are nonzero, it is theoretically impossible for this to happen.

How much duplication of eigenvalues occurs depends upon the eigenvalue gap structure within the original matrix. Repetitions however do not cause convergence problems. In fact if an eigenvalue of a Lanczos matrix is numerically multiple, then we can infer from that that the corresponding residual vector for that eigenvalue is small and that eigenvalue approximation has in fact converged. See Paige [1971]. In most cases the only problem which is incurred by such repetitions is that the Lanczos recursion has to be extended further to generate all of the desired eigenvalues. In extreme cases however, where there are very dominant eigenvalues with huge gaps, repetition can limit the convergence of small eigenvalues.

**2.9 Arithmetic Operation Counts Required.** In a Lanczos procedure with no reorthogonalization the Lanczos matrices required to compute the desired eigenvalues will typically be larger than the Lanczos matrix required to compute the same eigenvalues when selective reorthogonalization is used. However, selective reorthogonalization involves other costs, in particular reorthogonalizations of the Lanczos vectors and storing and processing both Lanczos vectors and Ritz vectors. Both types of procedures use the same residual estimates.

**2.10 Differences in Iteration Method.** In our Lanczos procedures with no reorthogonalization there may be no iteration in the ordinary sense. For a given eigenvalue problem, typically a Lanczos matrix of some size will be generated and an eigenvalue computation will be done on that matrix. If the desired convergence has not yet occurred, as can be determined by looking at the error estimates in the intervals of interest, that matrix is increased in size and the eigenvalue computation is repeated on the larger Lanczos matrix on subintervals not yet converged. Typically only a few different sized Lanczos matrices would be considered.

In a Lanczos procedure with selective reorthogonalization the recursion progresses continuously with the procedure stopping aperiodically to do Ritz vector and eigenvalue computations and certain reorthogonalizations. The procedure then continues the Lanczos recursion until other recursions monitoring the orthogonality indicate that a new eigenvalue has converged at which point in time the procedure pauses again, to do another eigenvalue and Ritz vector computation. This pattern repeats itself until the desired convergence is achieved. In such a procedure, the number of Lanczos matrices which must be considered is a function of the number of eigenvalues which converge prior to the convergence of the desired eigenvalues.

**2.11 Behavior of Lanczos Procedures with No Reorthogonalization.** It is important for a user to understand that Lanczos procedures with no reorthogonalization behave quite differently from procedures that incorporate reorthogonalization. If such a code is run on two different types of computers with the same original matrix $A$ and the same initial vector $v_1$, the computed results on the two computers are typically quite different due to the differences in the computer arithmetic. In practice, the two sets of Lanczos matrices may agree for a certain number of Lanczos steps but then begin to diverge upon the convergence of one or more of the eigenvalues of the Lanczos matrices to eigenvalues of the original matrix. If after a number of subsequent steps of the Lanczos recursion, we compare the entries in the two sets of Lanczos matrices, typically they are totally different.

Furthermore, if we compute the eigenvalues of the two sets of Lanczos matrices for various sizes and 'spurious' eigenvalues are present, then typically these spurious eigenvalues are quite different on the different computers and can even appear in different parts of the spectrum. Prior to the convergence of a given 'good' eigenvalue the values of the eigenvalue approximations to that eigenvalue generated on the different computers may differ. However, once a 'good' eigenvalue in either set of Lanczos matrices has converged, that converged 'good' eigenvalue will agree with a true eigenvalue of the original user-specified matrix. Moreover for a given level of arithmetic precision, the basic convergence patterns are the 'same' from computer to computer, meaning only that for a given size Lanczos matrix we see approximately the same amount of the spectrum converged.

We also observe that, as might be expected, if the precision of the computer arithmetic is varied, then the convergence of the eigenvalue approximations generated by the corresponding Lanczos procedure with no reorthogonalization is affected. If the precision is increased, the approximations converge more rapidly. If the precision is decreased, the approximations converge more slowly.

**2.12 Robustness of the Procedures.** It is interesting to note a certain nonrobustness of the Lanczos codes with no reorthogonalization which is actually useful in identifying bugs in user supplied subroutines. These Lanczos procedures will fall apart if there are inconsistencies in the user-supplied codes for the matrix-vector multiplies. For example, if something is being overwritten within such a subroutine so that each time that subroutine is called we get the equivalent of a small perturbation in the user supplied matrix, this will become evident in the nonconvergence of the spectrums of the associated Lanczos matrices. In one instance we had a user who was solving a partial differential equation by discretization and reduction to eigenvalue problems, who discretized incorrectly thereby generating unknowingly a slightly nonsymmetric matrix. The use of that subroutine in our real symmetric Lanczos codes produced partially correct eigenvalue approximations. Part of the spectrum was being approximated correctly, the rest was diverging, indicating quite clearly that something was wrong in the matrix setup and/or matrix vector subroutines. On more than one occasion this behavior has been useful in uncovering subtle bugs in user-supplied codes. However, it cannot help uncover bugs in codes which are consistently generating an incorrect matrix. As long as the information being provided is consistent, these Lanczos codes will consistently compute the eigenvalues of the matrix supplied.

**2.13 Complexity of the Codes.** A Lanczos procedure which uses selective reorthogonalization must include a tracking mechanism to determine when the Lanczos vectors are no longer sufficiently orthogonal and reorthogonalizations must be performed. It must keep all of the Lanczos vectors on some sort of readily-accessible storage along with the 'converged' Ritz vectors which are computed as the procedure progresses. The amount of such storage needed is not known a priori. Eigenvalues and eigenvectors of the Lanczos matrices must be computed as the procedure progresses.

On the other hand our Lanczos procedures with no reorthogonalization retain the simplicity of the original Lanczos recursion. No reorthogonalizations of the Lanczos vectors are performed and only the two most recent Lanczos vectors are needed as the procedure progresses. The 'complexity' is relegated to the eigenvalue portion of the computations. But as we said earlier that complexity consists of a simple tridiagonal, 'spurious' identification test which in the real symmetric case costs only a few multiplications per eigenvalue and is done as the eigenvalues are computed in the bisection subroutine.

**2.14 Theoretical Results.** The arguments for convergence of Lanczos procedures with selective reorthogonalization rest upon extensions of classical results to the case of 'sufficiently orthogonal' and hence linearly independent Lanczos vectors. See Parlett [1980], Chapter 13 for details. In his thesis [1971] and in a series of subsequent papers [1976], [1980], Paige presented several results which are key to the use of Lanczos procedures with no reorthogonalization. More will be said about these in the next Section. The theory however even for real symmetric procedures of this type is still not complete, and for such nonsymmetric Lanczos procedures there is only numerical evidence that they can produce good approximations to eigenvalues and eigenvectors of many nonsymmetric problems.

In the next Section we focus on our Lanczos procedures with no reorthogonalization and consider the differences between our procedures for real symmetric matrices and our procedures for nonsymmetric matrices.

## 3 Lanczos Procedures with No Reorthogonalization, Symmetric versus Nonsymmetric Procedures.

Lanczos procedures with no reorthogonalization have been devised for both real symmetric and nonsymmetric matrix eigenvalue problems. No reorthogonalization is done in either procedure. In these procedures the eigenvalue and eigenvector computations are separated. The eigenvalues are computed first. Once the eigenvalues are obtained, eigenvector approximations are obtained, but only for those eigenvalues specified by the user, by computing associated Ritz vectors as described in Section 2.1 for the real symmetric case.

Our nonsymmetric Lanczos procedure uses a variant of the following two-sided generalization of the Lanczos recursion given for example in Wilkinson [1965]. For $i = 1, 2, \ldots, m$ generate Lanczos vectors $W_m \equiv \{w_1, \ldots, w_m\}$ and $V_m \equiv \{v_1, \ldots, v_m\}$ and scalars $\gamma_{i+1}, \beta_{i+1}$ and $\alpha_i$ such that

$$\beta_{i+1} v_{i+1} = A v_i - \alpha_i v_i - \gamma_i v_{i-1} \equiv r_{i+1}$$
$$\gamma_{i+1} w_{i+1} = A^T w_i - \alpha_i w_i - \beta_i w_{i-1} \equiv t_{i+1} \qquad (3.1)$$

where

$$\alpha_i \equiv w_i^T A v_i, \quad \gamma_i \equiv w_{i-1}^T A v_i \quad \text{and} \quad \beta_i \equiv v_{i-1}^T A^T w_i. \qquad (3.2)$$

The coefficients in Equations (3.2) are chosen so that the resulting sets of vectors $V_m$ and $W_m$ are biorthogonal. The original matrix $A$ is mapped into a family of tridiagonal matrices

$$T_j \equiv \begin{bmatrix} \alpha_1 & \gamma_2 & & \\ \beta_2 & \alpha_2 & \ddots & \\ & \ddots & \ddots & \gamma_j \\ & & \beta_j & \alpha_j \end{bmatrix}, \quad 1 \leq j \leq m \qquad (3.3)$$

We observe first, that as in the real symmetric case, the original matrix $A$ (and $A^T$) is not modified by the recursions. We observe second, that the nonsymmetric matrix $A$ is not transformed into a Hessenberg matrix as in the standard eigenvalue procedures but is transformed into tridiagonal form. Wilkinson [1965] tells us that such tridiagonal transformations may not be stable for some nonsymmetric matrices, and in fact the recursions in Equations (3.1) and (3.2) will break down if $w_i^T v_i = 0$ for some $i$. In any of our discussions about the nonsymmetric procedure and how it behaves we are always assuming that such breakdown has not occurred. In practice, using complex arithmetic, we have not to our knowledge, encountered such a breakdown yet. However, others, for example, Parlett, Taylor and Liu [1985] who are working in real arithmetic, have constructed examples where breakdown does occur so this is a problem that cannot be ignored and must be understood. Breakdown however in complex arithmetic, is less likely to occur since it requires the simultaneous vanishing of two real inner products. More details about our nonsymmetric procedure can be found in Cullum and Willoughby [1986] and Cullum, Kerner, and Willoughby [1989].

In our discussion we assume $A$ is diagonalizable (nondefective) because defective eigenvalues cannot be computed accurately and the tolerances which our procedure uses assume that the eigenvalues of the Lanczos matrices can be computed accurately. In practice, if our procedure were used on a defective matrix, any nondefective eigenvalues would be computable. However, our tolerances for determining multiple or spurious eigenvalues of the Lanczos matrices would not be functional on any eigenvalue approximations corresponding to defective eigenvalues of $A$ and user intervention would be required to deal with such eigenvalue approximations. In practice, in finite precision arithmetic, we must introduce error terms in Equations (3.1) analogous to $E_m$ in Equation (2.3).

### 3.1 Theoretical Results and Assumptions.

For the real symmetric Lanczos recursion, Paige [1971, 1976, 1980] proved a number of theoretical results about the behavior of that recursion in finite arithmetic which we used to justify our Lanczos procedures for real symmetric problems. In Cullum and Willoughby [1985], Chapter 4, we used these results together with other results we derived to show that the local near-orthogonality of the Lanczos vectors demonstrated by Paige was sufficient to demonstrate that an associated sequence of conjugate gradient iterations has properties which we then related to the possible 'convergence' of our Lanczos procedure. (There is no actual convergence in the ordinary sense because the finite precision arithmetic prevents us from letting $m$ go to $\infty$). For more details see Chapter 4.

Paige also derived lower bounds on the norms of Ritz vectors obtained from the Lanczos matrices. Such lower bounds are needed to be able to infer from the size of the last component of an eigenvector of a Lanczos matrix that the residual vector for the corresponding Ritz vector is small. Paige also showed that the losses in orthogonality which occur when an eigenvalue approximation converges are along the directions of the corresponding converged Ritz vectors. This behavior is used to justify the use of Lanczos procedures with selective reorthogonalization which reorthogonalize the Lanczos vectors only along these directions. We do not however need such a result in our procedures. Another result Paige proved was that the spectrum of each Lanczos matrix is fully contained within the convex hull of the spectrum of the original matrix.

None of the preceding results have yet been extended to the nonsymmetric recursions. Currently the justification for our nonsymmetric Lanczos procedure with no reorthogonalization rests solely upon the results of our numerical experiments. See Cullum and Willoughby [1985, 1986] and Cullum et al [1989] for the results of extensive numerical experiments. In particular to justify our nonsymmetric procedure, we need first an argument that indeed the eigenvalues of the original matrix will appear in the spectrums of the Lanczos matrices if we make these matrices large enough. Currently, we have only numerical evidence that this does occur. We also need an analysis of the breakdown mechanism for these recursions. There is not yet published a matrix characterization of this breakdown. We also need to obtain lower bounds on the norms of associated Ritz vectors. In our discussions we will make the following assumptions for our nonsymmetric Lanczos procedure.

**Assumption 3.1.** For any nonsymmetric matrix $A$ considered and large enough $m$, any distinct eigenvalue of $A$ will appear in the spectrums of the Lanczos matrices $T_m$ generated by Recursions (3.1) and (3.4).

**Assumption 3.2.** The error terms which appear in Equations (3.1) when finite precision arithmetic is used stay uniformly small or grow very slowly as $m$ increases.

**Assumption 3.3.** The $\| v_j \|$ are 'not large' and for any isolated simple eigenvalue of a Lanczos matrix $T_m$ the norm $\| z \| = \| V_m u \|$ of a corresponding Ritz vector is 'not small'.

We need Assumption 3.2 for both our argument that our identification test makes sense and that the size of $| u(m) |$ for any $T_m u = \mu u$ is a viable estimate of the accuracy of $\mu$ when considered as an eigenvalue of $A$. Assumption 3.3 is also needed in justifying the use of $| u(m) |$. In addition we also need an assumption that is not a concern in the real symmetric case. In all of our discussions we have to assume that the nonsymmetric recursions do not encounter breakdown.

**Assumption 3.4.** For a given $A$ and for the particular starting vectors used, and the size $m$ Lanczos matrices considered, $w_i^T v_i \neq 0$ for $j = 1, \ldots, m$.

Thus, there are many open questions for our nonsymmetric Lanczos procedures. There are however also open questions for the real symmetric procedures. Paige's results in [1971, 1976, 1980] do not contain a convergence proof in any practical sense. He proved that for some $m < n+1$ that at least one of the eigenvalue approximations generated by the Lanczos procedure will be an accurate approximation to an eigenvalue of the original matrix. From that he showed that by $m = n^2$ at least $n/2$ eigenvalue approximations must have converged. For $n$ large, $n^2$ is very large. In practice, typically $n/2$ of the eigenvalues and often many more that that, are obtained for $m$ a very small multiple of $n$. The argument in Cullum and Willoughby [1985] provides a plausible explanation for this observed behavior but since we were not able to obtain bounds on some of the terms arising from the finite precision arithmetic, we were not able to get a complete proof.

Thus we are using both the symmetric and the nonsymmetric algorithms without proofs of convergence. However, in both cases, given computed quantities $(\mu, x)$ we can check the validity of these quantities by looking at the sizes of the associated residuals $\| Az - \mu z \|$. For a real symmetric problem this test is sufficient to guarantee that the computed quantities are accurate. For nonsymmetric problems this test is also sufficient if the quantities being computed have reasonable condition numbers. If the condition numbers are not reasonable, then the quantities cannot be computed accurately by any method.

**3.2 Orthogonality Becomes Biorthogonality.** The orthogonality of the Lanczos vectors in the real symmetric case is replaced by the biorthogonality of the two sets of vectors $V_m \equiv \{v_1, \ldots, v_m\}$ and $W_m = \{w_1, w_2, \ldots, w_m\}$ for some $m \leq n$. The corresponding projection matrices $T_m \equiv W_m^T A V_m$ are biorthogonal projections of the original matrix onto the Krylov subspaces $K_m \equiv \{v_1, A v_1, A^{m-1} v_1\}$ and $\bar{K}_m = \{w_1, A^T w_1, \ldots, (A^T)^{m-1} v_1\}$. See Saad [1980-1985] for a discussion of such oblique projections. This biorthogonality corresponds to the biorthogonality of the left and right eigenvectors of $Ax = \lambda x$ when $A$ is a nonsymmetric matrix.

**3.3 Real Symmetric Tridiagonal Matrices Become Complex Symmetric Tridiagonal Matrices.** There is some freedom in choosing the scaling factors $\beta_{i+1}$ and $\gamma_{i+1}$ in Equations (3.1). Only the product of these two quantities is specified by the biorthogonality conditions. We choose $\beta_{i+1} = \gamma_{i+1}$. With this choice the tridiagonal Lanczos matrices defined in Eqn(3.3) become symmetric but complex. Moreover, to improve numerical stability we use the following formulas for $\alpha_i$ and $\beta_{i+1}$.

$$\alpha_i \equiv (\alpha_i^v + \alpha_i^w)/2 \quad \text{and} \quad \beta_{i+1}^2 = \left(r_{i+1}^T t_{i+1}\right) \quad (3.4)$$
$$\text{where} \quad \alpha_i^v \equiv w_i^T(A v_i - \beta_i v_{i-1}) \quad \text{and} \quad \alpha_i^w \equiv v_i^T\left(A^T w_i - \beta_i w_{i-1}\right).$$

Complex symmetric tridiagonal matrices have several desirable properties which are discussed in Cullum and Willoughby [1985, 1986]. In particular, this choice allows us to justify easily the direct extension of our identification test for spurious eigenvalues to our nonsymmetric procedures. It is also easy to obtain a generalization to complex symmetric tridiagonal matrices of the implicit QL algorithm for computing the eigenvalues of a real symmetric tridiagonal matrix which requires no additional storage and which in terms of rates of convergence behaves very similarly to the real symmetric QL algorithm. This generalization allows us to work with large tridiagonal matrices and therefore to get quite a bit of information about the spectrum of the original nonsymmetric matrix.

**3.4 Identification of Spurious Eigenvalues.** The losses in orthogonality observed for the real symmetric case translate into losses in biorthogonality in the nonsymmetric case, and 'spurious' eigenvalues appear in the spectra of the Lanczos matrices. The identification test for 'spurious' eigenvalues is the same in both the real symmetric and the nonsymmetric procedures. In each case, for any Lanczos matrix $T_m$, a simple eigenvalue of that matrix which is also an eigenvalue of the corresponding submatrix $T(2, m)$ obtained by deleting the first row and column of the Lanczos matrix, is declared as spurious. A justification for this test is given in Cullum and Willoughby [1986]. However since this test is at the heart of our Lanczos procedures some of the discussion in that reference is repeated below. For both the real symmetric and the nonsymmetric Lanczos recursions we have that for any eigenvalue $\mu$ and corresponding eigenvector $u$ of any Lanczos matrix $T_m$ that the corresponding residual norm

$$\frac{\| Az - \mu z \|}{\| z \|} = \frac{| \beta_{m+1} | \; \| v_{m+1} \| \; | u(m) |}{\| z \|} + \bar{E}_m \tag{3.5}$$

where $z = V_m u$ is the corresponding Ritz vector, and $\bar{E}_m$ is bounded by the error term $\| E_m \| / \| z \|$.

For the real symmetric procedure, by construction, $\| v_j \| \equiv 1$ for all $j$. Furthermore, for that case Paige proved that if $\mu$ is any isolated simple eigenvalue of $T_m$, then the corresponding norm $\| z \| = \| V_m u \|$ is 'not small'. So for real symmetric problems the size of this residual in Equation (3.5) is controlled by the size of $| u(m) |$. For nonsymmetric problems we do not have an a priori bound on $\| v_{m+1} \|$ and there is no proof that $\| z \|$ is 'not small'. In fact we do not even have a bound on the error terms $E_m$. However numerical experiments indicate that results analogous to those obtained by Paige [1971, 1976] for the real symmetric case must extend to some significant class of nonsymmetric problems, and in the discussions which follow we are assuming that such extensions exist.

In both the real symmetric and the nonsymmetric cases, the Lanczos matrices are symmetric and we have that the $m$th component of an eigenvector of $T_m$ corresponding to a simple eigenvalue $\mu$ satisfies

$$| u(m) |^2 = \frac{| a_{m-1}(\mu) \hat{a}_2(\mu) |}{| a'_m(\mu) \hat{a}_2(\mu) |} = \frac{| \prod_{k=2}^{m+1} \beta_k |}{| a'_m(\mu) | \; | \hat{a}_2(\mu) |}. \tag{3.6}$$

In Equation (3.6), $\hat{a}_2(\mu)$ denotes the characteristic polynomial of $T(2, m)$ evaluated at $\mu$; and $a'_m(\mu)$ denotes the derivative of the characteristic polynomial of $T_m$ evaluated at $\mu$. Therefore for any isolated eigenvalue $\mu$ of $T_m$ the size of $| u(m) |$ and thus the size of the corresponding residual in Equation (3.5) is controlled by the size of $| \hat{a}_2(\mu) |$. Whenever $\mu$ is simultaneously in the spectrums of $T_m$ and $T(2, m)$, $| u(m) |$ will be pathologically large compared to its size for any $\mu$ that is not in the spectrum of $T(2, m)$.

If a Ritz pair $(\mu, z)$ has a pathologically large residual then it cannot be an eigenvalue, eigenvector pair of $Ax = \lambda x$ so we reject such $\mu$ (and $z$) as 'spurious'. All eigenvalues of $T_m$ that are not labelled 'spurious' are accepted as approximations to eigenvalues of $A$. In practice, for reasons not well understood, spurious eigenvalues are readily recognizable because the observed agreement between such eigenvalues as eigenvalues of $T_m$ and of $T(2, m)$ is to the accuracy to which those eigenvalues can be computed. In the real symmetric case we see such agreements to 10 or more digits.

The implementation of this identification test differs in the two procedures. In our real symmetric procedure this test is an integral part of the bisection eigenvalue computations and costs only a few multiplications per eigenvalue. This is accomplished by simultaneously computing the eigenvalues of the Lanczos matrix and of the submatrix $T(2, m)$ by using backward Sturm sequencing within the bisection subroutine. Unfortunately, we do not have an analogous bisection procedure for computing the eigenvalues of complex symmetric tridiagonal matrices. Therefore, in the nonsymmetric case we first compute the eigenvalues of $T_m$ and then use inverse iteration on $T(2, m)$. Each simple eigenvalue of $T_m$ is subjected to one step of a Rayleigh quotient, inverse iteration on that Lanczos matrix. The inverse iteration can simultaneously determine whether or not an eigenvalue is 'spurious' and compute an error estimate if the eigenvalue is 'good'. It is done by sweeping backwards up the matrix. If a particular simple eigenvalue of $T_m$ is also an eigenvalue of $T(2, m)$, the newly-computed Rayleigh quotient for that submatrix should be the same as the eigenvalue of the Lanczos matrix. Otherwise the quantities will differ. Thus, if the eigenvalue and the Rayleigh quotient agree, the eigenvalue is labelled as spurious. Otherwise the inverse iteration is completed to obtain an eigenvector of the Lanczos matrix which is then used to obtain an error estimate for that 'good' eigenvalue.

**3.5 Estimates of Convergence, Use of Residual Bounds.** In the real symmetric case an estimate of the norm of the residual vector typically provides a good estimate of the accuracy of the eigenvalue, although even in that case a residual bound is only an upper bound on the actual error. In the nonsymmetric case, however, the norm of the residual vector is not sufficient to estimate the accuracy of the corresponding eigenvalue approximation. The eigenvalues of a real symmetric matrix are perfectly conditioned, in the sense that if a small perturbation in the matrix occurs, then the eigenvalues of the perturbed matrix will differ from the eigenvalues of the original matrix by at most the size of the perturbation. This is not true for nonsymmetric matrices. Small perturbations in a nonsymmetric matrix may yield a matrix whose eigenvalues differ significantly from the eigenvalues of the original matrix. To obtain realistic estimates of the accuracy of computed eigenvalues of a nonsymmetric matrix it is necessary to also know the condition numbers of those eigenvalues which are a measure of the sensitivity of those eigenvalues to perturbations in the matrix. In practice however, these numbers are not known and all iterative eigenvalue procedures for large matrices rely on the sizes of residual norms as estimates of the accuracy of any computed eigenvalues (and eigenvectors). One could however use the approximations to the left and right eigenvectors to get an estimate of the condition number of those eigenvalues for which these eigenvectors are being computed.

In our nonsymmetric procedures, as in the real symmetric procedures, an estimate of any residual bound is obtained from the size of the last component of a corresponding eigenvector of the Lanczos matrix. Eigenvectors of complex symmetric tridiagonal matrix are easily computed using inverse iteration, as in the real symmetric case.

**3.6 Eigenvalue Computations.** For both our real symmetric and nonsymmetric Lanczos procedures it is often feasible to compute large portions of the spectrum of the matrix being used in the Lanczos recursions. In the real symmetric procedures we have the added flexibility that we can focus on any part of the spectrum which is desired without doing the Lanczos matrix eigenvalue computation on the entire spectrum of that matrix. The bisection subroutine for computing eigenvalues of real symmetric tridiagonal matrices allows us to confine those computations to any subintervals specified by the user. Of course it is still true that in the process of generating a Lanczos matrix large enough to contain the desired portion of the spectrum we may also have generated a matrix which contains many other parts of the spectrum which we were not interested in computing. Thus, the Lanczos matrix generation portion of the computations is not independent of the portion of the spectrum desired.

In our nonsymmetric Lanczos procedure we use a complex analog of the implicit QL algorithm IMTQL1 in EISPACK [1977]. We call this analog CMTQL1. The transformations $Q$ used in CMTQL1 are complex symmetric, and satisfy $Q^T Q = I$. CMTQL1 requires the existence of a $QL$ factorization for each tridiagonal matrix generated by that procedure. For more details see Cullum and Willoughby [1987]. In that reference we proved that under the assumption that the required $QL$ factorizations exist, CMTQL1, with all shifts set to zero, converges to a diagonal matrix containing the eigenvalues of the original tridiagonal matrix $T$ whenever these eigenvalues have distinct magnitudes. In practice however this procedure works well even on matrices with numerically-multiple eigenvalues. There are however many open questions related to this procedure.

**3.7 Ritz Vector Computations.** In both our procedures eigenvector approximations are obtained after the eigenvalues have been computed. For each eigenvalue $\mu$ considered, both procedures use inverse iteration on the Lanczos matrices $T_m$ to obtain a suitable size matrix $T_{M(\mu)}$ to use in a Ritz vector computation. The two procedures differ only in how they select the first guess at $M(\mu)$. In the real symmetric case Sturm sequencing can be used to obtain a guess at $M(\mu)$ which is often suitable without any further iterations on the sizes of the Lanczos matrices. The accuracy of the computed Ritz vectors as eigenvectors of $Ax = \lambda x$ is checked by computing the associated residuals $\| Az - \mu z \|$ where $T_{M(\mu)} u = \mu u$ and $z = V_{M(\mu)} u$.

**3.8 Computer Storage and Arithmetic Operations Required.** Both of our procedures have small storage requirements for the eigenvalue computations. The 6.5 double precision vectors of length $\max(m, n)$ required in our real symmetric Lanczos procedure approximately double in number and become double precision complex vectors in our nonsymmetric eigenvalue procedure. All of the computations in the nonsymmetric procedure are done in complex arithmetic even when the original matrix is real, and in the nonsymmetric procedure we have to work with vectors $Ax$ and $A^T y$.

The amount of storage required in the eigenvector computations is a function of how many eigenvectors are to be computed on each call to that program and how difficult it was to compute those eigenvalues. The Lanczos vectors are regenerated in the eigenvector computations so that whatever storage was required in the eigenvalue computations is again required here plus storage for the eigenvector of a Lanczos matrix associated with each eigenvalue considered and the corresponding Ritz vector for each such eigenvalue. Various tradeoffs can be made, depending upon the amount and type of computer storage available.

**3.9 Repetitions of Eigenvalues in the Spectrum of the Lanczos Matrices.** Neither our real symmetric nor our nonsymmetric Lanczos procedures can directly determine the multiplicity of the eigenvalues of the original matrix. These procedures compute only an approximation to each distinct eigenvalue of $A$. In some cases it is feasible to do auxiliary computations that allow you to determine whether or not a given computed eigenvalue is a multiple eigenvalue of $A$ but they still will not determine the actual multiplicity.

Because of the losses in orthogonality or biorthogonality simple or multiple eigenvalues of $A$ can be replicated in the spectrums of the Lanczos matrices. The amount of replication however has no relationship to the multiplicities of these eigenvalues in the spectrum of $A$.

**3.10 Observed Convergence Behavior.** The observed convergence behavior of our two Lanczos procedures is very similar. In both procedures numerical evidence indicates that losses in orthogonality correspond to convergence and that the rates of convergence are dominated by the overall eigenvalue gap structure, the dominance or nondominance of the desired eigenvalues, and their relative locations in the spectrum of the matrix being used in the Lanczos recursions.

Both our real symmetric and our nonsymmetric Lanczos procedures work well on many problems. As stated earlier the observed convergence depends upon the eigenvalue spectrum and upon what part of the spectrum is being computed. In any of the Lanczos algorithms with or without reorthogonalization, approximations to eigenvalues that are 'prominent' in the spectrum are obtained first as the iterations proceed. With a Lanczos procedure however, it is possible to get eigenvalues which are not prominent by extending the Lanczos recursion until those eigenvalues appear in the spectrums of the Lanczos matrices. In some cases, it is also possible to speed up this convergence by using a shift and invert strategy in combination with the basic procedure being used. The choice of appropriate shifts and the subsequent 'inversions' of the shifted matrix transform the desired eigenvalues into 'prominent' eigenvalues which can then be more readily computed by the Lanczos procedure.

In the next Section we propose such a shift and invert strategy for our Lanczos procedures with no reorthogonalization. The 'look-around' capabilities of these methods provide information which can be utilized to construct such a strategy. This idea could probably also be used in a Lanczos procedure with reorthogonalization but it may not be as obvious how to select the various parameters used in the constructions in the next Section.

The comments in Sections 2.11 and 2.13 apply equally well to our nonsymmetric Lanczos procedure.

## 4  A Shift and Invert Strategy for Lanczos Procedures.

In some applications, for example, in many structural analysis problems, instead of working with the original matrix it has been customary and in some cases necessary to work with a shifted and inverted version of the problem. In any problem where the desired eigenvalues are very small compared to other portions of the spectrum of the original matrix, iterative eigenvalue procedures have great difficulty computing such quantities. Standard eigenvalue procedures may also have difficulty computing such eigenvalues to the desired accuracy.

In such a situation, for either a real symmetric or a nonsymmetric matrix problem, the given problem

$$Ax = \lambda Bx \tag{4.1}$$

can be replaced by shifted and inverted problems:

$$(A - \sigma B)^{-1} Bx = \mu x \quad \text{where} \quad \mu = 1/(\lambda - \sigma) \tag{4.2}$$

and the shifts $\sigma$ are chosen judiciously. The objective is to choose $\sigma$ so that the desired eigenvalues $\lambda$ are transformed into large $\mu$ and any unwanted dominant $\lambda$ are simultaneously damped out. In practice a sequence of shifts $\sigma_i$, $1 \leq i \leq q$, may be used.

Explicit matrix inverses are not computed. The shifted matrices are factored, and the procedures use the factorizations $A - \sigma B = LDU$ to compute the required matrix-vector multiplies. The discussion in this Section assumes that such factorizations are feasible. $L$ denotes a unit lower triangular matrix. $U$ denotes a unit upper triangular matrix, and $D$ is a diagonal matrix. If $B$ is the identity matrix, problem (4.1) reduces to the standard eigenvalue problem.

We have devised shifted and inverted versions of our real symmetric and our nonsymmetric Lanczos codes, for both the standard and the generalized eigenvalue problems. In practice we replace the original problem Eqn. (4.1) by problems of the form

$$B(A - \sigma B)^{-1} Bx = \mu Bx \tag{4.3}$$

which are preferable numerically to the problems in Eqns. (4.2). In structural analysis problems often both matrices are real symmetric and the $B$ matrix is also positive semi-definite.

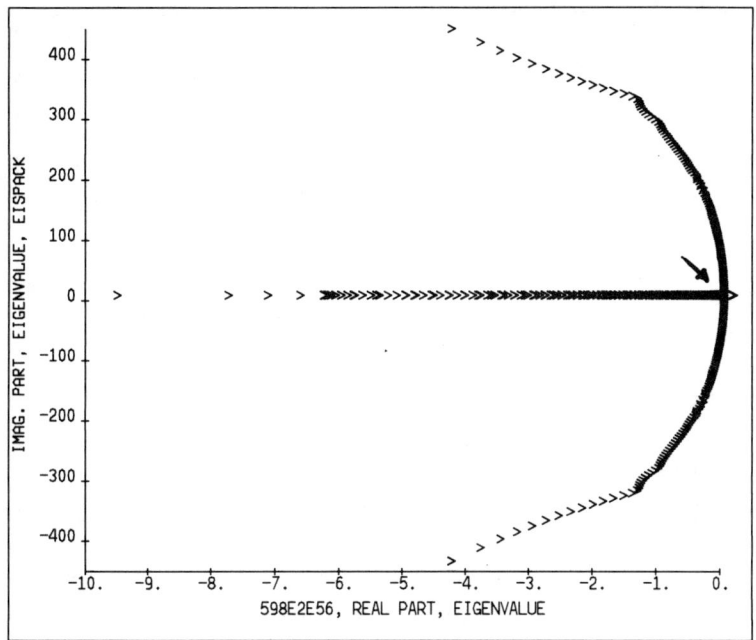

Figure 4.1: Spectrum, Magnetohydrodynamic example, equilibrium 2, $n = 598, \eta = 6*10^{-5}$.

For such problems all of the finite eigenvalues are real and furthermore, factorizations of shifted matrices can be used to count the numbers of eigenvalues in any subinterval. See Jennings [1977]. Such counts can be used to determine whether or not all of the desired eigenvalues have been computed, and to aid in the determination of appropriate shifts.

If however, neither the $A$ matrix nor the $B$ matrix is positive semi-definite or one of the matrices is nonsymmetric, then the eigenvalues need not be real and may be anywhere in the complex plane. In this situation, determining a shift and invert strategy is much more difficult. Typically for such a problem the user will specify a box of interest in the complex plane and want to compute the eigenvalues in that box. The eigenvalue procedure must then consider a sequence of shifts which move it through the box to find all of the eigenvalues in the box.

In Cullum, Kerner, and Willoughby [1989] we consider such a strategy for certain generalized nonsymmetric eigenvalue problems. The particular examples there arise in magnetohydrodynamics (MHD) problems but the procedure is applicable to other generalized problems. For these examples, the desired eigenvalues are the Alfven and sound mode spectrums. Figures 4.1 and 4.2 give the spectrum of a small MHD matrix, size $n = 598$, arising from a very crude discretization of the magnetohydrodynamic equations. See Cullum, Kerner, and Willoughby [1989] for details.

Figure 4.1 is a plot of the entire spectrum of this matrix obtained using codes from the EISPACK library [1977]. The eigenvalues of interest are marked in that Figure by the arrow. In these problems the desired eigenvalues are complex, are small in magnitude, have small gaps, are very interior to the overall spectrum, and are dominated by very large eigenvalues. As the discretization is refined, the magnitudes of the largest eigenvalues grow larger and this dominance increases. Figure 4.2 is a detailed plot of the eigenvalues of interest. Observe the different scalings in these two Figures. By choosing shifts near the desired eigenvalues and then working with the corresponding shifted and inverted problems, we can both amplify the desired eigenvalues and damp the large, unwanted eigenvalues. This

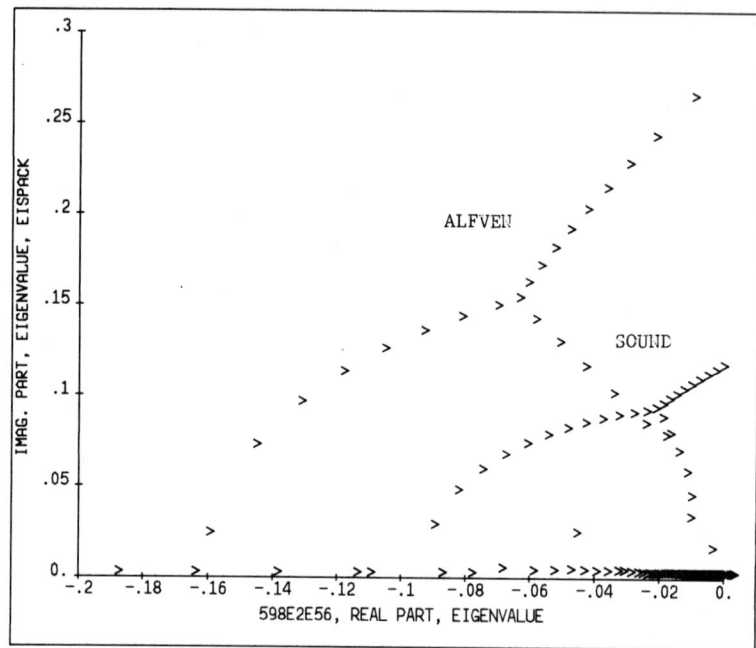

Figure 4.2: Alfven and Sound Spectra, Magnetohydrodynamic Example, Equilibrium 2, $n = 598$, $\eta = 6 * 10^{-5}$

example is treated in Section 5.

The basic problem to resolve is how to choose such shifts. There does not seem to be anything in the literature about shift selection when the eigenvalues are complex. In fact there are very few papers in the literature on algorithms for computing eigenvalues of large nonsymmetric matrices, with or without shifts.

As mentioned briefly in Section 2, our Lanczos procedures with no reorthogonalization have what we choose to call 'look-around' capabilities. This capability comes from our identification test which sorts the eigenvalues of each Lanczos matrix considered into 'good' and 'spurious' eigenvalues. Through it we identify not only eigenvalue approximations which have converged but also other eigenvalue approximations which may have only a few digits accuracy. In fact, we identify eigenvalue approximations independently of their accuracy since our test identifies the 'spurious' eigenvalues and then all remaining eigenvalues of the Lanczos matrix are accepted as eigenvalue approximations. Typically these 'good' eigenvalues have varying degrees of approximation.

This 'look-around' capability need not be unique to our Lanczos procedures with no reorthogonalization. In a Lanczos procedure with selective reorthogonalization or in a different Lanczos procedure with no reorthogonalization, the convergence tolerance could be relaxed to 'identify' more eigenvalue approximations which could then be used in a shift strategy. However, relaxing the convergence tolerance will not suffice, for example, in a Block Simultaneous Iteration procedure. In such a procedure convergence occurs in order of the magnitudes of the eigenvalues of the matrix used in the block procedure. If the convergence tolerance is relaxed, the additional part of the spectrum which will be seen is that part which is nearest in magnitude. That part may or may not be relevant for the desired computations. Being close in magnitude does not necessarily imply closeness in the complex plane.

We will describe our shifting strategy for our nonsymmetric Lanczos procedures. This strategy also works on real symmetric problems. It is equally applicable to any standard or generalized eigenvalue problem whenever it is feasible to factor shifted matrices $A - \sigma B$. Feasibility is measured by the amount of storage required and the number of operations required.

The examples given in Section 5 for the nonsymmetric case were generated by linearizing the compressible, resistive ($\eta$), MHD equations around various static equilibriums using cylindrical symmetry. For details on the construction see Kerner [1986]. As presented to us the problem was to determine the Alfven and sound spectra as a function of the resistivity parameter $\eta$. For this particular set of problems, as the resistivity $\eta$ is varied and for small enough $\eta$, all of the Alfven eigenvalues lie on the same curves. That fact however was not used in our procedures. We also give an example of the use of this strategy in a real symmetric problem.

The user must first specify a box in the complex plane. For a real symmetric problem these become intervals of interest on the real line. A starting shift must also be specified, along with the number of shifts which the user is willing to let the program consider. This determines the maximum number of factorizations which the procedure will do during the eigenvalue computations.

Our overall shift and invert procedures are actually two stage procedures, differing somewhat from our procedures which work directly with $A$ (and $B$). For details see Cullum, Kerner, and Willoughby [1989]. The first stage uses our shift and invert Lanczos procedure to compute eigenvalue approximations, using a sequence of shifts generated either automatically by our program or specified a priori by the user. After eigenvalue approximations have been computed using this first stage, the user selects a subset of these approximations to supply to a modified Rayleigh–Ritz inverse iteration procedure. This second stage improves the eigenvalue approximations determined in the first stage and computes eigenvector approximations. It is a two sided procedure so that it computes approximations to both the right and the left eigenvectors. In this paper we focus on the strategy for generating the shifts automatically.

It is important to note that in our algorithms, even when we are using shifting and inverting, we have separated the initial eigenvalue computations from the eigenvector computations, so that no Ritz vector computations are required in the initial eigenvalue computations.

To describe our shift and invert strategy we must first briefly describe the Lanczos procedure used for each shift which is considered. If we use the nonsymmetric Lanczos recursions in Equations (3.1) and (3.4), require $B$-biorthogonality, and then make a simple vector transformation, we obtain the following recursions.

$$\beta_{i+1}v_{i+1} = (A - \sigma B)^{-1} Bv_i - \alpha_i v_i - \beta_i v_{i-1} \equiv r_i$$

$$\beta_{i+1}w_{i+1} = (A - \sigma B)^{-T} Bw_i - \alpha_i w_i - \beta_i w_{i-1} \equiv p_i$$

(4.4)

with

$$\alpha_i \equiv \frac{\alpha_{iv} + \alpha_{iw}}{2}$$

$$\beta_{i+1} \equiv \sqrt{r_i^T p_i}, \quad v_1 = w_1$$

$$\alpha_{iv} = w_i^T B \left[ (A - \sigma B)^{-1} B v_i - \beta_i v_{i-1} \right]$$

$$\alpha_{iw} = v_i^T B \left[ (A - \sigma B)^{-T} B w_i - \beta_i w_{i-1} \right].$$

(4.5)

Recursions (4.4) and (4.5) map the given problem into a family of complex symmetric tridiagonal matrices as in Equations (3.3).

For each shift considered we want to accomplish two things: to obtain eigenvalue approximations and to identify candidates for the next shift. We must also make sure that each shifted matrix is numerically stable enough to use to generate a Lanczos matrix. If the shift selected is too close to an actual eigenvalue of the original matrix, then the resulting factorization may be too inaccurate to be a good reflection of the original problem. If that happens then obviously the eigenvalues of the associated Lanczos matrices will not be accurate reflections of the original problem. We therefore include a heuristic check on the sizes of the entries in the Lanczos matrices generated and reject a potential shift if a specified tolerance is exceeded. We note here that if we were simply doing inverse iteration to obtain an eigenvector approximation corresponding to a given eigenvalue approximation that Wilkinson [1965] has demonstrated that such ill conditioning does not hinder the accuracy of the computed vector. Such a computation would amount to doing one step in our Lanczos procedure. However, the additional steps which we require would introduce significant errors.

For each shift considered we use Equations (4.4) and (4.5) to compute a Lanczos matrix, $T_M$, of relatively small size. The numerical experiments described here used $M = 50, 30,$ and 90. We then compute the eigenvalues of the corresponding Lanczos matrix $T_M$, sort them, keep the 'good' eigenvalues and discard the 'spurious' eigenvalues and then compute error estimates for each of the 'good' eigenvalues.

We note that any errors in the factorization of a shifted problem correspond to introducing perturbations into the original problem, and the eigenvalues of these perturbed problems may differ somewhat from the eigenvalues of the original problem. Therefore, stage one of our procedure attempts only to obtain reasonably good eigenvalue approximations to the desired eigenvalues so it uses a weak 'convergence' tolerance in deciding which of the computed 'good' eigenvalues have 'converged'. The sizes of the last components of the associated eigenvectors of the Lanczos matrix are used in this determination because they are scale invariant.

In the numerical examples the weak convergence tolerance ERRTOL was set equal to $10^{-4}$. 'Good' eigenvalues with error estimates smaller than ERRTOL were declared 'weakly-converged' and accumulated in a file to be presented to the user prior to entering the second stage of our procedure. The remaining 'good' eigenvalues were ordered according to the sizes of their error estimates with those having the smallest error estimates listed first. Typically for each Lanczos matrix considered we see some 'good' eigenvalues with small error estimates that have 'weakly-converged' and other 'good' eigenvalue approximations with larger error estimates which we can consider as potential candidates for subsequent shifts.

We want to choose the next shift so that it is representative of a part of the spectrum of the original matrix which is of interest to us, but not too close to an actual eigenvalue of that matrix. We therefore, look through the list of good eigenvalues but focus on those 'good' eigenvalue approximations with poor error estimates. The average size of a component of a unit vector of length $M = 50$ is .14. Furthermore the error estimates are typically conservative. Therefore, if we use a shift tolerance, SHFTOL = .01 and select eigenvalue approximations with error estimates larger than this, we should obtain quantities which tell us something about the spectrum of the original matrix but are not too close to actual eigenvalues of that matrix.

Specifically, for each shift $\sigma$ considered we construct a list of potential next shifts by first listing those 'good' eigenvalues with error estimates larger than or equal to SHFTOL, and ordered such that their corresponding error estimates are increasing in magnitude. We then add to that list those 'good' eigenvalues whose error estimates are smaller than SHFTOL but larger than ERRTOL, ordered such that their corresponding error estimates are decreasing in magnitude. If for some $\sigma$ all of the eigenvalue approximations are 'weakly converged', then we get potential next shifts from the list generated for the previous shift used, selecting any entries on that previous list which were not already considered previously. As we said earlier, we do not consider any weakly converged eigenvalues as potential shifts because it is assumed that such values are 'too close' to actual eigenvalues of the original problem and would give us unreliable factorizations.

At each stage in our computations we check for redundancy using a tolerance CLSTOL and attempt to eliminate multiple copies of eigenvalue approximations from the list of 'weakly-converged' eigenvalues as well as duplicates in the lists of potential shifts. The list of potential shifts is initialized by incorporating five geometrically-constructed shifts in the user-specified box.

The shift selection process always starts with a user-supplied first shift. If for some reason that first shift is not appropriate, then the program terminates for the user to supply an alternative first shift. After the first successful shift, a successive shift is chosen from the current list of potential shifts. If a potential shift is rejected, the program simply proceeds to the next potential shift on the current list and attempts to use it. If at some point the list of potential shifts contains no acceptable shift, then the program terminates for user-intervention.

As we said earlier one can also specify a priori a set of shifts for the procedure to consider sequentially. In fact after any run with automatically generated shifts the user should examine a plot of the resulting eigenvalue approximations and identify a set of appropriate shifts which are then supplied to the procedure to check for and hopefully determine any missing eigenvalue approximations.

## 5  Numerical Examples.

In order to demonstrate the proposed shift strategy we present examples of its use on two matrices generated from the MHD application and on a difficult (but artificial) generalized real symmetric problem whose spectrum is both spread throughout a large interval and densely packed within a small subinterval of that large interval. We consider the effects of selecting different initial shifts, of using different size Lanczos matrices, and of using different values for SHFTOL.

Each of the resulting plots represents the results obtained by the first stage of our Lanczos procedure. In many of these plots there is some scatter of the eigenvalue approximations corresponding to duplications of approximations to some of the eigenvalues on different shifts. Any such duplications would be identified and eliminated by the second stage of our shift and invert Lanczos procedures. Examples of the effect of the second stages are given in Cullum et al [1989].

Table 1: MHD Example, $n = 598$, Equilibrium 2, $\eta = 6 * 10^{-5}$, 54 Eigenvalues

| Test | $\sigma_0$ | SHFTOL | $T_m$ | Defaults | Shifts in Small Box | Eval Approxs. | Evals Missing |
|---|---|---|---|---|---|---|---|
| 1 | (-.25, .25) | $10^{-2}$ | 51 | 0 | 13/14 | 54 | 0 |
| 2 | (-.03, .15) | $10^{-2}$ | 51 | 0 | 14/14 | 54 | 0 |
| 3 | (-.0125, .375) | $10^{-2}$ | 51 | 0 | 12/14 | 53 | 1 |
| 4 | (-.45, .125) | $10^{-2}$ | 51 | 0 | 13/14 | 54 | 0 |
| 5 | (-.4, .4) | $10^{-2}$ | 51 | 0 | 12/14 | 53 | 1 |
| 6 | (-.4, .4) | $10^{-1}$ | 51 | 0 | 4/14 | 36 | 18 |
| 7 | (-.4, .4) | $10^{-3}$ | 51 | 5 | 12/14 | 41 | 13 |

We consider the two nonsymmetric MHD examples first. For these particular examples the complex eigenvalues occur in conjugate pairs. Therefore it was only necessary to consider a box in the upper half plane. Furthermore, the shifts were explicitly kept away from the origin because the $A$-matrices were singular, and they were also kept away from the real line because the real eigenvalues were not of interest.

Tests were run on both a small example ($n = 598$), where the Lanczos eigenvalue approximations were compared with eigenvalues computed by EISPACK subroutines, and on a large example ($n = 4498$) where only the Lanczos generated values are available. In the plots for $n = 598$, the symbols + and > are used to denote respectively, eigenvalue approximations obtained from our Lanczos procedure and eigenvalues obtained using the EISPACK subroutines. Whenever a + and a > overlap we have 'convergence', at least pictorially, of the corresponding Lanczos approximation to an eigenvalue of the original problem. We also plot the Lanczos eigenvalue approximations by the number of the shift on which the approximations were first identified as 'weakly converged' and we plot the shifts used. For $n = 4498$ we have similar plots but without EISPACK eigenvalues.

Figures 4.1 and 4.2 in Section 4 are plots of the eigenvalues of the small MHD example considered. For details see Cullum et al [1989]. As stated earlier, the entire spectrum of this example is plotted in Figure 4.1 and the eigenvalues of interest, the Alfven and the sound spectra, are plotted in Figure 4.2.

We specified the box $S = \{[-.5, 0] \times [0, .5]\}$ in the complex plane and the objective was to compute all of the eigenvalues of these two examples in this box. Kerner used knowledge he had gained from the ideal ($\eta = 0$) Hermitian MHD case to tell him where to select his initial shift. In earlier work he computed the Alfven curves one eigenvalue at a time, by successive Rayleigh-quotient inverse iterations with extrapolation along the curves. With our Lanczos procedure we typically obtain a section or sections of this or the sound mode curve for each shift considered.

The results of the tests on these two MHD examples are summarized in Tables 1 and 2. In each case $\sigma_0$ denotes the initial shift specified, $T_m$ denotes the size of the Lanczos matrix used, and the number of defaults is the number of potential shifts which were considered and then rejected because entries in the corresponding Lanczos matrix exceeded the prespecified tolerance LTOL $= 10^6$. This test was used as a warning that perhaps the factors of $A - \sigma B$ were too ill-conditioned to be used to generate a Lanczos matrix.

In Tests 1 - 5 we varied the initial shift. The results of these computations are plotted in Figures 5.1 - 5.10. There were no additional eigenvalue approximations in the box $S$ and outside of the smaller box $\{[-.2, 0] \times [0, .3]\}$ used in the plots. Figures 5.1 - 5.5 are plots of the eigenvalue approximations obtained by our Lanczos procedure together with the

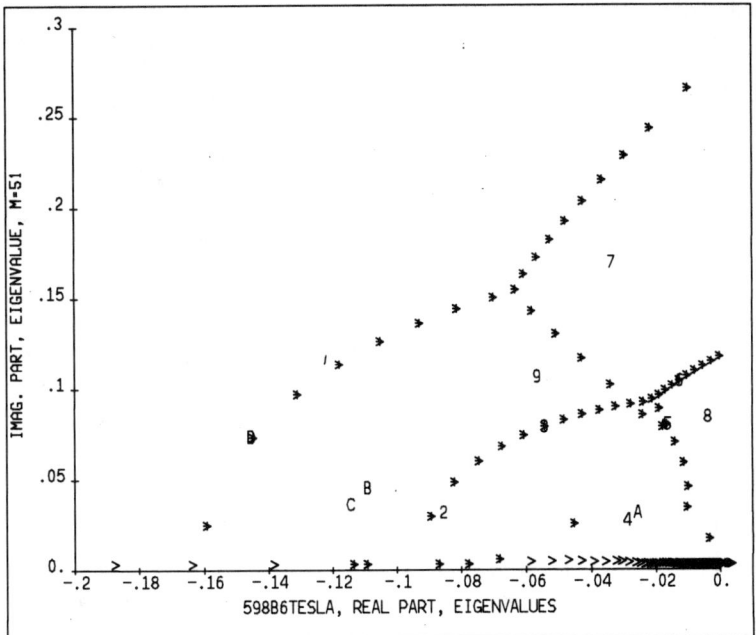

Figure 5.1: Test 1 in Table 1. MHD example, equilibrium 2, $n = 598, \eta = 6*10^{-5}$. EISPACK (>) and Lanczos (+) eigenvalues with shift locations plotted by number or letter. $\sigma_0 = (-.25,.25)$. $T_{51}$, SHFTOL $= 10^{-2}$.

eigenvalues obtained from EISPACK computations, and the locations of the shifts obtained in the small box $[-.2, 0] \times [0, .3]$. Figures 5.6 – 5.10 are plots of the Lanczos eigenvalue approximations by the number of the first shift on which each approximation was obtained. In these examples we see that although the choice of initial shift significantly altered the subsequent sequence of shifts generated, it did not make a big difference in the number of eigenvalues approximated.

In Tests 1 and 2, 53 of the 54 eigenvalues had been identified by shift seven. In Test 1 the remaining eigenvalue $\lambda = (-.1605, .0215)$ was not identified until shift eleven. In Test 2, ten shifts were required to identify all eigenvalues. Test 3 required only eight shifts to get the 53 it obtained. Test 4 got all 54 eigenvalues by shift eight. Test 5 required seven shifts to get its 53 eigenvalues. In Tests 3 and 5 the eigenvalue $(-.48 * 10^{-2}, .139 * 10^{-1})$ was not identified. However it lies on the tip of the lower vertical portion of the sound spectrum and would have been readily identified by a subsequent cursory check using the boundary points of the curves as prespecified shifts.

In each of these tests the shifts which were generated automatically by our procedure, were spread appropriately throughout the relevant portion of the user specified box, and allowed us to identify the relevant branches of the desired Alfven and sound mode curves. In Tests 2, 3 and 5 there is a small amount of redundancy at the branch point on the Alfven curve. The *'s in Figures 5.6 – 5.10 denote the locations of shifts used. The shifts are also labelled in Figures 5.1 – 5.5 using numbers and letters.

Similar plots are given for Tests 6 and 7 in Figures 5.11 to 5.14 which should be compared with Figures 5.1 to 5.10 to illustrate the effects of varying the size of SHFTOL for this MHD example. For this size problem when SHFTOL $= 10^{-1}$, there were very few eigenvalue approximations with error estimates larger than SHFTOL and not many of these few were in the smaller box containing the desired eigenvalues. Consequently only four

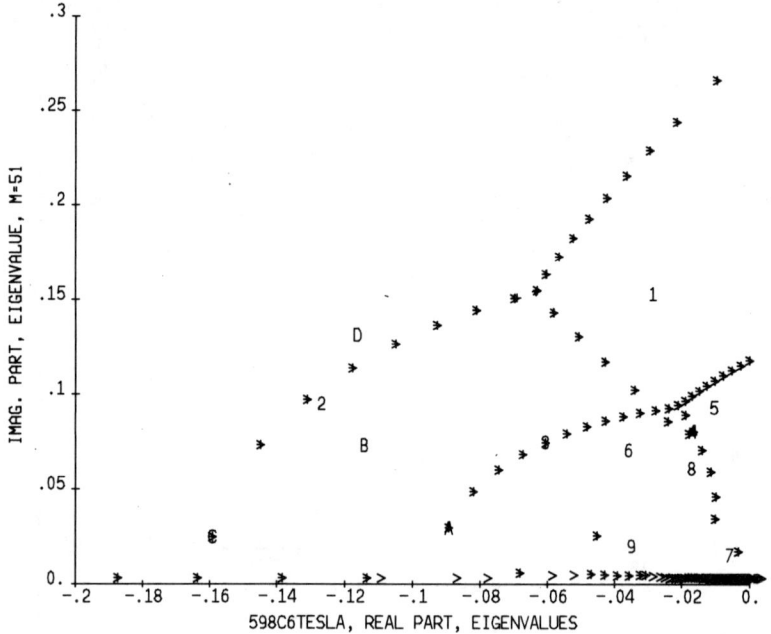

Figure 5.2: Test 2 in Table 1. MHD example, equilibrium 2. $n = 598, \eta = 6 * 10^{-5}$, Eispack (>) and Lanczos (+) eigenvalues with shift location plotted by number or letter. $\sigma_0 = (-.03, .15)$, $T_{51}$, SHFTOL $= 10^{-2}$.

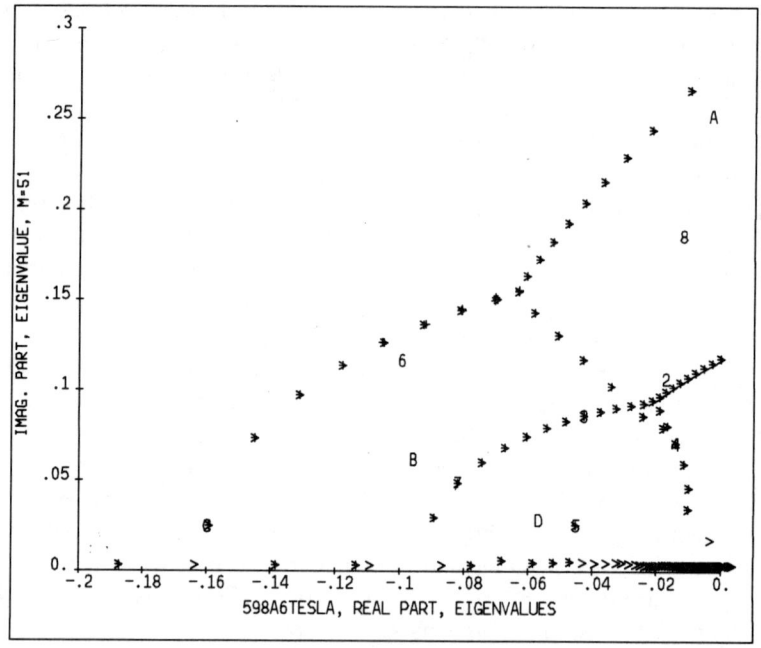

Figure 5.3: Test 3 in Table 1. MHD example, equilibrium 2, $n = 598, \eta = 6 * 10^{-5}$. EISPACK (>) and Lanczos (+) eigenvalues with shift locations plotted by number or letter. $\sigma_0 = (-.0125, .375)$, $T_{51}$, SHFTOL $= 10^{-2}$.

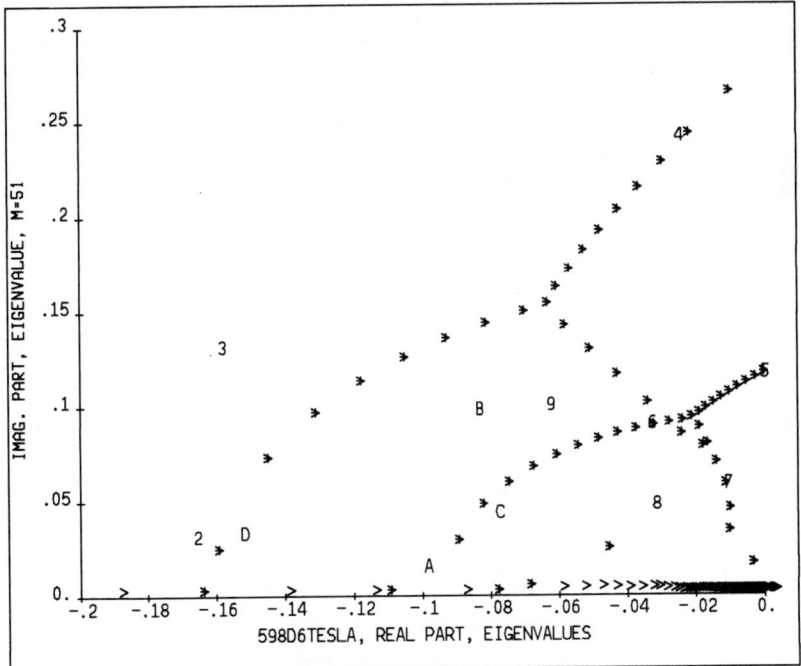

Figure 5.4: Test 4 in Table 1. MHD example, equilibrium 2. $n = 598, \eta = 6 * 10^{-5}$. EISPACK (>) and Lanczos (+) eigenvalues with shift locations indicated by number or letter. $\sigma_0 = (-.45, .125)$, $T_{51}$, SHFTOL $= 10^{-2}$.

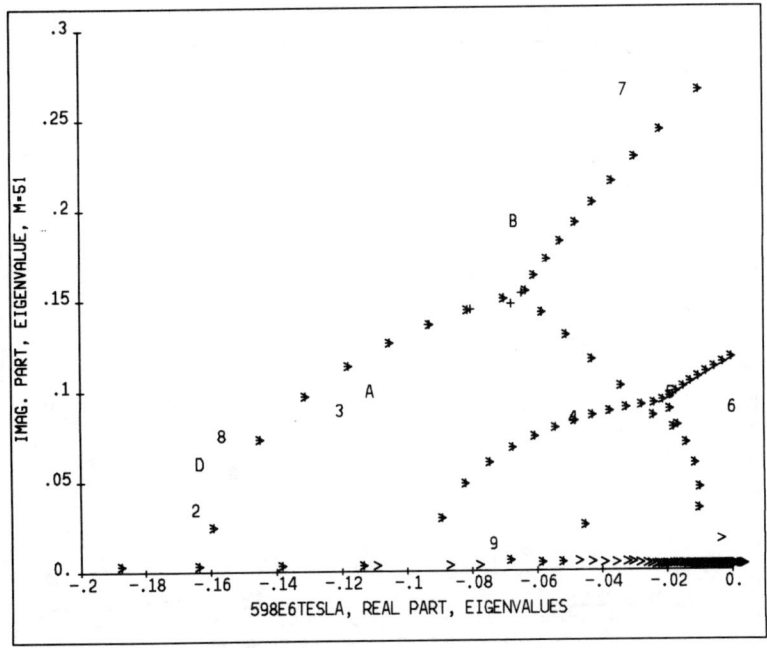

Figure 5.5: Test 5 in Table 1. MHD example, equilibrium 2. $n = 598, \eta = 6 * 10^{-5}$. EISPACK (>) and Lanczos (+) eigenvalues with shifts plotted by number or letter. $\sigma_0 = (-.4, .4)$, $T_{51}$, SHFTOL $= 10^{-2}$.

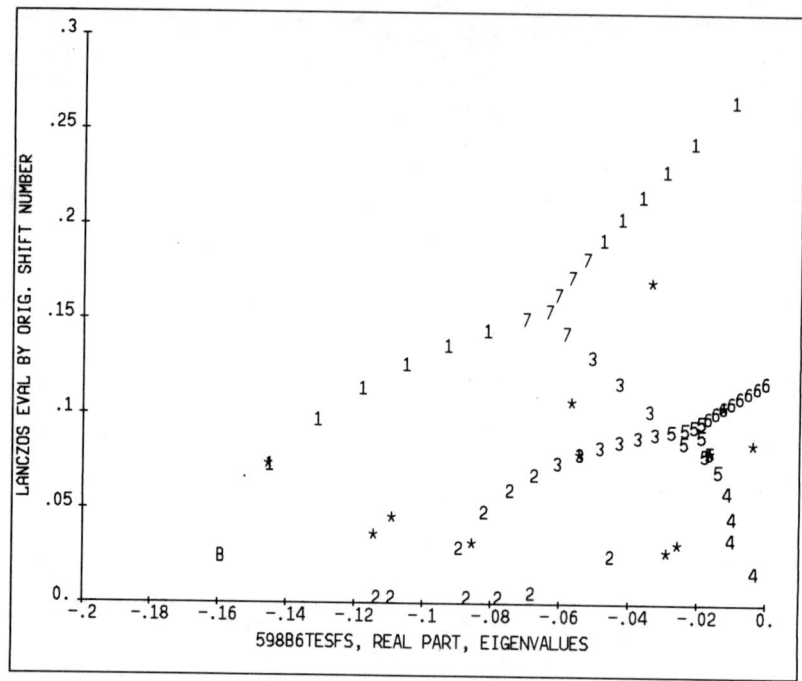

Figure 5.6: Test 1 in Table 1. MHD example, equilibrium 2, $n = 598$, $\eta = 6 * 10^{-5}$. Lanczos eigenvalue approximations plotted by number of first shift on which they were 'weakly converged'. $\sigma_0 = (-.25, .25)$. $*$ denotes a shift used. $T_{51}$, SHFTOL $= 10^{-2}$.

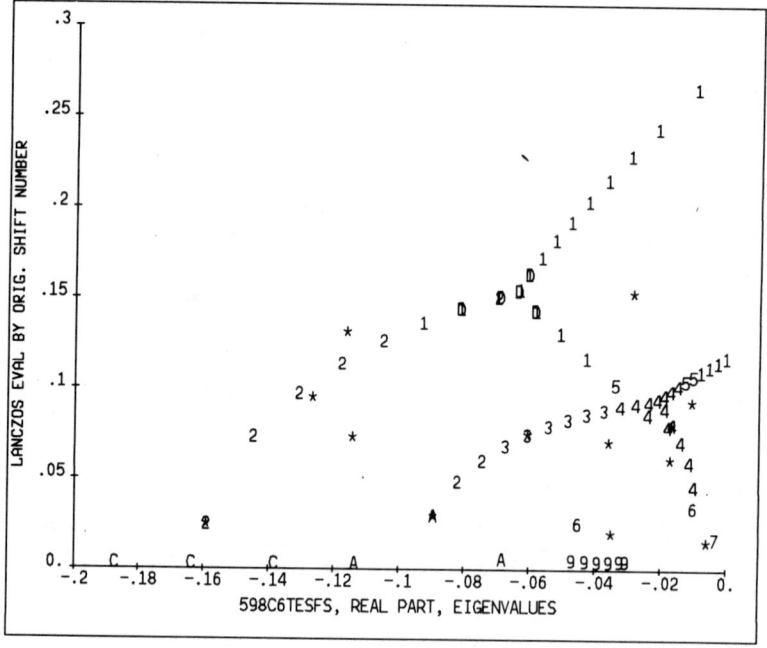

Figure 5.7: Test 2 in Table 1. MHD example, equilibrium 2, $n = 598$, $\eta = 6 * 10^{-5}$. Lanczos eigenvalue approximations plotted by number of first shift on which they were 'weakly converged'. $\sigma_0 = (-.03, .15)$. $*$ denotes a shift used. $T_{51}$, SHFTOL $= 10^{-2}$.

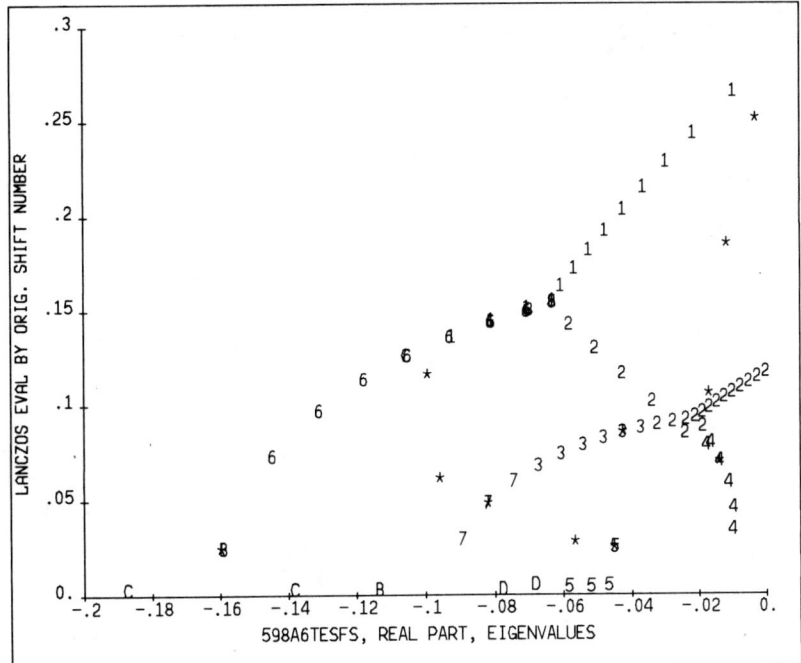

**Figure 5.8**: Test 3 in Table 1. MHD example, equilibrium 2, $n = 598, \eta = 6*10^{-5}$. Lanczos eigenvalue approximations plotted by number of first shift on which they were 'weakly-converged'. $\sigma_0 = (-.0125, .375)$. $*$ denotes a shift used. $T_{51}$, SHFTOL $= 10^{-2}$.

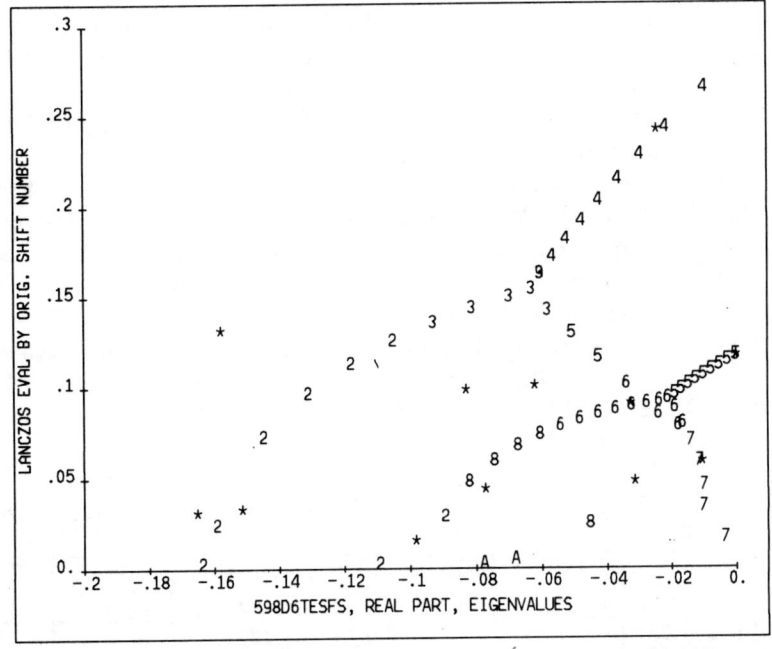

**Figure 5.9**: Test 4 in Table 1. MHD example, equilibrium 2, $n = 598, \eta = 6*10^{-5}$. Lanczos eigenvalue approximations plotted by number of first shift on which they were 'weakly-converged'. $\sigma_0 = (-.45, .125)$, $T_{51}$, SHFTOL $= 10^{-2}$.

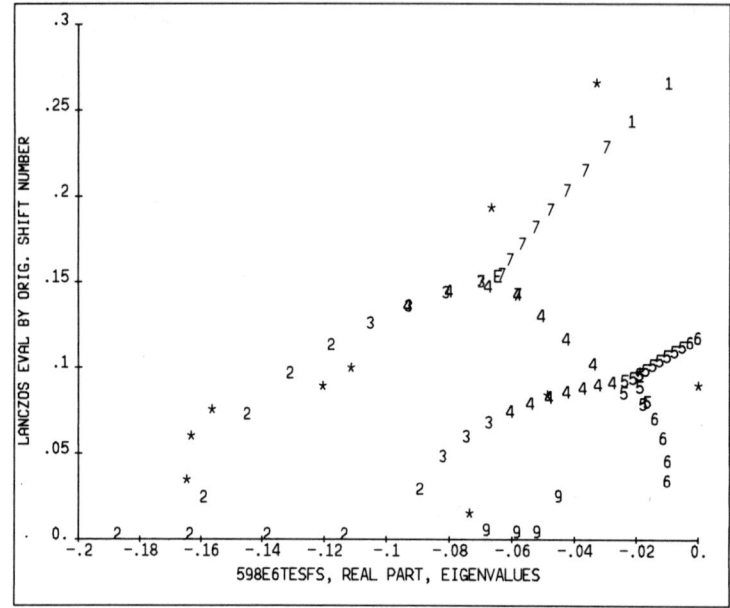

Figure 5.10: Test 5 in Table 1. MHD example, equilibrium 2, $n = 598, \eta = 6 * 10^{-5}$. Lanczos eigenvalue approximations plotted by number of first shift on which they were 'weakly-converged'. $\sigma_0 = (-.4, .4)$, $T_{51}$, SHFTOL $= 10^{-2}$.

shifts were located in the small box. When SHFTOL was reduced to $10^{-3}$, 12 shifts were identified in the box. However as we can see in Figure 5.12 the procedure did not 'look around' enough. Thus, in both Tests 6 and 7, although for different reasons, significant portions of the sound spectrum were not identified. See in particular Figures 5.11 and 5.12. In Test 7 there is some redundancy near the branch point of the Alfven curve.

When we increase the example size to $n = 4498$, we see quite different behavior. See in particular, Figures 5.17 to 5.20. With this increase in problem size the number of eigenvalue approximations with large error estimates also increases, improving the shift selection when SHFTOL $= 10^{-1}$. In these tests the initial shift was not varied, and the tests focussed on the effects of varying SHFTOL and the size of the Lanczos matrix. The results are summarized in Table 2. Corresponding plots are Figures 5.15 – 5.26. We note that the Alfven spectrums for the two MHD examples were identical. However, the sound spectrums varied considerably. EISPACK eigenvalues were not computed for this problem.

Table 2: MHD Example, $n = 4498$, Equilibrium 2, $\eta = 6 * 10^{-5}$, 57 eigenvalues

| Test | $\sigma_0$ | SHFTOL | $T_m$ | Defaults | Shifts in Small Box | Eval Approxs | Evals Missing |
|---|---|---|---|---|---|---|---|
| 1 | (-.25, .25) | $10^{-2}$ | 51 | 0 | 12/14 | 57 | 0 |
| 2 | (-.25, .25) | $10^{-1}$ | 51 | 0 | 13/14 | 57 | 0 |
| 3 | (-.25, .25) | $10^{-3}$ | 51 | 0 | 11/14 | 38 | 19 |
| 4 | (-.25, .25) | $10^{-2}$ | 91 | 0 | 12/14 | 57 | 0 |
| 5 | (-.25, .25) | $10^{-2}$ | 31 | 0 | 10/14 | 45 | 12 |
| 6 | (-.25, .25) | $10^{-2}$ | 31 | 0 | 17/21 | 51 | 6 |

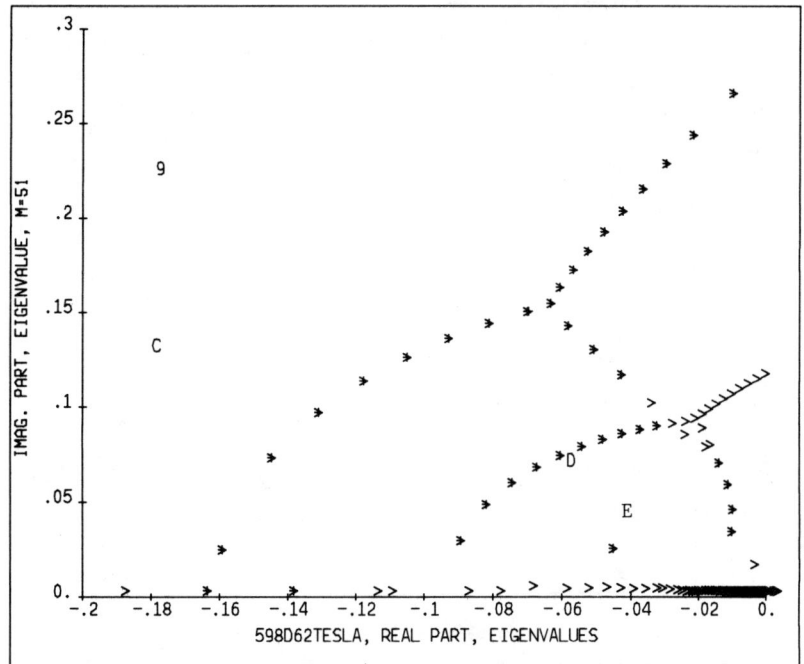

Figure 5.11: Test 6 in Table 1. MHD example, equilibrium 2, $n = 598, \eta = 6 * 10^{-5}$. EISPACK and Lanczos eigenvalues with shifts indicated by number or letter. $\sigma_0 = (-.4, .4)$, $T_{51}$, SHFTOL $= 10^{-1}$.

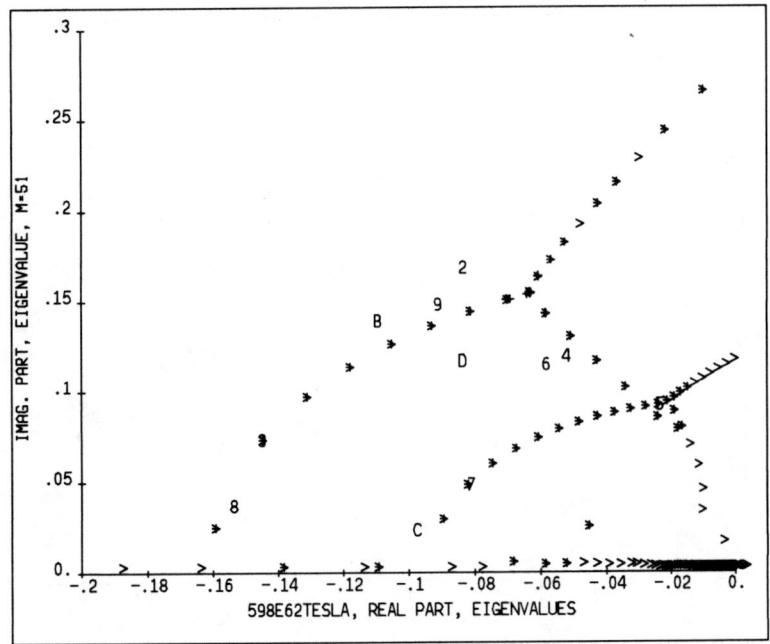

Figure 5.12: Test 7 in Table 1. MHD example, equilibrium 2, $n = 598, \eta = 6 * 10^{-5}$, Lanczos and EISPACK eigenvalues with shift locations indicated by number or letter. $\sigma_0 = (-.4, .4)$, $T_{51}$, SHFTOL $= 10^{-3}$.

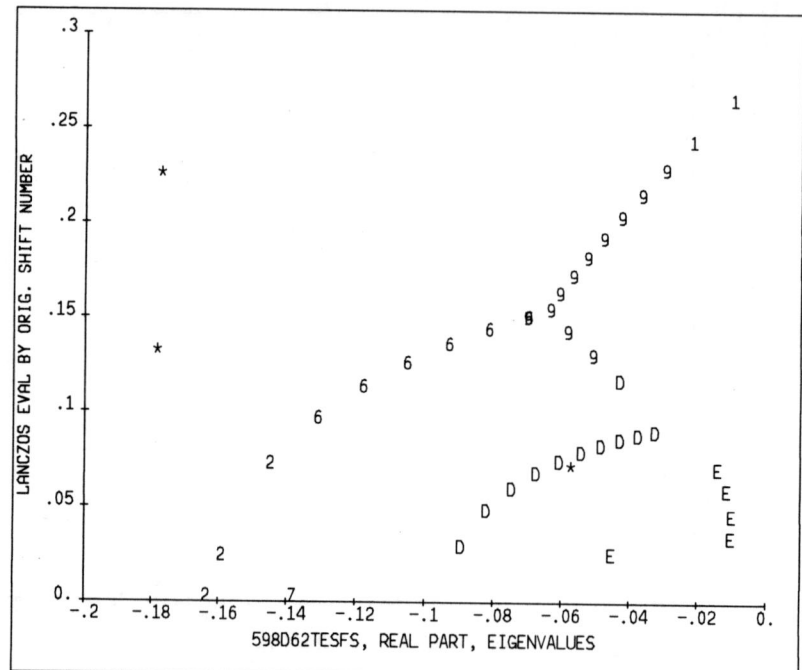

Figure 5.13: Test 6 in Table 1. MHD example, equilibrium 2, $n = 598$, $\eta = 6 * 10^{-5}$. Lanczos eigenvalue approximations plotted by number of first shift on which they were 'weakly-converged'. $\sigma_0 = (-.4, .4)$, $T_{51}$, SHFTOL $= 10^{-1}$.

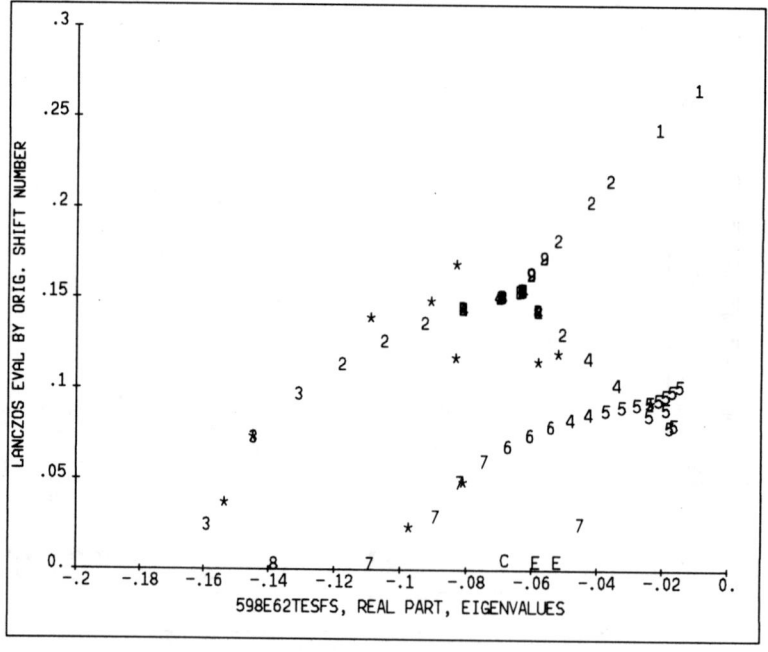

Figure 5.14: Test 7 in Table 1, MHD example, equilibrium 2, $n = 598$, $\eta = 6 * 10^{-5}$. Lanczos eigenvalue approximations plotted by number of first shift on which they were 'weakly-converged'. $\sigma_0 = (-.4, .4)$, $T_{51}$, SHFTOL $= 10^{-3}$.

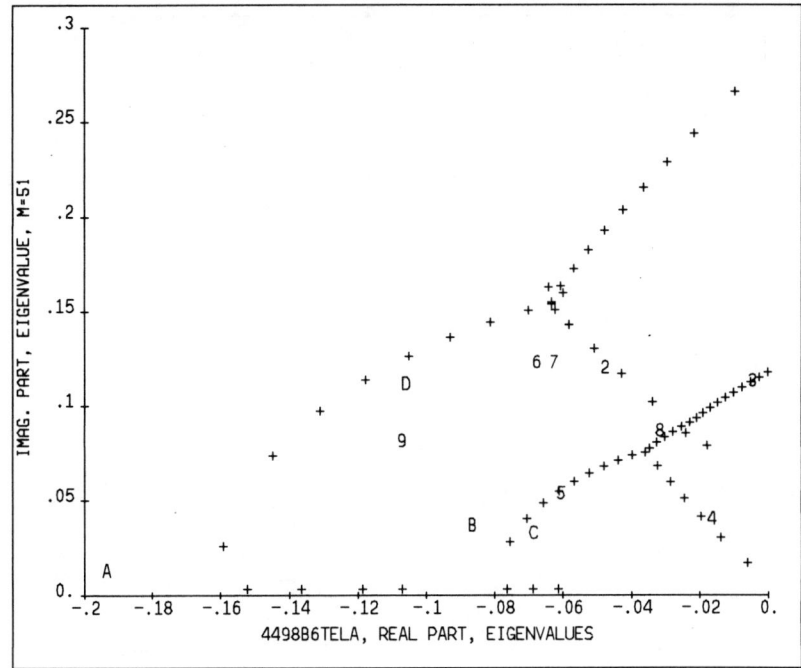

Figure 5.15: Test 1 in Table 2. MHD example, equilibrium 2, $n = 4498, \eta = 6 * 10^{-5}$. Lanczos (+) eigenvalue approximations. $\sigma_0 = (-.25, .25)$, $T_{51}$, SHFTOL = $10^{-2}$, and shifts plotted by number or letter.

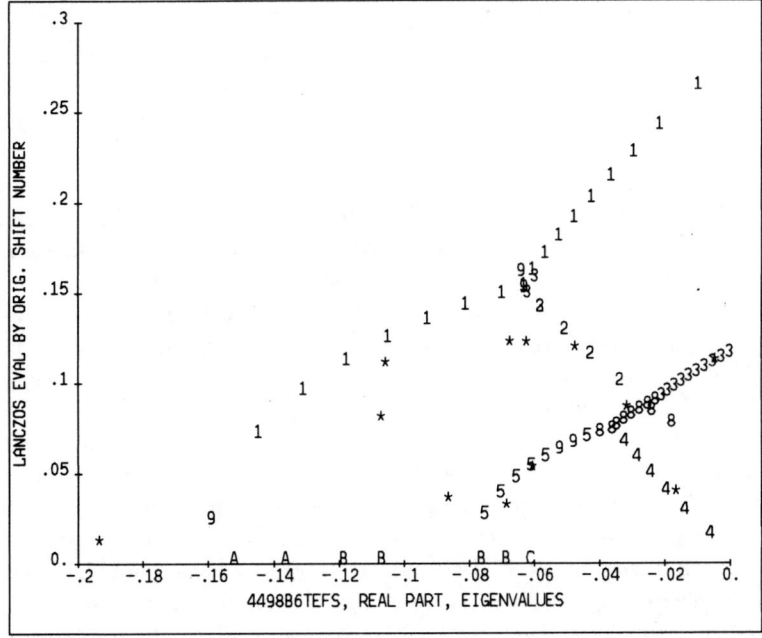

Figure 5.16: Test 1 in Table 2. MHD example, equilibrium 2, $n = 4498, \eta = 6 * 10^{-5}$. Lanczos eigenvalue approximations plotted by number of first shift on which they were 'weakly-converged'. $\sigma_0 = (-.25, .25)$, $T_{51}$, SHFTOL = $10^{-2}$. Shifts plotted as $*$.

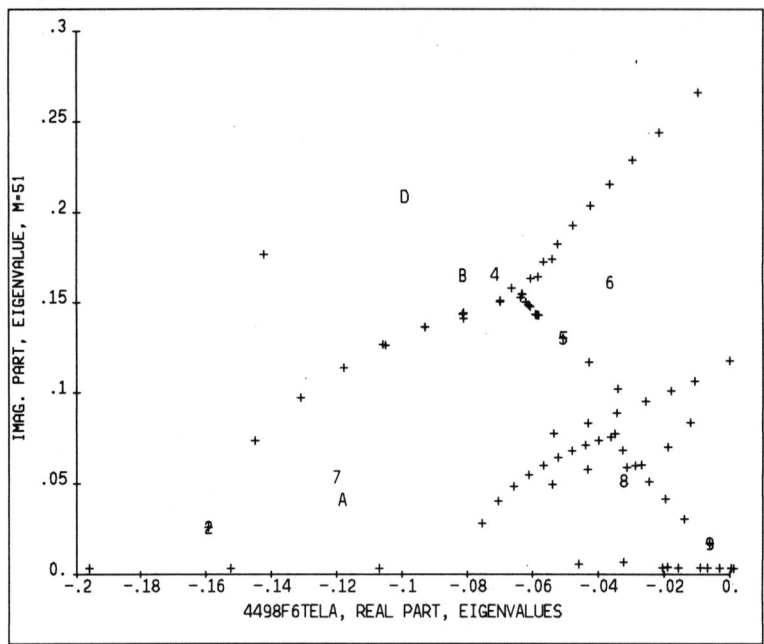

Figure 5.17: Test 3 in Table 2. MHD example, equilibrium 2, $n = 4498, \eta = 6 * 10^{-5}$. Lanczos (+) eigenvalue approximations with shifts plotted by number or letter. $\sigma_0 = (-.25, .25)$, $T_{51}$, SHFTOL $= 10^{-3}$.

In Tests 1 through 3, SHFTOL was varied. See Figures 5.15 – 5.20. With SHFTOL $= 10^{-1}$ or $10^{-2}$ all 57 eigenvalues were satisfactorily identified. There was a small amount of redundancy at the branch point on the Alfven spectrum in Test 1 corresponding to the ninth shift. However the second stage of our Lanczos procedure would easily resolve this redundance.

For SHFTOL $= 10^{-3}$ however, see Figures 5.17 and 5.18, there were real problems caused by shifts 5 and 9 which were too close to true eigenvalues of the original problem, agreeing to more than four digits. The corresponding eigenvalue approximations were very inaccurate. There is much scatter. But even in this case most of the Alfven curve and much of the sound spectrum were identified but are surrounded or buried in scatter and incorrect values. The computed upper right portion of the sound spectrum was totally incorrect due to the problems with shifts 5 and 9. This example clearly exhibits the consequences which can result from choosing shifts too close to true eigenvalues.

Tests 1, 4, 5 and 6 in Table 2, see Figures 5.15 and 5.16, and 5.21 to 5.26, can be used to compare the effects of changing the size of the Lanczos matrices used. Depending upon the relative costs of factoring $A - \sigma B$ versus the cost of the subsequent solves, it may or may not be more cost effective to generate larger Lanczos matrices and do fewer factorizations. For Test 4 with Lanczos matrices of size 91, see Figure 5.22, seven shifts were needed to identify all the desired eigenvalues. There is a small amount of scatter at the two branch points caused by later shifts. For Test 1 with Lanczos matrices of size 51, see Figures 5.15 and 5.16, nine shifts were needed.

Tests 5 and 6, see Figures 5.23 to 5.26, indicate that attempting to limit the size of the Lanczos matrices too much interferes with both the shift selection process and the computation of the desired eigenvalues. We need to use Lanczos matrices which are large enough to yield both some weakly-converged eigenvalue approximations and some approximations

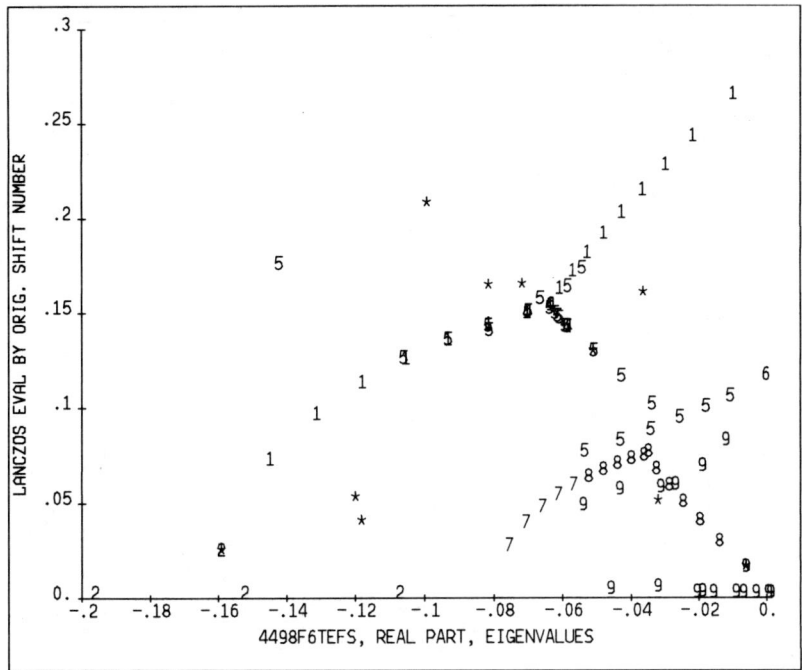

Figure 5.18: Test 3 in Table 2. MHD example, equilibrium 2, $n = 4498, \eta = 6 * 10^{-5}$, Lanczos (+) eigenvalue approximations plotted by number of first shift on which they had 'weakly-converged'. $\sigma_0 = (-.25, .25)$, $T_{51}$, SHFTOL = $10^{-3}$. Shifts 5 and 9 were too close to true eigenvalues.

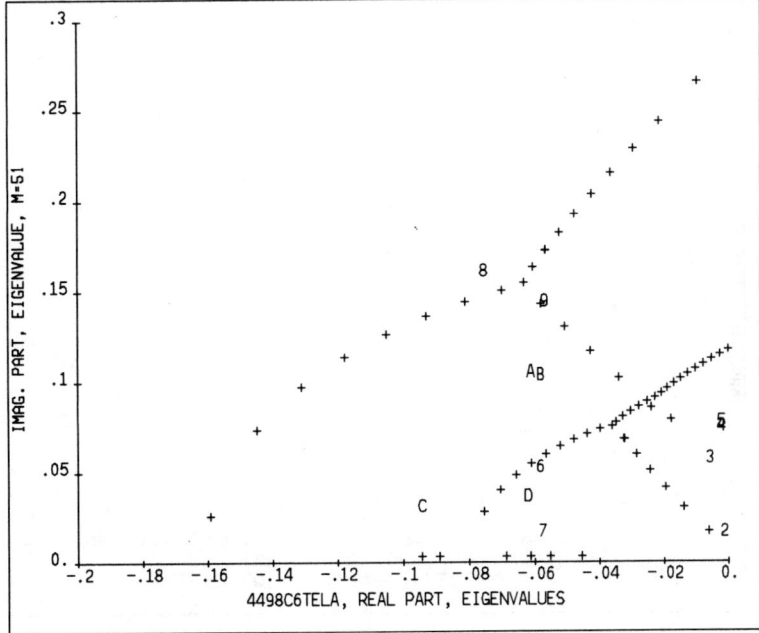

Figure 5.19: Test 2 in Table 2. MHD example, equilibrium 2, $n = 4498$, $\eta = 6 * 10^{-5}$. Lanczos (+) eigenvalue approximations with shifts plotted by number or letter. $\sigma_0 = (-.25, .25)$, $T_{51}$, SHFTOL = $10^{-1}$.

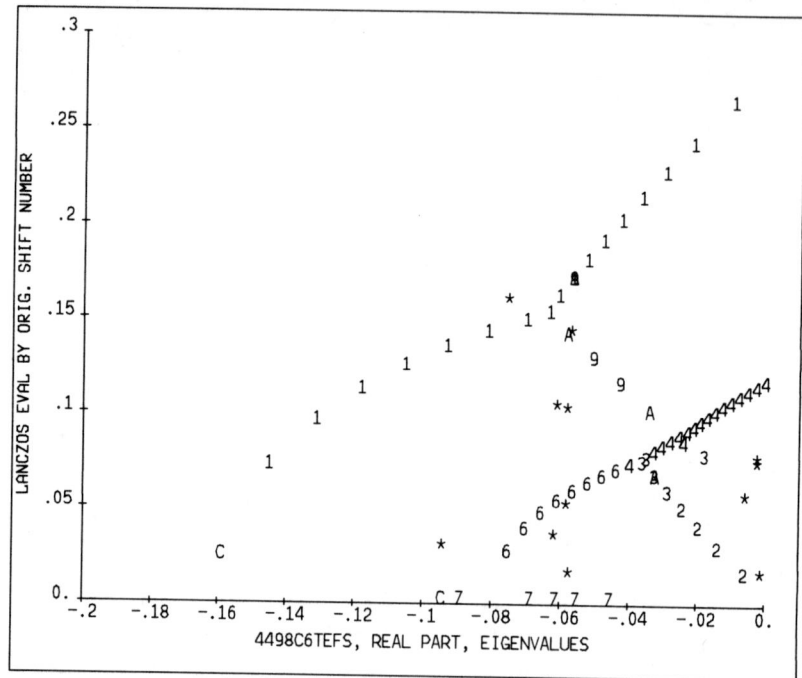

Figure 5.20: Test 2 in Table 2. MHD example, equilibrium 2, $n = 4498, \eta = 6 * 10^{-5}$. Lanczos (+) eigenvalue approximations plotted by number of first shift on which they were 'weakly-converged'. $\sigma_0 = (-.25, .25)$, $T_{51}$, SHFTOL $= 10^{-1}$. $*$ denotes a shift.

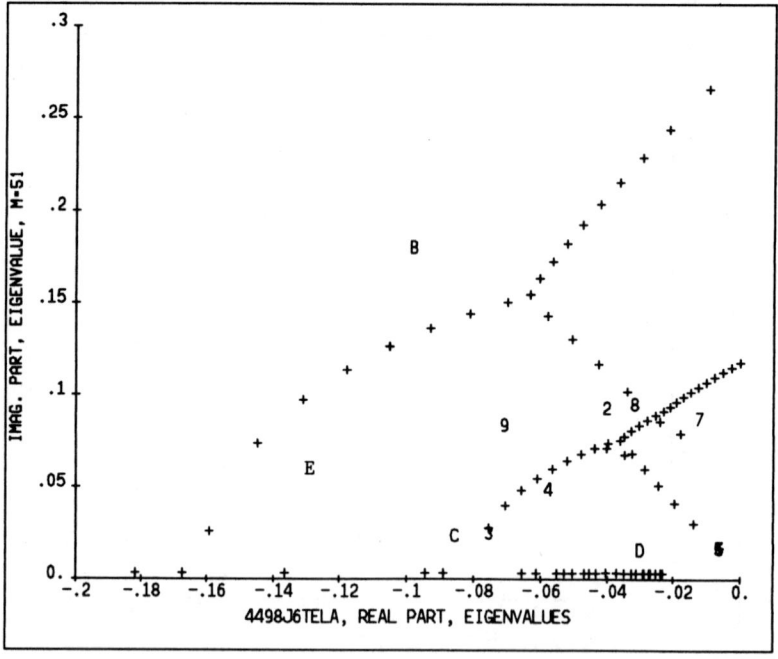

Figure 5.21: Test 4 in Table 2. MHD example, equilibrium 2, $n = 4498, \eta = 6 * 10^{-5}$. Lanczos (+) eigenvalue approximations with shifts plotted by number or letter. $\sigma_0 = (-.25, .25)$, $T_{91}$, SHFTOL $= 10^{-2}$.

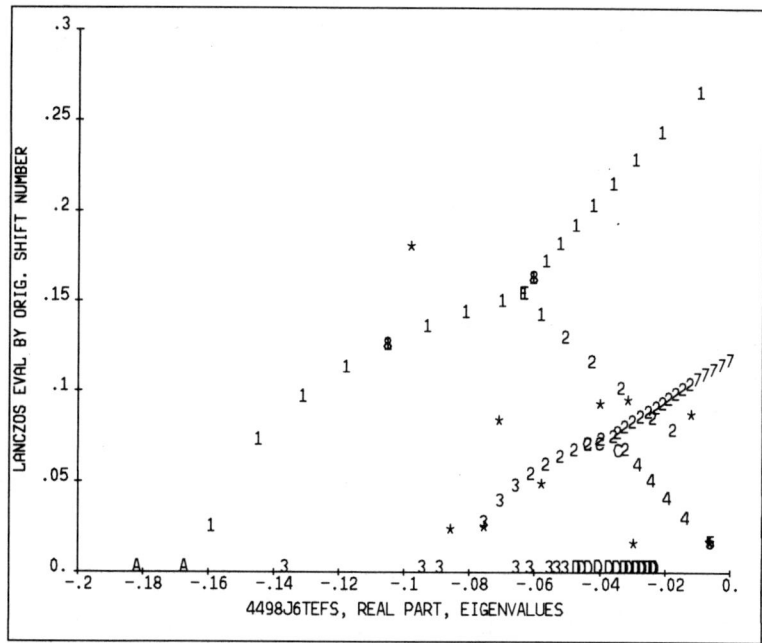

Figure 5.22: Test 4 in Table 2. MHD example, equilibrium 2, $n = 4498, \eta = 6 * 10^{-5}$. Lanczos (+) eigenvalue approximations plotted by number of first shift for which were they 'weakly-converged'. $\sigma_0 = (-.25,.25)$, $T_{91}$, SHFTOL $= 10^{-2}$.

for other eigenvalues 'around' the converged ones. In Test 5 with Lanczos matrices of size 31 and allowing 14 shifts, one Alfven and 11 sound mode eigenvalues were not found. The 45 eigenvalue approximations obtained were all obtained on shifts one through nine. Test 6 is the same as Test 5 except 21 shifts were allowed. Simply increasing the number of shifts did not solve the problem of the missing eigenvalues.

These tests demonstrate the relative insensitivity of our shift and invert strategy to the choice of initial shift and the importance of choosing the Lanczos matrices large enough to yield both weakly-converged eigenvalues, and eigenvalues that are not weakly-converged for use as potential shifts. They also illustrate the negative effects associated with choosing shifts that are 'too good', too close to actual eigenvalues, and the necessity for user interaction to assure complete resolution of the spectrum. Our Lanczos procedure should be run initially in the automatic mode with the user examining the resulting plots to determine appropriate shifts which our procedure can use to check for and locate any missing eigenvalues. Our Lanczos procedure would then be rerun using those shifts. This cycle could be repeated.

It is interesting to compare the behavior of our shift strategy in the nonsymmetric case with its behavior on a generalized real symmetric problem. We construct a difficult (but artificial) example with eigenvalues both spread throughout a large interval $[-815., 817.]$ and densely packed in a very small interval $[-.001, .001]$. Specifically we consider the problem $Ax = \lambda Bx$ where $A$ and $B$ are both diagonal matrices of order $n = 1000$. The diagonal entries in $A$ are chosen randomly from the interval $[-1, 1]$. The diagonal entries of $B$ are specified as follows

$$b_{jj} = j \quad \text{for} \quad 1 \leq j \leq .9n$$
$$= 1/j \quad \text{for} \quad .9n \leq j \leq n.$$

The overall eigenvalue distribution is given in Table 3.

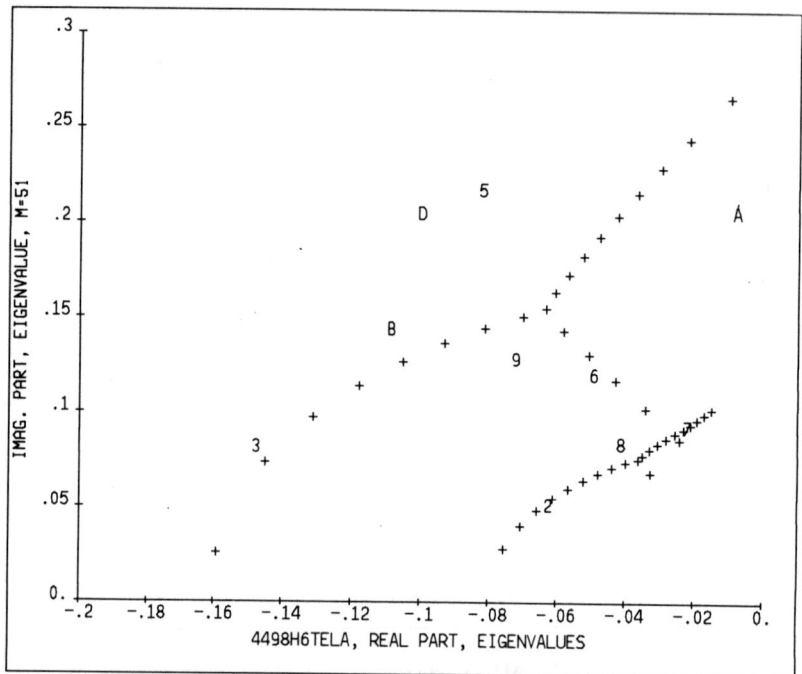

Figure 5.23: Test 5 in Table 2. MHD example, equilibrium 2, $n = 4498, \eta = 6 * 10^{-5}$. Lanczos (+) eigenvlaue approximations with shifts plotted by number or letter. $\sigma_0 = (-.25, .25)$, $T_{31}$, SHFTOL $= 10^{-2}$.

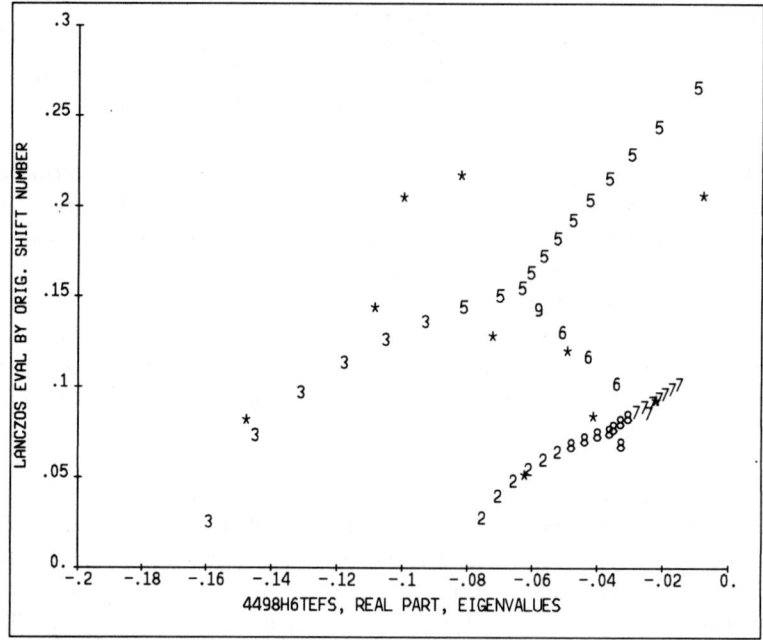

Figure 5.24: Test 5 in Table 2. MHD example, equilibrium 2, $n = 4498, \eta = 6 * 10^{-5}$. Lanczos (+) eigenvalue approximations plotted by number of shift for which they were 'weakly converged'. $\sigma_0 = (-.25, .25)$, $T_{31}$, SHFTOL $= 10^{-2}$.

LARGE MATRICES, LANCZOS ALGORITHMS, AND SHIFT AND INVERT 235

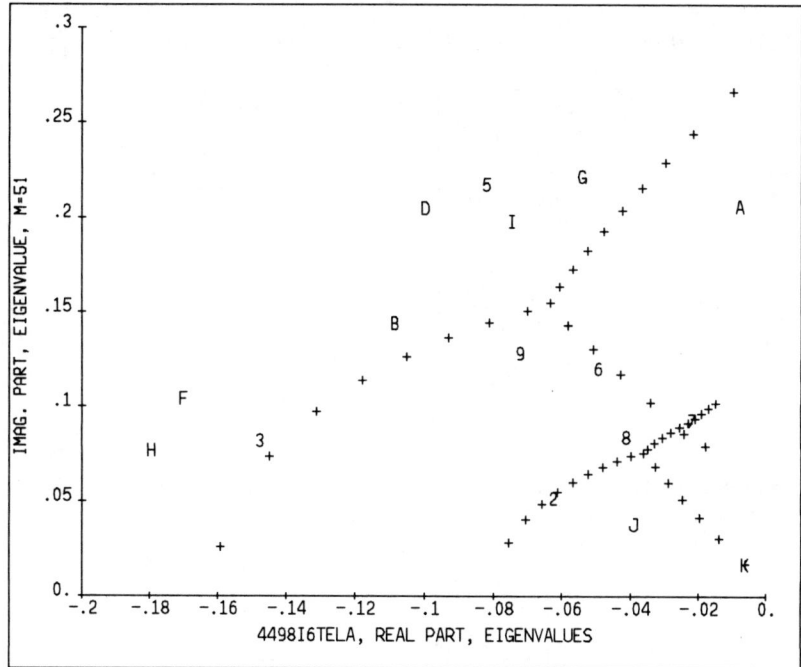

Figure 5.25: Test 6 in Table 2. MHD example, equilibrium 2, $n = 4498, \eta = 6 * 10^{-5}$. Lanczos (+) eigenvalue approximations with shifts plotted by number or letter. $\sigma_0 = (-.25, .25)$, $T_{31}$, SHFTOL $= 10^{-2}$, 21 shifts.

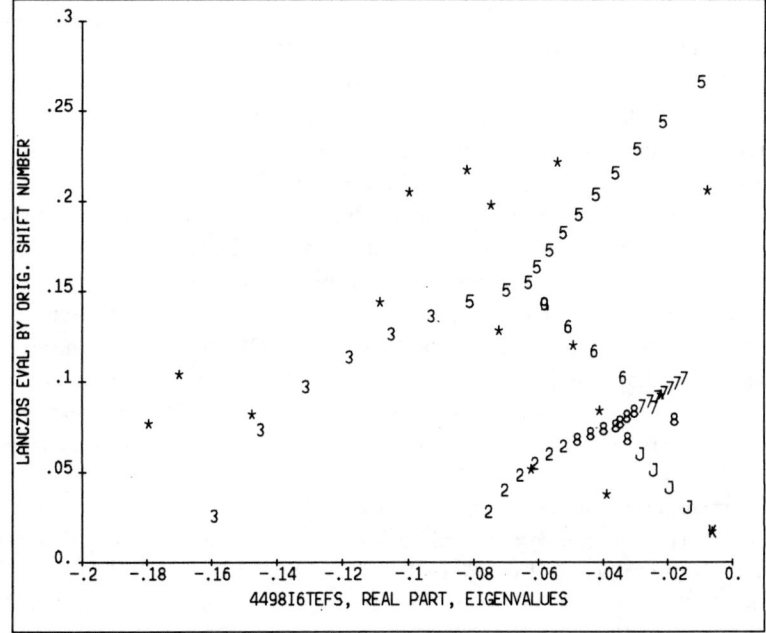

Figure 5.26: Test 6 in Table 2. MHD example, equilibrium 2, $n = 4498, \eta = 6 * 10^{-5}$. Lanczos (+) eigenvalue approximations plotted by number of first shift on which they were 'weakly-converged'. $\sigma_0 = (-.25, .25)$, $T_{31}$, SHFTOL $= 10^{-2}$, 21 shifts.

Table 3: Distribution of Eigenvalues, Real Symmetric Generalized Example, $n = 1000$

| Interval | Number of Eigenvalues |
|---|---|
| $[-815., -100.]$ | 287 |
| $[-98., -10.]$ | 146 |
| $[-9.95, -.73]$ | 20 |
| $[-.60, -.00102]$ | 6 |
| $[-.001, .001]$ | 95 |
| $[-.0011, .544]$ | 4 |
| $[1.7, 9.6]$ | 24 |
| $[10.12, 99.7]$ | 126 |
| $[101., 817.]$ | 292 |

Table 4: Samples of Gaps Between Eigenvalues, Real Symmetric Generalized Example, $n = 1000$

| Eigenvalue | Gap |
|---|---|
| $-736.9$ | 12 |
| $-287.3$ | $5 * 10^{-3}$ |
| $-78.7$ | $7.4 * 10^{-4}$ |
| $-8.9 * 10^{-4}$ | $1.03 * 10^{-7}$ |
| $2.4 * 10^{-5}$ | $8.5 * 10^{-7}$ |
| $756.27$ | 10.9 |

Tests were run on this example using a generalized, real symmetric version of our nonsymmetric, shift and invert Lanczos code. The shift strategies for both types of problems are identical except that here we want real shifts (and that is all we have), and we do not keep the shifts away from the origin. Restrictions to complex shifts and shifts not too near to the origin were incorporated in the nonsymmetric case because the desired eigenvalues were complex (the real eigenvalues were of no interest) and the $A$-matrix was singular. For any particular problem, users can add any additional restrictions on potential shifts that they want. These tests varied the choice of initial shift, the size of the tolerance SHFTOL, and the size of the Lanczos matrix used. The results of these tests are summarized in Table 5. Corresponding plots are in Figures 5.27 – 5.37. Not all tests are plotted.

Clearly the most densely packed subinterval is $[-.001, .001]$ which contains 95 eigenvalues. We asked the Lanczos procedure to compute all the eigenvalues in the interval (box) $[-.6, .6]$. Table 4 gives some indications of the large variations in gaps between neighboring eigenvalues. The gaps for the eigenvalues in the interval $[-.001, .001]$ vary from $1.04 * 10^{-7}$ to $4.7 * 10^{-5}$.

In this example all of the eigenvalues are real so we have chosen to plot the output slightly differently from the complex plots in the nonsymmetric case. Moreover, the high density of the eigenvalues in the interval $[-.001, .001]$ makes it difficult to plot the true eigenvalues and the Lanczos values in a single Figure. Therefore we have not tried to do that. However, all of the eigenvalue approximations which were obtained by our procedure and that are included in the counts in Table 5 agreed with the true eigenvalues to better than the pictorial resolution.

For each computation the box $[-.6, .6]$ was specified. However, each plot is re-

Table 5: Real Symmetric Generalized Example, $n = 1000$, box = $[-.6, 6]$, small box = $[-.0012, .0012]$

| Test | $\sigma_0$ | SHFTOL | $T_m$ | Defaults | Shifts in Small Box | Eval Approxs | Evals Missing |
|---|---|---|---|---|---|---|---|
| 1 | -.6 | $10^{-2}$ | 50 | 6 | 13/16 | 98 | 7 |
| 2 | 0. | $10^{-2}$ | 50 | 3 | 13/16 | 102 | 3 |
| 3 | .6 | $10^{-2}$ | 50 | 5 | 13/16 | 96 | 9 |
| 4 | .1 | $10^{-2}$ | 50 | 6 | 12/16 | 97 | 8 |
| 5 | 0. | $10^{-1}$ | 50 | 0 | 12/16 | 102 | 3 |
| 6 | 0. | $10^{-3}$ | 50 | 22 | 13/16 | 100 | 5 |
| 7 | 0. | $10^{-2}$ | 90 | 9 | 13/16 | 105 | 0 |
| 8 | 0. | $10^{-2}$ | 30 | 22 | 17/21 | 98 | 7 |
| 9 | -.6 | $10^{-2}$ | 30 | 11 | 17/21 | 77 | 28 |
| 10 | .6 | $10^{-2}$ | 30 | 13 | 17/21 | 88 | 17 |
| 11 | .1 | $10^{-2}$ | 30 | 7 | 14/21 | 73 | 32 |
| 12 | 0. | $10^{-2}$ | 50 | 6 | 18/21 | 102 | 3 |
| 13 | -.6 | $10^{-2}$ | 50 | 16 | 16/21 | 104 | 1 |
| 14 | .6 | $10^{-2}$ | 50 | 4 | 16/21 | 104 | 1 |
| 15 | .1 | $10^{-2}$ | 50 | 10 | 16/21 | 97 | 8 |

stricted to the small box, $[-.0012, .0012]$ because this is where almost all of the eigenvalues lie. For each test, we plot on the real axis those shifts which were in this small box. We then plot each weakly-converged eigenvalue approximation obtained by our Lanczos procedure on the vertical line through the first shift at which it was recognized as 'weakly-converged'. Typically for each shift we obtained several 'weakly-converged' eigenvalues. This however will not be apparent from the Figures because for some shifts no 'new' or 'additional' weakly-converged eigenvalues were identified, and therefore no eigenvalue approximations are plotted on the vertical lines through such shifts. The plots also do not include any shifts or 'weakly-converged' eigenvalue approximations which were in the big box $[-.6, .6]$ but not in the smaller box $[-.0012, .0012]$. Table 5 clearly illustrates the fact that for every case, almost all of the shifts generated automatically by our procedure fell in the small interval where all the eigenvalues are.

These plots can be used to illustrate the distribution of the shifts obtained in the smaller box, the numbers of such shifts needed to get the eigenvalue approximations listed in Table 5, and the corresponding distribution of those eigenvalue approximations over the different shifts. They do not however give any information on the errors in these approximations. In many of these plots, for the shift (or two shifts) nearest -.001, the plotting routine was unable to cope with the eigenvalue density and plotted two of the eigenvalue approximations alongside the vertical line (lines) corresponding to the shifts(s) on which they were obtained. See for example, Figure 5.27 and the eigenvalues corresponding to the eighth shift.

We decided to consider 16 shifts based purely on the nominal assumption that five or six new eigenvalue approximations might be obtained for each shift considered. In Tests 8 - 11, with Lanczos matrices of size 30, the number of shifts allowed was increased to 21. We also allowed 21 shifts in Tests 12 - 15 where we used Lanczos matrices of size 50 just to see if simply increasing the number of shifts would yield the eigenvalues which were still missing by the sixteenth shift. As Table 5 indicates this is not a good way to locate missing

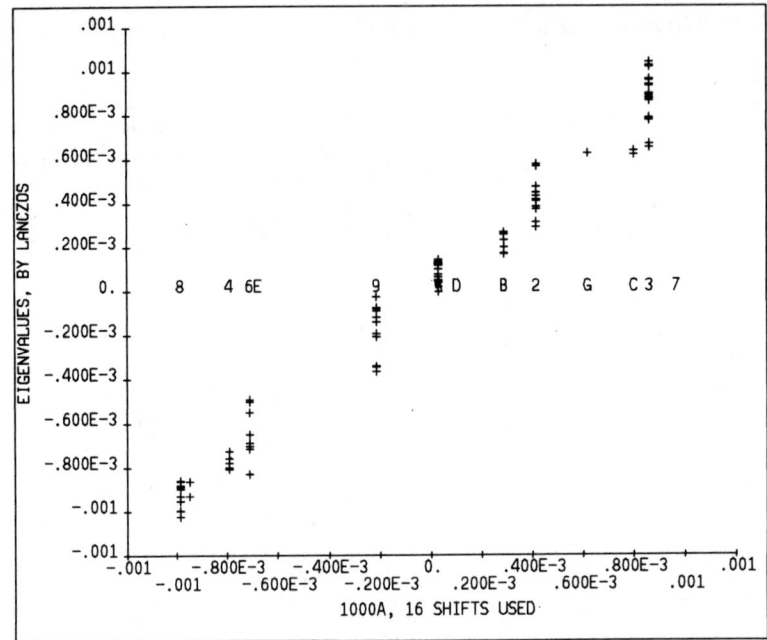

Figure 5.27: Test 1 in Table 5. Real symmetric generalized example, $n = 1000$. Lanczos (+) eigenvalues obtained on those shifts which were in smaller box $(-.0012, .0012)$. Corresponding shifts plotted on $x$-axis. $\sigma_0 = -.6$, $T_{50}$, SHFTOL $= 10^{-2}$.

eigenvalues.

Since $A$ and $B$ are real and symmetric with $B$ positive definite it is possible to use Sylvester's theorem, see for example Golub and Van Loan [1983, p. 308], to determine precisely how many eigenvalues of $Ax = \lambda Bx$ lie in any interval $(-\infty, \sigma)$ by simply counting the number of negative entries in the diagonal $D$ if the factorization $A - \sigma B = LDL^T$ is available. Since our procedure requires a factorization for each shift considered, counts on the intervals $(-\infty, \sigma_i)$ are available and could be used to identify subintervals in which eigenvalue approximations are missing. The user could then rerun our Lanczos procedure choosing a set of shifts based upon those intervals and these eigenvalue counts. We have not yet incorporated this check into our codes but we intend to do that.

Tests 1 – 4 in Table 5, see Figures 5.27–5.30, differ only in their choice of initial shift. None of these runs identified all 105 eigenvalues but they obtained respectively 98, 102, 96 and 97 eigenvalue approximations. If we take Test 2 for example with $\sigma_0 = 0$, then using the Sylvester counts on the eigenvalues in the intervals defined by the shifts, readily identifies the interval $(1.486 * 10^{-1}, 4.233 * 10^{-1})$ corresponding to shifts five and eleven as an interval containing the few eigenvalues which had not yet been approximated. These could be obtained by rerunning our code using a few a priori specified shifts in this small interval. Similar counts can be used in the other tests. These tests, as in the nonsymmetric case, indicate the relative insensitivity of our shift strategy to the particular choice of initial shift.

In Tests 2, 5 and 6, see Figures 5.28, 5.31 and 5.32, we examine the effect of varying SHFTOL, setting $\sigma_0 = 0$. Again a few eigenvalues are not identified. However, 100 or more are correctly identified and computed in spite of the fact that they are densely packed. In contrast to the scatter and redundancy observed for SHFTOL $= 10^{-3}$ on the $n = 4498$ nonsymmetric problem, on this real symmetric example the approximations and procedure

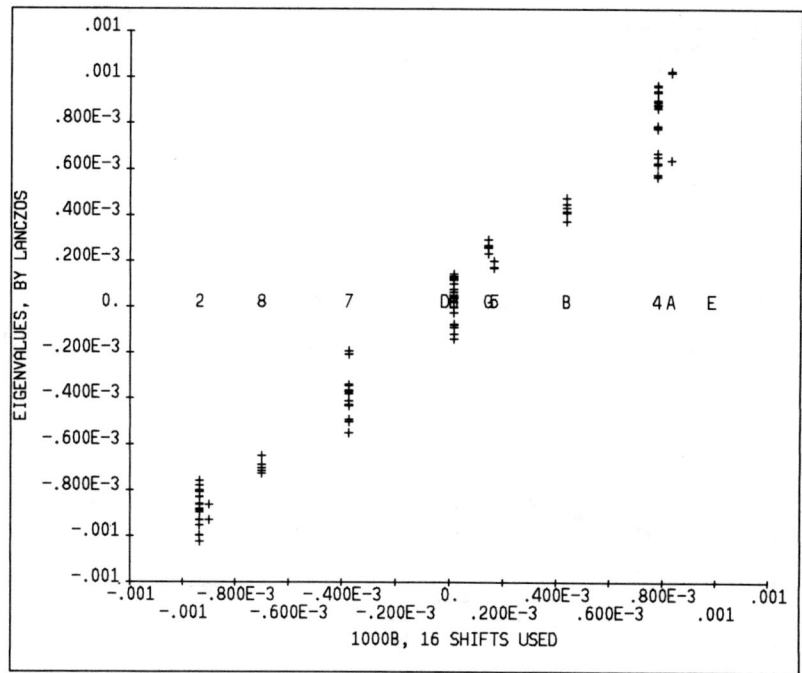

Figure 5.28: Test 2 in Table 5. Real symmetric generalized example, $n = 1000$. Lanczos (+) eigenvalues obtained on those shifts which were in smaller box $(-.0012, .0012)$. Corresponding shifts plotted on $x$-axis. $\sigma_0 = 0$, $T_{50}$, SHFTOL $= 10^{-2}$.

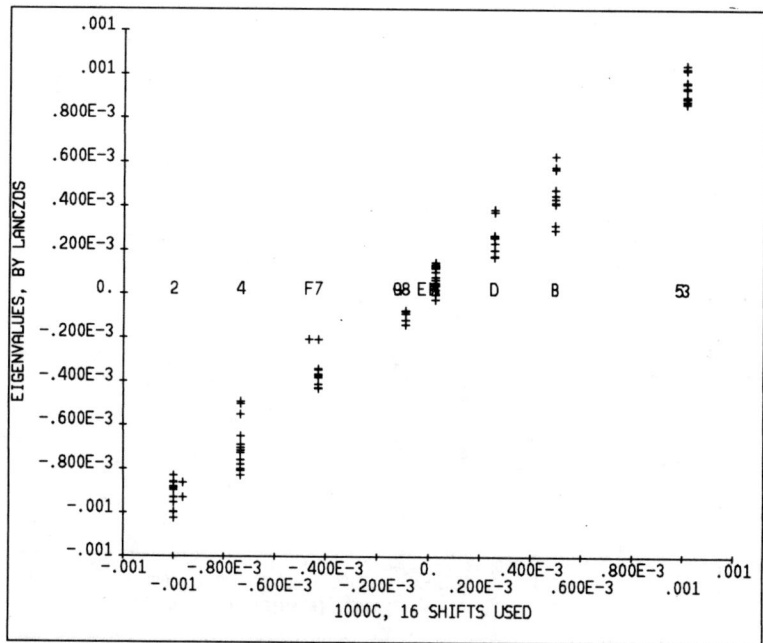

Figure 5.29: Test 3 in Table 5. Real symmetric generalized example, $n = 1000$. Lanczos (+) eigenvalues obtained on those shifts which were in smaller box $(-.0012, .0012)$. Corresponding shifts plotted on $x$-axis. $\sigma_0 = .6$, $T_{50}$, SHFTOL $= 10^{-2}$.

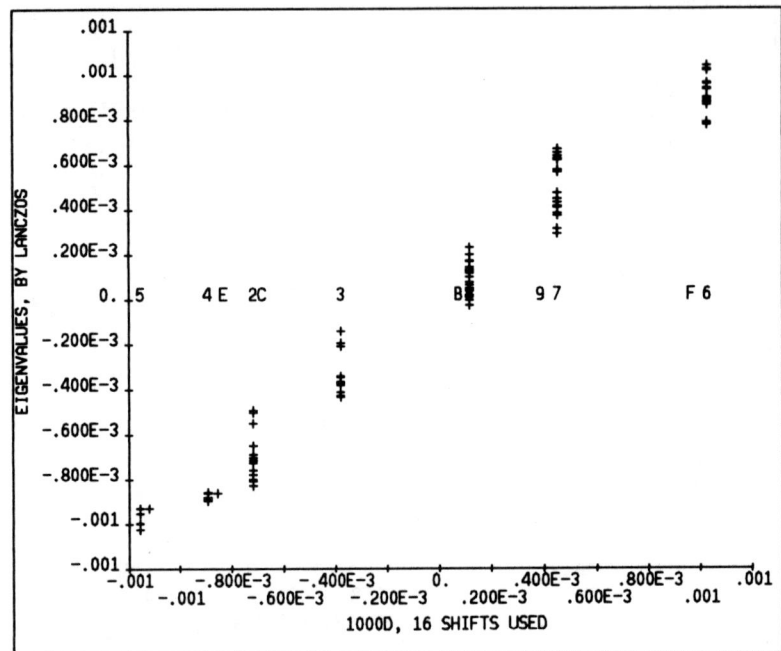

Figure 5.30: Test 4 in Table 5. Real symmetric generalized example, $n = 1000$. Lanczos (+) eigenvalues obtained on those shifts which were in the smaller box $(-.0012, .0012)$. Corresponding shifts plotted on $x$-axis. $\sigma_0 = .1$, $T_{50}$, SHFTOL $= 10^{-2}$.

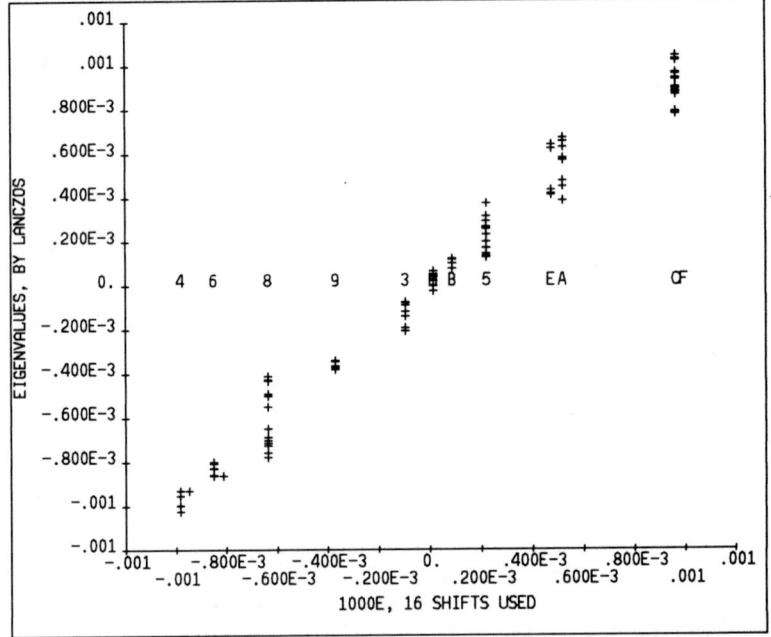

Figure 5.31: Test 5 in Table 5. Real symmetric generalized example, $n = 1000$. Lanczos (+) eigenvalues obtained on those shifts which were in the smaller box $(-.0012, .0012)$. Corresponding shifts plotted on $x$-axis. $\sigma_0 = 0$, $T_{50}$, SHFTOL $= 10^{-1}$.

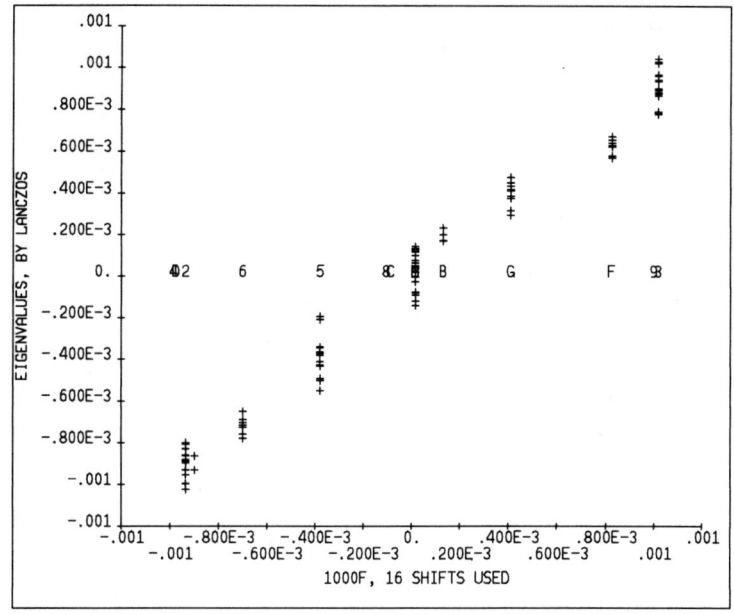

Figure 5.32: Test 6 in Table 5. Real symmetric generalized example, $n = 1000$. Lanczos (+) eigenvalues obtained on those shifts which were in the smaller box $(-.0012, .0012)$. Corresponding shifts plotted on $x$-axis. $\sigma_0 = 0$, $T_{50}$, SHFTOL $= 10^{-3}$.

continued to behave well even when SHFTOL $= 10^{-3}$. However, these results are misleading because for this example the factorizations are trivial since $A$ and $B$ are both diagonal matrices. The behavior on the nonsymmetric example with $n = 4498$ is much more representative of what we see in practice. On all three of these tests the automatically-generated shifts were appropriately distributed.

In Tests 2, 7 and 8, see Figures 5.28, 5.33 and 5.34, we can examine the effect for this Example of varying the size of the Lanczos matrices used. We considered three sizes, 30, 50 and 90. For Lanczos matrices of size 30 we also increased the number of shifts from 16 to 21. For Test 7, Figure 5.33, with Lanczos matrices of size 90 our procedure identified all 105 eigenvalues with redundancy on two eigenvalues. For Test 8, Figure 5.34, with Lanczos matrices of size 30 and 21 shifts, 98 eigenvalues were identified. There was however for shift 4, redundancy on two eigenvalues and two incorrect approximations. Shift 4 and two other shifts used in Test 8 were selected from lists of potential shifts that did not contain any eigenvalue approximations whose error estimates were larger than SHFTOL $= 10^{-2}$. Thus, for this example with the shift $\sigma_0 = 0$, there is no indication of any significant difference in behavior when we varied the size of the Lanczos matrix. We do see however, as expected that with a larger Lanczos matrix there is a corresponding increase in the numbers of weakly-converged eigenvalue approximations obtained on each shift. See Figure 5.33. By shift 8 approximations to all of the eigenvalues have been obtained. Correspondingly we see a similar decrease in the numbers of weakly-converged eigenvalue approximations obtained on each shift when we decrease the size of the Lanczos matrices and more approximations identified on later shifts. See Figure 5.34.

Tests 9, 10 and 11 however, see Figures 5.35-5.37, indicate clearly that the size of $T_m$ is important. For these tests, the initial shifts were $\sigma_0 = -.6, .6,$ and $.1$. Lanczos matrices of size 30 were used. Test 9 is typical. No shifts were identified in the in-

242   Cullum and Willoughby

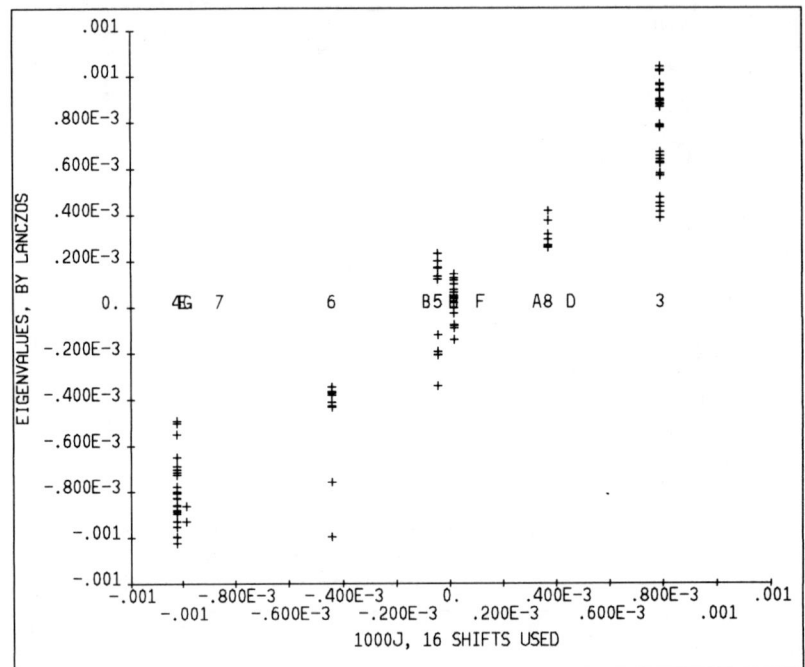

Figure 5.33: Test 7 in Table 5. Real symmetric generalized example, $n = 1000$. Lanczos (+) eigenvalues obtained on those shifts which were in the smaller box $(-.0012, .0012)$. Corresponding shifts plotted on $x$-axis. $\sigma_0 = 0$, $T_{90}$, SHFTOL $= 10^{-2}$.

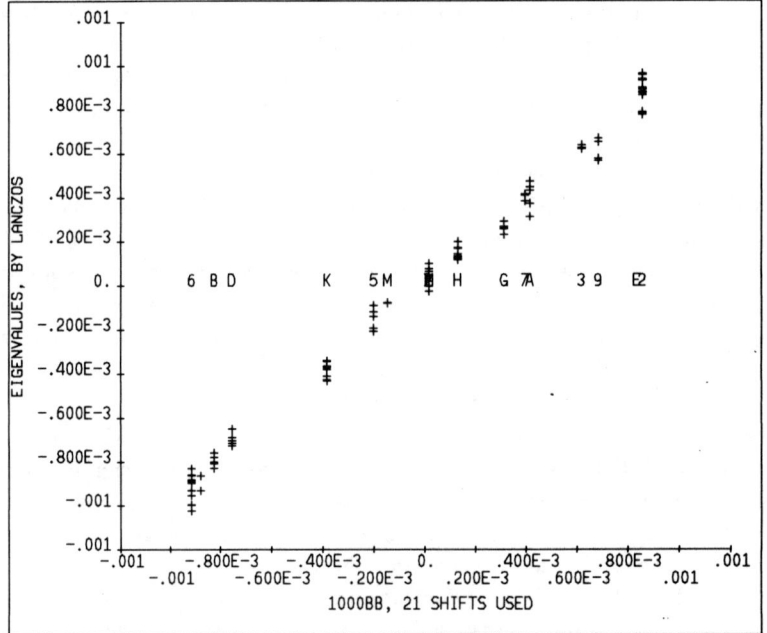

Figure 5.34: Test 8 in Table 5. Real symmetric generalized example, $n = 1000$. Lanczos (+) eigenvalues obtained on those shifts which were in the smaller box $(-.0012, .0012)$. Corresponding shifts plotted on $x$-axis. $\sigma_0 = 0$, $T_{30}$, SHFTOL $= 10^{-2}$.

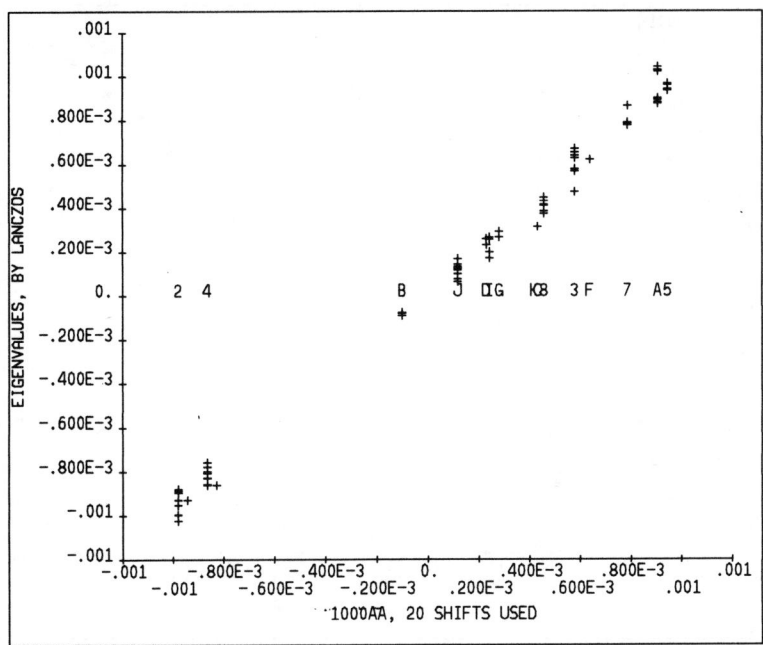

Figure 5.35: Test 9 in Table 5. Real symmetric generalized example, $n = 1000$. Lanczos (+) eigenvalues obtained on those shifts which were in the smaller box $(-.0012, .0012)$. Corresponding shifts plotted on $x$-axis. $\sigma_0 = -.6$, $T_{30}$, SHFTOL $= 10^{-2}$.

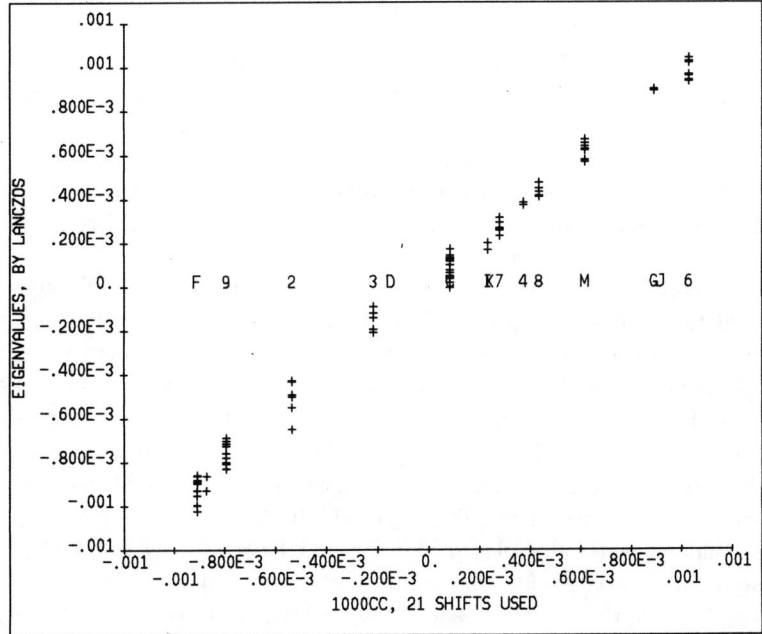

Figure 5.36: Test 10 in Table 5. Real symmetric generalized example, $n = 1000$. Lanczos (+) eigenvalue approximations obtained on those shifts which were generated in the smaller box $(-.0012, .0012)$. Corresponding shifts plotted on $x$-axis. $\sigma_0 = .6$, $T_{30}$, SHFTOL $= 10^{-2}$.

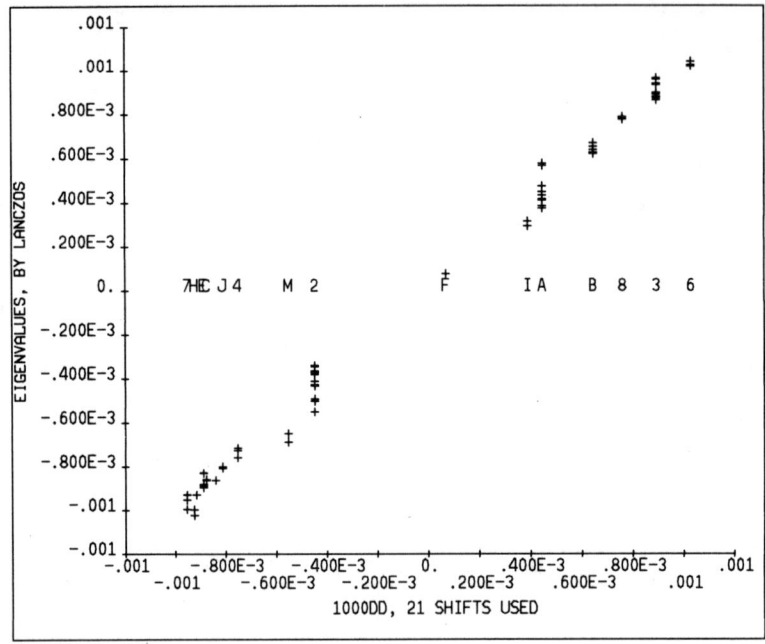

Figure 5.37: Test 11 in Table 5. Real symmetric generalized example, $n = 1000$. Lanczos (+) eigenvalue approximations obtained on those shifts which were in the smaller box $(-.0012, .0012)$. Corresponding shifts plotted on $x$-axis. $\sigma_0 = .1$, $T_{30}$, SHFTOL $= 10^{-2}$.

terval $(-.846*10^{-3}, -.118*10^{-3})$ and 22 of the missing eigenvalues were in the interval $(-.785*10^{-3}, -.116*10^{-3})$. The other six missing ones were in the interval $(-.102*10^{-3}, .37*10^{-4})$. A Lanczos matrix of size 30 did not provide enough information about the neighboring spectrum.

No plots are given for Tests 12 - 15. These tests were the same as Tests 1 - 4 with the number of allowable shifts increased to 21 from 16. As observed for the nonsymmetric MHD examples simply increasing the number of shifts which are allowed is not an effective way of locating any missing eigenvalues.

For any real symmetric problem we can use, as we said earlier, the Sylvester counts to verify that all of the required eigenvalues have been approximated. In the nonsymmetric case we do not have a similar test; and we must rely on exploring between gaps in the eigenvalue approximations, and in neighborhoods near eigenvalues with poor error estimates to determine if we have missed any eigenvalues. In the nonsymmetric case there is no known way of being certain that all of the desired eigenvalues have been computed.

One might want to argue that since for the real symmetric problems precise counts on the numbers of eigenvalues in any specified subinterval can be readily determined by factorizations, that geometrically-placed shifts would have produced similar results, i.e. there is no need for a strategy for generating shifts automatically. However, if for this example, we had taken the interval $[-.6, .6]$ and simply iteratively bisected it, and its successor intervals, doing counts on eigenvalues in each resulting subinterval, it would have taken quite a few bisections and associated factorizations to narrow the interval down to $(-.0012, .0012)$. Whereas, in each test, symmetric and nonsymmetric, our shift strategy almost immediately produced shifts within the relevant small 'box' and almost all of the shifts generated were in this small 'box'.

## 6 Conclusions.

We have summarized the key features in our Lanczos procedures for both real symmetric and nonsymmetric eigenvalue problems. Our shift and invert strategy was discussed in detail and several examples of its effectiveness were given. The numerical results in this paper and those in the earlier references Cullum and Willoughby [1985, 1986] and Cullum et al [1989] clearly indicate that our Lanczos algorithms can be used very effectively to solve very large problems. When factorizations are feasible, our shift and invert strategy focuses the computations on the region of interest.

Many open questions remain, not only for our shift and invert Lanczos procedures but even for our Lanczos procedures that do no shifting or inverting. Theoretical results have been difficult to obtain because the practical behavior of these procedures is completely entwined with the finite precision computer arithmetic. However, in the presence of strong numerical evidence together with ways of estimating the validity of the computed results, the lack of a convergence proof should not deter the use of these procedures. In fact, because real problems are typically more difficult than those problems which can be analyzed theoretically, many of the numerical algorithms in use today are being used in situations for which no convergence proof exists.

### References.

1. EISPACK Guide, *Matrix Eigensystem Routines — EISPACK Guide*, B.T. Smith, J.M. Boyle, B.S. Garbow, Y. Ikebe, V.C. Klema and C.B. Moler, eds. Lecture Notes in Computer Science, 6, Springer, New York, 1976.
2. EISPACK Guide, *Matrix Eigensystem Routines — EISPACK Guide Extension*, B.S. Garbow, J.M. Boyle, J.J. Dongarra and C.B. Moler, eds. Lecture Notes in Computer Science, 51, Springer, New York, 1977.
3. J. CULLUM AND R.A. WILLOUGHBY, *Lanczos Algorithms for Large Symmetric Eigenvalue Computations*, Volume 1 — Theory, Volume 2 — Programs; *Progress in Scientific Computing*, Volumes 3 and 4, eds. S. Abarbanel, R. Glowinski, G. Golub, P. Henrici, H.-O. Kreiss, Birkhäuser, Basel, 1985.
4. J. CULLUM AND R.A. WILLOUGHBY, A practical procedure for computing eigenvalues of large sparse nonsymmetric matrices, in *Large Scale Eigenvalue Problems*, J. Cullum and R.A. Willoughby, eds. Math. Stud., 127, North Holland, Amsterdam, 1986, pp. 193–240.
5. J. CULLUM, W. KERNER AND R.A. WILLOUGHBY, *A generalized nonsymmetric Lanczos procedure*, Computer Physics Communications, 53 (1989), pp. 19–48.
6. B.N. PARLETT AND D.S. SCOTT, *The Lanczos algorithm with selective reorthogonalization*, Math. Comp. 33 (1979), pp. 217–238.
7. B.N. PARLETT, *The Symmetric Eigenvalue Problem*, Prentice-Hall, Englewood Cliffs, NJ, 1980.
8. Y. SAAD, *Projection methods for solving large sparse eigenvalue problems*, Lecture Notes in Mathematics, 973 (1983), pp. 121–144.
9. Y. SAAD, *Chebyshev acceleration techniques for solving nonsymmetric eigenvalue problems*, Math. Comput. 42 (1984), pp. 567–588.
10. Y. SAAD, *The Rayleigh quotient iteration and some generalizations for nonnormal matrices*, Yale University Research Report U/DCS/RR-397, 1985.
11. Y. SAAD, *Variations on Arnoldi's method for computing eigenelements of large unsymmetric matrices*, Linear Algebra Appl. 34 (1980), pp. 269–295.
12. Y. SAAD, *Krylov subspace methods for computing eigenelements of large unsymmetric matrices*, Math. Comput. 37 (1981), pp. 105–126.

13. Y. SAAD, *The Lanczos biorthogonalization algorithm and other oblique projection methods for solving large unsymmetric linear systems*, SINUM 19 (1982), pp. 485–506.
14. DIEM HO AND F. CHATELIN, *Arnoldi-Tchebychev for large scale nonsymmetric matrices and its vectorizability*, IBM Science Center Report, Paris, France, 1988.
15. C.C. PAIGE, *The computation of eigenvalues and eigenvectors of very large sparse matrices*, Ph.D. Thesis, U. London, 1971.
16. HORST SIMON, *The Lanczos algorithm with partial reorthogonalization*, Mathematics of Computation 42 (1984), pp. 115–136.
17. J. CULLUM AND W.E. DONATH, *A block Lanczos algorithm for computing the q algebraically largest eigenvalues and a corresponding eigenspace of large, sparse, real symmetric matrices*, Proceedings of 1974 IEEE Conference on Decision and Control, Phoenix, Arizona, November 1974, pp. 505–509, IEEE Publications, Piscataway, NJ.
18. C.C. PAIGE, *Error analysis of the Lanczos algorithms for tridiagonalizing a symmetric matrix*, J. Inst. Math. Appl. 18 (1976), pp. 341–349.
19. C.C. PAIGE, *Accuracy and effectiveness of the Lanczos algorithm for the symmetric eigenproblem*, Linear Algebra Appl. 34 (1980), pp. 235–258.
20. J. CULLUM, M. LAKE AND R. WILLOUGHBY, *A Lanczos algorithm for computing eigenvalues and vectors of large matrices*, SIAM J. Sci. Stat. Comput. 4 (1983), pp. 197–251.
21. J.M. VAN KATS AND H.A. VAN DER VORST, *Numerical results of the Paige-style Lanczos method for the computation of extreme eigenvalues of large sparse matrices*, Acad. Comp. Centrum Report TR 3, U. Utrecht, 1976.
22. J.M. VAN KATS AND H.A. VAN DER VORST, *Automatic monitoring of Lanczos schemes for symmetric or skew symmetric generalized eigenvalue problems*, Acad. Comp. Centrum Report Utrecht TR 7, 1977.
23. B.N. PARLETT AND J.K. REID, *Tracking the progress of the Lanczos algorithm for large symmetric matrices*, IMA J. Numer. Anal. 1 (1981), pp. 135–155.
24. J.H. WILKINSON, *The Algebraic Eigenvalue Problem*, Oxford University Press, Oxford, England, 1965.
25. B.N. PARLETT, D.R. TAYLOR AND Z.A. LIU, *A look-ahead Lanczos algorithm for unsymmetric matrices*, Math. Comp. 44 (1985), pp. 105–124.
26. J. CULLUM AND R.A. WILLOUGHBY, *A QL algorithm for complex symmetric tridiagonal matrices*, IBM Research Report RC12835, Yorktown Heights, NY, 1987.
27. W. KERNER, *Computing the complex eigenvalue spectrum for resistive magnetohydrodynamics*, in Large Scale Eigenvalue Problems, J. Cullum and R.A. Willoughby eds., Math Stud. 127, North Holland, Amsterdam, 1986 pp. 241–266.

## CHAPTER 15

# Large-Scale Extended Linear-Quadratic Programming and Multistage Optimization*

## R. T. Rockafellar†

**Abstract.** Optimization problems in discrete time can be modeled more flexibly by extended linear-quadratic programming than by traditional linear or quadratic programming, because penalties and other expressions that may substitute for constraints can readily be incorporated and dualized. At the same time, dynamics can be written with state vectors as in dynamic programming and optimal control. This suggests new primal-dual approaches to solving multistage problems. The special setting for such numerical methods is described. New results are presented on the calculation of gradients of the primal and dual objective functions and on the convergence effects of strict quadratic regularization.

## 1. Introduction.

Large-scale problems in numerical optimization often arise from multistage models where decisions must be taken over a finite sequence of time periods. Such models may be deterministic or stochastic, and they may involve a process that naturally takes place in discrete time or the discretization of a process in continuous time. The computational challenges can be quite serious because of dimensionality, so the need to take advantage of specific problem structure has long been recognized. Different tactics can be followed in representing such structure, however, and this may be as important a key to eventual computational success as the refinement of standard algorithms for more general problems.

In work having its roots in the traditions of mathematical programming, the principal theme has been that of setting up problems as high-dimensional linear or quadratic programs and looking for helpful patterns of sparsity in the resulting matrices—staircase structure, and so forth. In work driven by the concepts and applications in areas like dynamic programming and optimal control, however, the emphasis is more often placed on a system of dynamical equations and the utilization of well known optimality conditions like the "maximum principle." In addition to the decision variables, termed "controls," there are auxiliary "state" variables around which the numerical procedures typically revolve.

---

*This work was supported by AFOSR grant 89-0081 and NSF grant DMS 881-9586.
†Department of Mathematics, University of Washington GN-50, Seattle, WA 98195.

Neither of these lines of thinking appears at present to derive full benefit from the other. In control, not enough attention has been paid to developing a first-level methodology for linear and quadratic types of problems in which inequality constraints can be handled alongside of equations, and in which the dual variables that emerge can be exploited on the basis of convexity. In mathematical programming, on the other hand, the modeling framework developed years ago for small-scale problems has perhaps been too readily accepted as appropriate for large-scale applications. The paradigm of just identifying a linear or quadratic objective function and writing down a list of exact linear constraints, then invoking a computational scheme that adapts to their special properties, may sometimes be unsuitable and unduly limiting.

The purpose of this paper is to describe an approach to multistage optimization that tries to bridge between the two camps, and to indicate the potential for numerical development that is thereby opened up. Problems are formulated in terms of "extended linear-quadratic programming," which differs from ordinary linear and quadratic programming mainly in allowing more flexible treatment of putative constraints, for instance in the direct incorporation of penalty terms. The problem models we draw on are ones developed recently in Rockafellar [1], [2], [3], [4], and Rockafellar and Wets [5]. We aim especially at explaining how the use of piecewise linear-quadratic penalty expressions along with state variables, generated from dynamical equations, alters the computational environment away from the familiar one in large-scale quadratic programming and linear complementarity but at the same time paves the way for new techniques like envelope methods [6], [7], and splitting methods [8]–[17].

## 2. Monitoring functions and extended linear-quadratic programming.

A problem of quadratic programming, as customarily defined, consists of minimizing a (linear or) quadratic convex objective function subject to a system of linear constraints. By introducing Lagrange multipliers for some of the constraints, one obtains a Lagrangian function for the problem and a saddle point characterization of optimality, which leads in turn to a dual problem. Solution methods may be "primal," which is to say roughly that they are molded by ideas of direct descent in the objective subject to the constraints, "dual," which means they involve ascent in an associated dual problem, or "primal-dual," which refers to a focus on solving for the primal and dual variables jointly through the Lagrangian saddle point condition.

Common in the case of primal methods have been "active set strategies" for identifying in a tentative way, to be updated as the procedure goes on, the set of linear inequality constraints likely to be active at the solution, and then treating these as equations. Such strategies often do not fit well with large-scale applications, however, because the dual dimension may be very high along with the primal dimension. Especially prominent among primal-dual methods have been those based on formulation of the saddle point condition as a problem in *linear complementarity*. We refer to Pang [18], Lin and Pang [19], for surveys of techniques in quadratic programming and linear complementarity. Recent developments based on Karmarkar's algorithm in linear programming can be found in Ye and Tse [20], Monteiro and Adler [21], and Goldfarb and Liu [22].

A problem of *extended linear-quadratic programming* can be described abstractly as consisting of the minimization of a piecewise linear-quadratic convex function subject to a system of linear constraints. A function is *piecewise linear-quadratic* if its domain can be expressed in principle (if not necessarily as a matter of convenience) by a union of polyhedral sets, on each of which the function is given by a linear or quadratic formula. This description of extended linear-quadratic programming obscures a fundamental difference in modeling concept, however, because the objective in this case may arise from some kind of mixing of what would be taken as the objective and constraints in a traditional approach.

In typical applications of optimization, the values of a variety of functions are of interest. The standard paradigm dictates that a single one of these functions must be chosen for minimization while the others are merely kept within certain bounds. The reality, of course, is that the choice of what should be the objective function versus what should be the constraint functions may be difficult because of trade-offs. The bounds to be used in writing down constraints may pose difficulties as well. In many situations black and white bounds on a function are not given, but just a desirable range of values that gradually merges into an undesirable or risky range. This is all the more true when the underlying application explicitly involves uncertainty and hedging. An insistence on exact constraints, except of a definitional variety (for instance, the natural nonnegativity of certain variables), makes little sense in such cases and can lead to poor models and unreliable "solutions."

There is nothing new in these observations, but their importance has grown with the scope of the problems being tackled with optimization methodology. Developments in stochastic programming as well as optimal control have made the deficiencies of mathematical models based on classical linear and quadratic programming quite apparent. Extended linear-quadratic programming, although not a panacea, offers a much more flexible approach to modeling which nonetheless retains the simple algebraic character of classical linear and quadratic programming *when viewed in primal-dual terms*. The function being minimized in the primal problem takes a more complicated form, but the Lagrangian saddle point problem characterizing optimality turns out still to be one that concerns just a quadratic function on a product of polyhedral sets.

In setting up a model in extended linear-quadratic programming, the notion of a "monitoring function" is central in obtaining from the given functions, either singly or in groups, the terms that make up the composite objective. Monitoring functions include penalty functions as a special case. For a start, let us consider a single function $h(u)$ from among those given in some application, where $u \in \mathbb{R}^k$ is the vector of "decision variables." The point of view is that the values of this function must be incorporated in the model, but not necessarily according to the standard paradigm: we are at a stage of formulation where no firm choice has yet been made about what should be a hard constraint or a contribution to the objective, at least as far as $h(u)$ is concerned.

In general we can think of placing a term $\rho(h(u))$ in the objective function being constructed for minimization—for some choice of $\rho : \mathbb{R} \to \overline{\mathbb{R}}$. Then $\rho$ is a *monitoring function* for $h(u)$. For instance, if we wish the problem to simply have the minimization of $h(u)$ as its objective, we can take $\rho(s) = s$. If instead we wish to minimize a weighted sum of expressions, of which $h(u)$ is just one, we can take $\rho(s) = \lambda s$ for some $\lambda > 0$. On the other hand, if our aim turns out to be that of enforcing the inequality $h(u) \leq 0$ as a hard constraint, this corresponds to taking $\rho$ to be the indicator of $\mathbb{R}_-$, i.e., $\rho(s) = 0$ when $s \leq 0$ but $\rho(s) = \infty$ when $s > 0$. In similar fashion, the equation $h(u) = 0$ corresponds to defining $\rho(0) = 0$, but $\rho(s) = \infty$ for all $s \neq 0$.

Penalty representations of a constraint like $h(u) \leq 0$ fit this picture very easily. A standard quadratic penalty term would arise in taking $\rho(s) = 0$ when $s \leq 0$ but $\rho(s) = \frac{1}{2}\mu s^2$ when $s > 0$ (where the penalty parameter $\mu$ is positive), while a so-called linear penalty would take the form $\rho(s) = 0$ when $s \leq 0$ but $\rho(s) = \mu s$ when $s > 0$. As is well known, a linear penalty term with $\mu$ high enough may suffice to represent a hard constraint

$h(u) \leq 0$ exactly. Exact penalty expressions can also be obtained through augmented Lagrangian terms of the kind where $\rho(s) = \lambda s + \frac{1}{2}\mu s^2$ when $s \geq -\lambda/\mu$, but $\rho(s) = -\lambda^2/2\mu$ when $s \leq -\lambda/\mu$ (see [23]). Further, one can make use of mixed penalties which start out quadratic but end up linear, as have turned out to be desirable in stochastic programming, cf. Rockafellar and Wets [24].

In all these examples the monitoring function $\rho$ is convex and piecewise linear-quadratic. This continues to be observed when several functions $h_i(u)$ are monitored together by a term $\rho(h(u))$ where $h(u)$ is a vector $(h_1(u), \ldots, h_l(u))$. For instance, an expression of the form $\max_i h_i(u)$ in the objective corresponds to the piecewise linear function $\rho(s) = \max_i s_i$, where $s = (s_1, \ldots, s_l)$. The hard constraint $h(u) \leq 0$ in vector terms corresponds to $\rho$ being the indicator of the nonpositive orthant of $\mathbb{R}^l$ (which is a polyhedral set), and so forth. Many other illustrations could be given in which the monitoring function acts like a weighted sum or max within the main region in which the vector $h(u)$ is desired to lie, but takes on penalty characteristics outside.

A dual form of representation turns out to characterize a large class of monitoring functions which includes the many examples indicated and is especially convenient to work with. A function in this class, on $\mathbb{R}^l$, is fixed by choosing a nonempty polyhedral set $V \subset \mathbb{R}^l$ (possibly all of $\mathbb{R}^l$) and a positive *semi*definite, symmetric matrix $Q \in \mathbb{R}^{l \times l}$ (which could be the zero matrix):

$$\rho_{VQ}(s) = \max_{v \in V}\{s \cdot v - \tfrac{1}{2} v \cdot Qv\}.$$

In the one-dimensional case ($l = 1$), the set $V$ is just a closed interval $[v^-, v^+]$ (perhaps unbounded) and $Q$ is specified by a single number $\beta \geq 0$, so that

$$\rho_{VQ}(s) = \max_{v^- \leq v \leq v^+}\{sv - \tfrac{1}{2}\beta s^2\}.$$

The maximum can then be calculated explicitly to obtain a more direct formula. In the *box-diagonal* case, by which we mean the case where $V$ is a box (a product of closed intervals $[v_i^-, v_i^+]$) and $Q$ is a diagonal matrix ($\mathrm{diag}\{\beta_1, \ldots, \beta_l\}$ with $\beta_i \geq 0$), the calculation can be carried out similarly, and $\rho_{VQ}(s)$ is a sum of separate terms in the components $s_i$ of $s$. On the other hand, the example of $\rho(s) = \max_i s_i$ corresponds to $V$ being the unit simplex in $\mathbb{R}^l$ (the set of vectors $(s_1, \ldots, s_l) \geq (0, \ldots, 0)$ with $\sum_{i=1}^l s_i = 1$) and $Q$ being the zero matrix. In Rockafellar and Wets [25] it is demonstrated that even the "recourse cost function" in a stochastic linear or quadratic programming problem can be written as $\rho_{VQ}(h(u))$ for some choice of $V$, $Q$, and a linear (or affine) mapping $h$.

Although composite objectives involving convex, possibly nonsmooth monitoring functions has been explored by many researchers in nonlinear programming in recent years, the fact that most of the monitoring expressions of interest fall in the $\rho_{VQ}$ class has not been recognized. The properties of this class make possible a treatment with a cleaner depiction of the role of generalized Lagrange multipliers in the form of dual variables.

These considerations lead to the adoption of the following *standard form* for a problem in extended linear-quadratic programming:

$$(\mathcal{P}_0) \qquad \text{minimize } p \cdot u + \tfrac{1}{2} u \cdot Pu + \rho_{VQ}(q - Ru) \text{ over } u \in U,$$

where the sets $U \subset \mathbb{R}^k$ and $V \subset \mathbb{R}^l$ are polyhedral and the matrices $P \in \mathbb{R}^{k \times k}$ and $Q \in \mathbb{R}^{l \times l}$ are symmetric and positive *semi*definite. In this, the set $U$ expresses fixed linear constraints not handled through "monitoring." The use of this form is discussed more fully in [1, §3]. While in many situations it would be right to think of the terms $p \cdot u + \frac{1}{2} u \cdot P u$ as the basic objective function, modified by a penalty function applied to the vector $q - Ru$, the form can also be used in other ways. For instance, in taking $V$ to be all of $\mathbb{R}^l$ and $Q$ to be the $l \times l$ identity matrix, one obtains $\frac{1}{2}|q - Ru|^2$ for the monitoring term. If $q = 0$, this yields $p \cdot u + \frac{1}{2} u \cdot (P + R^* R) u$ as the objective expression to be minimized, where $*$ denotes the transpose of a matrix. The problem then falls into the classical category of quadratic programming, but with the representation of additional structure that could aid in its solution.

An important advantage of the standard form is that it points specifically to the Lagrangian function to be used in characterizing optimality. The Lagrangian is

$$L(u,v) = p \cdot u + \tfrac{1}{2} u \cdot P u + q \cdot v - \tfrac{1}{2} v \cdot Q v - v \cdot R u \quad \text{on } U \times V. \tag{1}$$

We have demonstrated in [25] that *a vector $\bar{u}$ solves $(\mathcal{P}_0)$ if and only if there is a vector $\bar{v}$ such that the pair $(\bar{u}, \bar{v})$ furnishes a saddle point of $L$ relative to $U \times V$*. Especially to be noted here is the fact that the function $L$, unlike the objective function in $(\mathcal{P}_0)$, is truly linear-quadratic—not merely in a *piecewise* sense. The Lagrangian representation of $(\mathcal{P}_0)$ thus reveals an underlying simplicity in the problem format which might not immediately be evident. It also identifies the corresponding dual problem as

$$(\mathcal{Q}_0) \qquad \text{maximize } q \cdot v - \tfrac{1}{2} v \cdot Q v - \rho_{UP}(R^* v - p) \text{ over } v \in V.$$

The objective function in $(\mathcal{Q}_0)$ equals $\inf_{u \in U} L(u,v)$, whereas the objective function in $(\mathcal{P}_0)$ equals $\sup_{v \in V} L(u,v)$. The solutions to $(\mathcal{Q}_0)$ are the $\bar{v}$ components of the saddle points $(\bar{u}, \bar{v})$ for the minimax problem relative to (1) and may be interpreted as the optimal multiplier vectors associated with the monitoring expression in $(\mathcal{P}_0)$. When the monitoring expression represents hard constraints through infinite penalties, they are Lagrange multiplier vectors in the usual sense.

**3. Problem Models in Multistage Optimization.** With these modeling ideas, there are other ways of dealing with dynamical structure than just through the sparsity pattern in some constraint matrix. In problems involving a linear process operating in discrete time, the dynamics can generally be expressed by a system

$$x_t = A_t x_{t-1} + B_t u_t + b_t \text{ for } t = 1, \ldots, T, \text{ with } x_0 = B_0 u_0 + b_0, \tag{2}$$

where $u_t$ is a so-called *control* vector in $\mathbb{R}^{k_t}$ and $x_t$ is a *state* vector in $\mathbb{R}^{n_t}$. The control vectors give the true decision variables, while the state vectors may just be artificial constructs, but useful nevertheless. The state vectors stand for certain affine expressions in the control vectors as derived from (2):

$$x_1 = A_1(B_0 u_0 + b_0) + B_1 u_1 + b_1,$$
$$x_2 = A_2(A_1(B_0 u_0 + b_0) + B_1 u_1 + b_1) + B_2 u_2 + b_2, \ldots$$

and they could therefore be eliminated in principle from the formulation of any optimization problem with respect to $(u_0, u_1, \ldots, u_T)$. For a number of reasons, however, it seems best to keep them. Because the dimension $k_t$ of $u_t$ and the dimension $n_t$ of $x_t$ are allowed to vary with $t$, there is no loss of generality in taking (2) as the form for the dynamics: even if a problem may not seem to involve state vectors, they can always be introduced so as to get this type of expression. (This is a device of long standing; see [5] for details.)

The model problem we propose to consider for such a dynamical system is

(P) minimize subject to (2) with $u_t \in U_t$ for $t = 0, 1 \ldots, T$ the expression
$$f(u_0, \ldots, u_T) := p_0 \cdot u_0 + \tfrac{1}{2} u_0 \cdot P_0 u_0$$
$$+ \sum_{t=1}^{T} \left[ p_t \cdot u_t + \tfrac{1}{2} u_t \cdot P_t u_t + \rho_{V_t Q_t}(q_t - C_t x_{t-1} - D_t u_t) - c_t \cdot x_{t-1} \right]$$
$$+ \rho_{V_{T+1} Q_{T+1}}(q_{T+1} - C_{T+1} x_T) - c_{T+1} \cdot x_T,$$

where the sets $U_t \subset \mathbb{R}^{k_t}$ and $V_t \subset \mathbb{R}^{l_t}$ are polyhedral and the matrices $P_t \in \mathbb{R}^{k_t \times k_t}$ and $Q_t \in \mathbb{R}^{l_t \times l_t}$ are positive semidefinite. (The elements $V_t$, $Q_t$, $q_t$, $C_t$ and $c_t$ are indexed from $t = 1$ to $t = T + 1$ instead of from $t = 0$ to $t = T$ for the sake of achieving a symmetric formulation of duality, as will soon be seen.) Note that the objective is correctly written as $f(u_0, \ldots, u_T)$, because the vectors $x_t$ stand for affine expressions in $u_0, \ldots, u_T$ determined by the dynamics (2), as already mentioned:

$$(x_0, \ldots, x_T) = X(u_0, \ldots, u_T) \text{ for an affine mapping } X : \mathbb{R}^{k_0 + \cdots + k_T} \to \mathbb{R}^{n_0 + \cdots + n_T}.$$

Problem (P) was introduced in Rockafellar and Wets [5]. It was shown in that paper that (P) is the primal problem associated with the Lagrangian $L(u, v)$ on $U \times V$, where

$$\begin{aligned} U &:= U_0 \times \cdots \times U_T, & u &= (u_0, \ldots, u_T), \\ V &:= V_1 \times \cdots \times V_{T+1}, & v &= (v_1, \ldots, v_{T+1}), \\ L(u, v) &:= [p_0 \cdot u_0 + \tfrac{1}{2} u_0 \cdot P_0 u_0] + [q_{T+1} \cdot v_{T+1} - \tfrac{1}{2} v_{T+1} \cdot Q_{T+1} v_{T+1}] \\ & \quad + \sum_{t=1}^{T} \left[ p_t \cdot u_t + \tfrac{1}{2} u_t \cdot P_t u_t + q_t \cdot v_t - \tfrac{1}{2} v_t \cdot Q_t v_t - v_t \cdot D_t u_t \right] - l(u, v) \end{aligned} \quad (3)$$

and the expression $l(u, v)$ is affine separately in $u$ and $v$ and is defined from the dynamical equations (2) by

$$l(u, v) = \sum_{t=1}^{T+1} x_{t-1} \cdot (C_t^* v_t + c_t). \quad (4)$$

(As earlier, $*$ denotes the transpose of a matrix.)

**THEOREM 1** [5]. *A control sequence $\bar{u} = (\bar{u}_0, \ldots, \bar{u}_T)$ solves the multistage optimization problem (P) if and only if there is a dual control sequence $\bar{v} = (\bar{v}_1, \ldots, \bar{v}_{T+1})$ such that $(\bar{u}, \bar{v})$ is a saddle point of $L(u, v)$ relative to $U \times V$ in (3).*

The reason for calling $\bar{v}$ a "dual control sequence" is that it solves a dual problem of the same general kind. The dual dynamical system goes backward in time with the vectors $v_t \in \mathbb{R}^{l_t}$ as controls and certain vectors $y_t \in \mathbb{R}^{n_t}$ as states:

$$y_t = A_t^* y_{t+1} + C_t^* v_t + c_t \text{ for } t = T, \ldots, 1, \text{ with } y_{T+1} = C_{T+1}^* v_{T+1} + c_{T+1}. \quad (5)$$

The dual states $y_t$ thus represent linear expressions in the vectors $v_1, \ldots, v_{T+1}$:

$$(y_1, \ldots, y_{T+1}) = Y(v_1, \ldots, v_{T+1}) \text{ for an affine mapping } Y : \mathbb{R}^{l_1 + \cdots + l_{T+1}} \to \mathbb{R}^{n_1 + \cdots + n_{T+1}}.$$

It turns out (see [5]) that the Lagrangian dynamical term $l(u, v)$ defined in (4) can be written equivalently by means of (5) as

$$l(u, v) = \sum_{t=0}^{T} y_{t+1} \cdot (B_t u_t + b_t). \quad (6)$$

(Here we see some of the motivation for the indexing conventions.) From this one derives the dual problem as

(Q) maximize subject to (5) with $v_t \in V_t$ for $t = 1 \ldots, T+1$ the expression
$$g(v_1, \ldots, v_{T+1}) := q_{T+1} \cdot v_{T+1} - \tfrac{1}{2} q_{T+1} \cdot Q_{T+1} v_{T+1}$$
$$+ \sum_{t=1}^{T} \left[ q_t \cdot v_t - \tfrac{1}{2} v_t \cdot Q_t v_t - \rho_{U_t P_t}(B_t^* y_{t+1} + D_t^* v_t - p_t) - b_t \cdot y_{t+1} \right]$$
$$- \rho_{U_0 P_0}(B_0^* y_1 - p_0) - b_0 \cdot y_1.$$

From the dynamics (5) we have of course that

$(y_1, \ldots, y_{T+1}) = Y(v_1, \ldots, v_{T+1})$ for an affine mapping $Y : \mathbb{R}^{l_1 + \cdots + l_{T+1}} \to \mathbb{R}^{n_1 + \cdots + n_{T+1}}$.

Because $L$ in (3) is (at most) quadratic in $u$ and $v$, while the sets $U$ and $V$ are polyhedral, ($\mathcal{P}$) is in fact a problem of extended linear-quadratic programming in standard form, and so is its dual ($\mathcal{Q}$), except for a change of sign in converting to maximization. The following result then holds in consequence of the general theory of extended linear-quadratic programming in [25] (and [1]).

**THEOREM 2** [5]. *Problem* ($\mathcal{P}$) *has a solution if and only if its objective* $f(u)$ *is bounded below on* $U$ *and is finite for at least one* $u \in U$. *Then* $\min(\mathcal{P}) = \max(\mathcal{Q})$.

The saddle point condition in Theorem 1 can be expressed through the gradients of $L$ and the normal cones

$$\begin{aligned} N_U(\bar{u}) &:= \{r \mid r \cdot (u - \bar{u}) \leq 0 \text{ for all } u \in U\} \quad \text{(taken to be } \emptyset \text{ if } \bar{u} \notin U\text{)}, \\ N_V(\bar{v}) &:= \{s \mid s \cdot (v - \bar{v}) \leq 0 \text{ for all } v \in V\} \quad \text{(taken to be } \emptyset \text{ if } \bar{v} \notin V\text{)}, \end{aligned} \quad (7)$$

as the condition that

$$-\nabla_u L(\bar{u}, \bar{v}) \in N_U(\bar{u}), \qquad \nabla_v L(\bar{u}, \bar{v}) \in N_V(\bar{v}). \tag{8}$$

In terms of the affine mapping $M$ defined by

$$M(u, v) = (\nabla_u L(u, v), -\nabla_v L(u, v)) \tag{9}$$

this can be written in turn as

$$-M(\bar{u}, \bar{v}) \in N_{U \times V}(\bar{u}, \bar{v}), \tag{10}$$

a condition which is the same as the linear variational inequality

$$M(\bar{u}, \bar{v}) \cdot [(u, v) - (\bar{u}, \bar{v})] \leq 0 \text{ for all } (u, v) \in U \times V. \tag{11}$$

We shall not describe here the stochastic version in [5] of the multistage problem ($\mathcal{P}$), but it likewise corresponds to a quadratic convex-concave Lagrangian on a product of polyhedral sets much as in (3). The difference is just that $u$ and $v$ are then random vectors subjected to certain measurability constraints (expressing the state of information at any time $t$), and the sets $U$ and $V$ must be altered slightly to take this into account. An expectation must also be taken in the formula for the Lagrangian. The various data elements in the problem may be random variables along with the unknowns $u$ and $v$. See [5].

**4. Envelope Methods.** The formulas for the objective functions $f$ and $g$ in $(\mathcal{P})$ and $(\mathcal{Q})$ bring the linear-quadratic structure fully into view. They underline the fact that these functions are not necessarily smooth, because the monitoring terms may fail to be smooth. Even if $f$ and $g$ do have continuous first derivatives, they will generally fail to have continuous second derivatives, since $\rho_{V_t Q_t}$ and $\rho_{P_t U_t}$ fail to have such derivatives except in the very special cases. These circumstances make it important to look for new ways of utilizing the structure of $f$ and $g$ in terms of the primal and dual dynamics.

We shall focus on the *strictly quadratic* case of $(\mathcal{P})$ and $(\mathcal{Q})$, by which we mean the case where the matrices $P_t$ and $Q_t$ are *positive definite* (i.e., nonsingular). Then, as we shall demonstrate, there are special properties which strongly support the design of numerical methods. In particular, it turns out that we do have first-order smoothness of $f$ and $g$ in this case, although second-order smoothness is still lacking. While many problems of interest are not strictly quadratic, a basic device allows them to be approached by solving a sequence of auxiliary problems (of the same multistage form) that are strictly quadratic. This will be explained in the next section.

**THEOREM 3.** *In the strictly quadratic case, the objective function $f$ in $(\mathcal{P})$ is strongly convex with continuous first derivatives, and it can be expressed by*

$$f(u) = \max_{v \in V} L(u,v) = L(u, F(u)) \quad \text{with} \quad F(u) := \operatorname*{argmax}_{v \in V} L(u,v). \tag{12}$$

*Likewise, the objective function $g$ in $(\mathcal{Q})$ is in this case strongly concave with continuous first derivatives, and it can be expressed by*

$$g(v) = \min_{u \in U} L(u,v) = L(G(v), v) \quad \text{with} \quad G(v) := \operatorname*{argmin}_{u \in U} L(u,v). \tag{13}$$

*The mappings $F$ and $G$ are continuous and piecewise linear, and one has*

$$\nabla f(u) = \nabla_u L(u, F(u)), \qquad \nabla g(v) = \nabla_v L(G(v), v). \tag{14}$$

**Proof.** Formulas (12) and (13) follow right from the definitions of the monitoring functions $\rho_{V_t Q_t}$ and $\rho_{U_t P_t}$. The max and min are uniquely attained because of the assumed positive definiteness of the matrices giving the quadratic terms in (3). Indeed, $F(u)$ is the solution to a strictly quadratic programming program in $v$ in which only the linear part of the objective depends on $u$, from which it follows further (by the theory of parametric quadratic programming, cf. also the proof of our next result) that the mapping $F$ is continuous and piecewise linear; similarly for $G(v)$. The positive definiteness also yields the strong convexity of $f$ and the strong concavity of $g$, as is evident from the formulas for these functions in the statement of $(\mathcal{P})$ and $(\mathcal{Q})$. The gradient formulas have been proved in a more general context of extended linear-quadratic programming in [6, Prop. 3.2]. □

The expressions for $f$ and $g$ in (12) and (13) are called *envelope representations*: $f$ is the pointwise maximum of the collection of quadratic convex functions $\{L(\cdot, v)\}_{v \in V}$, whereas $g$ is the pointwise minimum of the collection of quadratic concave functions $\{L(u, \cdot)\}_{u \in U}$. Numerical schemes that make use of these representations in terms of various iterations on the mappings $F$ and $G$ are *envelope methods*. A basic theory of such methods has been developed by the author in [6].

A starting idea for envelope methods is to use a current point $u^0 \in U$ in $(\mathcal{P})$ and a current point $v^0 \in V$ in $(\mathcal{Q})$ to generate finitely many other points $u^1, \ldots, u^r$ in $U$ and $v^1, \ldots, v^r$ in $V$ by

$$v^k = F(u^{k-1}) \text{ and } u^k = G(v^{k-1}) \text{ for } k = 1, \ldots, r.$$

These points are seen as providing envelope information about the functions $f$ and $g$ through the fact that the finite envelope functions

$$\tilde{f}(u) := \max_{v \in \tilde{V}} L(u,v) \text{ with } \tilde{V} := \text{conv}\{v^0, v^1, \ldots, v^r\},$$
$$\tilde{g}(v) := \min_{u \in \tilde{U}} L(u,v) \text{ with } \tilde{U} := \text{conv}\{u^0, u^1, \ldots, u^r\},$$

satisfy

$$\begin{aligned} f(u) \geq \tilde{f}(u) \text{ for all } u, \text{ with equality for } u = u^k, k = 0, 1, \ldots, r-1, \\ g(v) \leq \tilde{g}(v) \text{ for all } v, \text{ with equality for } v = v^k, k = 0, 1, \ldots, r-1. \end{aligned} \quad (15)$$

One type of scheme based on this information has been worked out in some detail in [6] as a generalization of a method that has been successful in two-stage stochastic programming [25], [26], [27]. This calculates a new point for $(\mathcal{P})$ by minimizing $\tilde{f}(u)$ over $u \in \tilde{U}$ (instead of the true primal problem of minimizing $f(u)$ over $u \in U$) along with a new point in $(\mathcal{Q})$ by maximizing $\tilde{g}(v)$ over $v \in \tilde{V}$ (instead of maximizing $g(v)$ over $v \in V$). The calculation of these new $u$ and $v$ points, toward which one can then do line searches from the current primal and dual points, for instance, is equivalent to finding a saddle point of $L(u,v)$ relative to the polyhedral set $\tilde{U} \times \tilde{V}$. This can be converted to an explicit, low-dimensional problem in extended linear-quadratic programming which can be solved by way of standard codes (see [6]).

Another kind of envelope method is developed in the author's joint paper with Zhu [7]. It uses the generation scheme (15) with $r = 2$ to execute a sort of gradient projection method of simultaneous primal descent and dual ascent with restarts triggered by a certain feedback between the primal and dual processes.

In applying any envelope method to the multistage problems $(\mathcal{P})$ and $(\mathcal{Q})$, a critical issue is how to deal effectively with the mappings $F$ and $G$. The next two theorems provide the answer.

**THEOREM 4.** *In the strictly quadratic case, the following procedure calculates $F(u)$ and then $\nabla f(u)$ for any given $u$:*
(a) *Determine $x = (x_0, \ldots, x_T)$ from $u = (u_0, \ldots, u_T)$ via the primal dynamics (2).*
(b) *Calculate $\tilde{v} = (\tilde{v}_1, \ldots, \tilde{v}_{T+1}) \in V$ by*

$$\tilde{v}_t = \nabla \rho_{V_t Q_t}(q_t - C_t x_{t-1} - D_t u_t) \text{ for } t = 1, \ldots, T,$$
$$\tilde{v}_{T+1} = \nabla \rho_{V_{T+1} Q_{T+1}}(q_{T+1} - C_{T+1} x_T).$$

(c) *Determine $\tilde{y} = (\tilde{y}_1, \ldots, \tilde{y}_{T+1})$ from $\tilde{v} = (\tilde{v}_1, \ldots, \tilde{v}_{T+1})$ via the dual dynamics (5).*
(d) *Then $\tilde{v} = F(u)$, while $\nabla f(u)$ is given by*

$$\nabla_{u_t} f(u_0, u_1, \ldots, u_t) = \begin{cases} (p_t - B_t^* \tilde{y}_{t+1} - D_t^* \tilde{v}_t) + P_t u_t & \text{for } t = 1, \ldots, T, \\ (p_0 - B_1^* \tilde{y}_1) + P_0 u_0 & \text{for } t = 0. \end{cases}$$

**Proof.** From the definition of $F(u)$ in Theorem 3, with expression (4) used in the Lagrangian (3), one obtains that $F(u)$ is the vector $\tilde{v}'$ having components

$$\tilde{v}'_t = \underset{v_t \in V_t}{\mathrm{argmax}}\{v_t \cdot (q_t - C_t x_{t-1} - D_t u_t) - \tfrac{1}{2} v_t \cdot Q_t v_t\} \text{ for } t = 1, \ldots, T,$$

$$\tilde{v}'_{T+1} = \underset{v_{T+1} \in V_{T+1}}{\mathrm{argmax}}\{v_{T+1} \cdot (q_{T+1} - C_{T+1} x_T) - \tfrac{1}{2} v_{T+1} \cdot Q_{T+1} v_{T+1}\}.$$

Consider for each $t$ the function $\varphi_{V_t Q_t}(v_t)$, which has the value $\tfrac{1}{2} v_t \cdot Q_t v_t$ when $v_t \in V_t$ but equals $\infty$ when $v_t \notin V_t$. This function is strongly convex (because $Q_t$ is positive definite), and its conjugate is $\rho_{V_t Q_t}$. From this conjugacy, the formula just given can be interpreted in convex analysis (cf. [28, Thm. 23.5]) as saying that $\tilde{v}'_t$ is the unique subgradient of $\rho_{V_t Q_t}$ at the point $q_t - C_t x_{t-1} - D_t u_t$ (or, in the case of $t = T+1$, the point $q_{T+1} - C_{T+1} x_T$). When a convex function has a unique subgradient, that subgradient must actually be the gradient (cf. [28, Thm. 25.1]). Thus, $\tilde{v}'_t = \tilde{v}_t$. This proves the asserted formula for $F(u)$. We have then from Theorem 3 that $\nabla f(u) = \nabla_u L(u, \tilde{v})$, where in the definition (3) of $L(u, \tilde{v})$ one invokes (6) with the dual trajectory $\tilde{y}$. This yields the asserted formula for $\nabla_{u_t} f(u)$. □

**THEOREM 5.** *In the strictly quadratic case, the following procedure calculates $G(v)$ and then $\nabla g(v)$ for any given $v$:*
(a) *Determine $y = (y_1, \ldots, y_{T+1})$ from $v = (v_1, \ldots, v_{T+1})$ via the dual dynamics (5).*
(b) *Calculate $\tilde{u} = (\tilde{u}_0, \ldots, \tilde{u}_T) \in U$ by*

$$\tilde{u}_t = \nabla \rho_{U_t P_t}(B_t^* y_{t+1} + D_t v_t - q_t) \text{ for } t = 1, \ldots, T,$$

$$\tilde{u}_0 = \nabla \rho_{U_0 P_0}(B_0^* y_1 - p_0).$$

(c) *Determine $\tilde{x} = (\tilde{x}_0, \ldots, \tilde{x}_T)$ from $\tilde{u} = (\tilde{u}_0, \ldots, \tilde{u}_T)$ via the primal dynamics (2).*
(d) *Then $\tilde{u} = G(v)$, while $\nabla g(v)$ is given by*

$$\nabla_{v_t} g(v_1, \ldots, v_T, v_{T+1}) = \begin{cases} (q_t - C_t \tilde{x}_{t-1} - D_t^* \tilde{u}_t) - Q_t v_t & \text{for } t = 1, \ldots, T, \\ (q_{T+1} - C_{T+1} \tilde{x}_T) - Q_{T+1} u_{T+1} & \text{for } t = T+1. \end{cases}$$

**Proof.** The argument is parallel to the one for Theorem 4. □

**5. Strictly Quadratic Regularization.** The results stated in Theorems 3, 4 and 5 for the strictly quadratic case can be applied in a fundamental way even when problem $(\mathcal{P})$ is not strictly quadratic. One sets up a regularizing scheme of "outer" iterations which converts the solving of $(\mathcal{P})$ into the solving of a sequence of slightly modified, but strictly quadratic problems. An appropriate mechanism is the general *proximal point algorithm*, developed in Rockafellar [29].

The proximal point algorithm could be applied to our multistage model in a primal, dual, or primal-dual mode, following the pattern already known for its applications to standard problems in convex programming [30]. The primal-dual mode is the one of interest in the general case where neither $P_t$ nor $Q_t$ can be counted on as positive definite, and it is therefore the mode we concentrate on here. We can view the algorithm in this mode as operating on $(\mathcal{P})$ either in terms of the saddle point problem for the Lagrangian (3) or in the context of the variational inequality (10)-(11). Both ways, the implementation requires us to specify auxiliary matrices

$$\bar{P}_t \in \mathbb{R}^{k_t \times k_t}, \qquad \bar{Q}_t \in \mathbb{R}^{l_t \times l_t}, \qquad \text{symmetric and positive definite.}$$

Also required is a *bounded* sequence of values $\gamma_\nu > 0$, where $\nu = 1, 2, \ldots$, is the iteration index.

**THEOREM 6.** *Starting from any choice of elements $\bar{u}^0 \in U$ and $\bar{v}^0 \in V$, generate $\bar{u}^\nu$ and $\bar{v}^\nu$ iteratively for $\nu = 1, 2, \ldots$, by taking $\bar{u}^\nu$ to be the unique solution to the strictly quadratic problem $(\mathcal{P}^\nu)$ obtained from $(\mathcal{P})$ in replacing the elements $P_t$, $Q_t$, $p_t$ and $q_t$ by*

$$P_t^\nu := P_t + \gamma_\nu \bar{P}_t, \quad Q_t^\nu := Q_t + \gamma_\nu \bar{Q}_t, \quad p_t^\nu := p_t - \gamma_\nu \bar{P}_t \bar{u}^{\nu-1}, \quad q_t^\nu := q_t + \gamma_\nu \bar{Q}_t \bar{v}^{\nu-1}, \quad (16)$$

*and by taking $\bar{v}^\nu$ to be the unique solution to the corresponding dual problem $(\mathcal{Q}^\nu)$. If the original problem $(\mathcal{P})$ has a solution at all (cf. Theorem 2), then $\bar{u}^\nu$ converges to a solution $\bar{u}$ to $(\mathcal{P})$ (even though $(\mathcal{P})$ may have more than one solution), while $\bar{v}^\nu$ converges to a solution $\bar{v}$ to $(\mathcal{Q})$.*

**Proof.** The uniqueness of $\bar{u}^\nu$ and $\bar{v}^\nu$ follows from the objective function $f^\nu$ in $(\mathcal{P}^\nu)$ being strictly convex because $P^\nu$ is positive definite, and the objective function $g^\nu$ in $(\mathcal{Q}^\nu)$ being strictly concave because $Q^\nu$ is positive definite. From Theorem 1 as applied to $(\mathcal{P}^\nu)$ we know that the pair $(\bar{u}^\nu, \bar{v}^\nu)$ is the unique saddle point of $L^\nu(u,v)$ relative to $U \times V$, where $L^\nu$ is the Lagrangian for $(\mathcal{P}^\nu)$. In the norm notation

$$\|u\|_{\bar{P}} := \left(\sum_{t=0}^{T} u_t \cdot \bar{P}_t u_t\right)^{\frac{1}{2}}, \qquad \|v\|_{\bar{Q}} := \left(\sum_{t=1}^{T+1} v_t \cdot \bar{Q}_t v_t\right)^{\frac{1}{2}}, \quad (17)$$

this Lagrangian has the form

$$L^\nu(u,v) = L(u,v) + \tfrac{1}{2}\gamma_\nu \|u - \bar{u}^{\nu-1}\|_{\bar{P}}^2 - \tfrac{1}{2}\gamma_\nu \|v - \bar{v}^{\nu-1}\|_{\bar{Q}}^2 + \text{const.} \quad (18)$$

The norms (17) can be taken as inducing a Euclidean structure on the $u$ and $v$ spaces, and the procedure then fits the pattern of the minimax version of the proximal point algorithm in [29, Theorem 5]. In particular, one obtains the claimed convergence. □

The saddle point interpretation of the algorithm in Theorem 3, in terms of the pair $(\bar{u}^\nu, \bar{v}^\nu)$ being the unique saddle point of the Lagrangian $L^\nu(u,v)$ in (18) relative to $U \times V$, provides the connection with variational inequalities and leads to a result on the *rate* of convergence.

**THEOREM 7.** *The algorithm in Theorem 6 generates the pair $(\bar{u}, \bar{v})$ from $(\bar{u}^{\nu-1}, \bar{v}^{\nu-1})$ as the solution to the variational inequality*

$$M^\nu(\bar{u}^\nu, \bar{v}^\nu) \cdot [(u,v) - (\bar{u}^\nu, \bar{v}^\nu)] \leq 0 \quad \text{for all } (u,v) \in U \times V, \quad (19)$$

*where $M^\nu(u,v) = (\nabla_u L^\nu(u,v), -\nabla L^\nu(u,v))$, or equivalently in terms of the positive definite matrix $H = \text{diag}\{\bar{P}_0, \ldots, \bar{P}_T; \bar{Q}_1, \ldots, \bar{Q}_{T+1}\}$ and the affine mapping $M$ in (12),*

$$M^\nu(u,v) = M(u,v) + \gamma_\nu H\bigl[(u,v) - (\bar{u}^\nu, \bar{v}^\nu)\bigr]. \quad (20)$$

*Relative to the multivalued mapping $\bar{M}(u,v) := M(u,v) + N_{U \times V}(u,v)$, which is maximal monotone, this rule of generation can be written as*

$$(\bar{u}^\nu, \bar{v}^\nu) = (I + \gamma_\nu^{-1} H^{-1} \bar{M})^{-1}(\bar{u}^{\nu-1}, \bar{v}^{\nu-1}). \quad (21)$$

**Proof.** The reduction of the saddle point condition for $L^\nu$ in the proof of Theorem 6 to the variational inequality (19) follows through equivalence with the intermediate condition

$$-\nabla_u L^\nu(\bar{u}^\nu, \bar{v}^\nu) \in N_U(\bar{u}^\nu), \qquad \nabla_v L^\nu(\bar{u}^\nu, \bar{v}^\nu) \in N_V(\bar{v}^\nu).$$

The affine mapping $M$ is itself maximal monotone because $L(u,v)$ is convex in $u$ and concave in $v$ (see [31]). The maximal monotonicity of the multivalued mapping $\bar{M}$ is then follows because $U \times V$ is a closed convex set (see [32, Theorem 3].) The single-valuedness of the mapping on the right side of (21) is due to $(\bar{u}^\nu, \bar{v}^\nu)$ being uniquely determined as a saddle point of $L^\nu(u,v)$ on $U \times V$, which in turn is a consequence of the strictly quadratic terms in the definition of $L^\nu(u,v)$ in (18). □

**THEOREM 8.** *Suppose that* $(\mathcal{P})$ *and* $(\mathcal{Q})$ *have unique solutions* $\bar{u}$ *and* $\bar{v}$, *and let* $\gamma_\nu$ *be chosen such that* $\gamma_\nu \to \gamma_\infty \geq 0$. *Then, unless the algorithm in Theorem 6 actually terminates in finitely many steps (with* $(\bar{u}^\nu, \bar{v}^\nu) = (\bar{u}, \bar{v})$ *for all* $\nu$ *sufficiently large), there is a constant* $a \geq 0$ *such that*

$$\lim_{\nu \to \infty} \frac{\|(\bar{u}^\nu, \bar{v}^\nu) - (\bar{u}, \bar{v})\|_H}{\|(\bar{u}^{\nu-1}, \bar{v}^{\nu-1}) - (\bar{u}, \bar{v})\|_H} \leq \frac{a\gamma_\infty}{(1 + (a\gamma_\infty)^2)^{\frac{1}{2}}} < 1, \tag{22}$$

*where the norm is*

$$\|(u,v)\|_H := \sqrt{(u,v)\cdot H(u,v)} = \left(\sum_{t=0}^{T} u_t \cdot \bar{P}_t u_t + \sum_{t=1}^{T+1} v_t \cdot \bar{Q}_t v_t\right)^{\frac{1}{2}}. \tag{23}$$

**Proof.** Letting $K = H^{-\frac{1}{2}}$, we make the change of variables $\bar{w}^\nu = K^{-1}(\bar{u}^\nu, \bar{v}^\nu)$ to convert (21) into

$$\bar{w}^\nu = (I + \gamma_\nu^{-1}\tilde{M})^{-1}(\bar{w}^{\nu-1}) \quad \text{for } \tilde{M}(w) := K\bar{M}(Kw).$$

The mapping $\tilde{M}$ inherits maximal monotonicity from $\bar{M}$, and the procedure is thus converted to the fundamental form of the proximal point algorithm in Rockafellar [29]. Our hypothesis on the uniqueness of $\bar{u}$ and $\bar{v}$ as primal and dual solutions means that these elements uniquely satisfy $(0,0) \in \bar{M}(\bar{u}, \bar{v})$, and this translates to the pair $\bar{w} = (\bar{u}, \bar{v})$ being the unique solution to $0 \in \tilde{M}(\bar{w})$. According to Theorem 2 of the cited paper [29], we will have

$$\lim_{\nu \to \infty} \frac{\|(\bar{w}^\nu - \bar{w})\|}{\|\bar{w}^{\nu-1} - \bar{w}\|} \leq \frac{a\gamma_\infty}{(1 + (a\gamma_\infty)^2)^{\frac{1}{2}}}$$

if we can establish the existence of $a \geq 0$ and $\varepsilon > 0$ such that

$$\|w - \bar{w}\| \leq a\|w'\| \quad \text{when } w' \in \tilde{M}(w) \text{ and } \|w'\| \leq \varepsilon. \tag{24}$$

This will suffice for our result, because $\|\bar{w}^\nu - \bar{w}\| = \|(\bar{u}^\nu, \bar{v}^\nu) - (\bar{u}, \bar{v})\|_H$ from the definitions.

The mapping $\tilde{M}$ is polyhedral in the sense of Robinson [33]: its graph is the union of a finite collection of polyhedral sets. This follows from the definition of $\tilde{M}$ in terms of $\bar{M}$, which by (20) is the sum of the affine mapping $M$ and the normal cone mapping $(u,v) \mapsto N_{U \times V}(u,v)$. The latter is polyhedral because the sets $U$ and $V$ are polyhedral, see [33]. We know on the other hand that $\tilde{M}$ is maximal monotone with $\bar{w}$ the unique solution to $0 \in \tilde{M}(w)$. Therefore, the inverse $\tilde{M}^{-1}$ is maximal monotone and polyhedral as well as single-valued at 0 (with the value $\bar{w}$ there). The maximal monotonicity and single-valuedness at 0 imply that $\tilde{M}^{-1}(w') \neq \emptyset$ for all $w'$ in some neighborhood of 0; see [34, Theorem 1]. The polyhedral property then assures that $\tilde{M}^{-1}$ is actually single-valued and piecewise affine in some neighborhood of 0. Such a mapping in particular has, for some $a \geq 0$ and $\varepsilon > 0$, the property that

$$\|\tilde{M}^{-1}(w') - \tilde{M}^{-1}(0)\| \leq a\|w'\| \quad \text{when } \|w'\| \leq \varepsilon,$$

and this is equivalent to the desired statement (22). □

The version of the proximal point algorithm in Theorems 6, 7 and 8 calls for $\bar{u}^\nu$ and $\bar{v}^\nu$ to be the exact solutions to the strictly quadratic primal and dual subproblems in each iteration. Actually, the supporting theory in [29] do not require this. It is possible without great difficulty to develop a version in which approximate solutions suffice in each iteration, under a certain type of stopping condition.

Furthermore, the assumption in Theorem 8 that $(\mathcal{P})$ and $(\mathcal{Q})$ have unique solutions is not really needed. Making use of the refinements obtained by Luque [35] in the convergence properties of the general proximal point algorithm in [29], it can be demonstrated that (22) always holds in the broader sense of the distance (with respect to the norm induced by $H$) of $(\bar{u}^\nu, \bar{v}^\nu)$ from the set of all saddle points, rather than the distance from the unique saddle point.

**6. Splitting Methods and Lagrangian Decomposition.** The special formulation of the multistage problem $(\mathcal{P})$ also invites the application of so-called splitting methods. Such methods aim at solving a variational inequality like (11) by decomposing it into simpler conditions in the form of auxiliary variational inequalities. Iteratively these simpler conditions are solved relative to certain parameter elements, and their solutions are combined to get new parameter elements. Examples include the alternating direction method for convex programming [8], [9], [10], and the more general algorithm of Lions and Mercier [11]. Spingarn [12], [13], developed a class of splitting methods based on applying the proximal point algorithm [29] to the partial inverse of a maximal monotone mapping and showed that the alternating direction method and many other decomposition techniques were covered as a special case. Recently Eckstein [14], [15], has carried this further and provided an overview showing that even the Lions-Mercier algorithm falls essentially in this class.

Thus, a rich family of numerical methods for large-scale problems can be seen in terms of instances of the general proximal point algorithm. The properties of that algorithm have been invoked in Theorems 6, 7 and 8 with respect to strict quadratic regularization, but the larger question arises of whether splitting methods likewise have a special role to play in this framework of multistage optimization. Tseng [16], [17], has specifically applied a type of splitting method to $(\mathcal{P})$ and shown that this leads to a solution procedure allowing for massive parallelization. Many other possibilities can be explored, however, beyond the one discovered by Tseng. The structure of the Lagrangian in (3), with highly separable terms in $t$ supplemented by a bi-affine form expressing the dynamics, can be made the basis of a kind of decomposition in which, on the one hand, low-dimensional problems of extended linear-quadratic programming are repeatedly solved for each $t$, while on the other hand, the dynamics are treated through subproblems of ordinary linear-quadratic optimal control without constraints. This will be discussed elsewhere.

## REFERENCES

1. R. T. ROCKAFELLAR, *Linear-quadratic programming and optimal control*, SIAM J. Control Opt., 25 (1987), pp. 781–814.

2. R. T. ROCKAFELLAR, *On the essential boundedness of solutions to problems in piecewise linear-quadratic optimal control*, in: Analyse Mathématique et Applications, F. Murat and O. Pironneau (eds.), Gauthier-Villars, Paris, 1988, pp. 437–444.

3. R. T. ROCKAFELLAR, *Multistage convex programming and discrete-time optimal control*, Control and Cybernetics, 17 (1988), pp. 225–246.

4. R. T. ROCKAFELLAR, *Hamiltonian trajectories and duality in the optimal control of linear systems with convex costs*, SIAM J. Control Opt., 27 (1989), pp. 1007–1025.

5. R. T. ROCKAFELLAR and R. J-B WETS, *Generalized linear-quadratic problems of deterministic and stochastic optimal control in discrete time*, SIAM J. Control Opt., 28 (1990) (July).

6. R. T. ROCKAFELLAR, *Computational schemes for solving large-scale problems in extended linear-quadratic programming*, Math. Programming, 48 (1990).

7. C. ZHU and R. T. ROCKAFELLAR, *Finite-envelope gradient projection methods for extended linear-quadratic programming*, preprint.

8. D. GABAY and B. MERCIER, *A dual algorithm for the solution of nonlinear variational problems via finite element approximations*, Computers and Math. with Appl., 2 (1976), pp. 17–40.

9. M. FORTIN and R. GLOWINSKI, *On decomposition-coordination methods using an augmented Lagrangian*, in: Augmented Lagrangian Methods: Applications to the Solution of Boundary-Value Problems, M. Fortin and R. Glowinski (eds.), North-Holland, Amsterdam, 1983.

10. D. GABAY, *Applications of the method of multipliers to variational inequalities*, in: Augmented Lagrangian Methods: Applications to the Solution of Boundary-Value Problems, M. Fortin and R. Glowinski (eds.), North-Holland, Amsterdam, 1983.

11. P.-L. LIONS and B. MERCIER, *Splitting algorithms for the sum of two nonlinear operators*, SIAM J. Numer. Analysis, 16 (1979), 964–979.

12. J. E. SPINGARN, *Partial inverse of a monotone operator*, Appl. Math. Optimization, 10 (1983), pp. 247–265.

13. J. E. SPINGARN, *Applications of the method of partial inverses to convex programming*, Math. Programming, 32 (1985), pp. 199–223.

14. J. ECKSTEIN, *Splitting Methods for Monotone Operators with Applications to Parallel Optimization*, doctoral dissertation, M.I.T., 1989.

15. J. ECKSTEIN, *The Lions-Mercier algorithm and the alternating direction method are instances of the proximal point algorithm*, report LIDS-P-1769, Laboratory for Information and Decision Sciences, M.I.T., 1989.

16. P. TSENG, *Applications of a splitting algorithm to decomposition in convex programming and variational inequalities*, LIDS Report P-1836, M.I.T., Cambridge, MA, 1989.

17. P. TSENG, *Further applications of a splitting algorithm to decomposition in variational inequalities and convex programming*, LIDS Report P-1866, M.I.T., Cambridge, MA, 1989.

18. J.-S. PANG, *Methods for quadratic programming: a survey*, Computers and Chem. Engineering, 7 (1983), pp. 583–594.

19. Y.-Y. LIN and J.-S. PANG, *Iterative methods for large convex quadratic programs: a survey*, SIAM J. Control Opt., 25 (1987), pp. 383–411.

20. Y. YE and E. TSE, *A polynomial-time algorithm for convex quadratic programming*, preprint.

21. R. C. MONTEIRO and I. ADLER, *An $0(n^3 L)$ interior point algorithm for convex quadratic programming*, preprint.

22. D. GOLDFARB and S. LIU, *An $0(n^3 L)$ primal interior point algorithm for convex quadratic programming*, preprint.

23. R. T. ROCKAFELLAR, *Augmented Lagrange multiplier functions and duality in nonconvex programming*, SIAM J. Control, 12 (1974), pp. 268–285.

24. R. T. ROCKAFELLAR and R. J-B WETS, *Linear-quadratic problems with stochastic penalties: the finite generation algorithm*, in: Numerical Techniques for Stochastic Optimization Problems, Y. Ermoliev and R. J-B Wets (eds.), Springer-Verlag Lecture Notes in Control and Information Sciences No. 81, 1987, pp. 545-560.

25. R. T. ROCKAFELLAR and R. J-B WETS, *A Lagrangian finite generation technique for solving linear-quadratic problems in stochastic programming*, Math. Programming Studies, 28 (1986), pp. 63–93.

26. A. KING, *An implementation of the Lagrangian finite generation method*, in: Numerical Techniques for Stochastic Programming Problems, Y. Ermoliev and R. J-B Wets (eds.), Springer-Verlag, 1988.

27. J. M. WAGNER, *Stochastic Programming with Recourse Applied to Groundwater Quality Management*, doctoral dissertation, M.I.T, 1988.

28. R. T. ROCKAFELLAR, *Convex Analysis*, Princeton Univ. Press, Princeton, NJ, 1970.

29. R. T. ROCKAFELLAR, *Monotone operators and the proximal point algorithm*, SIAM J. Control Opt., 14 (1976), pp. 877–898.

30. R. T. ROCKAFELLAR, *Augmented Lagrangians and applications of the proximal point algorithm in convex programming*, Math. of Op. Research, 1 (1976), pp. 97–116.

31. R. T. ROCKAFELLAR, *Monotone operators associated with saddle functions and minimax problems*, in: Nonlinear Functional Analysis, Part 1, F. E. Browder (ed.), Symposia in Pure Math., vol. 18, Amer. Math. Soc., Providence, RI, 1970, pp. 397–407.

32. R. T. ROCKAFELLAR, *On the maximality of sums of nonlinear monotone operators*, Trans. Amer. Math. Soc., 149 (1970), pp. 75–88.

33. S. M. ROBINSON, *Some continuity properties of polyhedral multifunctions*, Math. Programming Study 14, 1981, pp. 206–214.

34. R. T. ROCKAFELLAR, *Local boundedness of nonlinear monotone operators*, Michigan Math. J., 16 (1969), pp. 397–407.

35. F. J. LUQUE, *Asymptotic convergence analysis of the proximal point algorithm*, SIAM J. Control Opt., 22 (1984), pp. 277–293.

# CHAPTER 16

A Finite Difference Approach to the
Kuramoto-Sivashinsky Equation

Alfredo Nicolás-Carrizosa*

**Abstract.** The Kuramoto–Sivashinsky equation is a nonlinear time dependent partial differential equation showing bifurcation phenomena, and is very sensitive to numerical accuracy. In one dimensional space, numerical results have been obtained using a sophisticated approach based on spectral methods, for which, a supercomputer was required. This work shows that those results can be obtained with a simpler approach based on classical finite difference methods, for which, a Vax computer is enough.

## 1. Introduction .

The Kuramoto-Sivashinsky (KS) equation in one space dimension is given by

$$(1.1) \quad \begin{cases} u_t + 4u_{xxxx} + \alpha[u_{xx} + \frac{1}{2}(u_x)^2] = 0, & 0 \le x \le 2\pi, \\ u(x,0) = u_0(x) + c, \\ \partial u_x^k(x+2\pi, t) = \partial u_x^k(x,t) \ k = 0,1,2,3, \end{cases}$$

where $\alpha > 0$ is the *bifurcation parameter*.

The KS equation models pattern formations in different physical contexts and is a paradigm of low dimensional behavior in solutions to partial differential equations (PDE's). KURAMOTO (1975, 1978), derived it in the context of angular-phase turbulence for a system of reaction-diffusion equations modeling the Belouzov-Zabotinskii reaction in three space dimensions. SIVASHINSKY (1977, 1980), derived it independently to model small thermal diffusive instabilities in laminar flame fronts.

The solution to the KS equation is characterized by a second-order unstable diffusion term, a fourth-order stabilizing viscosity and a quadratic non-

---

* Sección de Graduados, ESIME, IPN, México. Edif. 8, U.P. Zacatenco, México 14, D.F.

linear coupling term. This follows from the spectrum of problem (1.1) linearized about the trivial solution $u \equiv 0$. The eigenfunctions are the Fourier modes

(1.2) $\{\cos kx\}$ and $\{\sin kx\}$, $k = 0, 1, 2, \cdots$;

the corresponding eigenvalues are

(1.3) $\begin{cases} \Lambda_k &= 4k^4 - \alpha k^2 \\ &= k^2[4k^2 - \alpha], \end{cases}$

and the solution, formally, is

(1.4) $u(t) = a_0/2 + \sum_{k \geq 1} e^{-\Lambda_k t}(a_k \cos kx + b_k \sin kx),$

where $a_k$, $k \geq 0$, and $b_k$, $k \geq 1$ are the Fourier coefficients of $u_0(x)$ over $[0, 2\pi]$. We observe that $\Lambda_0 = 0$ is an eigenvalue with the constant function $\phi \equiv 1$ as the corresponding eigenfunction.

From (1.3) and (1.4), we have that the trivial solution is unstable as soon as $\alpha > 4$ ($\alpha = 4$ is *a bifurcation point*); for $\alpha > 4$ the component of the solution corresponding to the mode $k = 1$ is unstable; in general, there are $N = [\frac{1}{2}(\alpha)^{\frac{1}{2}}]$ positive eigenvalues governing the instability ( $[x]$ is the integer part of $x$). Thus, a nontrivial steady state exist for $\alpha > 4$. Since the eigenvalues increase as $k$ increases, in addition that there is an infinity of stable modes, they are extremely stable.

On the other hand, it has been proved by FOIAS, NICOLAENKO, SELL, and TEMAM (1985), that the KS equation is strictly equivalently to a low-dimensional dynamical system. That is, all orbits are attracted exponentially to a finite-dimensional, bounded, compact, smooth manifold, and the interesting dynamics take place on this "inertial" manifold; in a rough sense this manifold, say $\mathcal{M}$, plays the role of a "global center manifold" with complement (in the Sobolev space $H^1(0, 2\pi)$) corresponding to a global infinite-dimensional stable manifold (Fig. 1).

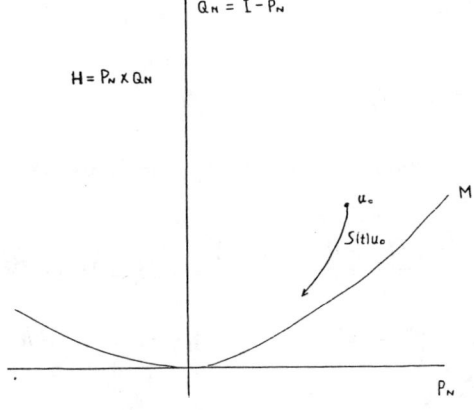

Numerically, problem (1.1) is *extremely sensitive* to numerical accuracy. This is due mainly to the fourth order derivative, and the *plus* sign of the second order derivative. The numerical work where this was observed is the one by J. M. HYMAN, B. NICOLAENKO (1986). Their computations confirm that $\|u_x(t)\|^2$ remains bounded in the $L^2$ norm. Thus, the linearly unstable low modes of the KS equation equation are stabilized by the strong nonlinear coupling $\frac{1}{2}(u_x)^2$ along with the extremely stable high modes.

To deal with the extreme sensitivity of the KS equation, Hyman and Nicolaenko used a *high-precision* pseudospectral approximation (discrete Fourier transform) to the spatial derivative and a variable time-step on a Cray XMP supercomputer (single precision: 14 digits).

In this work we show that the Hyman and Nicolaenko's numerical results can be obtained with a *simpler* approach based on *classical finite difference methods*, for which a VAX computer is enough (real*8 precision for steady state solutions and real*16 for solutions of oscillatory type: 16 and 32 digits, respectively).

## 2. Space Discretization.

Actually the problem which is solved numerically is the one obtained from problem (1.1) making

$$v(x,t) = u(x,t) - m(t)$$

where

$$m(t) = \frac{1}{2\pi}\int_0^{2\pi} u(x,t)dx,$$

then, with $u$ still denoting the variable, problem (1.1) is transformed into

$$(2.1) \quad \begin{cases} u_t + 4u_{xxxx} + \alpha[u_{xx} + \frac{1}{2}(u_x)^2] + \dot{m}(t) = 0, \quad 0 \leq x \leq 2\pi, \\ u(x,0) = u_0(x) + c, \\ \partial u_x^k(x+2\pi,t) = \partial u_x^k(x,t) \ k = 0,1,2,3, \end{cases}$$

where

$$c = -\frac{1}{2\pi}\int_0^{2\pi} u_0(x)dx,$$

$$(2.2) \quad \dot{m}(t) = -\frac{\alpha}{4\pi}\int_0^{2\pi} u_x^2(x,t)dx.$$

We approximate the second derivative, at time t, by its classical finite difference approximation

$$(2.3) \quad u_{xx}(x_i,t) \approx \frac{u_{i-1}(t) + u_{i+1}(t) - 2u_i(t)}{h^2}, i = 1,\cdots,N,$$

where, $u_i(t) = u(x_i,t)$, $x_i = ih, i = 0,1,2,\cdots,N+1$, $h = \frac{2\pi}{N+1}$. By periodicity,

(2.4) $u_{-1}(t) = u_N(t)$, $u_{N+1} = u_0$

Therefore, we have N+1 unknowns to be determined: $u_0(t), u_1(t), \cdots, u_N(t)$. From (2.2) and (2.4), we have that the corresponding (N+1)x(N+1) matrix is -A, where

$$(2.5) \quad A = \frac{1}{h^2} \begin{pmatrix} 2 & -1 & 0 & & -1 \\ -1 & 2 & -1 & & \\ & & \cdot & & \\ O & & \cdot & & O \\ & & & \cdot & -1 \\ -1 & & & 1 & 2 \end{pmatrix}_{(N+1)\times(N+1)}$$

The matrix $A$ is symmetric and positive semidefinite of rank N: the eigenvectors are

(2.6) $\{\cos kx_i\}_{i=0}^N$ and $\{\sin kx_i\}_{i=0}^N$, $k = 0, 1, 2, \cdots, N$,

with corresponding eigenvalues

(2.7) $\lambda_k = (4/h^2)\sin^2(kh/2)$;

for k = 0, $\lambda_0 = 0$, with eigenvector W = $\{1\}_{i=0}^N$. Then, $-A$ is symmetric and negative semidefinite of rank N. With $k$ fixed and $h$ small, (2.7) implies

(2.8) $\lambda_k \sim k^2$.

The matrix for the discrete fourth-order derivative is given by $A^2$, which is symmetric and positive semidefinite of rank N: the eigenvectors are the eigenvectors of $A$ and the corresponding eigenvalues, $\lambda_k^2$. Then for $A^2$, we have

(2.9) $\lambda_k^2 \sim k^4$.

For the discretization of the nonlinear term we consider the centered approximation

$$u_x^2(x_i, t) \approx \frac{(u_{i+1}(t) - u_{i-1}(t))^2}{4h^2}$$

To approximate the integral terms we use trapezoidal rule. Hence, after simplification, considering the periodic boundary conditions, we have the following:

i) From (2.1):

$$\dot{m}(t) \approx -[\frac{\alpha}{2}(N+1)] \sum_{i=0}^{N} u_x^2(x_i, t).$$

ii) Energy:

$$(2.10) \quad E(t) = \int_0^{2\pi} u_x^2(x,t)dx \approx h \sum_{i=0}^{N} u_x^2(x_i, t).$$

iii) Energy in mode 1 :

$$(2.11) \quad E_1(t) = \sqrt{A^2 + B^2},$$

where

$$A(t) = \frac{1}{\pi} \int_0^{2\pi} u(x,t) \cos(x) dx \approx [\frac{2}{(N+1)}] \sum_{i=0}^{N} u_i(t) \cos(x),$$

$$B(t) = \frac{1}{\pi} \int_0^{2\pi} u(x,t) \sin(x) dx \approx [\frac{2}{(N+1)}] \sum_{i=0}^{N} u_i(t) \sin(x).$$

Collecting the space discretization, the approximate problem, in space, of problem (1.1) is given by

$$(2.12) \quad \begin{cases} U_t + 4A^2 U - \alpha A U + B(U) = 0, \\ U(0) = U_0, \end{cases}$$

where

$$(2.13) \quad U = U(t) = \{u_i(t)\}_{i=0}^{N},$$

and B(U) contains the nonlinear term plus $\dot{m}(t)$.

Thus, taking into account (2.6) and (2.7), we have that for the spectrum of the linear problem of (2.12), the eigenvectors are given by (2.6) and the corresponding eigenvalues by

$$(2.14) \quad \lambda_k^* = 4(4/h^2)^2 \sin^4(kh/2) - \alpha(4/h^2) \sin^2(kh/2), k = 0, 1, \cdots, N.$$

Hence, by (2.8) and (2.9),

$$(2.15) \quad \begin{cases} \lambda_k^* \sim 4k^4 - \alpha k^2 \\ \phantom{\lambda_k^*} = k^2[4k^2 - \alpha]. \end{cases}$$

Then, (2.6), (2.14), and (2.15) are the discrete analogues of (2.1) and (2.2) for the continuous problem. Moreover,

$$(2.16) \quad U(t) = e^{-A(4A-\alpha I)t} U_0$$

is the exact solution for the linear problem associated to (2.12) (by forgetting momentarily the nonlinearity), which, in terms of the eigenvector basis, is given by

$$(2.17) \quad u_k(t) = e^{-\lambda_k(4\lambda_k - \alpha)t} u_0, k = 0, 1, \cdots, N.$$

Then, from (2.15) and (2.17) (or equivalently from (2.8)) it is clear that, for h small, the spectrum of problem (2.12), linearized about the trivial solution $U \equiv 0$, behaves like the corresponding spectrum of the continuous problem. In this case, instead of an infinity of stables modes we have a finite number:

$$(N+1) - [\frac{1}{2}(\alpha)^{\frac{1}{2}}].$$

## 3. Time Discretization. Solution of the Problem.

Now, we discretize problem (2.12) in time, and we solve the approximate problem. To this end we consider the following two methods.

### 3.1. First Method.

For the first-order derivative we use the following second-order accurate finite difference scheme

$$(3.1) \quad U_t(n\Delta t) \approx \frac{(1.5U^{n+1} - 2U^n + 0.5U^{n-1})}{\Delta t}, n \geq 1,$$

where $U^n = U(n\Delta t)$. Considering the nonlinear term from the previous time step, we have that problem (2.12) is transformed into the algebraic linear system

$$(3.2) \quad \begin{cases} (1.5I + 4\Delta t A^2 - \alpha \Delta t A)U^{n+1} = 2U^n - 0.5U^{n-1} - \\ \alpha \Delta t B(U^n), \quad n \geq 1. \end{cases}$$

With respect to the eigenvalues of the coefficient matrix of this system, recalling that the eigenvalues of $A$ are given by (2.7), we have that

$$(3.3) \quad 1.5 + 4\Delta t \lambda_k^2 - \alpha \Delta t \lambda_k = 1.5 + \Delta t \lambda_k (4\lambda_k - \alpha) > 0$$

for $\Delta t$ sufficiently small, regardless of $h$ and $k$; to be precise, (3.3) holds if $\Delta t < 24/\alpha^2$. Then, at each time step, regardless of the space mesh size, the coefficient matrix of (3.2) is symmetric and positive definitive for $\Delta t$ sufficiently small. By (2.5), this coefficient matrix has the structure given by

$$\begin{pmatrix} x & x & x & & & x & x \\ x & x & x & x & & & x \\ & & & \cdot & & & \\ O & & & \cdot & & O & \\ & & & \cdot & & & \\ x & & & x & x & x & x \\ x & x & & & x & x & x \end{pmatrix}_{(N+1) \times (N+1)}$$

Therefore, we may apply the Cholesky method to solve (3.2); for which, we need to store just 5 (N+1)-dimensional vectors, and the factorization is done once and for all.

By (2.8) and (2.9), we note that the condition number of the coefficient matrix, given by

$$(3.4) \quad \frac{1.5 + 4\Delta t \lambda_N^2 - \alpha \Delta t \lambda_N}{1.5},$$

is large, for h small.

As we see this is an easy method to implement in the computer to solve the KS equation. Unfortunately, it does not work in general. Keeping in mind that this is possibly due to the contribution of $-A$ and the large condition number given by (3.4), the aim of the following approach will be to avoid these difficulties using a factorization technique involving successive solution of discrete second order elliptic problems. Moreover, we consider the nonlinear term implicitly, since considering it explicitly, does not work. Hence, we must deal with a *fully implicit scheme*

## 3.2. Second Method.

Here we consider (3.1)-(3.2), via a fully implicit scheme; that is, we consider

$$(3.5) \quad \begin{cases} (1.5I + 4\Delta t A^2 - \alpha \Delta t A)U^{n+1} \alpha \Delta t B(U^{n+1}) \\ \qquad\qquad -2U^n + \frac{1}{2}U^{n-1} = 0, \quad n \geq 1. \end{cases}$$

Let's define $w$ and $S$ by $w = \sqrt{1.5\Delta t}$, and $S = \sqrt{1.5}I + 2\sqrt{\Delta t}A$ , then (3.5) yields

$$(3.6) \quad R(U^{n+1}) = 0, n \geq 1,$$

where

$$\begin{cases} R(U^{n+1}) = S^2 U^{n+1} - (\Delta t \alpha + 4w)AU^{n+1} + \alpha \Delta t B(U^{n+1}) - 2U^n + \\ \qquad \frac{1}{2}U^{n-1}. \end{cases}$$

To solve (3.6), we use the iterative process

$$(3.7) \quad X^{m+1} = X^m - \rho(S^2)^{-1}R(X^m), \; with \; 0 < \rho < 1; \; X^0 = U^n.$$

Then,

$$U^{n+1} = X^{m+1} \; if \; |X^{m+1} - X^m| < \varepsilon.$$

Assuming (3.7) converges, at each time step and for each iteration, we have to solve the linear system

$$(3.8) \quad S^2 X^{m+1} = S^2 X^m - \rho R(X^m),$$

equivalent to (3.7). The matrix S is symmetric and positive definite, independently of $\Delta t$. This matrix has the structure given by (2.5), hence we may solve (3.8) applying the Cholesky method twice:

$$\text{to solve } S^2 x = b, \; \text{we solve } Sy = b, \; \text{then } Sx = y.$$

In this case, we store only 3 (N+1)-dimensional vectors, and like the first method the factorization is done once and for all.

## 3.3 Numerical Results

Both methods successfully find the steady state solutions, giving the same results. When the solutions are of oscillatory type then the second one is effective, whereas the first one does not always converge. Also, the second method gives the steady state solutions with real*8 precision (16 digits).

The results, in the general case, were obtained using the second method with precision $10^{-10}$ in the convergence of the iterative process. The runs on a Vax computer required up to 3 hours of CPU time. We made runs of the program with $N = 99$, $\Delta t = .01$, used the initial condition

$$u_0(x) = 0.1(\cos x + \sin x),$$

and the values of

$$\alpha = 0, 2.5, 10, 18, 22, 22.3, 25, 35, 44, 45, 46, 52, 60, 80, 100.$$

Here we present three pictures which are in agreement with those of Hyman and Nicolaenko (1986), op. cit., (Figs. 3-5). Fig. 2 shows the corresponding ones from Hyman and Nicolaenko. (In NICOLAS-CARRIZOSA (1988), there are 18 more of such pictures). Finally, we point out that Hyman and Nicolaenko (1986), op. cit., include an analysis concerning the bifurcation phenomena of the results.

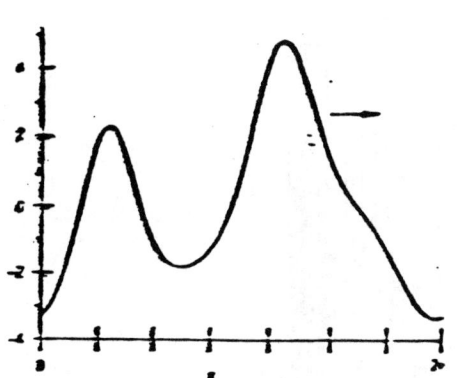
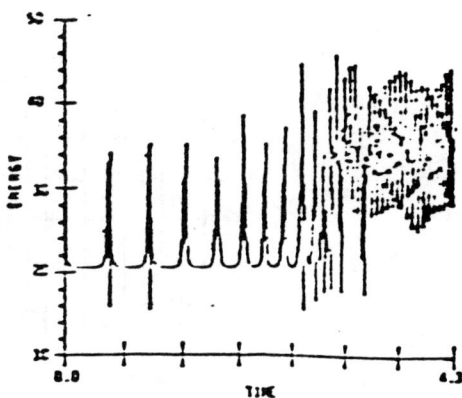

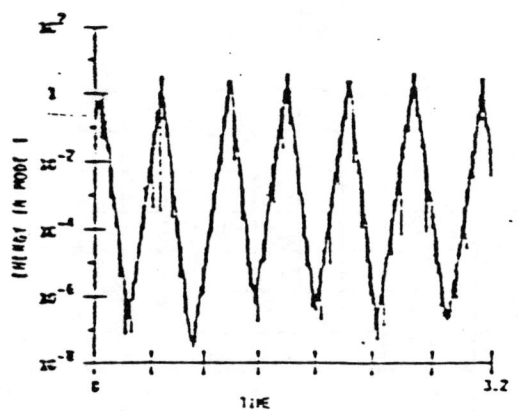

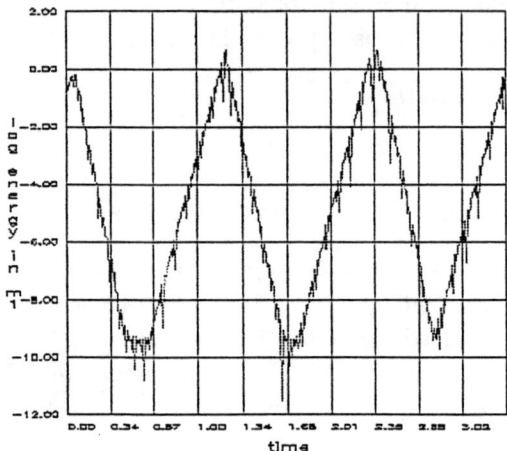

energy in m1 alpha=44

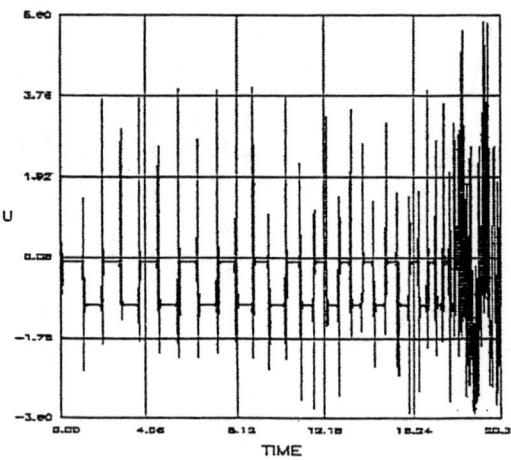

SOLUTION FOR ALPHA = 46.

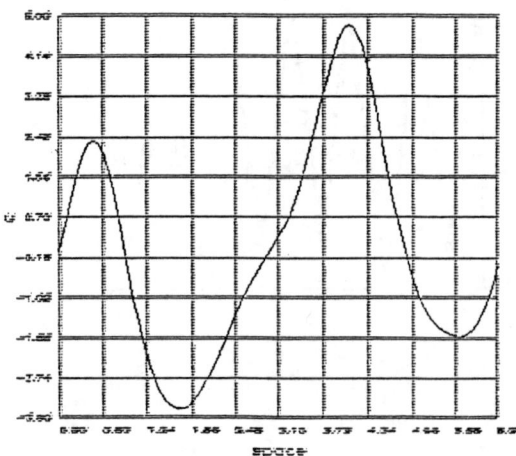

shape of sol alpha=52

*Acknowledgment*. I am grateful to have been involved with the solution of this problem through my collaboration with Professor R. Glowinski. I also thank Professor M. Golubitsky for suggesting us a numerical investigation of the Kuramoto-Sivashinsky equation.

*Note added*. We have obtained some numerical results for the KS equation in 2D using the same approach we discussed here, JUAREZ and NICOLAS (1990); further investigation is in progress. We have also applied successfully an iterative process like the one used here to solve the resultant nonlinear equation coming from a certain theoretic finite difference scheme for a semilinear wave equation, CASTILLO, JUAREZ, and NICOLAS (1990). However, about this latter problem, we realize later that a faster iterative process is given by the one used in the theoretic scheme to prove *existence*, which have let us to handle efficiently various semilinear wave equation problems; these results will published elsewhere.

## REFERENCES

F. CASTILLO, L. H. JUAREZ, A. NICOLAS (1990),"Un Estudio Teórico y Numérico de una Ecuación Hiperbólica Semilineal"; Reporte Interno No. 66 del Depto. de Matemáticas, CINVESTAV (México, D. F.).

C. FOIAS, G. R. SELL, R. TEMAM (1985), "Inertial Manifolds for Dissipative Partial Differential Equations"; C. R. Acad. Sc. Paris, t. 301, Serie I, **6**.

C. FOIAS, B. NICOLAENKO, G. R. SELL, R. TEMAM (1985), "Inertial Manifolds for the Kuramoto-Sivashinsky Equations"; C. R. Acad. Sc. Paris, t. 301, Serie I, **6**.

R. GLOWINSKI, Lecture Notes; Fall 1984 - Spring 1988, Dpt. of Math., University of Houston.

M. GOLUBITSKY, D. SCHAEFFER (1985), "Singularities and Groups in Bifurcation Theory"; Vol. I, Springer Verlag, N. Y.

J. GUCKENHEIMER, P. HOLMES (1984), "Nonlinear Oscillations, Dynamical Systems, and Bifurcation of Vector Fields"; Springer, Berlin.

J. M. HYMAN, B. NICOLAENKO (1986), "The Kuramoto-Sivashinsky Equation: A Bridge Between PDE's and Dynamical Systems"; Physica D, 113-126.

L. H. JUAREZ, A. NICOLAS (1990), "Solución Numérica de la Ecuación de Kuramoto-Sivashinsky en dos dimensiones"; Reporte Interno, UAM-Izt. (México, D, F.), (to appear).

Y. KURAMOTO, T. TSUZUKI (1975), "On the Formation of Dissipative Structures in Reaction-Diffusion Systems"; Prog. Theor. Phys. 54, 687-699.

Y. KURAMOTO (1978), "Diffusion-induced Chaos in Reactions Systems"; Supp. Prog. Theor. Phys. 64, 346-367.

A. NICOLAS-CARRIZOSA (1988), "Numerical Aspects of some Time Dependent Partial Differential Equation Problems"; PhD Thesis, Dpt. of Math., University of Houston.

G. SIVASHINSKY (1977), "Nonlinear Analysis of Hydrodynamic Instability in Laminar Flames, Part I. Derivation of basic equations"; Acta Astronautica 4, 1177-1206.

G. SIVASHINSKY (1980), "On Flame Propagation Under Condition of Stoichiometry"; SIAM J. Appl. Math. 39, 67-82.

… CHAPTER 17

# A Numerical Method for the Three Dimensional Inverse Acoustic Scattering Problem with Incomplete Data

Filippo Aluffi-Pentini*
Emanuele Cagliotit
Luciano Misicit
Francesco Zurilli§

**Abstract.** A numerical method for the three dimensional inverse acoustic scattering problem is presented. The far field pattern of the Helmholtz equation generated by an incoming plane wave incident on an obstacle $D$ is measured in several directions (15-30 directions). These measurements are repeated for the far fields corresponding to plane waves incoming in several directions (7-10 directions). From this data a closed surface $\partial D$ representing the boundary of an unknown obstacle $D$ is reconstructed. The data are incomplete in the sense that both the directions of the incoming waves and of the far fields measurements are taken for a fixed azimuthal angle.

**1. Introduction** Let $\mathbf{R}^3$ be the three dimensional euclidean space, $\underline{x} = (x,y,z) \in \mathbf{R}^3$ be a generic vector, $(.,.)$ will denote the euclidean scalar product and $\|.\|$ the euclidean norm. Let $D \subset \mathbf{R}^3$ be a bounded simply connected domain with smooth boundary. Let $u^i(\underline{x})$ be an incoming acoustic plane wave, that is:

$$u^i(\underline{x}) = e^{ik(\underline{x},\underline{\alpha})} \tag{1.1}$$

where $k > 0$ is the wave number and $\underline{\alpha} \in \mathbf{R}^3$ is a fixed unit vector. Let us denote with $u^s(\underline{x})$ the acoustic field scattered by the obstacle $D$ and with $u(\underline{x})$ the total acoustic field, that is:

$$u(\underline{x}) = u^i(\underline{x}) + u^s(\underline{x}) \tag{1.2}$$

The direct acoustic scattering problem for an acoustically soft obstacle $D$ is given by:

<u>Problem 1.1</u>: Find $u(\underline{x})$ defined for $\underline{x} \in \mathbf{R}^3 \backslash D$ such that

$$\triangle u + k^2 u = 0 \quad in \ \mathbf{R}^3 \backslash D \tag{1.3}$$

---

*Dipartimento di metodi e modelli matematici per le scienze applicate Università di Roma "La Sapienza" -00161 Roma- Italy

†Istituto Nazionale di Alta Matematica "F.Severi" Piazzale Aldo Moro 5 -00185 Roma- Italy. The research of this author has been made possible through the support and sponsorship of Elsag-Elettronica San Giorgio s.p.a.-Genova to the graduate fellowship program of the Istituto Nazionale di Alta Matematica "F. Severi" Roma.

‡Dipartimento di Matematica e Fisica Università di Camerino -62032 Camerino (MC)- Italy

§Dipartimento di Matematica "G. Castelnuovo" Università di Roma "La Sapienza" -00185 Roma- Italy

$$u = 0 \quad on \; \partial D \tag{1.4}$$

$$\lim_{r \to \infty} r\{\frac{\partial u^s}{\partial r} - iku^s\} = 0 \tag{1.5}$$

where $\triangle = \frac{\partial^2}{\partial x^2} + \frac{\partial^2}{\partial y^2} + \frac{\partial^2}{\partial z^2}$ is the laplacian and $r = (x^2 + y^2 + z^2)^{1/2}$. We note that (1.3) is the Helmholtz equation , (1.4) express the fact that the obstacle $D$ is acoustically soft and (1.5) is the Sommerfeld radiation condition at infinity. It can be shown [1] that the scattered field $u^s(\underline{x})$ when $r \to \infty$ has the following expansion

$$u^s(\underline{x}) = \frac{e^{ikr}}{r} F(\hat{\underline{x}}, k, \underline{\alpha}) + O(\frac{1}{r^2}) \; ; \; r \to \infty \tag{1.6}$$

where $\hat{\underline{x}} = \frac{\underline{x}}{\|\underline{x}\|}$ and $F(\hat{\underline{x}}, k, \underline{\alpha})$ is called far field pattern associated to the scattered field $u^s(\underline{x})$. Let $\lambda_n$ , $n = 1, 2, \ldots$ be the eigenvalues of the interior Dirichlet problem in $D$ for the Helmholtz equation, $B = \{\underline{x} \in \mathbf{R}^3 \mid \|\underline{x}\| < 1\}$ and $\partial B = \{\underline{x} \in \mathbf{R}^3 \mid \|\underline{x}\| = 1\}$ be the boundary of $B$. The inverse acoustic scattering problem for an acoustically soft obstacle $D$ is given by:

<u>Problem 1.2</u>: Let $k^2 \neq \lambda_n$ , $n = 1, 2, \ldots$ and $\Omega_1$ , $\Omega_2 \subseteq \partial B$ be two given subset of $\partial B$. From the knowledge of the far field patterns $F(\hat{\underline{x}}, k, \underline{\alpha})$ , $\forall \underline{\alpha} \in \Omega_1$ , $\forall \hat{\underline{x}} \in \Omega_2$ determine the boundary of the obstacle $\partial D$.

We observe that $\Omega_1$ is the set of directions of the incoming waves and $\Omega_2$ is the set of directions where the far field generated by the incoming waves is observed.

The inverse Problem 1.2 is of great importance in many applications such as underwater acoustic imaging, medical imaging, and has been recently considered by Colton and Monk in a series of very interesting papers [2], [3],[4],[5]. In [6] we have proposed a numerical method based on the work of Colton and Monk [5] to solve the inverse Problem 1.2. In this paper the method proposed in [6] is extended in order to handle the situation when $\Omega_1$, and $\Omega_2$ are finite sets, 7-30 points, with the same azimuthal angle.

We call these data incomplete since they are all taken on "one side" of the obstacle $D$. The reconstruction of $D$ from incomplete data is a difficult numerical problem, but its solution is necessary in the applications where the position of the obstacle is not known *a priori*.

In section 2 we illustrate our method, in section 3 some numerical experience obtained with it is shown.

**2. The numerical method** Given $D \subset \mathbf{R}^3, k > 0, \Omega_1 = \{\underline{\alpha}_i \in \partial B \mid i = 1, 2, \cdots, N\}$ be the set of directions of the incoming waves and $\Omega_2 = \{\hat{\underline{x}}_i \in \partial B \mid i = 1, 2, \cdots, M\}$ be the set of directions where the far fields are measured moreover let us assume that, for $i = 1, 2$ the directions contained in $\Omega_i$ are distinct. The data of our problem are the measurements of the far-field patterns $F(\hat{\underline{x}}_j, k, \underline{\alpha}_i)$ , $F_{ij}, i = 1, 2, \cdots, N$, $j = 1, 2, \cdots, M$. The data $F_{ij}$ , $i = 1, 2, \cdots, N$, $j = 1, 2, \cdots, M$ are obtained by solving numerically the direct Problem 1.1.

Let $(\theta, \phi)$ be the polar angles so that

$$\hat{\underline{x}}(\theta, \phi) = (\sin \theta \cos \phi, \; \sin \theta \sin \phi, \; \cos \theta) \tag{2.1}$$

and $U_{lm}(\hat{\underline{x}}) = P_l^m(\cos \theta) \cos m\phi$, $V_{lm}(\hat{\underline{x}}) = P_l^m(\cos \theta) \sin m\phi$ be the spherical harmonics, $P_l$ are the Legendre polynomials and $P_l^m$ is the $m^{th}$ derivative of $P_l$.

From these data our computations proceeds in four steps:

<u>Step 1</u> From the measurements $F_{ij}$ , $j = 1, 2, \cdots, M$ to the Fourier coefficients of $F(\hat{\underline{x}}, k, \underline{\alpha}_i)$.

Given $L_{max} \geq 0$ we assume that the far field $F(\hat{\underline{x}}, k, \underline{\alpha}_i)$ can be approximated by a truncated Fourier series, that is:

$$F(\hat{\underline{x}}, k, \underline{\alpha}_i) = \sum_{l=0}^{L_{max}} \sum_{m=0}^{l} F_{lm1}^i \, \gamma_{lm} \, U_{lm}(\hat{\underline{x}}) + \sum_{l=1}^{L_{max}} \sum_{m=1}^{l} F_{lm2}^i \, \gamma_{lm} \, V_{lm}(\hat{\underline{x}}) \tag{2.2}$$

where $\gamma_{lm}$ are normalization factors in $L_2(\partial B)$. We observe that the Fourier coefficients are complex numbers.

In order to determine the $\frac{(L_{max}+1)(L_{max}+2)}{2} + \frac{L_{max}(L_{max}+1)}{2}$ Fourier coefficients $\{F^i_{lmk}\}_{k=1,2}$ of the far field we impose that:

$$\sum_{l=0}^{L_{max}}\sum_{m=0}^{l} F^i_{lm1}\gamma_{lm}U_{lm}(\hat{\underline{x}}_j) + \sum_{l=1}^{L_{max}}\sum_{m=1}^{l} F^i_{lm2}\gamma_{lm}V_{lm}(\hat{\underline{x}}_j) = F_{ij} \quad j = 1,2,\cdots,M \quad (2.3)$$

So that in order to determine the Fourier coefficients $\{F^i_{lmk}\}_{k=1,2}$ we need at least $M \geq \frac{(L_{max}+1)(L_{max}+2)}{2} + \frac{L_{max}(L_{max}+1)}{2}$ measurements. Since the linear system (2.3) is ill conditioned it may be useful to split (2.3) in two linear systems; the first one will be used to determine the coefficients $\{F^i_{lm1}\}$ and the second one will be used to determine the coefficients $\{F^i_{lm2}\}$. This can be obtained for example by choosing $\Omega_2 = \Omega'_2 \cup \Omega''_2$ where

$$\Omega'_2 = \{\hat{\underline{x}}_i \in \partial B |\ \hat{\underline{x}}_i = \hat{\underline{x}}(\theta_i, 0)\ ,\ i = 1,2,\cdots,M_1\}$$
$$\Omega''_2 = \{\hat{\underline{x}}_i \in \partial B\ |\ \hat{\underline{x}}_i = \hat{\underline{x}}(\theta_i, \phi_0)\ ,\ 0 < \phi_0 < \frac{\pi}{2}\ ,\ \sin m\phi_0 \neq 0\ ,$$
$$m = 1,2,\cdots,L_{max}\ ,\ i = M_1 + 1, M_1 + 2, \cdots, M\}$$

The linear system (2.3) becomes:

$$\sum_{l=0}^{L_{max}}\sum_{m=0}^{l} F^i_{lm1}\gamma_{lm}U_{lm}(\hat{\underline{x}}_j) = F_{ij}\ ,\ \hat{\underline{x}}_j \in \Omega'_2 \quad (2.4)$$

and

$$\sum_{l=1}^{L_{max}}\sum_{m=1}^{l} F^i_{lm2}\gamma_{lm}V_{lm}(\hat{\underline{x}}_j) = F_{ij} - \sum_{l=0}^{L_{max}}\sum_{m=0}^{l} F^i_{lm1}\gamma_{lm}U_{lm}(\hat{\underline{x}}_j)\ ,\ \hat{\underline{x}}_j \in \Omega''_2 \quad (2.5)$$

The linear system (2.4) has $\frac{(L_{max}+1)(L_{max}+2)}{2}$ unknowns and the linear system (2.5) has $\frac{L_{max}(L_{max}+1)}{2}$ unknowns so that these are constraints on $M_1$ and $M - M_1$. Since the linear systems (2.4), (2.5) are ill conditioned and their condition number increases dramatically with $L_{max}$ they are solved by "parabolic regularization".

Let

$$A\underline{\xi} = \underline{\beta} \quad (2.6)$$

be a linear system such that $A$ is non-singular and let $A^*$ be the hermitian conjugate of $A$. Let $\underline{\xi}(t)$ be the solution of the differential equation:

$$\frac{d\underline{\xi}}{dt} = -A^*A\underline{\xi} + A^*\underline{\beta} \quad (2.7)$$

with the initial condition

$$\underline{\xi}(0) = \underline{0} \quad (2.8)$$

It is easy to see that $\lim_{t\to\infty} \underline{\xi}(t) = \underline{\xi}_*$, where $\underline{\xi}_*$ is the unique solution of (2.6). For solution of (2.6) via "parabolic regularization" we mean obtaining $\underline{\xi}_*$ by integrating numerically (2.7), (2.8) with an L-stable method such as backward Euler.

The previous procedure can easily be adapted to the case of non square systems solving in the least squares sense as long as the solution is unique.

The procedure previously described is performed for $i = 1, 2, \cdots, N$ so that the Fourier coefficients of $F(\hat{\underline{x}}, k, \underline{\alpha}_i)$, $\underline{\alpha}_i \in \Omega_1$ are obtained.

<u>Step 2</u> From the far field patterns to the Herglotz kernel.

Let $g(\hat{x}) \in \mathbf{L}_2(\partial B)$ be the Herglotz kernel associated to the domain $D$, we have [5]:

$$\int_{\partial B} F(\hat{x}, k, \underline{\alpha})\overline{g(\hat{x})}ds(\hat{x}) = 1/k \quad \forall \underline{\alpha} \in \partial B \tag{2.9}$$

where $ds(\hat{x})$ is the surface element of $\partial B$ and $\overline{g}$ the complex conjugate of $g$. We assume for $g$ the expression of a truncated Fourier expansion:

$$g(\hat{x}) = \sum_{l=0}^{L_g} \sum_{m=0}^{l} g_{lm1} \gamma_l^m U_{lm}(\hat{x}) + \sum_{l=1}^{L_g} \sum_{m=1}^{l} g_{lm2} \gamma_l^m V_{lm}(\hat{x}) \tag{2.10}$$

where $0 \leq L_g \leq L_{max}$. Imposing (2.9) between (2.2) and (2.10) and using the hortogonality properties of the spherical harmonics we have:

$$\sum_{l=0}^{L_g} \sum_{m=0}^{l} F_{lm1}^i \overline{g}_{lm1} + \sum_{l=1}^{L_g} \sum_{m=1}^{l} F_{lm2}^i \overline{g}_{lm2} = 1/k \quad i = 1, 2, ..., N \tag{2.11}$$

So that the Fourier coefficients $\{g_{lm1}\}, \{g_{lm2}\}$ of $g$ are determined by solving (2.11) The linear system (2.11) has $\frac{(L_g+1)(L_g+2)}{2} + \frac{L_g(L_g+1)}{2}$ unknowns, that is in order to determine the $\{g_{lmk}\}_{k=1,2}$ we need $N \geq \frac{(L_g+1)(L_g+2)}{2} + \frac{L_g(L_g+1)}{2}$.

However for obstacles $D$ such that $\partial D$ is symmetric with respect to the $z$-axis, that is $\partial D = \{\underline{x} = f(\theta)\hat{x}(\theta, \phi) \mid 0 < a < f(\theta) < b < \infty, \; 0 \leq \theta \leq \pi\}$ we can assume that $g(\hat{x})$ has the same symmetry that is $g_{lm1} = 0$ if $m > 0$ and $g_{lm2} = 0$ so that the linear system (2.11) has only $L_g + 1$ unknowns. Moreover if $\partial D$ is also symmetric with respect to the equator $(\theta = \frac{\pi}{2})$, that is $f(\theta)$ is an even function of $\theta - \frac{\pi}{2}$, we can assume that $g(\hat{x})$ has the same symmetry, that is, $g_{lm1} = 0$ if $m = 0$ and $l$ odd or $m > 0$, and $g_{lm2} = 0$ so that the linear system (2.11) has only $[\frac{L_g}{2}] + 1$ unknowns, where with $[\frac{L_g}{2}]$ we mean the integer part of $\frac{L_g}{2}$.

The use of these symmetries gives us the possibilty of using a small set $\Omega_1$ (i.e. $N = L_g + 1$ or $N = [\frac{L_g}{2}] + 1$) so that the linear system (2.11) is easy to handle. Moreover the solution of the linear system (2.5) can be avoided since the Fourier coefficients $\{F_{lm2}^i\}$ are not used in (2.11) since the $\{g_{lm2}\}$ are zero.

The Step 3 and 4 of our computation are identical to the ones performed in [6] and are reported here only for convenience of the reader.

<u>Step 3</u> From the Herglotz Kernel $g$ to the Herglotz wave function $v$. Let $\underline{y} \in \mathbf{R}^3$, the Herglotz wave function $v(k\underline{y})$ is the Fourier transform of the Herglotz kernel $g(\hat{x})$ that is:

$$v(k\underline{y}) = \int_{\partial B} g(\hat{x}) e^{ik(\hat{x}, \underline{y})} ds(\hat{x}) \tag{2.12}$$

Since $g(\hat{x})$ is given by (2.10) we have:

$$v(k\underline{y}) = \sum_{l=0}^{L_g} \sum_{m=0}^{l} g_{lm1} \gamma_l^m Q_{lm}(k\underline{y}) + \sum_{l=1}^{L_g} \sum_{m=1}^{l} g_{lm2} \gamma_l^m R_{lm}(k\underline{y}) \tag{2.13}$$

where

$$Q_{lm}(k\underline{y}) = \int_{\partial B} U_{lm}(\hat{x}) e^{ik(\hat{x}, \underline{y})} ds(\hat{x}) \tag{2.14}$$

$$R_{lm}(k\underline{y}) = \int_{\partial B} V_{lm}(\hat{x}) e^{ik(\hat{x}, \underline{y})} ds(\hat{x}) \tag{2.15}$$

The integrals (2.14), (2.15) can be evaluated in terms of spherical Bessel's functions of the first kind $j_m(r)$, and their derivatives. We note that the spherical Bessel's functions are elementary functions (for exemple $j_0(r) = \frac{\sin r}{r}$). Moreover we note that we have obtained an analytic expression for $v$ and that in a similar way analytic expressions can be obtained for $\frac{\partial v}{\partial r}, \frac{\partial v}{\partial \theta}, \frac{\partial v}{\partial \phi}$

that is the derivatives of $v$ with respect to the polar variables. For later convenience we split $v$ in real and immaginary part and use polar variables

$$v(k\underline{y}) = v^R(kr,\theta,\phi) + iv^I(kr,\theta,\phi) \qquad (2.16)$$

**Step 4** From the Herglotz wave function $v$ to $\partial D$. We assume that the origin $O \in D$ and that exists $a > 0, b > 0$ and a function $f(\theta,\phi)$ with $a \le f \le b$ such that $\partial D = \{(r,\theta,\phi) \in \mathbf{R}^3 \mid r = f(\theta,\phi) \; 0 \le \theta < \pi, 0 \le \phi < 2\pi\}$. If $D$ is an Herglotz domain we have [5]:

$$v(kf,\theta,\phi) + \frac{e^{-ikf}}{kf} = 0 \quad 0 \le \theta < \pi \,,\, 0 \le \phi < 2\pi \qquad (2.17)$$

Equation (2.17) defines $f$ implicitly as a function of $\theta$ and $\phi$. We note that the set where (2.17) is satisfied is a rather complicated set that contains $\partial D$ as a subset. From (2.17) differentiating with respect to $\theta$ and $\phi$ we have:

$$\left[k\frac{\partial v}{\partial r} - (if + \frac{1}{k})\frac{e^{-ikf}}{f^2}\right]\frac{df}{d\theta} + \frac{\partial v}{\partial \theta} = 0 \quad 0 \le \theta < \pi \,,\, 0 \le \phi < 2\pi \qquad (2.18)$$

$$\left[k\frac{\partial v}{\partial r} - (if + \frac{1}{k})\frac{e^{-ikf}}{f^2}\right]\frac{df}{d\phi} + \frac{\partial v}{\partial \phi} = 0 \quad 0 \le \theta < \pi \,,\, 0 \le \phi < 2\pi \qquad (2.19)$$

In order to obtain $f(\theta,\phi)$ we proceed as follows:

(i) for fixed $\theta$ and $\phi$, let $\theta = \phi = 0$, we solve (2.17) that is we solve

$$v^R(kf,0,0) + \frac{\cos kf}{kf} = 0 \qquad (2.20)$$

and we verify that the solution found $f_{00}$ satisfy

$$v^I(kf_{00},0,0) - \frac{\sin kf_{00}}{kf_{00}} = 0 \qquad (2.21)$$

We note that (2.20) is a non-linear equation in one unknown

(ii) given $f_{00}$ we solve for $0 \le \theta < \pi$ the differential equation for $\frac{df(\theta,0)}{d\theta}$ obtained by taking the real part of equation (2.18) with initial condition

$$f(0,0) = f_{00} \qquad (2.22)$$

Let $f_0(\theta) \; 0 \le \theta < \pi$ be the solution found we verify than for $f_0(\theta)$ we have

$$v^I(kf_0(\theta),\theta,0) - \frac{\sin kf_0(\theta)}{kf_0(\theta)} = 0 \quad 0 \le \theta < \pi \qquad (2.23)$$

(iii) given $f_0(\theta)$ we solve for $0 \le \phi < 2\pi$ the differential equation for $\frac{df}{d\phi}$ obtained by taking the real part of (2.19) with initial condition

$$f(\theta,0) = f_0(\theta) \qquad (2.24)$$

and we verify than for $f(\theta,\phi)$ we have

$$v^I(kf,\theta,\phi) - \frac{\sin kf}{kf} = 0 \quad 0 \le \theta < \pi \,,\, 0 \le \phi < 2\pi \qquad (2.25)$$

The differential problems considered in (ii),(iii) are initial values problems for a scalar differential equations and are solved numerically via a Runge-Kutta method. The problem considered in (iii) is performed only for a finite number of $\theta$. We note that if the obstacle is cylindrically symmetric with respect to the $z$-axis (iii) is not needed and the problem is solved after performing (i),(ii).

Moreover (i),(ii),(iii) are fully parallelizable in fact: in (i) many roots of (2.20) that satisfy (2.21) can be found indipendently from different initial guesses. Since the set of zeroes of (2.17) is a complicated set that contains $\partial D$ in proper sense this may be useful. In (ii) the differential equation considered can be integrated indipendently from the initial conditions obtained in (i). Only the trajectories that exists for $0 \leq \theta < \pi$ must be considered for (iii). Finally in (iii) the differential equation considered can be integrated indipendently for different values of $\theta$ and different initial conditions obtained in (ii). Only the trajectories that generate a closed surface should be considered.

**3. The numerical experience** In Fig. 1,2,3,4 the numerical results obtained on some simple geometries with the numerical method described in section 2 are shown.

In the reconstructions shown in these figures the symmetries of the obstacles are always exploited in the way suggested in Step 2 of section 2. The relative $L_2$ error, $E_{L_2}$ shown in the figures is given by

$$E_{L_2} = \left\{ \frac{\sum_{j=0}^{36}[f(\theta_j,0) - f_c(\theta_j,0)]^2}{\sum_{j=0}^{36}(f(\theta_j,0))^2} \right\}^{\frac{1}{2}} \quad (3.1)$$

where $r = f(\theta,\phi)$ is the surface of the obstacle, $f_c(\theta_j,0)$ is the computed value of this surface with the procedure described in section 2 and $\theta_j = \frac{j\pi}{36}$, $j = 0,1,\cdots,36$.

The set $\Omega_1$ is chosen between the following ones:

$$A_1 = \{(\theta_j,\phi_k) \mid \theta_j = \arccos(\frac{j}{4} - 1), j = 1, \cdots, 7; \phi_k = \frac{2\pi k}{7}, k = 1, \cdots, 7\} \cup \{(0,0),(\pi,0)\}$$

$$A_2 = \{(\theta_j,0) \mid \theta_j = \arccos(\frac{j}{3} - 1), \ j = 0,1,\cdots,6\}$$

$$A_3 = \{(\theta_j,0) \mid \theta_j = \arccos(\frac{\sqrt{2}}{2}(\frac{j}{3} - 1)), \ j = 0,1,\cdots,6\}$$

moreover the set $\Omega_2$ is chosen between the following ones:

$$B_1 = \partial B$$

$$B_2 = \{(\theta_j,0) \mid \theta_j = \frac{j\pi}{14}, \ j = 0,1,\cdots,14\}$$

$$B_3 = \{(\theta_j,0) \mid \theta_j = \frac{(7+j)\pi}{28}, \ j = 0,1,\cdots,14\}$$

When $\Omega_2 = B_1 = \partial B$ we assume that a truncated Fourier expansion, that is $\{F^i_{lmk}\}_{k=1,2}$, of the relevant far fields is known as has been assumed in [5], [6].

The far fields data $F_{ij}$ of (2.3) are substituted by $F_{ij} + \varepsilon\varsigma$ where $\varsigma$ is a random number uniformly distributed between $-0.5$ and $0.5$. Finally we observe that in our numerical experience the far field data, $F_{ij}$, are roughly of magnitude one. The figures 1,2,3,4 show how the reconstruction deteriorates when the data are taken in a smaller portion of the unit sphere (i.e. $A_1, A_2, A_3$ and $B_1, B_2, B_3$) or when the data are affected by an increasing random error (i.e. increasing values of $\varepsilon$).

Overall the reconstruction shown in fig. 1,2,3,4 seem to be very satisfactory. In particular the figures 1a),2a),3a),4a) represent the original obstacle $D$ and the equation of $\partial D$, the figures 1b),2b),3b),4b) represent the reconstructed obstacles when $\Omega_1 = A_1$ and $\Omega_2 = B_1$, that is the maximum amount of data allowed in our reconstructions is used and the figures 1c),2c),3c),4c),1d),2d),1e),2e) represent the reconstructed obstacles when incomplete data are used. In these figures $L_m = L_g$.

Further work is in progress to extend our method to non axially symmetric obstacles and to boundary conditions more general than the Dirichlet boundary condition (1.4) considered here.

The computations are performed on an IBM-3090 with VM operating system and the figures are drawn using the GRAFMATIC package on a personal computer AT-IBM with ega-color graphic card.

## Prolate Ellipsoid

a)

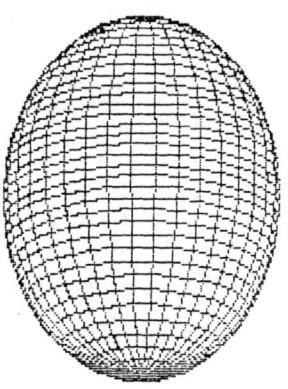

$$x^2 + y^2 + \left(\tfrac{2}{3}z\right)^2 = 1$$

b)

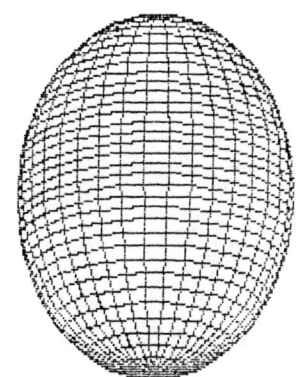

$L_m = 8, L_g = 8, K = 3, \varepsilon = 0$
$\Omega_1 = A_1, \Omega_2 = B_1, E_{L_2} = 0.000019$

c)

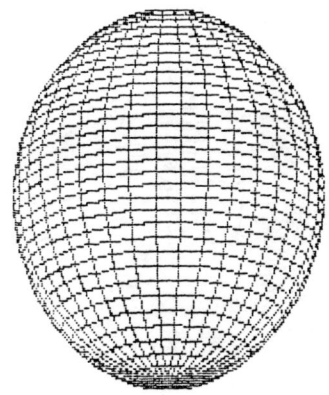

$L_g = 4, K = 3, \varepsilon = 0.001$
$\Omega_1 = A_2, \Omega_2 = B_2, E_{L_2} = 0.079$

d)

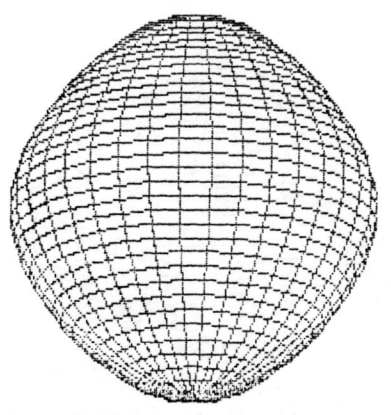

$L_g = 4, K = 3, \varepsilon = 0.001$
$\Omega_1 = A_3, \Omega_2 = B_3, E_{L_2} = 0.16$

e)

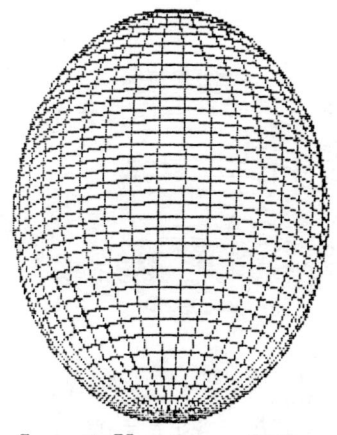

$L_g = 4, K = 3, \varepsilon = 0.02$
$\Omega_1 = A_2, \Omega_2 = B_2, E_{L_2} = 0.13$

Fig. 1

## Horizontal Platelet

a)

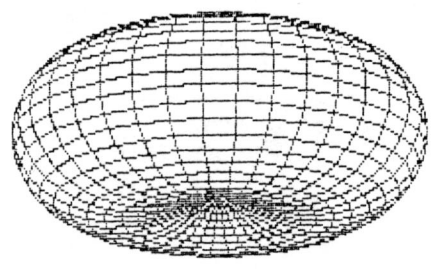

$r(\theta) = 1 - 0.5 \cos 2\theta$

b)

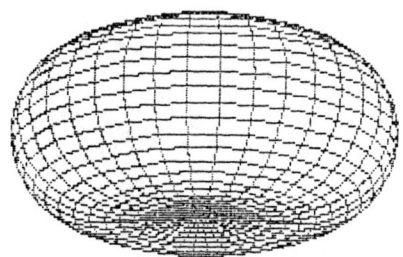

$L_m = 8, L_g = 8, K = 3.5, \varepsilon = 0$
$\Omega_1 = A_1, \Omega_2 = B_1, E_{\mathbf{L_2}} = 0.043$

c)

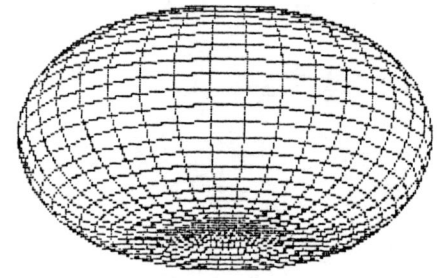

$L_g = 4, K = 3.5, \varepsilon = 0.001$
$\Omega_1 = A_2, \Omega_2 = B_2, E_{\mathbf{L_2}} = 0.20$

d)

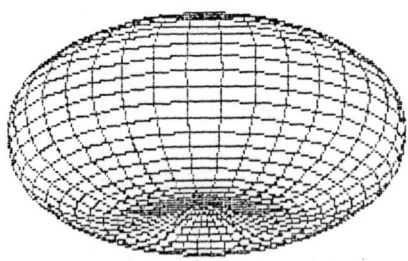

$L_g = 4, K = 3.5, \varepsilon = 0.001$
$\Omega_1 = A_3, \Omega_2 = B_3, E_{\mathbf{L_2}} = 0.085$

e)

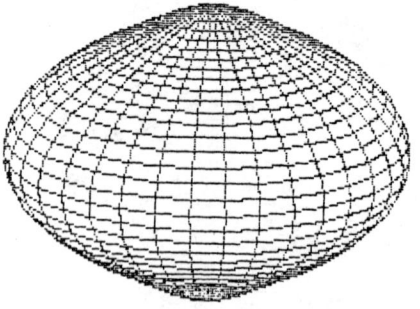

$L_g = 4, K = 3.5, \varepsilon = 0.02$
$\Omega_1 = A_2, \Omega_2 = B_2, E_{\mathbf{L_2}} = 0.36$

Fig. 2

## Short Cylinder

a)

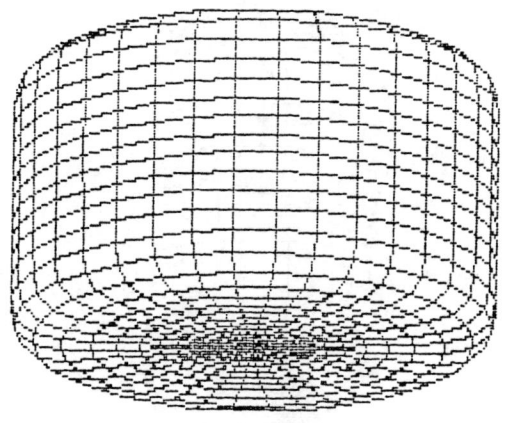

$$((\tfrac{2}{3}x)^2 + (\tfrac{2}{3}y)^2)^5 + z^{10} = 1$$

b)

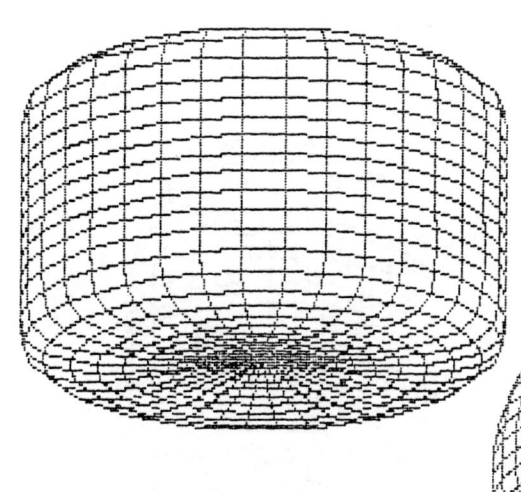

c)

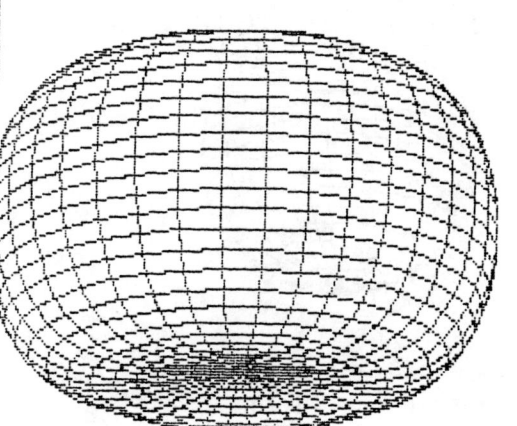

$L_m = 8, L_g = 8, K = 3, \varepsilon = 0$
$\Omega_1 = A_1, \Omega_2 = B_1, E_{L_2} = 0.018$

$L_g = 6, K = 3, \varepsilon = 0.001$
$\Omega_1 = A_2, \Omega_2 = B_2, E_{L_2} = 0.066$

Fig. 3

## Vogels Peanut

a)

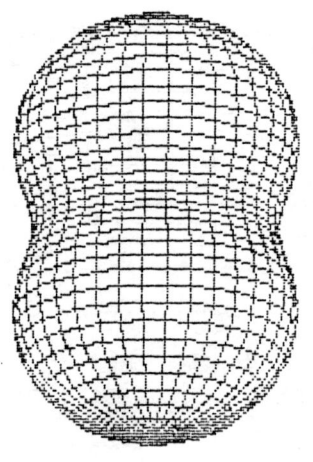

$$r^2(\theta) = \tfrac{9}{4}(\cos^2\theta + \tfrac{1}{4}\sin^2\theta)$$

b)

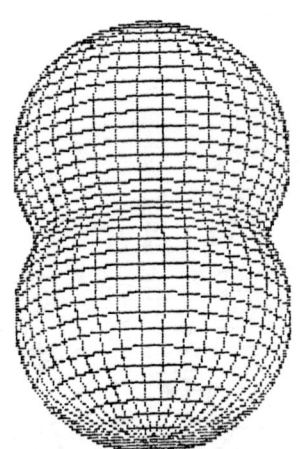

$L_m = 8, L_g = 8, K = 4, \varepsilon = 0$
$\Omega_1 = A_1, \Omega_2 = B_1, E_{L_2} = 0.0066$

c)

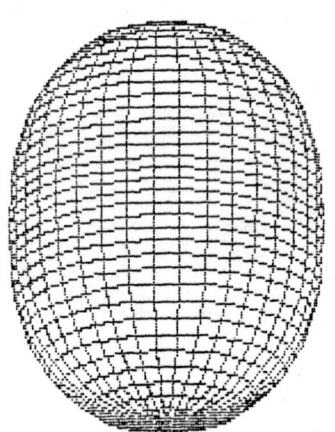

$L_g = 4, K = 3.4, \varepsilon = 0.001$
$\Omega_1 = A_2, \Omega_2 = B_2, E_{L_2} = 0.088$

Fig. 4

**Acknowledgements** One of us (F.Z.) gratefully acknowledge the help of Prof. D. Colton and Prof. P. Monk that have given to him a copy of the computer program used to obtain the numerical experience described in [5].

## REFERENCES

1. D. COLTON, R. KRESS, *Integral equation methods in scattering theory*, J.Wiley & Sons, New York, 1983.

2. D. COLTON, *The inverse scattering problem for time harmonic acoustic waves*, SIAM Review 26, (1984), pp. 323-350.

3. D. COLTON, P. MONK, *A novel method for solving the inverse scattering problem for time harmonic acoustic waves in the resonance region*, SIAM J.Appl.Math. 45, (1985), pp. 1039-1053.

4. D. COLTON, P. MONK, *A novel method for solving the inverse scattering problem for time harmonic acoustic waves in the resonance region II*, SIAM J.Appl.Math 46, (1986), pp. 506-523.

5. D. COLTON, P. MONK, *The numerical solution of the three dimensional inverse scattering problem for time harmonic acoustic waves*, SIAM J.Sci.Stat.Comput. 8, (1987), pp. 278-291.

6. F. ALUFFI-PENTINI, E. CAGLIOTI, L. MISICI, F. ZIRILLI, *A parallel algorithm for a three dimensional inverse acoustic scattering problem*, in Parallel Computing, D. J. Evans, C. Sutti ed., Philadelphia, 1988, pp. 193-200.

# CHAPTER 18

## The Application of Globally Convergent Homotopy Methods to Nonlinear Optimization*

Layne T. Watson**
Raphael T. Haftka†
Frederick H. Lutze†
Raymond H. Plautt‡
Philip Y. Shin§

**Abstract.** Probability-one homotopy methods are a class of algorithms for solving nonlinear systems of equations that are accurate, robust, and converge from an arbitrary starting point almost surely. These new globally convergent homotopy techniques have been successfully applied to solve Brouwer fixed point problems, polynomial systems of equations, discretizations of nonlinear two-point boundary value problems based on shooting, finite differences, collocation, and finite elements, and finite difference, collocation, and Galerkin approximations to nonlinear partial differential equations. This paper surveys some recent major advances in globally convergent homotopy algorithms for unconstrained and constrained optimization, with applications to the optimal design of composite laminated plates and fuel-optimal orbital satellite maneuvers, previously unsolved by locally convergent iterative techniques.

**1. Introduction.** Continuation is a well known and established procedure in numerical analysis. The idea is to continuously deform a simple (easy) problem into the given (hard) problem, while solving the family of deformed problems. The solutions to the deformed problems are related, and can be tracked as the deformation proceeds. The function describing the deformation is called a *homotopy map*. Homotopies are a traditional

---

*The work of L.T. Watson was supported in part by AFOSR Grant 85-0250, and that of R.T. Haftka by NASA grant NAG-1-168.
**Department of Computer Science, Virginia Polytechnic Institute & State University, Blacksburg, VA 24061.
†Department of Aerospace and Ocean Engineering, Virginia Polytechnic Institute & State University, Blacksburg, VA 24061.
‡Department of Civil Engineering, Virginia Polytechnic Institute & State University, Blacksburg, VA 24061.
§Department of Mechanical Engineering, Naval Postgraduate School, Monterey, CA 93940.

part of topology, and have found significant application in nonlinear functional analysis and differential geometry. Similar ideas, such as incremental loading, are also widely used in engineering.

These traditional continuation algorithms have serious deficiencies, which have been removed by modern homotopy algorithms. The differences, however, are subtle and mathematically deep, and the mathematical proofs of the statements in this article are beyond the scope of the presentation here. To explain the differences between the old and new homotopy techniques, a more detailed discussion is required. Suppose the given problem is to find a root of the nonlinear equation $f(x) = 0$, and that $s(x) = 0$ is a simple version of the given problem with an easily obtainable unique solution $x_0$. Then a homotopy map could be, e.g.,

$$H(\lambda, x) = \lambda f(x) + (1-\lambda) s(x), \quad 0 \leq \lambda \leq 1.$$

The family of problems is $H(\lambda, x) = 0$, $0 \leq \lambda \leq 1$, and the idea would be to track the solutions of $H(\lambda, x) = 0$, starting from $(\lambda, x) = (0, x_0)$, as $\lambda$ goes from 0 to 1. If everything worked out well, this would lead to a point $(\lambda, x) = (1, \bar{x})$, where $f(\bar{x}) = 0$. The "standard" approach is to start from a point $(\lambda_i, x_i)$ with $H(\lambda_i, x_i) = 0$, and solve the problem $H(\lambda_i + \Delta\lambda, x) = 0$ for $x$, with $\Delta\lambda$ being a sufficiently small, fixed, positive number. The bad things that can happen are:
1) The points $(\lambda_i, x_i)$ may diverge to infinity as $\lambda \to 1$.
2) The problem $H(\lambda_i + \Delta\lambda, x) = 0$ may be singular at its solution, causing numerical instability.
3) There may be no solution of $H(\lambda_i + \Delta\lambda, x) = 0$ near $(\lambda_i, x_i)$.

The modern approach to homotopy methods is to construct a homotopy map $\rho_a(\lambda, x)$, involving additional parameters in the vector $a$, such that 1), 2), and 3) never occur or never cause any difficulty. The details of how this is done are given in Section 2. Section 3 summarizes basic homotopy results for optimization, and makes the connection between nonlinear equations, homotopies, and optimization. Examples of the globally convergent homotopy techniques applied to optimization are given in Sections 4–7.

**2. Homotopy theory.** The theoretical foundation of all probability one globally convergent homotopy methods is given in the following differential geometry theorem:

DEFINITION. Let $E^n$ denote n-dimensional real Euclidean space, let $U \subset E^m$ and $V \subset E^n$ be open sets, and let $\rho : U \times [0, 1) \times V \to E^n$ be a $C^2$ map. $\rho$ is said to be transversal to zero if the Jacobian matrix $D\rho$ has full rank on $\rho^{-1}(0)$.

PARAMETRIZED SARD'S THEOREM [5]. *If $\rho(a, \lambda, x)$ is transversal to zero, then for almost all $a \in U$ the map*

$$\rho_a(\lambda, x) = \rho(a, \lambda, x)$$

*is also transversal to zero; i.e., with probability one the Jacobian matrix $D\rho_a(\lambda, x)$ has full rank on $\rho_a^{-1}(0)$.*

The import of this theorem is that the zero set $\rho_a^{-1}(0)$ consists of smooth, nonintersecting curves in $[0, 1) \times V$. These curves are either closed loops, or have endpoints in $\{0\} \times V$ or $\{1\} \times V$, or go to infinity. Another important consequence is that these curves have finite arc length in any compact subset of $[0, 1) \times V$. The recipe for constructing a globally convergent homotopy algorithm to solve the nonlinear system of equations

$$F(x) = 0, \tag{1}$$

where $F : E^n \to E^n$ is a $C^2$ map, is as follows: For an open set $U \subset E^m$ construct a $C^2$ homotopy map $\rho : U \times [0,1) \times E^n \to E^n$ such that

1) $\rho(a, \lambda, x)$ is transversal to zero,
2) $\rho_a(0, x) = \rho(a, 0, x) = 0$ is trivial to solve and has a unique solution $x_0$,
3) $\rho_a(1, x) = F(x)$,
4) $\rho_a^{-1}(0)$ is bounded.

Then for almost all $a \in U$ there exists a zero curve $\gamma$ of $\rho_a$, along which the Jacobian matrix $D\rho_a$ has rank $n$, emanating from $(0, x_0)$ and reaching a zero $\bar{x}$ of $F$ at $\lambda = 1$. This zero curve $\gamma$ does not intersect itself, is disjoint from any other zeros of $\rho_a$, and has finite arc length in every compact subset of $[0,1) \times E^n$. Furthermore, if $DF(\bar{x})$ is nonsingular, then $\gamma$ has finite arc length.

The general idea of the algorithm is now apparent: just follow the zero curve $\gamma$ emanating from $(0, x_0)$ until a zero $\bar{x}$ of $F(x)$ is reached (at $\lambda = 1$). Of course it is nontrivial to develop a viable numerical algorithm based on that idea, but at least conceptually, the algorithm for solving the nonlinear system of equations $F(x) = 0$ is clear and simple. The homotopy map (usually, but not always) is

$$\rho_a(\lambda, x) = \lambda F(x) + (1 - \lambda)(x - a), \qquad (2)$$

which has the same form as a standard continuation or embedding mapping. However, there are two crucial differences. In standard continuation, the embedding parameter $\lambda$ increases monotonically from 0 to 1 as the trivial problem $x - a = 0$ is continuously deformed to the problem $F(x) = 0$. The present homotopy method permits $\lambda$ to both increase and decrease along $\gamma$ with no adverse effect; that is, turning points present no special difficulty. The second important difference is that there are never any "singular points" which afflict standard continuation methods. The way in which the zero curve $\gamma$ of $\rho_a$ is followed and the full rank of $D\rho_a$ along $\gamma$ guarantee this.

The scheme just described is known as a probability-one globally convergent homotopy algorithm. The phrase "probability-one" refers to the almost any choice for $a$, and the "global convergence" refers to the fact that the starting point $x_0$ need not be anywhere near the solution $\bar{x}$. It should be mentioned that the form of the homotopy map $\rho_a(\lambda, x)$ in (2) is just a special case used here for clarity of exposition. The more general theory can be found in [24–31], and practical engineering problems requiring a $\rho_a$ nonlinear in $\lambda$ are in [35] and [36]. Below are some typical theorems for various classes of problems.

The computation of Brouwer fixed points represents one of the first successes for both simplicial [1, 19] and continuous homotopy methods [5, 24, 34]. Brouwer fixed point problems can be very nasty, and often cause locally convergent iterative methods a great deal of difficulty.

THEOREM [24]. *Let* $B = \{x \in E^n \mid \|x\|_2 = 1\}$ *be the closed unit ball, and* $f : B \to B$ *a* $C^2$ *map. Then for almost all* $a \in \text{int } B$ *there exists a zero curve* $\gamma$ *of*

$$\rho_a(\lambda, x) = \lambda(x - f(x)) + (1 - \lambda)(x - a),$$

*along which the Jacobian matrix* $D\rho_a(\lambda, x)$ *has full rank, emanating from* $(0, a)$ *and reaching a fixed point* $\bar{x}$ *of* $f$ *at* $\lambda = 1$. *Furthermore,* $\gamma$ *has finite arc length if* $I - Df(\bar{x})$ *is nonsingular.*

Typically a mathematical problem (such as a partial differential equation) reduces to a finite dimensional nonlinear system of equations, and what is desired are conditions on the original problem, not on the final discretized problem. Thus the results in this section are used to derive, working backwards, useful conditions on the original problem, whatever it might be. The following four lemmas, which follow from the results of [5, 24–31], are used for that purpose.

LEMMA 1. *Let $g : E^p \to E^p$ be a $C^2$ map, $a \in E^p$, and define $\rho_a : [0,1) \times E^p \to E^p$ by*

$$\rho_a(\lambda, y) = \lambda g(y) + (1 - \lambda)(y - a).$$

*Then for almost all $a \in E^p$ there is a zero curve $\gamma$ of $\rho_a$ emanating from $(0, a)$ along which the Jacobian matrix $D\rho_a(\lambda, y)$ has full rank.*

LEMMA 2. *If the zero curve $\gamma$ in Lemma 1 is bounded, it has an accumulation point $(1, \bar{y})$, where $g(\bar{y}) = 0$. Furthermore, if $Dg(\bar{y})$ is nonsingular, then $\gamma$ has finite arc length.*

LEMMA 3. *Let $F : E^p \to E^p$ be a $C^2$ map such that for some $r > 0$, $x F(x) \geq 0$ whenever $\|x\| = r$. Then $F$ has a zero in $\{x \in E^p \mid \|x\| \leq r\}$, and for almost all $a \in E^p$, $\|a\| < r$, there is a zero curve $\gamma$ of*

$$\rho_a(\lambda, x) = \lambda F(x) + (1 - \lambda)(x - a),$$

*along which the Jacobian matrix $D\rho_a(\lambda, x)$ has full rank, emanating from $(0, a)$ and reaching a zero $\bar{x}$ of $F$ at $\lambda = 1$. Furthermore, $\gamma$ has finite arc length if $DF(\bar{x})$ is nonsingular.*

Lemma 3 is a special case of the following more general lemma.

LEMMA 4. *Let $F : E^p \to E^p$ be a $C^2$ map such that for some $r > 0$ and $\tilde{r} > 0$, $F(x)$ and $x - a$ do not point in opposite directions for $\|x\| = r$, $\|a\| < \tilde{r}$. Then $F$ has a zero in $\{x \in E^p \mid \|x\| \leq r\}$, and for almost all $a \in E^p$, $\|a\| < \tilde{r}$, there is a zero curve $\gamma$ of*

$$\rho_a(\lambda, x) = \lambda F(x) + (1 - \lambda)(x - a),$$

*along which the Jacobian matrix $D\rho_a(\lambda, x)$ has full rank, emanating from $(0, a)$ and reaching a zero $\bar{x}$ of $F$ at $\lambda = 1$. Furthermore, $\gamma$ has finite arc length if $DF(\bar{x})$ is nonsingular.*

These theoretical algorithms have been implemented in sophisticated mathematical software packages such as PITCON [18], CONKUB [10], and HOMPACK [32]. The latter is an extensive collection of FORTRAN 77 routines implementing three different tracking algorithms for both dense and sparse problems, and containing high level drivers for special classes of problems.

**3. Basic optimization homotopies.** Consider first the unconstrained optimization problem

$$\min_x f(x). \tag{3}$$

THEOREM [26]. *Let $f : E^n \to E$ be a $C^3$ convex map with a minimum at $\tilde{x}$, $\|\tilde{x}\|_2 \leq M$. Then for almost all $a$, $\|a\|_2 < M$, there exists a zero curve $\gamma$ of the homotopy map*

$$\rho_a(\lambda, x) = \lambda \nabla f(x) + (1 - \lambda)(x - a),$$

*along which the Jacobian matrix $D\rho_a(\lambda, x)$ has full rank, emanating from $(0, a)$ and reaching a point $(1, \bar{x})$, where $\bar{x}$ solves (3).*

A function is called uniformly convex if it is convex and its Hessian's smallest eigenvalue is bounded away from zero. Consider next the constrained optimization problem

$$\min_{x \geq 0} f(x). \tag{4}$$

This is more general than it might appear because the general convex quadratic program reduces to a problem of the form (4).

THEOREM [26]. *Let $f : E^n \to E$ be a $C^3$ uniformly convex map. Then there exists $\delta > 0$ such that for almost all $a \geq 0$ with $\|a\|_2 < \delta$ there exists a zero curve $\gamma$ of the homotopy map*

$$\rho_a(\lambda, x) = \lambda K(x) + (1 - \lambda)(x - a),$$

*where*

$$K_i(x) = -\left|\frac{\partial f(x)}{\partial x_i} - x_i\right|^3 + \left(\frac{\partial f(x)}{\partial x_i}\right)^3 + x_i^3,$$

*along which the Jacobian matrix $D\rho_a(\lambda, x)$ has full rank, connecting $(0, a)$ to a point $(1, \bar{x})$, where $\bar{x}$ solves the constrained optimization problem (4).*

Given $F : E^n \to E^n$, the nonlinear complementarity problem is to find a vector $x \in E^n$ such that

$$x \geq 0, \quad F(x) \geq 0, \quad x^t F(x) = 0. \tag{5}$$

At a solution $\bar{x}$, $\bar{x}$ and $F(\bar{x})$ are "complementary" in the sense that if $\bar{x}_i > 0$, then $F_i(\bar{x}) = 0$, and if $F_i(\bar{x}) > 0$, then $\bar{x}_i = 0$. This problem is difficult because there are linear constraints $x \geq 0$, nonlinear constraints $F(x) \geq 0$, and a combinatorial aspect from the complementarity condition $x^t F(x) = 0$. It is interesting that homotopy methods can be adapted to deal with nonlinear constraints and combinatorial conditions.

Define $G : E^n \to E^n$ by

$$G_i(z) = -\left|F_i(z) - z_i\right|^3 + \left(F_i(z)\right)^3 + z_i^3, \quad i = 1, \ldots, n, \tag{6}$$

and let

$$\rho_a(\lambda, z) = \lambda G(z) + (1 - \lambda)(z - a).$$

THEOREM [25]. *Let $F : E^n \to E^n$ be a $C^2$ map, and let the Jacobian matrix $DG(z)$ be nonsingular at every zero of $G(z)$. Suppose there exists $r > 0$ such that $z > 0$ and $z_k = \|z\|_\infty \geq r$ imply $F_k(z) > 0$. Then for almost all $a > 0$ there exists a zero curve $\gamma$ of $\rho_a(\lambda, z)$, along which the Jacobian matrix $D\rho_a(\lambda, z)$ has full rank, having finite arc length and connecting $(0, a)$ to $(1, \bar{z})$, where $\bar{z}$ solves (5).*

THEOREM [25]. *Let $F : E^n \to E^n$ be a $C^2$ map, and let the Jacobian matrix $DG(z)$ be nonsingular at every zero of $G(z)$. Suppose there exists $r > 0$ such that $z \geq 0$ and $\|z\|_\infty \geq r$ imply $z_k F_k(z) > 0$ for some index $k$. Then there exists $\delta > 0$ such that for almost all $a \geq 0$ with $\|a\|_\infty < \delta$ there exists a zero curve $\gamma$ of $\rho_a(\lambda, z)$, along which the Jacobian matrix $D\rho_a(\lambda, z)$ has full rank, having finite arc length and connecting $(0, a)$ to $(1, \bar{z})$, where $\bar{z}$ solves (5).*

Homotopy algorithms for convex unconstrained optimization may not be computationally competitive with other approaches, but it is reassuring that the globally convergent homotopy techniques can theoretically be directly applied. For constrained optimization the homotopy approach offers some advantages, and, especially for the nonlinear complementarity problem, is competitive with other algorithms See [33] for an application of homotopy techniques to the linear complementarity problem. Constrained optimization is addressed in the next few sections.

## 4. Expanded Lagrangian Homotopy.

The expanded Lagrangian homotopy method of Poore [16, 17] is applicable to the general nonlinear programming problem

$$\min \theta(x)$$
$$\text{subject to} \quad g(x) \leq 0,$$
$$h(x) = 0,$$

where $x \in E^n$, $\theta$ is real valued, $g$ is an $m$-dimensional vector, and $h$ is a $p$-dimensional vector. In the most general situation the formulation and solution algorithm for the expanded Lagrangian homotopy are rather complicated. The method will be illustrated by applying it to a special case, namely the linear complementarity problem:

$$w - Mz = q,$$
$$w \geq 0, \quad z \geq 0, \quad w^t z = 0$$

where $M$ is a given real $n \times n$ matrix and $q \in E^n$ is given; the unknowns are $w \in E^n$ and $z \in E^n$.

The expanded Lagrangian approach [17] may be described as an optimization/continuation approach and has in its simplest form two main steps.

Step 1. (Optimization phase).

At $r = r_0 > 0$ solve the unconstrained minimization problem

$$\min_{w,z} P(w, z, r)$$

where

$$P(w, z, r) = \frac{1}{2r}\|w - Mz - q\|_2^2 + \frac{1}{2r}\langle w, z\rangle^2 - r\sum_{i=1}^n \ln z_i - r\sum_{i=1}^n \ln w_i.$$

Step 2A. (Switch to expanded system).

A (local) solution of $\min P$ must satisfy

$$0 = \nabla_{(w,z)} P = \begin{pmatrix} I \\ -M^t \end{pmatrix} \frac{(w - Mz - q)}{r} + \begin{pmatrix} z \\ w \end{pmatrix} \frac{\langle w, z\rangle}{r} - r\left(\frac{1}{w_1}, \ldots, \frac{1}{w_n}, \frac{1}{z_1}, \ldots, \frac{1}{z_n}\right)^t.$$

Introduce the following variables:

$$\beta = \frac{w - Mz - q}{r},$$
$$\theta = \frac{\langle w, z\rangle}{r},$$
$$\mu_i = \frac{r}{w_i}, \quad i = 1, \ldots, n,$$
$$\eta_i = \frac{r}{z_i}, \quad i = 1, \ldots, n,$$

which ultimately represent the Lagrange multipliers. This helps to remove the inevitable ill-conditioning associated with penalty methods for small $r$ and we thus obtain our equivalent but expanded system:

$$\begin{pmatrix} I \\ -M^t \end{pmatrix} \beta + \begin{pmatrix} z \\ w \end{pmatrix} \theta - \begin{pmatrix} \mu \\ \eta \end{pmatrix} = 0,$$

$$w - Mz - q - r\beta = 0,$$

$$\langle w, z \rangle - r\theta = 0,$$

$$\mu_i w_i - r = 0, \quad i = 1, \ldots, n,$$

$$\eta_i z_i - r = 0, \quad i = 1, \ldots, n.$$

(Remark. As a result of the optimization phase and the initial starting point with $r_0 > 0$, the solution $(w^{(0)}, z^{(0)})$ of min $P(w, z, r_0)$ satisfies $z^{(0)} > 0$ and $w^{(0)} > 0$. As a consequence, $\mu^{(0)} > 0$ and $\eta^{(0)} > 0$ from the definitions of $\mu$ and $\eta$. They remain positive until $r = 0$ where we formally have

$$\begin{pmatrix} I \\ -M^t \end{pmatrix} \beta + \begin{pmatrix} z \\ w \end{pmatrix} \theta - \begin{pmatrix} \mu \\ \eta \end{pmatrix} = 0,$$

$$w - Mz - q = 0,$$

$$\langle w, z \rangle = 0,$$

$$\mu_i w_i = 0, \quad i = 1, \ldots, n,$$

$$\eta_i z_i = 0, \quad i = 1, \ldots, n,$$

$$w, z, \theta, \mu, \eta \geq 0,$$

which implies that we have solved the problem.)

In practice we do not solve the optimization problem min $P$ to high accuracy since a highly accurate solution may have only a digit or two in common with the final answer. However, it is imperative that $\nabla P$ be reasonably small in magnitude, say less than $r_0/10$. The expanded system is converted to a homotopy map by letting $r = r_0(1 - \lambda)$ and modifying the first equation to obtain:

$$\begin{pmatrix} I \\ -M^t \end{pmatrix} \beta + \begin{pmatrix} z \\ w \end{pmatrix} \theta - \begin{pmatrix} \mu \\ \eta \end{pmatrix} - \frac{r}{r_0} \nabla P(w^{(0)}, z^{(0)}, r_0) = 0,$$

$$w - Mz - q - r\beta = 0,$$

$$\langle w, z \rangle - r\theta = 0,$$

$$\mu_i w_i - r = 0, \quad i = 1, \ldots, n,$$

$$\eta_i z_i - r = 0, \quad i = 1, \ldots, n.$$

Write this system of $5n + 1$ equations in the $5n + 2$ variables $\lambda, w, z, \beta, \theta, \mu, \eta$ as

$$\Upsilon(\lambda, w, z, \beta, \theta, \mu, \eta) = 0.$$

Step 2B. (Track the zero curve of $\Upsilon$ from $r = r_0$ to $r = 0$.)

Starting with arbitrary $r_0 > 0$, $w^{(0)} > 0$ and $z^{(0)} > 0$, the rest of the initial point $(0, w^{(0)}, z^{(0)}, \beta^{(0)}, \theta_0, \mu^{(0)}, \eta^{(0)})$ is given by

$$\beta^{(0)} = \frac{w^{(0)} - Mz^{(0)} - q}{r_0},$$

$$\theta_0 = \frac{\langle w^{(0)}, z^{(0)} \rangle}{r_0},$$

$$\mu_i^{(0)} = \frac{r_0}{w_i^{(0)}}, \quad i = 1, \ldots, n,$$

$$\eta_i^{(0)} = \frac{r_0}{z_i^{(0)}}, \quad i = 1, \ldots, n.$$

This approach requires careful attention to implementation details. For example, the linear algebra and globalization techniques with dynamic scaling are critically important in the optimization phase. For degenerate problems the path can still be long. One possible resolution is the use of shifts and weights as developed in the method of multipliers [3], but holding $r = r_0$ fixed. (This approach is currently under investigation in the context of linear programming [16].) However, in keeping with the philosophy of the "pure" homotopy approach of the current work, we do not solve the optimization problem (Step 1.), but instead use the above equations $\Upsilon(\lambda, w, z, \beta, \theta, \mu, \eta) = 0$ as a "pure" homotopy.

**5. An example of a special purpose homotopy.** The optimization problem that we consider here is to maximize the lowest buckling load of a structure for a given amount of resources. The structure is discretized by finite elements. Expressing the lowest buckling load with Rayleigh's quotient, the problem is written as

$$\max_v \min_u \frac{u^T K u}{u^T K_G u}$$
$$\text{such that} \quad c^T v - \theta = 0 \tag{7}$$
$$\text{and} \quad v_{i\,min} \leq v_i \leq v_{i\,max} \quad \text{for } i = 1, \ldots, M,$$

where $v$ is a vector of design variables with components $v_i$, $u$ is the displacement vector, $K$ and $K_G$ (depending on $v$) are the stiffness matrix and the geometric stiffness matrix, respectively, $c$ is a positive cost vector, and $\theta$ is the amount of available resources. The $M$ design variables are subject to upper and lower bounds, $v_{i\,max}$ and $v_{i\,min}$, respectively.

A typical optimization method, applied to solve this problem, starts from a given design and continuously searches for better designs until it finds an optimum design. The trial designs along the path are of no value. The proposed method instead proceeds along a path of optimal designs for increasing amounts of resource $\theta$. The resource $\theta$ is varied between the minimum $\theta_{min}$ required to satisfy the lower bound constraints and a maximum $\theta_{max}$ when all variables are at their upper bounds.

The path consists of several smooth segments, each segment being characterized by a set $I_A$ of variables which are at their upper or lower bounds. Along each segment, some inequality constraints can be treated as equality constraints,

$$v_j = v_{j\,min} \quad \text{or} \quad v_j = v_{j\,max} \quad \text{for } j \in I_A, \tag{8}$$

so that these variables can be eliminated from the optimization problem, while the other variables do not have to be constrained. The optimization problem along a segment can, therefore, be written as

$$\max_{v_i} \min_u \frac{u^T K u}{u^T K_G u} \quad \text{for } i \notin I_A \tag{9}$$

$$\text{such that} \quad c^T v - \theta = 0.$$

The solution of the problem consists of three related problems: solving the optimization problem along a segment, locating the end of the segment where the set $I_A$ changes, and finding the set $I_A$ for the next segment.

It is common practice to normalize the displacement vector u such that the denominator of Rayleigh's quotient is unity and to treat this as an equality constraint. Then, using Lagrange multipliers $\eta$ and $\mu$, the augmented function $P^*$ is formed:

$$P^* = u^T K u - \eta \left[ u^T K_G u - 1 \right] - \mu \left[ c^T v - \theta \right]. \tag{10}$$

The following stationary conditions are obtained by taking the first derivative of $P^*$ with respect to $v_i$, $u$, $\eta$, and $\mu$, and setting it equal to zero:

i) Optimality conditions

$$u^T \frac{\partial K}{\partial v_i} u - \eta u^T \frac{\partial K_G}{\partial v_i} u - \mu c_i = 0 \quad \text{for } i \notin I_A. \tag{11}$$

ii) Stability conditions

$$K u - \eta K_G u = 0. \tag{12}$$

iii) Normalization constraint

$$1 - u^T K_G u = 0. \tag{13}$$

iv) Total resource constraint

$$\theta - c^T v = 0. \tag{14}$$

Equations (11)-(14) form a system of nonlinear equations to be solved for $v_i$, $u$, $\eta$, and $\mu$. A homotopy method is used to find the solutions of these equations as a function of $\theta$.

In certain ranges of structural resources, the optimal solution is known to be bimodal, i.e., the lowest buckling load is a repeated eigenvalue. The formulation for bimodal solutions is given in the appendix of [21]. The existence of bimodal solutions also introduces additional transitions (bimodal to unimodal and vice versa) along the path of optimum solutions.

The homotopy method as described here earlier is intended to solve a *single* nonlinear system of equations, and converge from an arbitrary starting point with probability one. In this context $\theta \in [0, 1]$, and the zero curve $\gamma$ is bounded and leads to the (single) desired solution at $\theta = 1$. The $a$ vector, viewed as an artificial perturbation of the problem, plays a crucial role. In the version of the method employed here, $\theta \in (\theta_0, \theta_1)$, *each* point along $\gamma$ has physical significance, and $a$ is fixed at zero (no perturbation). Because $a$ is not random, the claimed properties for $\gamma$ hold only in subintervals $(\theta_0, \theta_1)$ of $[0, \infty)$. Detecting and dealing with these subinterval transition points is the essence of the modification of the homotopy method used in this section.

**Switching from one segment to the next.** There are four types of events which end a segment and start a new one:

Type 1: a bound constraint becoming active (i.e., being satisfied as an equality);
Type 2: a bound constraint becoming inactive;
Type 3: transition from a unimodal solution to a bimodal solution;
Type 4: transition from a bimodal solution to a unimodal solution.

To switch from one segment to the next, we first need to locate the transition point. At a transition point there are a number of solution paths which satisfy the stationary equations, and we need to choose the optimum path.

Transition points are located by checking the bound constraints and the optimality conditions. The bound constraints

$$v_{i\,min} \leq v_i \leq v_{i\,max} \quad \text{for } i = 1, \ldots, M \tag{15}$$

are checked to detect a transition point of type 1.

Optimality of the solution is checked by the Kuhn-Tucker conditions and the second-order conditions discussed below. The solution satisfies the Kuhn-Tucker conditions when all Lagrange multipliers are nonnegative. So a transition of type 2 is detected by checking the positivity of the Lagrange multipliers associated with the bound constraints. These multipliers are obtained by adding the bound constraints to the formulation (9) and replacing the augmented function $P^*$ by

$$P^* = u^T K u - \eta \left[ u^T K_G u - 1 \right] - \mu \left[ c^T v - \theta \right] - \sum_{i \in I_A} \lambda_{1i} \left[ v_{i\,min} - v_i \right] - \sum_{i \in I_A} \lambda_{2i} \left[ v_i - v_{i\,max} \right]. \tag{16}$$

Taking the first derivative of $P^*$ with respect to $v_i$ gives

$$u^T \frac{\partial K}{\partial v_i} u - \eta u^T \frac{\partial K_G}{\partial v_i} u - \mu c_i + \lambda_{1i} - \lambda_{2i} = 0 \quad \text{for } i \in I_A. \tag{17}$$

Since $\lambda_{1i}$ is 0 for $v_i \neq v_{i\,min}$ and $\lambda_{2i}$ is 0 for $v_i \neq v_{i\,max}$ for the above equations, $\lambda_{1i}$ and $\lambda_{2i}$ are given by

$$\begin{aligned}
\lambda_{1i} &= -u^T \frac{\partial K}{\partial v_i} u + \eta u^T \frac{\partial K_G}{\partial v_i} u + \mu c_i \quad \text{for } v_i = v_{i\,min} \\
\lambda_{2i} &= u^T \frac{\partial K}{\partial v_i} u - \eta u^T \frac{\partial K_G}{\partial v_i} u - \mu c_i \quad \text{for } v_i = v_{i\,max}.
\end{aligned} \tag{18}$$

A type 2 transition is detected by a Lagrange multiplier becoming nonpositive. Similar equations for the bimodal case are given in the appendix of [21].

The bimodal formulation replaces $\eta$ by $\eta_1$ and $\eta_2$ which are the Lagrange multipliers for the normalization constraints on the two buckling modes. When one of them becomes negative, the corresponding mode should be removed for the optimum design, so that we have a transition of type 4 from bimodal to unimodal design.

For a transition of type 3, we need to check if there is another buckling mode associated with a lower buckling load. This can be accomplished by checking the second-order optimality conditions for the buckling mode variables u given by

$$r^T \left[ \nabla_u^2 P^* \right] r > 0 \quad \text{for every } r \text{ such that} \quad \nabla_u h^T r = 0 \tag{19}$$

where

$$[\nabla_u^2 P^*] = \left[\frac{\partial^2 P^*}{\partial u_s \partial u_t}\right]$$

$$\nabla_u h = \left[\frac{\partial h}{\partial u_s}\right] \quad (20)$$

$$h = u^T K_G u - 1.$$

Alternatively we can solve the buckling problem (12) for the current design and check whether the buckling load obtained from the stationary conditions is truly the lowest one. The transition of type 3 is detected by checking if

$$p \neq p_1 \quad (21)$$

where $p$ is the buckling load obtained from the stationary conditions while $p_1$ is the first buckling load obtained by solving the stability conditions (12) for the given structure.

Once a transition point is located, we need to choose a path which satisfies the optimality conditions. Choosing an optimum path constitutes finding a set of active bound constraints for type 1 and 2 transitions and the correct buckling modes for type 3 and 4 transitions. These are obtained by using the Lagrange multipliers of the previous path and the sensitivity calculation on the buckling load. The procedure is explained further in [21].

**Summary.** A typical optimization method starts from a given design and continuously searches for better designs until it finds an optimum design. The trial designs along the path are of no value. Here a strategy for tracing a path of optimum solutions parameterized by an amount of available resources was discussed. Equations for the optimum path were obtained using Lagrange multipliers, and were solved by a homotopy method.

The solution path has several branches due to changes in the active constraint set and transition from unimodal to bimodal solutions. The Lagrange multipliers and the second-order optimality conditions were used to detect branching points and to switch to the optimum solution path.

In [21] this procedure was applied to the design of a foundation which supports a column for maximum buckling load, where the total available foundation was used as a homotopy parameter. Starting from a minimum foundation which satisfies the lower bound (in this case zero), a set of optimum foundation designs was obtained for the full range of total foundation stiffness.

**6. Example of a smooth envelope function for nonlinear constraints.** The two previous sections presented ways to deal with inequality constraints. Both are theoretically "correct" and computationally "practical". However, there are numerous practical difficulties in dealing with them, and the implementation and tuning details become absolutely crucial. For example, with the expanded Lagrangian formulation, line searches may generate negative arguments for the ln functions, and the homotopy zero curve may diverge if the Step 1 solution is not good enough. For the approach in Section 5, the detection and switching criteria for transition points may become extremely cumbersome and inefficient. This section suggests an alternate way of dealing with inequality constraints.

Consider inequality constraints of the form

$$g_i(x) \leq 0, \quad i = 1, \ldots, m, \quad (22)$$

where each $g_i : E^n \to E$ is $C^2$. For a constant $\rho > 0$, the Kreisselmeier-Steinhauser [7] envelope function for (22) is

$$K(x) = \frac{1}{\rho} \ln \left[ \sum_{i=1}^{m} \exp(\rho g_i(x)) \right]. \tag{23}$$

$K(x)$ is a cumulative measure of the satisfaction or violation of the constraints (22). Let $g_{max}(x) = \max\{g_1(x), \ldots, g_m(x)\}$, and observe that

$$K(x) = g_{max}(x) + \frac{1}{\rho} \ln \left[ \sum_{i=1}^{m} \exp\left(\rho(g_i(x) - g_{max}(x))\right) \right], \tag{24}$$

from which it directly follows that

$$g_{max}(x) \leq K(x) \leq g_{max}(x) + \frac{1}{\rho} \ln m. \tag{25}$$

Thus the envelope $K(x)$ follows the maximum constraint, more closely for large $\rho$. In particular, (22) could be replaced by

$$K(x) \leq 0 \tag{26}$$

with an error of no more than $(\ln m)/\rho$.

The choice of $\rho$ involves a tradeoff between modelling the maximum constraint (large $\rho$ preferred) and avoiding large gradients (small $\rho$ preferred). If the practical criterion for an active constraint is $|g_i| \leq \epsilon$, then a choice for $\rho$ which has worked well in practice is

$$\rho = \frac{\ln m}{\epsilon}. \tag{27}$$

Observe that $K(x)$ is $C^2$ and defined *everywhere*, a decided advantage over barrier functions. Furthermore, (26) is a *single* nonlinear constraint, which makes any active set strategy very simple. (26) has been successfully used in large scale structural optimization [2] and optimal control [7].

**7. Probability-one homotopy for Kuhn-Tucker optimality conditions.** Section 6 explained why the approaches of Sections 4 and 5 are not always entirely adequate. The cumulative constraint function (23) is however decidedly unnatural, extremely nonlinear and ill conditioned for large $\rho$, and does not take advantage of a known solution to a related problem. Consider again the general nonlinear programming problem of Section 4:

$$\begin{aligned} \min \;& \theta(x) \\ \text{subject to} \;& g(x) \leq 0, \\ & h(x) = 0, \end{aligned} \tag{28}$$

under the same assumptions mentioned before. The Kuhn-Tucker necessary optimality conditions for (28) are

$$\nabla \theta(x) + \beta^t \nabla h(x) + \mu^t \nabla g(x) = 0,$$
$$h(x) = 0,$$
$$g(x) \leq 0, \qquad (29)$$
$$\mu \geq 0,$$
$$\mu^t g(x) = 0,$$

where $\beta \in E^p$ and $\mu \in E^m$. Following Mangasarian [9] and Watson [25], the complementarity conditions $\mu \geq 0$, $g(x) \leq 0$, $\mu^t g(x) = 0$ are replaced by the equivalent nonlinear system of equations

$$W(x, \mu) = 0, \qquad (30a)$$

where

$$W_i(x, \mu) = -\left|\mu_i + g_i(x)\right|^3 + \mu_i^3 - (g_i(x))^3, \quad i = 1, \ldots, m. \qquad (30b)$$

Thus the optimality conditions (29) take the form

$$F(x, \beta, \mu) = \begin{pmatrix} [\nabla \theta(x) + \beta^t \nabla h(x) + \mu^t \nabla g(x)]^t \\ h(x) \\ W(x, \mu) \end{pmatrix} = 0. \qquad (31)$$

With $z = (x, \beta, \mu)$, the proposed homotopy map is

$$\rho_a(\lambda, z) = \lambda F(z) + (1 - \lambda)(z - a), \qquad (32)$$

where $a \in E^{n+p+m}$. Simple conditions on $\theta$, $g$, and $h$ guaranteeing that the above homotopy map $\rho_a(\lambda, z)$ will work are unknown, although this map has worked very well on some difficult fuel optimal orbital rendezvous problems [23].

Frequently in practice the functions $\theta$, $g$, and $h$ involve a parameter vector $c$, and a solution to (28) is known for some $c = c^{(0)}$. Suppose that the problem under consideration has parameter vector $c = c^{(1)}$. Then

$$c = (1 - \lambda)c^{(0)} + \lambda c^{(1)} \qquad (33)$$

parametrizes $c$ by $\lambda$ and $\theta = \theta(x; c) = \theta(x; c(\lambda))$, $g = g(x; c(\lambda))$, $h = h(x; c(\lambda))$. The optimality conditions in (31) become functions of $\lambda$ as well, $F(\lambda, x, \beta, \mu) = 0$, and

$$\rho_a(\lambda, z) = \lambda F(\lambda, z) + (1 - \lambda)(z - a) \qquad (34)$$

is a highly implicit nonlinear function of $\lambda$. If $F(0, z^{(0)}) = 0$, a good choice for $a$ in practice has been found to be $a = z^{(0)}$. A natural choice for a homotopy would be simply

$$F(\lambda, z) = 0, \qquad (35)$$

since the solution $z^{(0)}$ to $F(0, z) = 0$ (the problem corresponding to $c = c^{(0)}$) is known. However, for various technical reasons, (34) is much better than (35) [23].

The homotopy (34) was used in [23] to solve a fuel-optimal orbital rendezvous problem, and for such optimal control problems appears to be far superior to other known algorithms.

## REFERENCES

[1] E. ALLGOWER AND K. GEORG, *Simplicial and continuation methods for approximating fixed points*, SIAM Rev., 22 (1980), pp. 28–85.

[2] J.-F. M. BARTHELEMY AND M. F. RILEY, *Improved multi-level optimization approach for the design of complex engineering systems*, AIAA J., 26 (1988), pp. 353–360.

[3] D. P. BERTSEKAS, *Constrained Optimization and Lagrange Multiplier Methods*, Academic Press, New York, 1982.

[4] S. C. BILLUPS, *An augmented Jacobian matrix algorithm for tracking homotopy zero curves*, M.S. Thesis, Dept. of Computer Sci., VPI & SU, Blacksburg, VA, Sept., 1985.

[5] S. N. CHOW, J. MALLET-PARET, AND J. A. YORKE, *Finding zeros of maps: Homotopy methods that are constructive with probability one*, Math. Comput., 32 (1978), pp. 887–899.

[6] G. H. ELLIS AND L. T. WATSON, *A parallel algorithm for simple roots of polynomials*, Comput. Math. Appl., 10 (1984), pp. 107–121.

[7] G. KREISSELMEIER AND R. STEINHAUSER, *Systematic control design by optimizing a vector performance index*, Proc. IFAC Symp. on Computer Aided Design of Control Systems, Zurich, Switzerland, (1979), pp. 113–117.

[8] M. KUBICEK, *Dependence of solutions of nonlinear systems on a parameter*, ACM Trans. Math. Software, 2 (1976), pp. 98–107.

[9] O.L. MANGASARIAN, *Equivalence of the complementarity problem to a system of nonlinear equations*, SIAM J. Appl. Math., 31 (1976), pp. 89–92.

[10] R. MEJIA, *CONKUB: A conversational path-follower for systems of nonlinear equations*, J. Comput. Phys., 63 (1986), pp. 67–84.

[11] A. P. MORGAN, *A transformation to avoid solutions at infinity for polynomial systems*, Appl. Math. Comput., 18 (1986), pp. 77–86.

[12] ――――, *A homotopy for solving polynomial systems*, Appl. Math. Comput., 18 (1986), pp. 87–92.

[13] ――――, *Solving polynomial systems using continuation for engineering and scientific problems*, Prentice-Hall, Englewood Cliffs, NJ, 1987.

[14] A. P. MORGAN AND L. T. WATSON, *A globally convergent parallel algorithm for zeros of polynomial systems*, Nonlinear Anal., 13 (1989), pp. 1339–1350.

[15] W. PELZ AND L. T. WATSON, *Message length effects for solving polynomial systems on a hypercube*, Parallel Comput., 10 (1989), pp. 161–176.

[16] A. B. POORE AND D. SORIA, *Continuation algorithms for linear programming*, manuscript.

[17] A. B. POORE AND Q. AL-HASSAN, *The expanded Lagrangian system for constrained optimization problems*, SIAM J. Control Optim., 26 (1988), pp. 417–427.

[18] W. C. RHEINBOLDT AND J. V. BURKARDT, *Algorithm 596: A program for a locally parameterized continuation process*, ACM Trans. Math. Software, 9 (1983), pp. 236-241.

[19] H. SCARF, *The Computation of Economic Equilibria*, Yale University Press, (1973).

[20] L. F. SHAMPINE AND M. K. GORDON, *Computer Solution of Ordinary Differential Equations: The Initial Value Problem*, W. H. Freeman, San Francisco, 1975.

[21] Y. S. SHIN, R. T. HAFTKA, L. T. WATSON, AND R. H. PLAUT, *Tracing structural optima as a function of available resources by a homotopy method*, Comput. Methods Appl. Mech. Engrg., 70 (1988), pp. 151–164.

[22] ———, *Design of laminated plates for maximum buckling load*, J. Composite Materials, 23 (1989), pp. 348–369.

[23] G. VASUDEVAN, L. T. WATSON, AND F. H. LUTZE, *A homotopy approach for solving constrained optimization problems*, Tech. Rep. 88-50, Dept. of Computer Sci., VPI&SU, Blacksburg, VA, 1988.

[24] L. T. WATSON, *A globally convergent algorithm for computing fixed points of $C^2$ maps*, Appl. Math. Comput., 5 (1979), pp. 297–311.

[25] ———, *Solving the nonlinear complementarity problem by a homotopy method*, SIAM J. Control Optim., 17,1 (1979), pp. 36–46.

[26] ———, *Computational experience with the Chow-Yorke algorithm*, Math. Programming, 19 (1980), pp. 92–101.

[27] ———, *An algorithm that is globally convergent with probability one for a class of nonlinear two-point boundary value problems*, SIAM J. Numer. Anal., 16 (1979) 394–401.

[28] ———, *Solving finite difference approximations to nonlinear two-point boundary value problems by a homotopy method*, SIAM J. Sci. Stat. Comput., 1 (1980) 467–480.

[29] L. T. WATSON AND M. R. SCOTT, *Solving spline-collocation approximations to nonlinear two-point boundary-value problems by a homotopy method*, Appl. Math. Comput., 24 (1987) 333–357.

[30] L. T. WATSON AND L. R. SCOTT, *Solving Galerkin approximations to nonlinear two-point boundary value problems by a globally convergent homotopy method*, SIAM J. Sci. Stat. Comput., 8 (1987) 768–789.

[31] L. T. WATSON, *Numerical linear algebra aspects of globally convergent homotopy methods*, Tech. Report TR-85-14, Dept. of Computer Sci., VPI&SU, Blacksburg, VA, 1985, and SIAM Rev., 28 (1986), pp. 529–545.

[32] L. T. WATSON, S. C. BILLUPS, AND A. P. MORGAN, *HOMPACK: A suite of codes for globally convergent homotopy algorithms*, Tech. Rep. 85-34, Dept. of Industrial and Operations Eng., Univ. of Michigan, Ann Arbor, MI, 1985, and ACM Trans. Math. Software, 13 (1987), pp. 281–310.

[33] L. T. WATSON, J. P. BIXLER, AND A. B. POORE, *Continuous homotopies for the linear complementarity problem*, Tech. Report TR-87-38, Dept. of Computer Sci., VPI&SU, Blacksburg, VA, 1987.

[34] L. T. WATSON AND D. FENNER, *Chow-Yorke algorithm for fixed points or zeros of $C^2$ maps*, ACM Trans. Math. Software, 6 (1980), pp. 252–260.

[35] L. T. WATSON AND C. Y. WANG, *A homotopy method applied to elastica problems*, Internat. J. Solids Structures, 17 (1981), pp. 29–37.

[36] L. T. WATSON AND W. H. YANG, *Optimal design by a homotopy method*, Applicable Anal., 10 (1980), pp. 275–284.

# CHAPTER 19

Relation Between the Regularization and the
Multipliers Method for the Min-Max Problem

Cristina Gígola*
Susana Gómez**

**Abstract.** The regularized function is a lower differentiable approximation to the maximum of a finite number of functions. We use in a previous work, this approximation to solve the min-max problem.

We show that the regularized function can be viewed as a type of augmented Lagrangian, where the vector of Lagrange multipliers correspond to what we call the smoothing parameter $v$. We then present different ways to update this parameter using the ideas of the Lagrange multipliers updates.

To incorporate second order information, we solve QPP in the dual space, and we show that these quadratics correspond to using Newton's Method to solve an ordinary Lagrange formulation of the min-max problem.

## 1. Introduction

The finite min-max problem is stated as follows

$$\min ( \varphi(x) \mid x \in \mathbb{R}^n) \tag{1.1}$$

where

$$\varphi(x) = max(f_i(x) \mid i = 1,\ldots,m) \tag{1.2}$$

We will show here that the regularized function proposed in our previous work, Gígola and Gómez (1990 a), to solve this problem, is a type of augmented Lagrangian for problem (1.1), and this allows to compare it with the multiplier method.

We will give here several ways to update the smoothing parameter which can be viewed as the Lagrange multipliers. We will use first and second order information to update this parameter.

---

\* Departamento de Matemáticas, Escuela Superior de Física y Matemáticas, COFAA, IPN, México.

\*\* Instituto de Investigaciones en Matemáticas Aplicadas y en Sistemas, UNAM, México.

## 2. The Regularized Function

In Gígola and Gómez (1990 a) we showed that the function $\varphi(x)$ can also be written

$$\varphi(x) = \sup_{u \in U} u^T f(x) \qquad (2.1)$$

where $f = (f_1, \ldots f_m)^T$ and

$$U = \{u \in \mathbb{R}^m | \sum_{i=1}^{m} u_i = 1, u_i \geq 0\}. \qquad (2.2)$$

This formulation of the problem, allows us to define the regularized function $\varphi_v$, *

$$\varphi_v(x) = \sup_{u \in U} (u^T f(x) - \frac{1}{2}\|u - v\|^2), \quad v \in U. \qquad (2.3)$$

In Gígola and Gómez (1990 a) we propose a method that uses (2.3) to generate a sequence of differentiable subproblems.

For every subproblem, we minimize the function $\varphi_v$ to obtain a new approximation to the solution. Parameter $v$ must be updated in order to have convergence.

Later on, in Sections 7 and 8, we will describe a general form of the method and several update formulae for parameter $v$.

## 3. Properties

We will now consider some important properties of the regularized function $\varphi_v$ and the parameters $u(x, v)$ and $v$. To simplify notation, we will write the arguments of $u(x, v)$, only when it is necessary.

**P1.** The regularized function $\varphi_v(x)$, is a lower bounded approximation of $\varphi(x)$.

$$\varphi(x) - \frac{1}{2}(1 + \|v\|^2) \leq \varphi_v \leq \varphi(x); \qquad (3.1)$$

**P2.** We can always obtain the original function $\varphi(x)$ from the regularized approximation

$$\varphi(x) = \sup_{v \in U} \varphi_v(x); \qquad (3.2)$$

---

* through out this paper $\| \cdot \|$ denotes the Euclidean norm

**P3.** $\varphi(x)$ is differentiable everywhere.

$$\nabla_x \varphi_v(x) = \sum_{i=1}^{m} u_i(x,v) \nabla f_i(x) = J^T(x) u(x,v); \quad (3.3)$$

Where $J(x)$ is the jacobian whose elements are $J_{ij} = \frac{\partial f_i}{\partial x_j}$

**P4.** The value of the vector $u(x,v)$ can be explicitly computed.

$$\text{if } u(x,v) = \arg \sup_{u \in U} \left( u^T f(x) - \frac{1}{2} \|u - v\|^2 \right) \text{ then}$$

$$u(x,v) = (u_1(x,v)), \ldots u_m(x,v)) \text{ where}$$

$$u_i(x,v) = \begin{cases} f_i(x) + v_i - \delta & \text{if } f_i(x) + v_i - \delta > 0 \\ 0 & \text{otherwise} \end{cases} \quad (3.4)$$

and $\delta$ is such that $\sum_{i=1}^{m} u_i(x,v) = 1$

The proofs of properties **P1** to **P4**, are in Gígola and Gómez (1990 a).

Let us define now the following sets,

$$I(x) = \{i = 1, \ldots, m | f_i(x) = \varphi(x)\} \quad (3.5)$$

called the Active Set, and

$$N(x) = \{i = 1, \ldots, m | u_i(x,v) > 0\} \quad (3.6)$$

The variables $u$ and $v$, both in the dual space, satisfy several relations that will be useful later on.

**P5.**

If $j \in I(x)$ then $u_j \geq v_j$

**Proof**

Suppose that $u_j < v_j, j \in I(x)$; now, from the definition of $u(x,v)$, (2.7) $f_j(x) + v_j - \delta \leq u_j$, then $f_j(x) - \delta \leq u_j - v_j < 0$ explicitly and $\varphi(x) - \delta < 0$ which implies $f_i(x) - \delta < 0$ $\forall i = 1, \ldots m$. Then

$$1 = \sum_{i=1}^{m} (f_i(x) + v_i - \delta)^+ = \sum_{i \in N(x)} (f_i(x) + v_i - \delta) < \sum_{i \in N(x)} v_i \leq 1$$

which is a contradiction. Here $(\cdot)^+ = \max(0, \cdot)$

**P6.**

If $u \not\equiv v$, then there exists $i_o \in \{1, \cdots, m\}$ such that $u_{i_o} < v_{i_o}$

**Proof**

Suppose that $u_j \geq v_j \ \forall \ j = 1, \ldots, m$ with at least one $j_o$ such that $u_{j_o} > v_{j_o}$ (otherwise $u \equiv v$). Then $1 = \sum_{i=1}^{m} u_i > \sum_{i=1}^{m} v_i = 1$ which is a contradiction.

**P7.**

If there exists a $j \in I(x)$ such that $j \notin N(x)$, then $u \equiv v$.

**Proof**

From **P5** if $u_j = 0, j \in I(x)$ then $v_j = 0$ and $u_j - v_j = 0$
If $u_i > 0$, then $u_i - v_i = f_i(x) - \delta \leq \varphi(x) - \delta$;
Since $\varphi(x) - \delta \leq u_j - v_j$, then $u_i - v_i \leq u_j - v_j = 0 \ \forall \ u_i > 0$
If $u_i = 0$ then always $0 = u_i \leq v_i$, and we have that $u_i \leq v_i, \ \forall \ i = 1, \ldots, m$;
Then we necessarily have $u \equiv v$.

**P8.**

For each $i \in N(x)$, we have $u_i - v_i \leq u_j - v_j$, for all $j \in I(x)$

**Proof**

The proof follows from P7.

**P9.**

If $j \in I(x)$, then $u_j = f_j(x) + v_j - \delta \geq 0$

**Proof**

If $j \in I(x)$ and $f_j(x) + v_j - \delta < 0$
then $u_j = 0$ and from **P7** $v_j = 0$, which implies $\varphi(x) - \delta < 0$ and we have a contradiction.

**P10.**

We always have that $\varphi(x) \geq \delta$

**Proof**

The proof follows from **P9**.

## 4. Smoothing Effect of the Regularized Function

We will show now in a simple nonconvex example (one dimension) how the regularized function smoothes at points where the original problem is not differentiable.

Let us take

$$f_1(x) = x^3 - x$$
$$f_2(x) = x^2 - k, \qquad k = \frac{2+\sqrt{3}}{3\sqrt{3}} + 1$$

The max function have two minimum points at the same level.

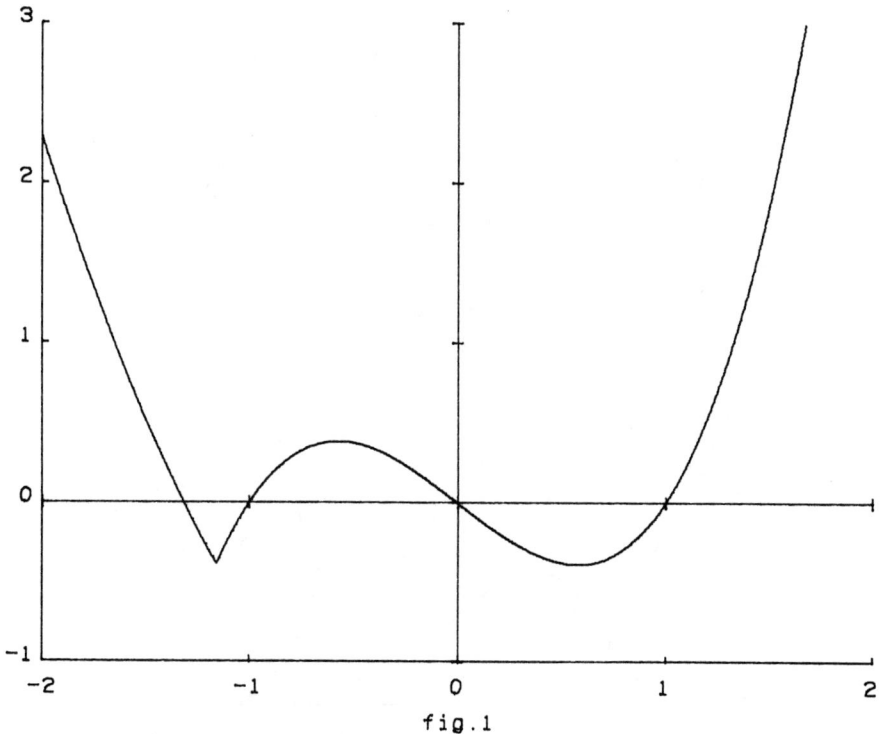

fig.1

Original function $\varphi(x)$.

If we take $v = (Ro, 1 - Ro), Ro \in [0,1]$
then

$$\varphi_v(x) = \begin{cases} x^3 - x - (1 - Ro)^2 & \text{if } x^3 - x^2 - x \geq 2(1 - Ro) - k \\ x^2 - k - Ro^2 & \text{if } x^3 - x^2 - x \leq -2Ro - k \\ \frac{x^3 - x^2 - x + k}{2} + \frac{Ro}{2}(x^3 - x) + (1 - Ro)(x^2 - k) \\ \qquad \text{if } -2Ro - k \leq x^3 - x^2 - x \leq 2(1 - Ro) - k \end{cases}$$

In the following figures we show $\varphi_v$ as an approximation to $\varphi$, for different values of $R_o$.

Note that for $v = (1,0)$ (Fig. 3), we get the optimum $v$ for the minimum at the right, and for $v = (0.43, 057)$ (Fig. 4) we get the optimum $v$ for the minimum at the left.

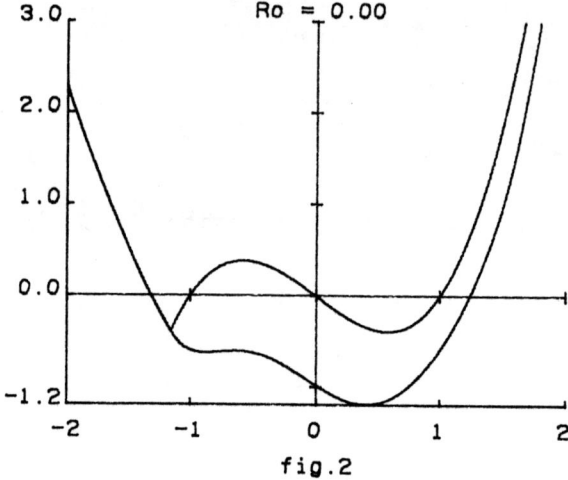

fig.2

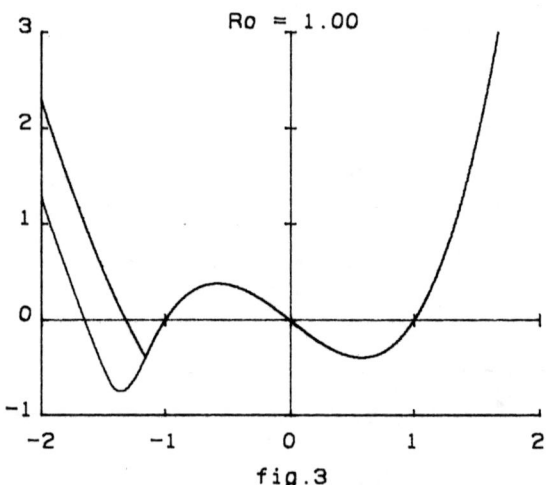

fig.3

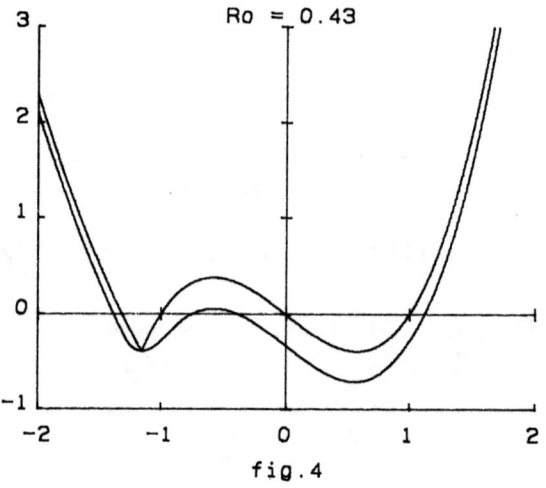

fig.4

## 5. The Regularized Function and The Augmented Lagrangian

We can write problem (1.1) as a general nonlinear programming problem. Adding an extra variable $\delta$, it can be stated as

$$\min_{x,\delta} \quad \delta \tag{5.1}$$

$$\text{s.t.} \quad f_i(x) \leq \delta, \quad i = 1,\ldots,m$$

where $x \in \mathbb{R}^n, \delta \in \mathbb{R}$. In order to have an equality constrained problem, we introduce the slack variables $w_i$. The problem now has the form

$$\min_{x,\delta,w} \quad \delta \tag{5.2}$$

$$\text{s.t.} \quad f_i(x) + w_i^2 - \delta = 0, \quad i = 1,\ldots,m$$

where $w \in \mathbb{R}^m$. The Lagrangian function for problem (5.2) is

$$L_0(x,\delta,w,v) = \delta + \sum_{i=1}^{m} v_i(f_i(x) + w_i^2 - \delta) \tag{5.3}$$

and we can also define the augmented Lagrangian (see Bertsekas (1975))

$$L(x,\delta,w,v,r) = \delta + \sum_{i=1}^{m} v_i(f_i(x) + w_i^2 - \delta) + \frac{r}{2}\sum_{i=1}^{m}(f_i(x) + w_i^2 - \delta)^2. \tag{5.4}$$

But

$$\operatorname*{Inf}_{w} L(x,\delta,w,v,r) = \delta + \sum_{i=1}^{m} \operatorname*{Inf}_{w}\left[v_i(f_i(x) - w_i^2 - \delta) + \frac{r}{2}(f_i(x) + w_i^2 - \delta)^2\right].$$

Where

$$\operatorname*{Inf}_{w}\left[v_i(f_i(x) + w_i^2 - \delta) + \frac{r}{2}(f_i(x) + w_i^2 - \delta)^2\right]$$

is reached at $\quad w_i^* = max(0, -(f_i(x) + \frac{v_i}{r} - \delta))\quad$ and

$$L(x,\delta,w^*,v,r) = \delta + \sum_{i=1}^{m}\left[v_i \max\left(f_i(x) - \delta, -\frac{v_i}{r}\right) + \frac{r}{2}\max^2\left(f_i(x) - \delta, -\frac{v_i}{r}\right)\right]$$

$$= \delta + \sum_{i=1}^{m} \begin{cases} v_i(f_i(x)-\delta) + \frac{r}{2}(f_i(x)-\delta)^2 & \text{if } f_i(x)+\frac{v_i}{r}-\delta \geq 0 \\ -\frac{v_i^2}{2r} & \text{otherwise} \end{cases}$$

$$L(x,\delta,w^*,v,r) = \delta + \sum_{i=1}^{m}\left[v_i \max\left(f_i(x)-\delta,-\frac{v_i}{r}\right) + \frac{r}{2}\max^2\left(f_i(x)-\delta,-\frac{v_i}{r}\right)\right]$$

$$= \delta + \sum_{i=1}^{m}\begin{array}{ll} v_i(f_i(x)-\delta)+\frac{r}{2}(f_i(x)-\delta)^2 & \text{if } f_i(x)+\frac{v_i}{r}-\delta\geq 0 \\ -\frac{v_i^2}{2r} & \text{otherwise} \end{array}$$

See Rockafellar (1973),

$$L(x,\delta,w^*,v,r) = \delta + \frac{r}{2}\sum_{i=1}^{m}\max^2\left(0, f_i(x)+\frac{v_i}{r}-\delta\right) - \left(\frac{v_i}{r}\right)^2 \qquad (5.5)$$

We can also see that the $\operatorname*{Inf}_{\delta} L(x,\delta,w^*,v,r)$ is reached at $\delta = \delta^*$ for which $\sum (f_i(x)+\frac{v_i}{r}-\delta^*)^+ = \frac{1}{r}$.

If we also note that

$$\hat{u}_i = \begin{cases} r(f_i(x)+\frac{v_i}{r}-\delta^*) & \text{if } f_i(x)+\frac{v_i}{r}-\delta^* \geq 0 \\ 0 & \text{otherwise} \end{cases}$$

with $\delta^*$ such that $\sum \hat{u}_i = 1$   then

$$\hat{u} \in U = \{u \in R^m | \sum u_i = 1 \quad u_i \geq 0\} \quad \text{and}$$

$$\operatorname*{Inf}_{\delta,w} L(x,\delta,w,v,r) = \sum_{i=1}^{m}\left[\hat{u}_i\delta^* + \frac{1}{2r}(\hat{u}_i^2 - v_i^2)\right]$$

$$= \sum_{i\in J}\frac{\hat{u}_i}{r}(-\hat{u}_i + v_i + rf_i(x)) + \sum_{i=1}^{m}\frac{1}{2r}(\hat{u}_i^2 - v_i^2)$$

$$= \sum_{i\in J}\hat{u}_i f_i(x) - \frac{1}{2r}\|\hat{u} - v\|^2 \qquad (5.6)$$

$$= \operatorname*{Sup}_{u\in U}\left(u^T f - \frac{1}{2r}\|u-v\|^2\right) = \varphi_{v,r}(x)$$

We now see that the regularized function, that is

$$\varphi_v(x) = \varphi_{v,1}(x) = \operatorname*{Inf}_{\delta,w} L(x,\delta,w,v,1) \qquad (5.7)$$

is the minimum in $\delta$ of the augmented Lagrangian for Problem (5.2) with the penalty parameter $r = 1$.

This result establishes a relationship between the multiplier method, minimizing the augmented Lagrangian (5.4) with $r = 1$, and the regularization method, that minimizes $\varphi_v(x)$. The parameter vector $v$ can be viewed as a Lagrange multiplier.

## 6. The Geometric Interpretation of the Regularized Function and the Penalty Parameter r

In order to see how the penalty parameter $r$, changes the behavior of the regularized function we offer the following figures.

The function is:

$$\varphi_{v,r}(x) = \begin{cases} x^3 - x - \frac{1}{r}(1 - Ro)^2 & \text{if } x^3 - x^2 - x \geq \frac{2(1-Ro)}{r} - k \\ x^2 - k - \frac{Ro^2}{r} & \text{if } x^3 - x^2 - x \leq -\frac{2\,Ro}{r} - k \\ \frac{r}{2}(x^3 - x^2 - x + k) + \frac{Ro}{2}(x^3 - x) + (1 - Ro)(x^2 - k) & \text{otherwise} \end{cases}$$

where the parameter $v$ is fixed at the optimum value for the minimum at the left, $Ro = 0.43$

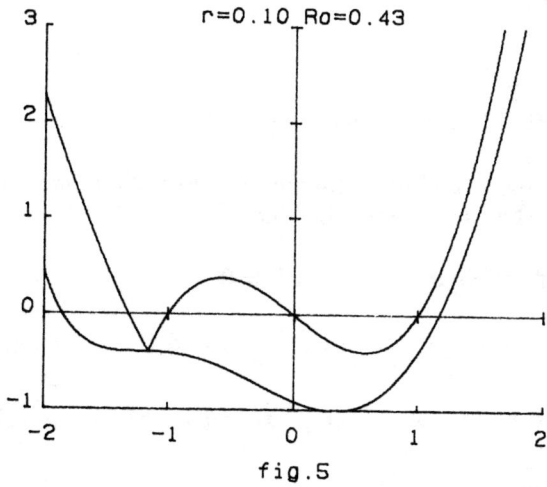
fig.5

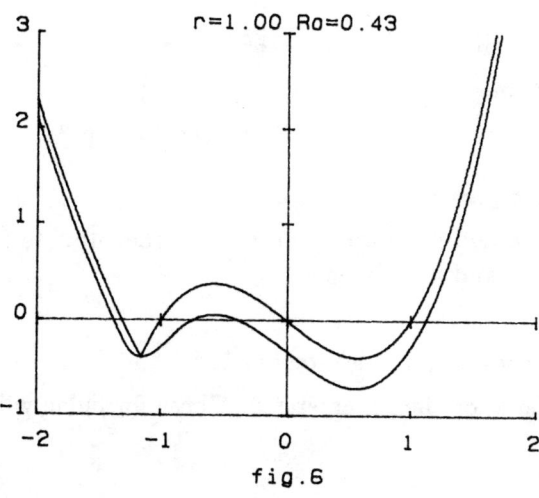
fig.6

From the figures the effect of increasing $r$ should be clear, *i.e* the regularized function gets closer to the original function as $r$ increases.

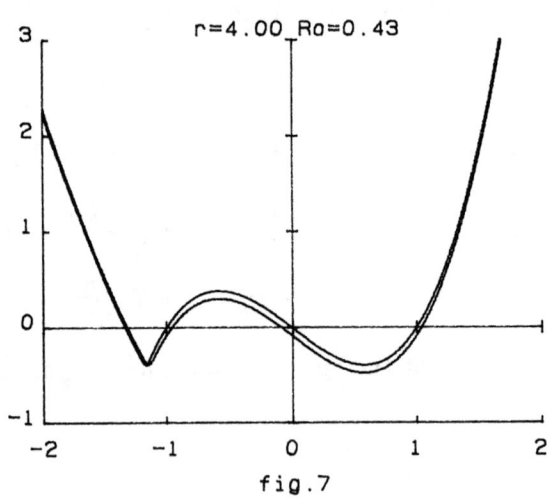

fig.7

## 7. A General Regularization–Multiplier Method

After we have should that $\varphi_v$ is an augmented Lagrangian, we can also described the regularization method as a multiplier method.

0. Let $(x^\circ, v^\circ)$ be an initial pair, $\epsilon > 0$ and $k = 0$

1. At the $k - th$ iteration we have a pair $(x^k, v^k)$ and the $k - th$ element of a decreasing sequence $Gnorm = \|\nabla_x \varphi_v^{k-1}(x)\|$

We then decrease the function $\varphi_{v^k}$, from $x^k$ to get a $x^{k+1}$ that satisfies

$$\|\nabla_x \varphi_{v^k}(x^{k+1})\|^2 \leq Gnorm$$

and go to step 2.

2. If $\|\nabla_x \varphi_{v^k}(x^{k+1})\| \leq \epsilon$ and $\|u(x^{k+1}, v^k) - v^k\|^2 \leq \epsilon$
STOP, exit has been attained.
Otherwise update $Gnorm = \|\nabla_x \varphi_{v^k}(x^{k+1})\|^2$ and go to step 3.

3. Update the smoothing parameter $v$
If $\|u(x^{k+1}, v^k) - v^k\| < \epsilon$ and if we are using a first order update (see (8.3)), then take $v^{k+1} = v^k, k = k + 1$ and go to step 1.
otherwise take
$$v^{k+1} = \Lambda(x^{k+1}, v^k), \quad k = k + 1 \text{ and go to step 1};$$

where $\Lambda$ is an updating formula for the parameter $v$. These formulae will be given in the next section.

Definition. An updating formula for the parameter $v$ is a function

$$\Lambda : \mathbb{R}^n \times \mathbb{R}^m \to \mathbb{R}^m$$

such that if the pair $(x^*, v^*)$ is a stationary pair of problem (2.1), then $v^* = \Lambda(x^*, v^*)$, see Tapia (1977).

We will present now in the following section different updating formula. The algorithm depends strongly on the formula used. Each formula is developed with a different criteria that we will give in every case.

## 8. Update Formulae for Parameter $v$

The function $\varphi_v$ is differentiable with respect the variable $v$, and we have, see Gígola and Gómez (1990 a),

$$\nabla_v \varphi_v(x) = u(x, v) - v \tag{8.1}$$

$$\varphi(x) = \underset{v \in V}{Sup}\ \varphi_v(x) \tag{8.2}$$

these properties suggests updating the parameter $v$ so that the regularized function moves towards the original function (1.2), moving along the gradient direction (8.1). These formulae will then be of first order.

**First Order Updates.** We define first order updates by

$$v(\lambda) = v + \lambda(u(x, v) - v) \tag{8.3}$$

where $\lambda$ is the step length, and will be chosen according to several criteria.

If we take

$$\lambda_1 = 1 \tag{8.4}$$

we have

$$\Lambda(x, v) = u(x, v) \tag{8.5}$$

this update is equivalent to the one proposed by Hestenes (1969) and Powell (1969), in the multipliers method and was also used by the authors (1990 a).

Along the gradient direction $(u(x, v) - v)$, we could find the step length $\lambda_2$ such that

$$\lambda_2 = \arg\left(\min_{\lambda \geq 0} v^T(\lambda)\ J(x)\ J(x)^T v(\lambda)\right) \tag{8.6}$$

where $v(\lambda) = v + \lambda(u(x, v) - v) \in U$, and $J(x)$ is the Jacobian defined in Property P3.

In order to justify the use of the quadratic in (8.6) we need the following proposition

**Proposition 1.** The quadratic function $h_v(\lambda) = v^T(\lambda) J(x) J(x)^T v(\lambda)$ is the norm of the gradient in the variable $x$ of the following regularized function

$$\varphi_{v,\lambda}(x) = \underset{u \in U}{Sup} \left( v^T f(x) - \frac{1}{2\lambda} \|u - v\|^2 \right) \tag{8.7}$$

**Proof.**     Let $\nabla_x \varphi_{v,\lambda}(x) = J^T(x) v(\lambda)$
where

$$v(\lambda) = \arg \underset{\eta}{Sup} \left( \eta^T f(x) - \frac{1}{2\lambda} \|\eta - v\|^2 \right)$$

that is

$$v_i(\lambda) = \begin{cases} \lambda \left( f_i(x) + \frac{v_i}{\lambda} - \delta \right) & \text{if } > 0 \\ 0 & \text{otherwise} \end{cases} \tag{8.8}$$

Now using property **P6**, we can define a $\lambda$ that keeps $v$ in the set $U$, i.e

$$\lambda \max = \min \left( \frac{-v_j}{u_j - v_j} \;\middle|\; u_j - v_j < 0, u_j > 0 \right) \tag{8.9}$$

We easily obtain.

$$v_i(\lambda) = v_i + \lambda(u_i - v_i) \qquad \text{if } 0 \leq \lambda \leq \lambda \max$$

which proves the proposition.

Note that (8.7) is the function obtained in (5.6) with $\lambda$ instead of $r$.

Then the step length parameter $\lambda$ makes the regularization function to get closer the original function. See fig. (2, 3, 4).

In order to find $\lambda_2$, we solve the problem

$$\min h_v(\lambda) = \underset{0 \leq \lambda \leq \lambda \max}{\min} v^T(\lambda) J(x) J^T(x) v(\lambda). \tag{8.10}$$

However, $v(\lambda)$ is updated along $(u - v)$ using (8.3), so omitting the arguments $(x, v)$

$$h_v(\lambda) = \lambda^2 (u - v)^T J J^T (u - v) + 2\lambda v^T J J^T (u - v) + v^T J J^T v;$$

This function has its unconstrained minimum at

$$\lambda_2^* = -\frac{v^T J J^T (u - v)}{(u - v)^T J J^T (u - v)}; \tag{8.11}$$

The updating formula will then be as follow

-If $(u-v)^T JJ^T(u-v) > 0$ then

$$\lambda_2 = \begin{cases} \lambda_2^* & \text{if } v^T JJ^T(u-v) < 0 \text{ and } \lambda_2^* \leq \lambda\max \\ \lambda\max & \text{if } v^T JJ^T(u-v) < 0 \text{ and } \lambda_2^* > \lambda\max \\ 0 & \text{otherwise} \end{cases}$$

-If $(u-v)^T JJ^T(u-v) = 0$ then

$$\lambda_2 = \begin{cases} \lambda\max & \text{if } v^T JJ^T(u-v) < 0 \\ 0 & \text{otherwise} \end{cases}$$

Now because $\nabla_v \varphi_v(x) = u(x,v) - v$, the function $\varphi_v(x)$ increases along $(u(x,v) - v)$. More precisely we have

$$\varphi_v(x) + \frac{1-\lambda^2}{2}\|u-v\|^2 \leq \varphi_{v(\lambda)}(x) \leq \varphi_v(x) + \|u-v\|^2 \quad (8.12)$$

Formula (8.11) was proposed in the context of the multiplier method by Miele et al. (1971).

Another update could be developed when trying to find $v(\lambda)$ so that the norm of the gradient of the regularized function using $v(\lambda)$, has a minimum at $x$. That is, we have to find $\lambda \epsilon [0, \lambda\max]$ such that $\|\nabla_x \varphi_{v(\lambda)}(x)\|^2$ is minimized.

We need for this purpose to have an explicit expression of $u(x, v(\lambda)) = u(\lambda)$. To find this expression, let us suppose that

$$u_i(x) = \begin{cases} f_i(x) + v_i - \delta_o & \text{if } f_i(x) + v_i - \delta_o > 0 \\ 0 & \text{otherwise} \end{cases}$$

If we suppose that $f_1 + v_1 \geq f_2 + v_2 \geq \ldots \geq f_m + v_m$

$$\delta_o = \frac{\sum\limits_{j=1}^{j_o} f_j + v_j - 1}{j_o}$$

where $j_o$ is the last index with $f_{j_o} + v_{j_o} - \delta_o > 0$. Let us call

$$K_{j_o} = \frac{\sum\limits_{j=1}^{j_o}(u_j - v_j)}{j_o} \quad (8.14)$$

Note that $K_{j_o} = 0$ if $u_i(x,v) > 0$ for all $i$. In general $0 \leq K_{j_o} \leq 1$.

In order to be sure that $u(\lambda) \in U$ we need to find the maximum step length allowed. We then have the following result

**Proposition 2.** If

$$\beta \max = \min\{\frac{-u_j}{(u_j - v_j) - K_{j_0}} | u_j > 0, u_j - v_j - K_{j_0} < 0\} \qquad (8.15)$$

$\beta \max$ is the step length that forces $u(\lambda) \in U$. If $0 \leq \lambda \leq \beta \max$, then

$$u_i(\lambda) = u_i(x, v(\lambda)) = \begin{cases} u_i + \lambda[(u_i - v_i) - K_{j_0}] & \text{for } i \in N(x) \\ 0 & \text{otherwise} \end{cases} \qquad (8.16)$$

where the set $N(x)$ is defined in (3.6), and $K_{j_0}$ is as in (8.14).

**Proof.** Let $u_i = u_i(x, v) > 0$ and let us take $\delta(\lambda) = \lambda K_{j_0} + \delta_o$ then

$$u_i + \lambda[(u_i - v_i) - K_{j_0}] = f_i + v_i - \delta_o + \lambda(u_i - v_i) - \lambda K_{j_0}$$

$$= f_i(x) + v_i + \lambda(u_i - v_i) - \delta(\lambda)$$

$$= f_i(x) + v_i(\lambda) - \delta(\lambda) = u_i(x, v(\lambda))$$

We also have

$$\sum_{i \in N(x)} u_i + \lambda[(u_i - v_i) - K_{j_0}] = \sum_{i \in N(x)} u_i + \lambda \left[ \sum_{i \in N(x)} (u_i - v_i) - \sum_{i \in N(x)} K_{j_0} \right]$$

$$= \sum_{i \in N(x)} u_i + \lambda \left( \sum_{i \in N(x)} (u_i - v_i) - j_0 K_{j_0} \right) = \sum_{i \in N(x)} u_i = 1$$

Since the set $N(x)$ has cardinality $j_0$

-If $i \in N(x)$ and $i \in I(x)$ from **P5**, then we have that $u_i > v_i$ and $u_i - v_i \geq K_{j_0}$ then

$$u_i(x, v(\lambda)) = u_i + \lambda[(u_i - v_i) - K_{j_0}] > 0 \text{ for all } \lambda \geq 0.$$

-If $i \in N(x)$ but $i \notin I(x)$,

$$u_i + \lambda[(u_i - v_i) - K_{j_0}] > 0 \qquad for \qquad 0 \leq \lambda \leq \beta \max$$

which implies that $u(\lambda) \in U$. So the proof is completed.

We also note that if $0 \leq \lambda \leq \beta\max, u_i > 0$ then $u_i(\lambda) > 0$, that is

$$N(x) = N_\lambda(x) = \{i | u_i(x, v(\lambda)) > 0\}$$

Also if $\beta\max = \frac{-u_{i_o}}{u_{i_o} - v_{i_o} - K_{j_o}}$ for one $i_o$, then $u_{i_o}(x, v(\lambda)) = 0$, where $u_{i_o}(x, v)$ is a positive multiplier with $i_o \notin I(x)$.

Now we can determine a new updating formula minimizing the following function:

$$h_u(\lambda) = u^T(x, v(\lambda))J(x)J^T(x)u(x, v(\lambda)) \tag{8.17}$$

We note that $u(x, v) - v - K_{j_o} \not\equiv 0$, because otherwise $u_i - v_i = K_{j_o} \geq 0$ for all $i \in N(x)$, then $u_i \geq v_i$ for all $i$, which implies that $u_i = v_i$ for all $i$.

The unconstrained minimum of $h_u(\lambda)$ is attained at

$$\lambda_3^* = -\frac{\eta^T(\lambda)JJ^T u}{\eta^T(\lambda)JJ^T \eta(\lambda)} \tag{8.18}$$

where

$$\eta_i(\lambda) = \begin{cases} u_i - v_i - K_{j_o} & \text{if } i \in N(x) \\ 0 & \text{otherwise} \end{cases}$$

The updating procedure will now be

$$\Lambda_3(x, v) \equiv v(\lambda_3) = v + \lambda_3(u(x, v) - v)$$

and $\lambda_3$ has the following form:

-If $\eta^T(\lambda)J(x)J^T(x)\eta(\lambda) > 0$

$$\lambda_3 = \begin{cases} \lambda_3^* & \text{if } 0 \leq \lambda_3^* \leq \beta\max \\ 0 & \text{if } \lambda_3^* \leq 0 \\ \beta\max & \text{if } \lambda_3* \geq \beta\max \end{cases} \tag{8.19}$$

-If $\eta^T(\lambda)J(x)J^T(x)\eta(\lambda) = 0$

$$\lambda_3 = \begin{cases} \beta\max & \text{if } \eta^T(\lambda)J(x)J^T(x)u < 0 \\ 0 & \text{otherwise} \end{cases}$$

In the following proposition we show how the gradient of $\varphi_v$ in variable $v$ changes when $v$ is updated along the gradient direction $(u(x, v) - v)$.

**Proposition 3.** If $0 \leq \lambda \leq \beta \max$, we have

$$\|u(x,v(\lambda)) - v(\lambda)\|^2 = \|u(x,v) - v\|^2 + \lambda(\lambda - 2)W_{j_o}$$

where $W_{j_o} = [j_o K_{j_o}^2 + \sum_{i \notin N(x)} v_i^2]$.

**Proof.** Substituting $u(\lambda)$ from eq. (8.16) and $v(\lambda)$ from eq. (8.3), and because $u_i(\lambda) = 0$ if $i \in N(x)$ then

$$\|u(\lambda) - v(\lambda)\|^2 = \sum_{i \in N(x)} (u_i + \lambda[(u_i - v_i) - K_{j_o}] - v_i - \lambda(u_i - v_i)) + \sum_{i \notin N(x)} (\lambda - 1)^2 v_i^2$$

with some algebra we finally get

$$\|u(\lambda) - v(\lambda)\|^2 = \|u - v\|^2 + \lambda(\lambda - 2) \sum_{i \notin N(x)} (u_i - v_i)^2 - 2\lambda\, K_{j_o}^2\, j_o + j_o \lambda^2 K_{j_o}^2$$

since

$$\sum_{i \notin N(x)} u_i - v_i = -j_o\, K_{j_o}$$

then

$$\|u(\lambda) - v(\lambda)\|^2 = \|u - v\|^2 + \lambda(\lambda - 2)[j_o\, K_{j_o}^2 + \sum_{i \notin N(x)} v_i^2]$$

which proves the proposition.

Note: If $K_{j_o} = 0$ then

$$\|u(\lambda) - v(\lambda)\|^2 = \|u - v\|^2$$

then the gradient in variable $v$ does not change with $\lambda$.

However $h_v(\lambda)$ and $h_u(\lambda)$ are different, and this can be seen in the following example: for

$$f_1 = x^2$$
$$f_2 = -x^2 + 1$$

at $x = 1$, we have $\lambda \max = 2, \beta \max = 1, K_{j_o} = 0$ and

$$J(x)J^T(x) = 4\begin{bmatrix} 1 & -1 \\ -1 & 1 \end{bmatrix}$$

$$u^T(\lambda)J(x)J^T(x)u(\lambda) = \begin{cases} \frac{\lambda^2}{4} & \text{if } 0 \leq \lambda \leq beta\max = 1 \\ \frac{1}{4} & \text{if } 1 \leq \lambda \leq \lambda\max = 2 \end{cases}$$

$$v^T(\lambda)J(x)J^T(x)v(\lambda) = \tfrac{1}{4}(\lambda-1)^2 \qquad \text{if } 0 \leq \lambda \leq \lambda\max = 2$$

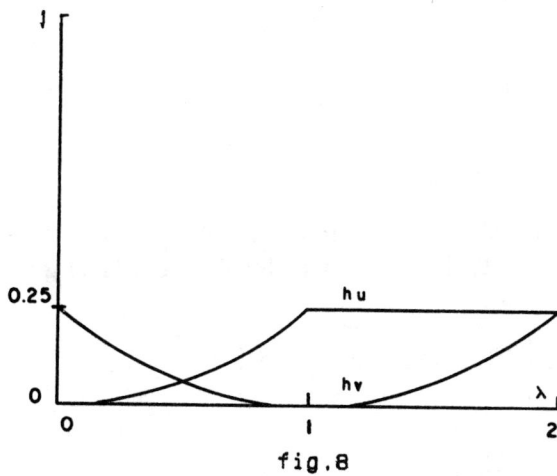

fig.8

## 9. Second Order Updates

In order to use second order information to update $v$ we will take

$$u_q = \Lambda(x,u)$$

where $u_q$ is the solution of the quadratic problem

$$u_q = \arg\left( \sup_{u \in U} \left( u^T f(x) - \frac{1}{2} u^T J(x) H(x) J^T(x) u \right) \right) \qquad (9.1)$$

or

$$u_q = \arg\left( \sup_{u \in U} \left( u^T f(x) - \frac{1}{2} (u-v)^T J(x) H(x) J^T(x) (u-v) \right) \right) \qquad (9.2)$$

where $H(x)$ is a positive definite matrix.

Now in the context of multiplier methods, these update formulae are equivalent to the updates that Tapia present in (1977) for equality constrained problems.

The Kuhn–Tucker conditions for (9.1) can be expressed, taking into account the equality constrained formulation given in (5.2), as follows

$$\begin{cases} J(x)H(x)J^T(x)u_q = f(x) + w^2 - \delta_q \\ \sum_{i=1}^{m} u_{q_i} = 1 \\ u_{q_i} \geq 0 \quad i = 1,..,m \\ w_i^2\, u_{q_i} = 0 \end{cases} \qquad (9.3)$$

with $u_{q_i}$ and $w_i$ being the $i-th$ component of $u_q$ and $w$. These conditions are equivalent to the Kuhn-Tucker conditions for problem (1.1) given in Han (1981):

$$\begin{cases} B(x)p_q = -\sum_{i=1}^{m} u_{q_i}\nabla f_i(x) \\ \sum_{i=1}^{m} u_{q_i} = 1 \\ u_{q_i} \geq 0 \\ u_{q_i}^T(f_i(x) + \nabla f_i^T(x)p_q - \delta_q) = 0 \\ f_i(x) + \nabla f_i^T(x)p_q \leq \delta_q \quad i = 1,\ldots,m \end{cases} \qquad (9.4)$$

with $B = H^{-1}$.

The solution of problem (9.3) for $u_q, w, \delta$ is equivalent to solving the system (omitting the argument $x$)

$$(J\,H\,J^T + \mathbf{1}_{m \times m} + D_{4w^2})\ u_q = \mathbf{1}_{m \times 1} + (f + w^2 - \delta_q) \qquad (9.5)$$

where $\mathbf{1}_{m \times 1}$ is a m–vector of ones, $\mathbf{1}_{m \times m}$ is a matrix of ones, $D_{4w^2}$ is a diagonal matrix with elements $4w_i^2$.

Now (9.5) can be also obtained by applying Newton's Method to solve problem

$$\begin{aligned} \nabla_{x,\delta,w,v}\ L(x,\delta,w,v) &= 0 \\ f(x) + w^2 - \delta &= 0 \end{aligned} \qquad (9.6)$$

where $L(x,\delta,w,v)$ is the ordinary Lagrangian defined in (5.3).

To be able to work with (9.6) we need to define the matrix

$$J_1(x,w) = \begin{bmatrix} J(x) \\ 1_{m \times 1} \\ D_{2w} \end{bmatrix} \qquad (9.7)$$

and the matrix

$$H_1 = \begin{bmatrix} H & 0 & 0 & J_1^T \\ 0 & 1 & 0 & \\ 0 & 0 & I & \\ J_1^T & & & 0 \end{bmatrix} \qquad (9.8)$$

To take a Newton step we solve the following system,

$$H_1 \begin{bmatrix} \Delta x \\ \Delta \delta \\ \Delta w \\ \Delta v \end{bmatrix} = - \begin{bmatrix} \nabla_x L(x,\delta,w,v) \\ \nabla_\delta L(x,\delta,w,v) \\ \nabla_w L(x,\delta,w,v) \\ \nabla_v L(x,\delta,w,v) \end{bmatrix} \qquad (9.9)$$

Expressing the solution of (9.9) in terms of $\Delta v$, we get (9.5). Then $u_q$ is the solution of (9.5) for $x^*, w^*, \delta^*$ where the ordinary Lagrangian (5.3) has a critical point.

When solving the quadratic (9.2) instead of the matrix $H_1$ we now use

$$H_2 = \begin{bmatrix} H & 0 & 0 & 0 \\ 0 & 1 & 0 & \\ 0 & 0 & I & \\ 0 & & & -JHJ^T \end{bmatrix} \qquad (9.10)$$

and in terms of $\Delta v$, we get the system

$$(JHJ^T)u_q = (f + w^2 - \delta) + JHJ^T v \qquad (9.11)$$

Note: Following these ideas we can also see the first order update $v = u(x, v)$ as the solution of the quadratic

$$u = \arg(\sup_{u \in U} (u^T f(x) - \frac{1}{2}(u-v)^T I(u-v))) \tag{9.12}$$

and it corresponds to solving a system of equations using $H_2$ with the identity on the right bottom.

## 10. Conclusions

We have shown in this work, that the regularized function can be viewed as a type of augmented Lagrangian, where the vector of Lagrange multipliers correspond to what we call the smoothing parameter $v$. We then presented different ways to update this parameter using first and second order information.

We have shown that solving the Quadratic Programming Problem (9.1) is equivalent to using Newton's method to solve (9.6). When we use the augmented Lagrangian to find the equivalent quadratic problem, we found that we get the same updates than when using the ordinary Lagrangian for the cases (9.2) and (9.12). These updates are then equivalent to Buys update (1972), and the Hestenes and Powell (1969). However we have been unable so far to find the associated quadratic resulting when applying Newton's Method to the augmented Lagrangian formulation.

With this second order information obtained through the dual quadratic problem, we will present two ways to incorporate this information into what we call second order methods to solve the finite min–max problem. These methods and the numerical results obtained, will be given in another paper in this volume, Gígola and Gómez (1990 b).

A future paper with convergence results for these methods is now in preparation.

Because of the similarity between the way we approximate the original non-differentiable min–max problem, and the multiplier methods applied to an inequality constrained nonlinear programming problem, we think that our results could help to define similar updates when we apply the multiplier method to the general inequality constrained problem.

## REFERENCES

Bertsekas D.P. (1975), "Combined primal-dual and penalty methods for constrained minimization" SIAM J. Control Vol. 13, No. 3 pp. 521-544.

Buys. L.D., (1972), "Dual algorithms for constrained optimization", Rijks Universiteit de Leiden, The Netherlands, Ph. D. Thesis.

Gígola C. and Gómez, S., (1990), "A regularization method for solving the finite convex min–max problem". To appear in SIAM Journal on Numerical Analysis.

Gígola C. and Gómez S., (1990), "Two second order regularization methods to solve the finite min–max problem". Proceedings of the Fifth Mexico–United States Numerical Analysis Workshop, SIAM.

Gígola C. and Gómez S. "Convergence results for the regularization methods for the min–max problem", In preparation.

Han. S.P., (1981), "Variable metric methods for minimizing a class of non differentiable functions" Math. Programming 20, pp. 1-13.

Hestenes, M. R., (1969) "Multiplier and gradient methods" Journal of Opt. Theory and Appl. Vol. 4, pp. 303-320.

Miele A., Cragg E. and Levy, A.V., (1971), "Use of the augmented penalty function in mathematical programming problems", Part I, Journal of Opt. Theory and Appl. Vol. 8, pp. 115-130.

Powell, M.J.D.,(1969), "A method for nonlinear constraints in minimization problems", Optimization, Edited by R. Fletcher, Academic Press, London, England.

Rockafellar R.T., (1973). "A dual approach to solving nonlinear programming problems by unconstrained optimization" Math. Programming, 5, pp. 354-373.

Tapia R.A.,(1977), "Diagonalized multiplier methods and quasi–newton methods for constrained optimization" J. Optim. Theory Appl. No. 22, pp. 135-194.

# CHAPTER 20

Two Second Order Regularization Methods
to Solve the Finite Min-Max Problem

Cristina Gígola*
Susana Gómez**

**Abstract.** We want to solve the problem so called finite min–max problem, which is not differentiable.

In a previous work, we propose an approximation to the original problem which makes it differentiable; that is why we call this approximation the regularized function.

We present here several ways to generate second order methods using the regularized function as the merit function. The smoothing parameter used in the regularization, is updated solving QPP.

Numerical results with these methods and a comparison with Han's method are presented.

## 1. Introduction

The problem we want to solve is a non differentiable optimization problem, and can be stated as follows

$$\min_{x} \ \max \ (f_i(x) \mid i = 1, \ldots m) = \min_{x} \varphi(x) \tag{1.1}$$

where $x \in \mathbb{R}^n$.

In a previous paper Gígola and Gómez (1990 a) we proposed a first order method to solve problem (1.1). We showed in another paper in this volume, Gígola and Gómez (1990 b) that this can be viewed as a multiplier method.

We want to present here, how to use second order information on the dual variables presented in Gígola and Gómez (1990 b), in order to generate second order methods to solve problem (1.1).

We will present numerical results for these methods and compare them to the results obtained with our implementation of Han's method (1981).

---

\* Departamento de Matemáticas, Escuela Superior de Física y Matemáticas, y COFAA, IPN, México.

\*\* Instituto de Investigaciones en Matemáticas Aplicadas y en Sistemas, UNAM, México.

## 2. The Regularization Technique

One way of solving the problem (1.1) is by approximating the max function $\varphi(x)$, which is not differentiable everywhere, by the differentiable function, (see Gígola and Gómez (1990 a)),

$$\varphi_v(x) = \underset{u \in U}{Sup}(u^T f(x) - \frac{1}{2}\|u - v\|^2), \qquad v \in U \tag{2.1}$$

where $U = \{u \in \mathbb{R}^m | \sum u_i = 1, u_i \geq 0\}$ and $f = (f_1 \ldots f_m)^T$. We call $\varphi_v$ a Regularized function for $\varphi$.

In Gígola and Gómez (1990 a, b) we suggest minimizing $\varphi_v$ in order to approximate the optimum of $\varphi$.

## 3. General Regularization Second Order Methods

The method we are presenting here for solving the min-max problem (1.1), has the structure of an SQP method.

For a given pair $(x^k, v^k)$ we minimize function $\varphi_v(x^k)$, defined in (2.2), by generating a sequence

$$x^{k+1} = x^k + \alpha_k p^k \tag{3.1}$$

where $p^k$ is a descent direction generated using the formula

$$p^k = -H(x)J^T(x)\eta^k, \qquad \eta^k \in U \tag{3.2}$$

Here $J(x)$ is the Jacobian matrix whose columns are the gradients $\nabla f_i(x)$, and $H(x)$ is an approximation of the Hessian $H(x) = \sum_{i=1}^{m} u_i \nabla^2 f_i(x)$. The step length $\alpha_k$ is found using a one dimensional search. We use quadratic interpolation until the following descent condition is satisfied

$$\varphi_v(x + \alpha p) \leq \varphi_v(x) + \alpha \omega p^T \nabla \varphi_v(x). \tag{3.3}$$

In this work we have used $\omega = 0.1$.

Then we update $v^{k+1} = \Lambda(x^{k+1}, v^k)$ where $\Lambda(x, v)$ is an updating formula (see Gígola and Gómez (1990 b).

We obtain different results, depending on how we update $v^k$ and how we choose $\eta^k$. We have to prove then, that for every $\eta^k, p^k$ is a descent direction for $\varphi_{v^k}$.

In order to incorporate second order information to update $v^k$ and $\eta^k$, we will solve the following quadratic programming problem

$$u_{q^k} = \arg ( Sup(u^T f(x^k) - \frac{1}{2} u^T J(x^k) H(x^k) J^T(x^k) u)) \tag{3.4}$$

which satisfies the Kuhn-Tucker conditions

$$\begin{cases} H^{-1}(x)p + \sum_{i=1}^{m} u_i \nabla f_i(x) = 0 \\ \\ \sum_{i=1}^{m} u_i = 1 \\ \\ u_i \geq 0 \quad i = 1, \ldots m \\ \\ f_i(x) + \nabla^T f_i(x)p - \delta_q \leq 0 \quad i = 1, \ldots m \\ \\ u_i(f_i(x) + \nabla^T f_i(x)p - \delta_q) = 0 \quad i = 1, \ldots m \end{cases} \quad (3.5)$$

where $p = -H(x)J^T(x)u_q$.

**Method 1.**

We can take the updates

$$\eta^k = u_{q^k} \quad (3.6)$$

$$v^k = u_{q^k}$$

**Proposition 3.** Given $x \in \mathbb{R}^n$, the direction

$$p_q = -H(x)J^T(x)u_q \quad (3.7)$$

is a descent direction for the regularized function $\varphi_{u_q}$ and also for the original function $\varphi$.

**Proof.** The gradient in $x$ of $\varphi_{u_q}(x)$ is

$$\nabla_x \varphi_{u_q}(x) = J^T(x)u(x, u_q)$$

and the directional derivative has the form

$$\nabla_x \varphi_{u_q}^T(x) p_q = -u^T(x, u_q) J(x) H(x) J^T(x) u_q \quad (3.8)$$

Now omitting the arguments of $J(x), H(x), B(x)$ and observing that

$$Bp_q = -J^T u_q, \quad B = H^{-1} \quad (3.9)$$

we have
$$p_q^T B p_q = u_q^T J H J^T u_q \tag{3.11}$$
and
$$\begin{aligned}\nabla_x \varphi_{u_q}^T(x) p_q &= -p_q^T B p_q + u_q^T J H J^T u_q - u^T(x, u_q) J H J^T u_q \\ &= -p_q^T B p_q + (u(x, u_q) - u_q)^T J p_q\end{aligned}$$

From the Kuhn–Tucker conditions (3.5) we see that
$$\nabla^T f_i(x) p_q \leq \delta_q - f_i(x) \qquad \forall i \tag{3.10}$$
with equality if $u_{q_i} > 0$. Hence
$$\nabla_x \varphi_{u_q}^T(x) p_q \leq -p_q^T B p_q + \sum_{i=1}^m (u_i(x, u_q) - u_{q_i})(\delta_q - f_i(x)). \tag{3.12}$$
Also from the definition of the regularized function
$$\varphi_{u_q}(x) = u^T(x, u_q) f(x) - \frac{1}{2}\|u(x, u_q) - u_q\|^2 \geq u_q^T f(x). \tag{3.13}$$
so
$$(u(x, u_q) - u_q)^T f(x) \geq \frac{1}{2}\|u(x, u_q) - u_q\|^2 \tag{3.14}$$
and we also have
$$\nabla_x \varphi_{u_q}^T(x) p_q^T \leq -p_q^T B p_q - \frac{1}{2}\|u(x, u_q) - u_q\|^2 < 0. \tag{3.15}$$
Hence we have a descent direction for $\varphi_{u_q}(x)$.

Now in order to see that $p_q$ is also a descent direction for the original max function $\varphi(x)$, (1.1), we have that the directional derivative along direction $p_q$ is defined by
$$\varphi'(x; p_q) = \max_{i \in I(x)} (\nabla f_i^T(x) p_q) \tag{3.16}$$
where
$$I(x) = \{i | \varphi(x) = f_i(x)\}$$
(see Han (1981)). From (3.11)
$$\varphi'(x; p_q) \leq \max_{i \in I(x)} (\delta_q - f_i(x)) = \delta_q - \varphi(x) \tag{3.17}$$

But

$$-p_q^T B p_q = \sum_{i=1}^m u_{q_i} \nabla f_i^T(x) p_q = \sum_{u_{q_i}>0} u_{q_i} \nabla f_i^T(x) p_q =$$
$$= \sum_{u_{q_i}>0} u_{q_i}(\delta_q - f_i(x)) \geq \delta_q - \varphi(x) \tag{3.18}$$

which shows that

$$\varphi'(x; p_q) \leq -p_q^T B p_q \tag{3.19}$$

That is, $p_q$ is a descent direction for $\varphi(x)$.

If in the minimization phase we use $\varphi$ as the merit function we get Han's method (1981). If we use $\varphi_{u_q}$ as the merit function the method seems to be more like the diagonalized multiplier method of Tapia (1977), (see Fontecilla (1988)) applied to a general constrained problem.

**Method 2.**

Now if we take

$$\eta^k = u(x^k, u_q^k) \tag{3.20}$$

$$v^{k+1} = u_q^k$$

we have that the direction

$$p_u = -H(x) J^T(x) u(x, u_q) \tag{3.21}$$

is a descent direction for $\varphi_{u_q}(x)$ but not necessarily for $\varphi(x)$. This is so because

$$\nabla_x \varphi_{u_q}^T(x) p_u = -u(x, u_q) J(x) H(x) J^T(x) u(x, u_q) = -p_u^T B\, p_u \tag{3.22}$$

For the first order method with $v(\lambda) = v + \lambda_3(u(x,v) - v)$ given in Gígola and Gómez (1990 b), we can show in a similar way as we did in Proposition 3, that $p = -H\, J^T v(\lambda_3)$ is a descent direction for $\varphi_{v(\lambda_3)}(x)$ and $\varphi(x)$.

From the point of view of convergence, the following Proposition will allow to prove convergence only showing that the norm of the gradient of $\varphi_v$ at $x^k$ tends to zero. It shows that the norm of the gradient of $\varphi_v$ in variable v, given by (u(x,v) - v) is less or equal than the H-norm of the gradient in variable x. This is true for $v = u_q$ or $v = v(\lambda_3)$.

## Proposition 4.

$$\|u(x,v) - v\|^2 \leq u(x,v) J(x) \ H(x) \ J^T(x) u(x,v) - v^T J(x) \ H(x) \ J^T(x) v$$

**Proof**

By definition of the regularized function $\varphi_v(x)$, and $v = u_q$ or $v = v(\lambda_3)$, we have that, (omitting the argument $x$ in $J$ and $H$).

$$\varphi_v(x) = u^T(x,v) f(x) - \frac{1}{2}\|u(x,v) - v\|^2 \geq v^T f(x)$$
$$\geq v^T f(x) - \frac{1}{2} v^T J \ H \ J^T v$$
$$\geq u^T(x,v) f(x) - \frac{1}{2} u^T(x,v) J \ H \ J^T u(x,v)$$

that is

$$(u(x,v) - v)^T f(x) \leq \frac{1}{2}(u^T(x,v) J \ H \ J^T u(x,v)) - \frac{1}{2} v^T J \ H \ J^T v$$

and

$$(u(x,v) - v)^T f(x) \geq \frac{1}{2}\|u(x,v) - v\|^2$$

which proves the proposition.

## Method 3.

There is another way to incorporate the second order information using $u_q$. Suppose we move in a Newton fashion on both variables $x$ and $v$, using for $x$ the direction $p_q = -HJ^T u_q$ and for variable $v$ a direction $\rho = u_q - v$.

Now in order to have a general direction

$$d = \begin{bmatrix} p_q \\ \rho \end{bmatrix} \tag{3.23}$$

being a descent direction for the regularized function $\varphi_v(x)$, we need to use a penalty parameter in the definition of the regularized function. Let us then define

$$\varphi_r(x,v) = \underset{u \in U}{\text{Sup}} \left( u^T f(x) - \frac{1}{2r}\|u - v\|^2 \right) \tag{3.24}$$

In this context we think that the proper notation yields $x$ and $v$ as variables and not as before taking $v$ as a fixed parameter. We can show as in Proposition 3, that $p_q$ is a descent direction for $\varphi_r(x,v)$ in the variable $x$ for any r.

Now for the general directional derivative in both variables, we have the result that

$$\nabla_{x,v}\varphi_r^T(x,v)d \leq -p_q^T B p_q + \frac{1}{r}(u_q - v)^T(u(x,v) - v) \qquad (3.25)$$

and we can determine $r$ so that $d$ is a descent direction for both variables in $\varphi_r(x,v)$.
We then obtain

$$r \geq \frac{(u_q - v)^T(u(x,v) - v)}{p_q^T B p_q} \qquad (3.26)$$

Following Schittkowski's ideas, (1981), we use $\varphi_r(x,v)$ as the merit function and the method will minimize $\varphi_r(x,v)$ using $d$ as a descent direction, moving at the same time in both variables $x, v$

$$\begin{pmatrix} x^{k+1} \\ v^{k+1} \end{pmatrix} = \begin{pmatrix} x^k \\ v^k \end{pmatrix} + \lambda \begin{bmatrix} p_q \\ u_q - v \end{bmatrix} \qquad (3.27)$$

for $0 \leq \lambda \leq \lambda \max$.

Convergence results for these methods will be given in a future paper Gígola and Gómez (In preparation).

## 4. Implementation

In this Section we will first describe, several variants of the methods introduced in Section 2.

We will call Method 1, M1, when the variable $x$ is the following

$$p = -H(x)J^T(x)u_q \qquad (4.1)$$

In Method 2, M2, we use the direction.

$$p = -H(x)J^T(x)u(x, uq) \qquad (4.2)$$

Method 3, M3, is the resulting method when taking $H(x) = I$ that is, we take

$$p = -J^T(x)u_{q_I} \qquad (4.3)$$

where $u_{q_I}$ is obtained when solving the quadratic

$$u_{q_I} = \arg\left(\sup_{u \in U}(u^T f(x) - \frac{1}{2}u^T J(x)J^T(x)u)\right) \qquad (4.4)$$

We decided to report the results obtained when using M3, because they are much better than the results obtained using first order updates, (see Gígola and Gómez (1990 a)). We will discuss this fact later on.

For these last methods M1, M2 and M3, we move $x$ using $p$ as a descent direction. Then we update the smoothing parameter $v = u_q$, where

$$u_q = \arg \left( \underset{u \in U}{Sup}(u^T f(x) - \frac{1}{2} u^T J(x) H(x) J^T(x) u) \right) \qquad (4.5)$$

for methods M1 and M2, and if $H(x) \equiv I$ for method M3.

We call Method 4, M4, the method that we obtain when moving along $x$ and $v$ at the same time using $d = \begin{bmatrix} p_q \\ \rho \end{bmatrix}$ as the general descent direction as described in Section 2, and

$$\begin{aligned} p &= -H(x) J^T(x) u_q \\ \rho &= u_q - v \end{aligned} \qquad (4.6)$$

In order to solve the quadratic programming problem (4.4) and (4.5), we make a special modification of the algorithm due to Lemke, (see Ravindran (1972)). This special modification is due to the fact that in our problem we only have one linear equality constraint

$$\sum_{i=1}^{m} u_i = 1$$

We can then reformulate the quadratic problem and solve it in a reduced space $\mathbb{R}^{m-1}$. We describe it as follows:

Let the general quadratic program be,

$$\begin{aligned} \min\ & c^T x + x^T Q x \\ s.t.\ & Ax \geq b \\ & x \geq 0 \end{aligned}$$

where A is an (m x n) matrix, Q is an (n x n) matrix, c is an (n x 1) vector, and $b$ is an (m x 1) vector.

An optimum may be obtained by solving the linear complementary problem: find $w$ and $z$ such that

$$\begin{cases} w = Mz + q \\ w, z \geq 0 \\ w^T z = 0 \end{cases} \qquad (4.7)$$

where

$$M = \begin{bmatrix} 2Q & -A^T \\ A & 0 \end{bmatrix} \text{ and } q = \begin{bmatrix} c \\ -b \end{bmatrix}$$

In our case we have a quadratic programming problem of the form:

$$\min d^T u + \frac{1}{2} u^T D u$$

$$\sum_{i=1}^{m} u_i = 1$$

$$u_i \geq 0 \quad i = 1, \ldots, m$$

where D is a (m x m) matrix, $D = J(x) H(x) J^T(x)$ and d is a (m x 1) vector, $d = -f(x)$.

In order to profit from the special form of the constraints (a single equality constraint), we may solve the equivalent quadratic problem

$$\min \hat{d} u + \frac{1}{2} u \hat{D} u$$

$$\sum_{i=1}^{m-1} u_i \leq 1$$

$$u_i \geq 0 \quad i = 1, \ldots m-1$$

where

$\hat{D}$ is a $(m-1) \times (m-1)$ matrix such that $\hat{D}ij = Dij - Dim - Dmj + Dmm$ $i, j = 1, \ldots m-1$, $\hat{d}$ is a $(m-1)$ vector such that $\hat{d}i = dm - di + 2(Dim - Dmm)$ $i = 1, \ldots, m-1$ and $y_m = 1 - \sum_{i=1}^{m} y_i$

by solving a linear complementary problem of the form (4.7) where

$$M = \begin{bmatrix} \hat{D} & \mathbf{1}_m \\ -\mathbf{1}_m & 0 \end{bmatrix}$$

is a (m x m) matrix, $\mathbf{1}_m$ a (m x 1) column vector of ones, and

$$q = \begin{bmatrix} \hat{d} \\ 1 \end{bmatrix}$$

is a (m x 1) column vector.

For the line search, which can be a difficult problem when solving nonconvex problems, we used the line search that Han (1981) suggests. However we have been working in adapting the line search proposed by Pierre and Lowe (1975).

## 5. Numerical Results

We will present here, the numerical results obtained when using methods M1, M2, M3 and M4, described in the last section, on a set of convex and nonconvex problems.

The problems we are solving are the following,

Problem 1.    (Demyanov and Malozemov (1971))
$$f_1 = -5x_1 + x_2$$
$$f_2 = 4x_2 + x_1^2 + x_2^2$$
$$f_3 = 5x_1 + x_2$$

Problem 2 .    (Charalambous and Conn (1978))
$$f_1 = x_1^2 + x_2^4$$
$$f_2 = (2 - x_1)^2 + (2 - x^2)^2$$
$$f_3 = 2e^{-x_1 + x_2}$$

Problem 3.    (Charalambous and Conn (1978))
$$f_1 = x_1^4 + x_2^2$$
$$f_2 = (2 - x_1)^2 + (2 - x_2)^2$$
$$f_3 = 2e^{-x_1 + x_2}$$

The following are nonconvex problems that we introduce in this paper

Problem 4.
$$f_1 = \sin x_1 + x_1 + \sin x_2 + x_2$$
$$f_2 = \cos x_1 + \cos x_2$$
$$f_3 = x_1^2 + x_2^2 - 10$$

Problem 5.
$$f_1 = e^{\sqrt{x_1^2 + x_2^2}} \sin\left(\sqrt{x_1^2 + x_2^2}\right)$$
$$f_2 = (x_1^2 + x_2^2)/12$$
$$f_3 = e^{\sqrt{x_1^2 + x_2^2}} \cos\left(\sqrt{x_1^2 + x_2^2}\right)$$

We compare the results obtained when solving these problems with the methods M1, M2, M3 and M4 described in the last section and the method proposed by Han (1981).

We ran the experiments using a UNYSIS A-12, with TOL $=10^{-8}$ as the stopping tolerance.

All problems have three functions $m = 3$, and two variables $n = 2$. We have used $(2n + 1)^n$ equidistant initial points in the square $[-5, 5] \times [-5, 5]$

For the nonconvex problems convergence to a Kuhn-Tucker point is attained.

We report in the following tables the average function evaluations and the average number of times that we solve quadratic problems.

| METHOD/PROBLEM | 1 | 2 | 3 | 4 | 5 |
|---|---|---|---|---|---|
| M1  | 6.08 | 8.33  | 7.87  | 8     | 5.33 |
| M2  | 7.02 | 58.20 | 40.16 | 358*  | 11.8 |
| M3  | 6.75 | 12.00 | 8.66  | 6.29  | 8.33 |
| M4  | 6.12 | 9.29  | 7.87  | 8     | 5.33 |
| Han | 6.08 | 8.37  | 7.87  | 8.20* | 5.33 |

Table 1. Average function evaluations

| METHOD/PROBLEM | 1 | 2 | 3 | 4 | 5 |
|---|---|---|---|---|---|
| M1  | 15.50 | 28.75  | 22.58 | 21.18   | 19.33 |
| M2  | 68.04 | 106.58 | 50.70 | 381.88* | 29.50 |
| M3  | 17.95 | 33.58  | 24.04 | 20.20   | 29.00 |
| M4  | 15.79 | 29.75  | 22.75 | 21.12   | 19.33 |
| Han | 15.50 | 29.58  | 22.58 | 21.79   | 19.33 |

Table 2. Average number of quadratic problems solved

At points where $H$ is not positive definite, we use the Identity to generate a descent direction.

## 6. Conclusions

We have presented several ways to generate second order methods to solve the finite min max problem, using the regularized function as the merit function.

From the numerical results we conclude that these methods are fast, robust and easy to implement.

It is interesting to see how well works the method M3 with the identity used instead of the Hessian, specially for the nonconvex problems. We might conclude from these results with method M3, that having a good approximation to the multipliers (in our case the smoothing parameter), is so important that in these problems the curvature information seems less relevant.

These methods can be used to solve large problems, although we have only solved here low dimensional problems.

Convergence results for these methods is the subject of ongoing research.

# REFERENCES

Charalambous C. and Conn A. (1978), "An efficient method to solve min-max problem directly". SIAM Num. Anal. 15, No. 1, pp. 162-187.

Demyanov V. F. and Malozemov V. N. (1971), "On the theory of non-linear min-max problems" Russian Math. Surveys 26, No. 3, pp. 57-115.

Fontecilla R. (1988), "Local convergence of secant method for non linear constrained optimization" SIAM J. Num. Anal. Vol. 25, No. 3, pp. 692-712.

Gígola C. and Gómez, S. (1990 a), "A regularization method for solving the finite convex min-max problem". To appear in SIAM Journal on Numerical Analysis.

Gígola C. and Gómez S. (1990 b), "Relation between the regularization and the multiplier methods for the min-max problem" Proceeding of the Fifth IIMAS Numerical Analysis Workshop SIAM.

Gígola C. and Gómez S. "Convergence results of regularization methods for the min-max problem" In preparation.

Han, S.P. (1981), "Variable metric method for minimizing a class of non-differentiable functions" Math. Programming 20, pp. 1-13.

Pierre D. and Lowe M. (1975), "Mathematical programming via augmented Lagrangians" Addison Wesley.

Ravindran A. (1972), "A computer routine for quadratic and linear programming problems communication of ACM," Vol. 15, No. 9, pp. 818-820.

Schittkowski, K. (1981), "The non linear programming method of Wilson, Han and Powell with an augmented Lagrangian type line search function" Part 1 Numerische Mathematik Vol. 38, pp. 83-114.

Tapia R.A. (1977), "Diagonalized multiplier method and quasi-newton methods for constrained optimization" JOTA Vol. 22, No.2, pp. 135-194.

# CHAPTER 21

## Quasi-Newton Methods for Maximum-Likelihood Estimation*

John D. Gonglewski†
Homer F. Walker‡

**Abstract.** General unconstrained maximum-likelihood estimation problems have special structure which parallels that found in nonlinear least-squares problems. We consider quasi-Newton algorithms for maximum-likelihood estimation which are analogues of successful algorithms for nonlinear least-squares problems and which are intended to exploit this special structure. Particular applications are to the finite mixture estimation problem.

**Key words.** maximum-likelihood estimation, quasi-Newton methods, method of scoring, mixture densities, mixture estimation, nonlinear least-squares problems

**1. Introduction.** The fundamental problem of interest here is a statistical parameter estimation problem, which we phrase as follows:

PROBLEM 1.1. *Given a random (independent, identically distributed) sample $\{x_k\}_{k=1,\ldots,N} \subset \mathbf{R}^n$ on a random variable $X$ with probability density function (PDF) $p(x|\Phi_*) \in \{p(x|\Phi) : \Phi \in \Omega\}$, estimate $\Phi_*$.*

There are a variety of estimation methods which one might bring to bear on this problem. We focus on the *method of maximum likelihood*. In this, one determines a *maximum-likelihood estimate (MLE)* $\hat{\Phi}$ of $\Phi_*$, defined roughly to be a value of $\Phi$ which maximizes the *log-likelihood function*

$$L(\Phi) = \frac{1}{N} \sum_{k=1}^{N} \log p(x_k|\Phi). \tag{1}$$

---

* Received by the editors August, 1989.
† U. S. Air Force's Weapons Laboratory, Optical Phased Array Branch, Kirtland Air Force Base, New Mexico 87117. The work of this author was supported in part by the United States Department of Energy under Contract Number DE-FG02-86ER25018 with Utah State University. Many of the results in this paper first appeared in this author's doctoral dissertation at the University of Houston.
‡ Department of Mathematics and Statistics, Utah State University, Logan, Utah 84322-3900. The work of this author was supported by United States Department of Energy Grant Number DE-FG02-86ER25018, Department of Defense/Army Grant Number DAAL03-88-K, and National Science Foundation Grant Number DMS-0088995, all with Utah State University.

We say "roughly" because this definition is not always adequate, as we note further below. Also, the factor $1/N$ is included not for any essential reason but merely because it results in a cleaner treatment of certain things in the sequel.

Maximum-likelihood estimation is widely used because it typically enjoys a number of desirable statistical and perhaps other properties, both in theory and (usually) in practice. Of course, "desirable" means "desirable relative to the alternatives" and does not mean that MLE's are always as well-behaved, well-defined, or easily obtained as one might wish. Indeed, in some applications, maximum-likelihood estimation may involve considerable statistical, mathematical, and numerical difficulties: The bias and variance of MLE's may be undesirably large for practical sample sizes. The log-likelihood function may have multiple local and global maximizers or singularities near which it is arbitrarily large; in particular, it may not have a unique global maximizer or, indeed, any global maximizer at all (but see below). Except in familiar elementary cases in which $p(x|\Phi)$ is a member of one of the well-known parametric families, one cannot obtain MLE's analytically and must approximate them numerically, and the numerical approximation of MLE's can be quite challenging. The existence of multiple local or global maximizers or singularities of $L$ is obviously a complicating factor; in addition, there may be expensive function evaluations, ill-conditioning, and various other numerical difficulties.

Since there is not always a unique global maximizer of the log-likelihood function, MLE's are not always well-defined by the rough definition above. However, we note that under mild assumptions there is with probability one a unique local maximizer of $L$ in any sufficiently small neighborhood of $\Phi_*$ whenever $N$ is sufficiently large; furthermore, with probability one, the value of $L$ at this maximizer eventually exceeds the value of $L$ on any compact set which does not contain a point $\Phi$ such that $p(x|\Phi) = p(x|\Phi_*)$ for almost all $x$. See Redner and Walker [25, pp. 210–212] for a detailed discussion. We refer to this maximizer as the *(unique) strongly consistent MLE* and denote it by $\hat{\Phi}_N$. Of course $\hat{\Phi}_N$ depends on the sample observations as well as the sample size, but there is no need to denote this. For future reference, we note explicitly that $\lim_{N \to \infty} \hat{\Phi}_N = \Phi_*$.

A very important context in which all of the difficulties
of maximum-likelihood estimation occur, and which has provided us with motivation for the developments in this paper, is that of *(finite) mixture estimation*. In this context, $p(x|\Phi)$ is a *(finite) mixture density*, i. e., a PDF of the form

$$(2) \qquad p(x|\Phi) = \sum_{i=1}^{m} \alpha_i p_i(x|\Phi_i),$$

where $\Phi = (\alpha_1, \ldots, \alpha_m, \Phi_1, \ldots, \Phi_m)$, the $\alpha_i$'s are nonnegative and satisfy $\sum_{i=1}^{m} \alpha_i = 1$, and each *component density* $p_i(x|\Phi_i)$ is itself a PDF, typically from one of the familiar parametric families. A mixture density (2) arises naturally—and can naturally be interpreted—as a PDF associated with a statistical population which is a mixture of $m$ component populations with associated densities $\{p_i\}_{i=1,\ldots,m}$ and mixing proportions $\{\alpha_i\}_{i=1,\ldots,m}$. Thus mixture densities appear often in the statistical modeling of chemical mixtures, biological populations, economic processes, etc. They also occur as fundamental models in areas such as statistical pattern recognition, classification, and clustering. Some specific applications in which mixture densities play a central role are the following, cited along with many more applications in Titterington, Smith, and Makov [26]: tracking in a multi-target environment, in which noisy signals are classified into the categories of noise alone, false alarm, and the desired target; interpretation and classification of remotely-sensed satellite data (see also the special issue of *Communications in Statistics* [5]); resolution of chemical mixtures via absorption spectroscopy, chromatographic scanning, electrophoresis, and other methods; and medical diagnosis and prognosis. The simplest case of the mixture estimation problem, which is implicitly the case considered throughout the following, is just Problem 1.1 with $p(x|\Phi)$ of the form (2). Other cases involve samples in which some of the observations have labels indicating their component populations of origin. For recent general references on mixture densities and mixture estimation problems, we suggest Everitt and Hand [11], McLachlan and Basford[21], and Redner and Walker [25] as well as Titterington, Smith, and Makov [26].

To give a concrete illustration of the role of mixture densities and mixture estimation in applications, we outline a famous example of Hosmer [19].

EXAMPLE 1.2. According to the International Halibut Commission of Seattle, Washington, the length distribution of halibut of a given age is closely approximated by a mixture of two univariate normal PDF's corresponding to the length distributions of the male and female subpopulations. Thus an appropriate PDF for lengths of the overall halibut population of a given age has the form (2) with $m = 2$,

$$(3) \qquad p_i(x|\Phi_i) = \frac{1}{\sqrt{2\pi}\sigma_i} e^{-\frac{(x-\mu_i)^2}{2\sigma_i^2}}, \qquad \text{with} \qquad \Phi_i = (\mu_i, \sigma_i^2)^T \in \mathbf{R}^2$$

for $i = 1, 2$, and $\Phi = (\alpha_1, \alpha_2, \Phi_1, \Phi_2)$. Once satisfactory parameter estimates have been obtained, such a mixture density can be used for a variety of purposes, e. g., in-the-field classification of fish according to sex on the basis of length. If one had a large sample of halibut lengths which were labeled according to sex, then it would be trivial to obtain very satisfactory parameter estimates. Unfortunately, the sex of halibut is not easily determined in the field, at least by humans, and as a practical matter one is forced to deal with samples which are at least partially unlabeled with respect to sex and thereby led to a mixture estimation problem. In the case of a totally unlabeled sample, the estimation problem is just Problem 1.1 with $p(x|\Phi)$ given by (2) and (3).

The estimation problem for a mixture of two univariate normal PDF's arising in this example is perhaps the simplest mixture estimation problem of widespread interest and was first outlined in 1894 by Pearson in [23], which appears to be the earliest published work relating to mixture estimation problems of any kind. In [23] and in other work on mixture estimation problems through the first half of this century, only very simple mixtures and relatively simple mixture estimation methods were considered because of the limitations of available analytical methods. Maximum-likelihood estimation could not be treated analytically at all and, although the subject of some early wishful thinking and experimentation in very simple cases, began to be seriously considered as a widely applicable approach to mixture estimation only as the use of computers became widespread in the 1960's. At present, maximum-likelihood estimation appears to be the generally preferred approach for mixture estimation problems of all types, and the best studied mixture estimation problem is unquestionably that of obtaining MLE's for a mixture of two or more univariate or multivariate normal PDF's.

One can see the difficulties which arise in maximum-likelihood estimation for mixtures even in the simple case of a mixture of two univariate normal PDF's. The log-likelihood function has singularities; indeed, if $\mu_i = x_k$ for some $i$ and $k$, then $L$ blows up as $\sigma_i^2 \to 0$. Thus $L$ has no global maximum, a fact which has led some investigators to consider such devices as constraints on the parameters (Hathaway [17]) or augmenting $L$ with penalty terms (Redner [24]). Furthermore, "label switching", i. e., interchanging $(\alpha_1, \Phi_1)$ and $(\alpha_2, \Phi_2)$ does not change the value of $L$, and so there is never a unique local maximizer. Still, Theorems 3.1 and 3.2 of [25] imply the existence of the (unique) strongly consistent MLE $\hat{\Phi}_N$ with probability one for sufficiently large samples, and one might take finding $\hat{\Phi}_N$

to be the goal of the estimation procedure. However, if the component PDF's are "poorly separated" in the sense that $\Phi_1 \approx \Phi_2$, then the variance of $\hat{\Phi}_N$ will be undesirably large for samples of reasonable size and, furthermore, the numerical problem of determining it will be ill-conditioned as well; see [7].

Our interest here is in the numerical aspects of maximum-likelihood estimation and in particular in *quasi-Newton methods* for approximating MLE's. In order to keep the discussion as much to the point as possible, we consider a quasi-Newton method only in the general basic form

$$(4) \qquad \Phi^{(j+1)} = \Phi^{(j)} - B_j^{-1} \nabla L(\Phi^{(j)}), \qquad B_j \approx \nabla^2 L(\Phi^{(j)}).$$

In practice, of course, one would have to modify this basic form with safeguards which modify the step $-B_j^{-1} \nabla L(\Phi^{(j)})$ if necessary to ensure progress toward a solution. Such safeguards are incorporated in the codes used in the numerical experiments described below, but we need not consider them at this point. Since these safeguards measure progress only by the increase in $L$ that results from a step, they may lead to local maximizers which are not of interest or to singularities. This is a serious matter, but we do not consider it here both because it is not clear how to resolve it effectively and because consideration of it would detract from the issues of major interest.

Various well-known optimization methods have the form (4). Newton's method is obtained with $B_j = \nabla^2 L(\Phi^{(j)})$, and the Davidon-Fletcher-Powell (DFP) and Broyden-Fletcher-Goldfarb-Shanno (BFGS) methods result if, for each $j$, one obtains $B_{j+1}$ through the respective update formulas

$$(5) \qquad B_{j+1} = B_j + \frac{(y^{(j)} - B_j s^{(j)}) y^{(j)T} + y^{(j)} (y^{(j)} - B_j s^{(j)})^T}{y^{(j)T} s^{(j)}} - \frac{s^{(j)T} (y^{(j)} - B_j s^{(j)}) y^{(j)} y^{(j)T}}{(y^{(j)T} s^{(j)})^2}$$

and

$$(6) \qquad B_{j+1} = B_j + \frac{y^{(j)} y^{(j)T}}{y^{(j)T} s^{(j)}} - \frac{B_j s^{(j)} s^{(j)T} B_j}{s^{(j)T} B_j s^{(j)}},$$

where $s^{(j)} = \Phi^{(j+1)} - \Phi^{(j)}$ and $y^{(j)} = \nabla L(\Phi^{(j+1)}) - \nabla L(\Phi^{(j)})$. See Dennis and Schnabel [9] for more on these updates and as a general reference on all aspects of quasi-Newton methods.

These well-known general optimization methods can certainly be brought to bear on maximum-likelihood estimation problems, but our specific interest is in certain special quasi-Newton methods which are suggested by the special structure of the log-likelihood function. In §2, we first develop this special structure of the log-likelihood function, which closely parallels the well-known structure of the residual sum of squares in nonlinear least-squares problems. We then formulate analogues for maximum-likelihood estimation of certain well-known quasi-Newton methods for nonlinear least-squares problems. In §3, we offer the results of two numerical experiments in which these and other methods are use to obtain MLE's for a mixture of two univariate normal PDF's.

Recent work which is closely related to that described here has been done by Bunch [2,3] for probabilistic choice models and by Gay and Welsch [15] for nonlinear exponential family regression models. Also, the special updating methods described in §2 can be considered a special case of the very general algorithms for nonlinear fitting proposed by Dennis [6].

Finally, while this paper was in the final stages of preparation, we became aware of very recent work of Bunch [4] which deals directly with the algorithms and applications of interest here, and the reader is especially referred to [4].

**2. Special quasi-Newton methods.** We begin by examining the special structure of the log-likelihood function. For convenience, we assume that $\Phi = (\xi_1, \ldots, \xi_\nu)^T \in \mathbf{R}^\nu$, where the $\xi_i$'s are unconstrained, mutually independent variables. This assumption is valid, e. g., in typical mixture estimation problems after minor reparametrization. Assuming sufficient differentiability, one has

$$(7) \qquad \nabla L(\Phi) = \frac{1}{N} \sum_{k=1}^{N} \frac{\nabla_\Phi p(x_k|\Phi)}{p(x_k|\Phi)} = \frac{1}{N} J(\Phi)^T \vec{1},$$

where

$$(8) \qquad J(\Phi) = \begin{pmatrix} \frac{\nabla_\Phi p(x_1|\Phi)^T}{p(x_1|\Phi)} \\ \vdots \\ \frac{\nabla_\Phi p(x_N|\Phi)^T}{p(x_N|\Phi)} \end{pmatrix} \quad \text{and} \quad \vec{1} = \begin{pmatrix} 1 \\ \vdots \\ 1 \end{pmatrix}.$$

Further differentiation gives

$$(9) \qquad \nabla^2 L(\Phi) = -\frac{1}{N} J(\Phi)^T J(\Phi) + \frac{1}{N} \sum_{k=1}^{N} \frac{\nabla_\Phi^2 p(x_k|\Phi)}{p(x_k|\Phi)}.$$

The special structure of interest is evident in (9): The first term on the right is easily computed, since $J(\Phi)$ must be computed anyway in evaluating $\nabla L(\Phi)$. Furthermore, it is always negative semi-definite and is likely to be negative-definite in typical applications for reasonable sample sizes. The second term on the right is likely to be relatively expensive to evaluate. Furthermore, in most applications one can expect it to be a relatively insignificant part of $\nabla^2 L(\Phi)$ if the sample size is large and $\Phi$ is near $\Phi_*$, as we show below.

The special structure of $L$ reflected in (9) is completely analogous to the special structure of the residual sum of squares in nonlinear least-squares problems. Indeed, if we write a residual sum of squares in a variable $x$ as $f(x) = \frac{1}{2} \sum_{k=1}^{N} r_k(x)^2$, then, assuming sufficient differentiability, we have

$$(10) \qquad \nabla f(x) = \sum_{k=1}^{N} r_k(x) \nabla r_k(x) = R'(x)^T R(x),$$

where

$$(11) \qquad R'(x) = \begin{pmatrix} \nabla r_1(x)^T \\ \vdots \\ \nabla r_N(x)^T \end{pmatrix} \quad \text{and} \quad R(x) = \begin{pmatrix} r_1(x) \\ \vdots \\ r_N(x) \end{pmatrix},$$

and

$$(12) \qquad \nabla^2 f(x) = R'(x)^T R'(x) + \sum_{k=1}^{N} r_k(x) \nabla^2 r_k(x).$$

The similarity of the special structure reflected in (12) to that in (9) is clear: The first term on the right is easily computed; furthermore, it is always positive semi-definite and is likely to be positive-definite in typical applications. The second term may be expensive to evaluate but is likely to be relatively small near the minimizer in the small-residual or nearly-linear case.

The special structure of $f$ reflected in (12) has inspired two special methods of the quasi-Newton form

$$(13) \qquad x^{(j+1)} = x^{(j)} + B_j^{-1} \nabla f(x^{(j)})$$

which are of interest here. The first is the classical *Gauss-Newton method*, in which $B_j = R'(x^{(j)})^T R'(x^{(j)})$ in (13). Under mild assumptions, this method exhibits local $q$-linear convergence to a minimizer $\hat{x}$ if $f(\hat{x})$ is small or if $R$ is not too nonlinear near $\hat{x}$, and it exhibits local convergence to $\hat{x}$ of $q$-order $(1+p)$ if $f(\hat{x}) = 0$ and both $\nabla^2 f$ and $\sum_{k=1}^N r_k \nabla^2 r_k$ are Hölder continuous with exponent $p$ at $\hat{x}$. For a detailed local convergence analysis for the Gauss-Newton method in the case $p = 2$, see Dennis and Schnabel [9, §10.2].

The Gauss-Newton method achieves simplicity and success in some applications by ignoring the term $\sum_{k=1}^N r_k(x) \nabla^2 r_k(x)$ in $\nabla^2 f(x)$, but it loses its effectiveness in situations when this term cannot be safely ignored. The second special method of interest here is intended to retain its effectiveness in such situations. This is the basic method underlying the NL2SOL algorithm of Dennis, Gay, and Welsch [7,8]. The idea is to choose

$$(14) \qquad B_j = R'(x^{(j)})^T R'(x^{(j)}) + A_j$$

in (13), where the augmentation $A_j$ of the Guass-Newton approximate Hessian is maintained by updating. The hope is that the resulting method will enjoy rapid local convergence at significantly less cost per iteration than would result if $B_j$ were the full Hessian of $L$. The update originally used in NL2SOL to maintain $A_j$ was a DFP-like update; it follows from results of Dennis and Walker [10] that the iteration (13), (14) with $A_j$ maintained by this update enjoys local $q$-superlinear convergence to a minimizer $\hat{x}$ under mild assumptions. We understand from Gay [14] that the update currently used in NL2SOL is an update derived from the BFGS update as in Al-Baali and Fletcher [1]. The NL2SOL algorithm has been quite successful in general practice.

Given the similarity of the structure of the log-likelihood function $L$ and the residual sum of squares $f$, as reflected in (9) and (12), respectively, it is reasonable to consider analogues for maximum-likelihood estimation of the Gauss-Newton method and NL2SOL-type methods for nonlinear least-squares problems. We now do this.

**2.1. The method of scoring.** The analogue for maximum-likelihood estimation of the Gauss-Newton method is the *method of scoring*. This method is attributable to Fisher [12,13], and its classical form is the quasi-Newton form (4) with $B_j = -I(\Phi^{(j)})$, where $I(\Phi)$ is the *Fisher information matrix* given by

$$(15) \qquad I(\Phi) = \int_{\mathbf{R}^n} [\nabla_\Phi \log p(x|\Phi)] [\nabla_\Phi \log p(x|\Phi)]^T p(x|\Phi) \, d\mu,$$

where $\mu$ denotes the underlying measure on $\mathbf{R}^n$ appropriate for $p(x|\Phi)$. In many applications, evaluating $I(\Phi)$ is undesirably expensive or impractical, and an effective alternative is the sample approximation

$$(16) \qquad B_j = -\frac{1}{N} J(\Phi^{(j)})^T J(\Phi^{(j)}),$$

where $J(\Phi)$ is given by (8). We consider the method of scoring in this form, i. e., (4) with $B_j$ given by (16). The similarity to the Gauss-Newton method is clear.

In Theorems 2.2 and 2.4 below, we give a local $q$-linear convergence analysis for the method of scoring which is adequate for the present purposes. A more detailed local convergence analysis which parallels the treatment of the Gauss-Newton method in Dennis and Schnabel [9, §10.2] is given in Gonglewski [16]. Theorem 2.2 gives general conditions under which the method of scoring exhibits local $q$-linear convergence to a maximizer of $L$. Theorem 2.4 exploits the probabilistic properties of the problem and the method to draw more refined conclusions about the behavior of the method near the strongly consistent MLE. The upshot of this analysis is that under mild assumptions, with probability one, the method of scoring enjoys arbitrarily fast local $q$-linear convergence to the strongly consistent MLE whenever the sample size is sufficiently large.

Here and throughout the following, we assume a norm of interest is given on $\mathbf{R}^\nu$, and we denote both it and its induced matrix norm on $\mathbf{R}^{\nu \times \nu}$ by the same symbol $|\cdot|$. For convenience we write

$$\nabla^2 L(\Phi) = C(\Phi) + A(\Phi), \tag{17}$$

where

$$C(\Phi) = -\frac{1}{N} J(\Phi)^T J(\Phi) \quad \text{and} \quad A(\Phi) = \frac{1}{N} \sum_{k=1}^{N} \frac{\nabla_\Phi^2 p(x_k|\Phi)}{p(x_k|\Phi)}. \tag{18}$$

We also formulate the following standard hypothesis typically used in the local convergence analysis of Newton's method and quasi-Newton methods.

HYPOTHESIS 2.1. *Let $\hat{\Phi} \in \mathbf{R}^\nu$ be a point for which $\nabla L(\hat{\Phi}) = 0$, and suppose $\nabla L(\Phi)$ is differentiable in an open convex neighborhood $\Omega$ of $\hat{\Phi}$, $C$ is continuous at $\hat{\Phi}$, and there are $\gamma \geq 0$ and $p \in (0,1]$ such that*

$$|\nabla^2 L(\Phi) - \nabla^2 L(\hat{\Phi})| \leq \gamma |\Phi - \hat{\Phi}|^p \tag{19}$$

*for all $\Phi \in \Omega$.*

THEOREM 2.2. *Under Hypothesis 2.1, if $C(\hat{\Phi})$ is invertible and $|C^{-1}(\hat{\Phi}) A(\hat{\Phi})| \equiv r < 1$, then there is an $\epsilon > 0$ such that if $|\Phi^{(0)} - \hat{\Phi}| < \epsilon$, then the iterates $\{\Phi^{(j)}\}_{j=0,1,...}$ produced by the method of scoring converge to $\hat{\Phi}$, and*

$$\limsup_{j \to \infty} \frac{|\Phi^{(j+1)} - \hat{\Phi}|}{|\Phi^{(j)} - \hat{\Phi}|} \leq r, \tag{20}$$

*provided $\Phi^{(j)} \neq \hat{\Phi}$ for all $j$.*

*Proof.* For $\Phi \in \Omega$ sufficiently near $\hat{\Phi}$ that $C(\Phi)$ is invertible, we set $\Phi_+ = \Phi - C(\Phi)^{-1} \nabla L(\Phi)$. Then

$$\Phi_+ - \hat{\Phi} = C(\Phi)^{-1} \left\{ \nabla^2 L(\Phi)(\Phi - \hat{\Phi}) - \nabla L(\Phi) \right\} - C(\Phi)^{-1} A(\Phi)(\Phi - \hat{\Phi}). \tag{21}$$

By standard arguments,

$$\left| \nabla^2 L(\Phi)(\Phi - \hat{\Phi}) - \nabla L(\Phi) \right| \leq \frac{\gamma}{1+p} |\Phi - \hat{\Phi}|^{1+p}, \tag{22}$$

and so the first term on the right of (21) is $O(|\Phi - \hat{\Phi}|^{1+p})$. Since $|C^{-1}(\hat{\Phi}) A(\hat{\Phi})| \equiv r < 1$ and $C$ and (hence) $A$ are continuous at $\hat{\Phi}$, it follows that for any $\rho \in (r, 1)$ there is an $\epsilon > 0$ such that if $|\Phi - \hat{\Phi}| < \epsilon$, then $|\Phi_+ - \hat{\Phi}| \leq \rho |\Phi - \hat{\Phi}|$. It follows immediately that if $|\Phi^{(0)} - \hat{\Phi}| < \epsilon$, then the iterates $\{\Phi^{(j)}\}_{j=0,1,...}$ produced by the method of scoring converge $q$-linearly to $\hat{\Phi}$, and (20) follows from (21) and (22). □

For Theorem 2.4 below, we need a stronger hypothesis than Hypothesis 2.1. Hypothesis 2.3 below is mainly necessary to ensure with probability one the proper limiting behavior of various things as the sample size $N$ approaches infinity, but as it happens the differentiability conditions on $p(x|\Phi)$ in Hypothesis 2.3 imply with probability one that for $\hat{\Phi} \in \Omega$ such that $\nabla L(\hat{\Phi}) = 0$, the conditions of Hypothesis 2.1 hold with $p = 1$ in (19). The first two conditions of Hypothesis 2.3 may look forbidding, but they amount to rather mild regularity conditions on $p(x|\Phi)$ that are met in most applications. It is possible that these conditions could be tightened somewhat, since they all relate to the regularity of $p(x|\Phi)$ in $\Phi$, but we prefer to give them in the form below because it is most convenient for the proof of Theorem 2.4.

HYPOTHESIS 2.3. *Let $\Omega$ be an open convex neighborhood of $\Phi_*$ in which the following hold:*

1. *For all $\Phi \in \Omega$, for almost all $x \in \mathbf{R}^n$, and for all $i$, $j$, $k = 1, \ldots, \nu$, the partial derivatives $\partial p/\partial \xi_i$, $\partial^2 p/\partial \xi_i \partial \xi_j$, and $\partial^3 p/\partial \xi_i \partial \xi_j \partial \xi_k$ exist and satisfy*

$$(23) \quad \left|\frac{\partial p(x|\Phi)}{\partial \xi_i}\right| \leq f_i(x), \quad \left|\frac{\partial^2 p(x|\Phi)}{\partial \xi_i \partial \xi_j}\right| \leq f_{ij}(x), \quad \left|\frac{\partial^3 \log p(x|\Phi)}{\partial \xi_i \partial \xi_j \partial \xi_k}\right| \leq f_{ijk}(x),$$

*where $f_i$ and $f_{ij}$ are integrable and $f_{ijk}$ satisfies*

$$(24) \quad \int_{\mathbf{R}^n} f_{ijk}(x) p(x|\Phi_*) \, d\mu < \infty.$$

2. *For all $\Phi \in \Omega$ and for almost all $x \in \mathbf{R}^n$,*

$$(25) \quad \left|\frac{\nabla_\Phi p(x|\Phi) \nabla_\Phi p(x|\Phi)^T}{p(x|\Phi)^2} - \frac{\nabla_\Phi p(x|\Phi_*) \nabla_\Phi p(x|\Phi_*)^T}{p(x|\Phi_*)^2}\right| \leq g_1(x) o_1(|\Phi - \Phi_*|)$$

*and*

$$(26) \quad \left|\frac{\nabla_\Phi^2 p(x|\Phi)}{p(x|\Phi)} - \frac{\nabla_\Phi^2 p(x|\Phi_*)}{p(x|\Phi_*)}\right| \leq g_2(x) o_2(|\Phi - \Phi_*|),$$

*where for $i = 1, 2$,*

$$(27) \quad \lim_{\Phi \to \Phi_*} o_i(|\Phi - \Phi_*|) = 0 \quad \text{and} \quad \int_{\mathbf{R}^n} g_i(x) p(x|\Phi_*) \, d\mu < \infty.$$

3. *The Fisher information matrix $I(\Phi)$ given by (15) is well-defined and positive definite at $\Phi_*$.*

THEOREM 2.4. *Under Hypothesis 2.3, with probability one the strongly consistent MLE $\hat{\Phi}_N$ is well-defined whenever $N$ is sufficiently large and the following holds: For any $r \in (0, 1)$, there are $N_r$ and $\epsilon_r > 0$ such that if $N > N_r$ and $|\Phi^{(0)} - \hat{\Phi}_N| < \epsilon_r$, then the iterates $\{\Phi^{(j)}\}_{j=0,1,\ldots}$ produced by the method of scoring converge to $\hat{\Phi}_N$, and*

$$(28) \quad \limsup_{j \to \infty} \frac{|\Phi^{(j+1)} - \hat{\Phi}_N|}{|\Phi^{(j)} - \hat{\Phi}_N|} \leq r,$$

*provided $\Phi^{(j)} \neq \hat{\Phi}_N$ for all $j$.*

*Proof.* Under conditions 1 and 3 of Hypothesis 2.3, it follows from Theorem 3.1 of Redner and Walker [25] that with probability one the strongly consistent MLE $\hat{\Phi}_N$ is well-defined whenever $N$ is sufficiently large and $\lim_{N \to \infty} \hat{\Phi}_N = \Phi_*$.

To complete the proof, we suppose $r \in (0,1)$ is given and note that with probability one

$$
\begin{aligned}
|C(\Phi) - I(\Phi_*)| &\leq |C(\Phi) - C(\Phi_*)| + |C(\Phi_*) - I(\Phi_*)| \\
&\leq \left[\frac{1}{N}\sum_{k=1}^{N} g_1(x_k)\right] o_1(|\Phi - \Phi_*|) + |C(\Phi_*) - I(\Phi_*)|
\end{aligned}
\tag{29}
$$

and

$$
\begin{aligned}
|A(\Phi)| &\leq |A(\Phi) - A(\Phi_*)| + |A(\Phi_*)| \\
&\leq \left[\frac{1}{N}\sum_{k=1}^{N} g_2(x_k)\right] o_2(|\Phi - \Phi_*|) + |A(\Phi_*)|.
\end{aligned}
\tag{30}
$$

It follows from the Strong Law of Large Numbers (see Loève [20, p. 239]) that with probability one

$$
\lim_{N\to\infty} \frac{1}{N}\sum_{k=1}^{N} g_i(x_k) = \int_{\mathbf{R}^n} g_i(x) p(x|\Phi_*)\, d\mu < \infty, \quad i = 1, 2,
\tag{31}
$$

$$
\begin{aligned}
\lim_{N\to\infty} C(\Phi_*) &= \lim_{N\to\infty} \frac{1}{N}\sum_{k=1}^{N} \nabla_\Phi \log p(x_k|\Phi_*) \nabla_\Phi \log p(x_k|\Phi_*)^T \\
&= I(\Phi_*),
\end{aligned}
\tag{32}
$$

and

$$
\begin{aligned}
\lim_{N\to\infty} A(\Phi_*) &= \lim_{N\to\infty} \frac{1}{N}\sum_{k=1}^{N} \frac{\nabla_\Phi^2 p(x_k|\Phi_*)}{p(x_k|\Phi_*)} \\
&= \int_{\mathbf{R}^n} \nabla_\Phi^2 p(x|\Phi_*)\, d\mu = 0.
\end{aligned}
\tag{33}
$$

One sees from (29) – (33) that with
probability one there are $\bar{N}_r, \bar{\epsilon}_r > 0$, and $M$ such that if $N > \bar{N}_r$ and $|\Phi - \Phi_*| < \bar{\epsilon}_r$, then $C(\Phi)$ is invertible with $|C(\Phi)^{-1}| \leq M$ and $|C(\Phi)^{-1} A(\Phi)| \leq r/2$.

We note further that it follows from the third inequality in (23) that with probability one there exists a $\gamma \geq 0$ independent of $N$ for which (19) holds with $p = 1$ for all $\Phi, \hat{\Phi} \in \Omega$ and for which (22) also holds with $p = 1$ whenever $\nabla L(\hat{\Phi}) = 0$. In particular, for such a $\gamma \geq 0$, (19) and (22) hold with $p = 1$ for $\Phi \in \Omega$ and $\hat{\Phi} = \hat{\Phi}_N$.

We now choose $N_r \geq \bar{N}_r$ and $\epsilon_r > 0$ so that if $N > N_r$ and $|\Phi - \hat{\Phi}_N| < \epsilon_r$, then $|\Phi - \Phi_*| < \bar{\epsilon}_r$ and also

$$
\frac{M\gamma}{1+p}|\Phi - \hat{\Phi}_N|^p + |C(\Phi)^{-1}A(\Phi)| \leq r.
\tag{34}
$$

It follows from (21) and (22) with $\hat{\Phi} = \hat{\Phi}_N$ and from (34) that if $N > N_r$ and $|\Phi^{(0)} - \hat{\Phi}_N| < \epsilon_r$, then the iterates $\{\Phi^{(j)}\}_{j=0,1,...}$ produced by the method of scoring converge to $\hat{\Phi}_N$ and (28) holds. □

## 2.2. MLE updating methods.
We now consider analogues for maximum-likelihood estimation of NL2SOL-type methods for nonlinear least-squares problems, which we refer to as *MLE updating methods*. These have the quasi-Newton form (4) with

$$(35) \qquad B_j = C(\Phi^{(j)}) + A_j, \qquad A_j \approx A(\Phi^{(j)}),$$

where $C(\Phi)$ and $A(\Phi)$ are given by (18) and $A_j$ is maintained by updating. The updating is done to satisfy a *secant condition*

$$(36) \qquad A_{j+1} s^{(j)} = y^{(j)},$$

where $s^{(j)} = \Phi^{(j+1)} - \Phi^{(j)}$ and

$$(37) \qquad y^{(j)} \approx A(\Phi^{(j+1)}) s^{(j)} = \left[ \frac{1}{N} \sum_{k=1}^{N} \frac{\nabla_\Phi^2 p(x_k | \Phi^{(j+1)})}{p(x_k | \Phi^{(j+1)})} \right] s^{(j)}.$$

There are at least several reasonable choices of $y^{(j)}$ in (37) and also a number of applicable update formulas which yield (36). We consider choices of $y^{(j)}$ together with a particular DFP-like update formula which yield a locally $q$-superlinearly convergent method according to the results of Dennis and Walker [10]. Another update which we have not yet tried but which seems especially promising is a BFGS-like update analogous to that in Al-Baali and Fletcher [1] which is currently used in NL2SOL. Guidelines for making an admissible choice of $y^{(j)}$ are given in [10], and three readily available choices which fit these guidelines are

$$(38) \qquad y_1^{(j)} = \sum_{k=1}^{N} \frac{\nabla_\Phi p(x_k | \Phi^{(j+1)}) - \nabla_\Phi p(x_k | \Phi^{(j)})}{p(x_k | \Phi^{(j+1)})},$$

$$(39) \qquad y_2^{(j)} = \nabla L(\Phi^{(j+1)}) - \nabla L(\Phi^{(j)}) + J(\Phi^{(j+1)})^T J(\Phi^{(j+1)}) s^{(j)},$$

$$(40) \qquad y_3^{(j)} = \nabla L(\Phi^{(j+1)}) - \nabla L(\Phi^{(j)}) + J(\Phi^{(j)})^T J(\Phi^{(j)}) s^{(j)}.$$

The first choice is the analogue of the choice originally reported in [7,8] to be the most effective in NL2SOL and is the choice used in the numerical experiments discussed in §3 below. The second and third choices are obvious "default" and are included here for completeness.

The particular update formula which we consider is a DFP-like update analoguous to that originally used in NL2SOL, viz.,

$$(41) \qquad A_{j+1} = A_j + \frac{(y^{(j)} - A_j s^{(j)}) v^{(j)T} + v^{(j)}(y^{(j)} - A_j s^{(j)})^T}{v^{(j)T} s^{(j)}} - \frac{s^{(j)T}(y^{(j)} - A_j s^{(j)}) v^{(j)} v^{(j)T}}{(v^{(j)T} s^{(j)})^2},$$

where $v^{(j)} = \nabla L(\Phi^{(j+1)}) - \nabla L(\Phi^{(j)})$.

The resulting MLE updating method which we consider is summarized as

$$(42) \qquad \begin{aligned} B_j &= C(\Phi^{(j)}) + A_j, \text{ with } C \text{ given by (18)}, \\ \Phi^{(j+1)} &= \Phi^{(j)} - B_j^{-1} \nabla L(\Phi^{(j)}), \\ A_{j+1} &\text{ given by (41), with} \\ y^{(j)} &\in \{y_1^{(j)}, y_2^{(j)}, y_3^{(j)}\} \text{ given by (38) – (40).} \end{aligned}$$

The following local $q$-superlinear convergence result for this method is an immediate corollary of Theorems 4.2 and 4.3 of [10].

THEOREM 2.5. *Under Hypothesis 2.1 and the additional assumptions that $\nabla^2 L(\hat{\Phi})$ is invertible and there is a $\gamma_C \geq 0$ such that*

(43) $$|C(\Phi) - C(\hat{\Phi})| \leq \gamma_C |\Phi - \hat{\Phi}|^p,$$

*there are $\epsilon > 0$ and $\delta > 0$ such that if $|\Phi^{(0)} - \hat{\Phi}| < \epsilon$ and $|A_0 - A(\hat{\Phi})| < \delta$, then the iterates produced by (42) are well-defined and converge $q$-superlinearly to $\hat{\Phi}$. Furthermore, $\{|B_j|\}_{j=0,1,...}$ and $\{|B_j^{-1}|\}_{j=0,1,...}$ are uniformly bounded.*

**3. Numerical experiments.** In this section we report on two numerical experiments in which the special quasi-Newton methods outlined in §2 and other methods were applied to mixture estimation problems involving a mixture of two univariate normal PDF's.

For these experiments we constructed an experimental code MLESOL, in which the basic quasi-Newton form (4) is augmented with various options for determining $B_j$, several choices of "global convergence" safeguards intended to ensure progress toward a solution, several sophisticated stopping tests, some of which are tailored to the maximum-likelihood estimation context, and a variety of other features. We refer the reader to Gonglewski [16] for a more detailed description of the features of MLESOL but note that the "global convergence" procedures follow routines from Appendix A of Dennis and Schnabel [9] and that many practical suggestions for the coding, especially for the stopping tests, were taken from the programming comments on NL2SOL [7,8].

In the experiments described here, we used the following ways of determining $B_j$ at each step: finite-difference evaluation of $\nabla^2 L(\Phi^{(j)})$, standard BFGS updating according to (6), the method of scoring choice $B_j = C(\Phi^{(j)})$ given by (18), and MLE updating as in (42), (41), with $y^{(j)} = y_1^{(j)}$ given by (38). The "global convergence" procedure used in all cases was that which determined the "locally constrained optimal" ("hook") step described in [9, §6.4.1] and based on Hebden [18] and Moré [22]. All computing was done in single precision on a Digital Equipment Corporation VAX 11/780 using the VMS Fortran compiler. All random samples on mixtures of two univariate normal PDF's were generated by using the RAN(ISEED) function from VMS Fortran first to generate a uniform-[0,1] random number which determined a component PDF and then to generate a pair of uniform-[0,1] random numbers $r_1, r_2$ which determined an appropriately normally distributed sample point $x_k$ by the standard Box-Muller transformation $x_k = \mu_i + \sigma_i \sqrt{-2 \log r_1} \cos(2\pi r_2)$. Mixtures of two univariate normal PDF's were viewed as depending on five parameters: one proportion in $[0,1]$, two means, and two positive variances. These were treated as unconstrained parameters by MLESOL, except that the iterations were stopped if any constraint was reached or violated.

In the first experiment, we considered a sample of 200 observations from a mixture with true parameters $\alpha_1^* = \alpha_2^* = .5$, $\mu_1^* = 0$, $\mu_2^* = 2.5$, and $\sigma_1^{2*} = \sigma_2^{2*} = 1$. With these parameters, the component PDF's are not well-separated and a sample of this size is rather small for reliable estimates; however, there is no reason to expect numerical problems such as ill-conditioning near

the strongly consistent MLE. (See Redner and Walker [25, §3].) For each of eight different starting points, we used MLESOL to find a maximizer of $L$ for this sample using each of the four ways of determining $B_j$ described above. Table 1 gives the average numbers of iterations required by the four methods for six of these starting points which we feel gave meaningful results. One of the left-out starting points resulted in the iterates produced by all methods quickly reaching a constraint boundary point which did not represent a

| | |
|---|---|
| Finite-Difference Hessian | 10.8 |
| MLE Updating | 12.3 |
| Method of Scoring | 16.7 |
| BFGS Updating | 23.2 |

TABLE 1
*Average numbers of iterations over six starting points for a sample of 200 observations.*

maximizer of $L$. The other of the left-out starting points resulted in the finite-difference Hessian method inexplicably taking an anomolously large number of iterations, viz., 83, for convergence, while the other three methods behaved more or less in line with their averages. We also note that of the six starting points included in the average, one resulted in the iterates produced by all methods converging to a "spurious" maximizer of $L$ which was not the strongly consistent MLE, while the other five resulted in the iterates produced by all methods converging to the strongly consistent MLE in each case. See [16] for more details.

The results in Table 1 indicate that the two special quasi-Newton methods performed very effectively in comparison to the other two methods. The MLE updating method and the method of scoring require the same function-evaluation work per iteration as the BFGS method, although they also require some additional arithmetic, but they gave convergence in significantly fewer iterations in this experiment. On the average, the MLE updating method required only a few more iterations for convergence than the finite-difference Hessian method in this experiment, but the latter requires significantly more function-evaluation work per iteration.

In the second experiment the object was not only to compare the performance of the four methods used in the first experiment but also to assess the sensitivity of these methods to the separation of the component PDF's in the mixture. Our interest

in this issue arises as follows: The *EM algorithm* is a well-known method which is widely used for maximumum-likelihood mixture estimation. For mixtures of normal PDF's, as well as for many other mixtures, the EM algorithm has the form of a fixed-point iteration which, in general, exhibits local $q$-linear convergence to an MLE. The actual speed of this $q$-linear convergence is known to depend strongly on the separation of the component PDF's in the mixture and, in particular, deteriorates badly under increasingly poor separation of the component PDF's. See Redner and Walker [25] as a general reference on the EM algorithm and its properties, especially in the context of mixture estimation. If the quasi-Newton methods of interest here show relatively little deterioration of performance under increasingly poor component separation, then this gives them an important advantage over the EM algorithm on poorly separated mixtures.

We considered three samples of 1000 observations each from each of several mixtures with true proportions $\alpha_1^* = .3$, $\alpha_2^* = .7$ and true variances $\sigma_1^{2*} = \sigma_2^{2*} = 1$ and with varying $\Delta \mu = \mu_1^* - \mu_2^*$. We observed the number of iterations required in each case to meet both loose and tight bounds on the maximum error component, i. e., on $|\Phi^{(j)} - \hat{\Phi}|_\infty$. The starting point in each case was given by $\alpha_i = .5$, $\sigma_i^2 = .5$, and $\mu_i = 1.5\mu_i^*$ for $i = 1, 2$. The averages of the iteration numbers are given in Tables 2 and 3. For completeness, we have also included in these tables iteration numbers observed for the EM algorithm in a similar experiment reported in [25]. However, we caution the reader that these results for the EM algorithm were obtained using only a single sample of 1000 points for each value of $\Delta \mu$, and that sample was not one of the ones used in our experiment here. In Tables 2 and 3, the

| $\Delta\mu$ | FDH | MLE | MOS | BFGS | EM |
|---|---|---|---|---|---|
| 6.0 | 5 | 9 | 11 | 15 | 1 |
| 4.0 | 4 | 7 | 8 | 14 | 1 |
| 2.0 | 5 | 7 | 9 | 15 | 126 |
| 1.0 | 12 | 14 | 17 | 33 | 349 |
| 0.4 | 30 | 37 | 33 | 70 | 710 |
| 0.2 | 43 | 49 | 47 | 77 | 2078 |

TABLE 2

*Average numbers of iterations required to make $|\Phi^{(j)} - \hat{\Phi}|_\infty \leq 10^{-1}$ over three samples of 1000 observations for each $\Delta\mu$.*

| $\Delta\mu$ | FDH | MLE | MOS | BFGS | EM |
|---|---|---|---|---|---|
| 6.0 | 11 | 13 | 14 | 24 | 5 |
| 4.0 | 7 | 13 | 13 | 22 | 44 |
| 2.0 | 8 | 10 | 15 | 26 | 883 |
| 1.0 | 23 | 22 | 27 | 46 | 777 |
| 0.4 | 37 | 42 | 38 | 76 | 1381 |
| 0.2 | 51 | 54 | 59 | 87 | 3095 |

TABLE 3

*Average numbers of iterations required to make $|\Phi^{(j)} - \hat{\Phi}|_\infty \leq 10^{-6}$ over three samples of 1000 observations for each $\Delta\mu$.*

column headings "FDH", "MLE", "MOS", and "BFGS" denote, respectively, the methods using finite-difference Hessians, MLE updating, method-of-scoring approximate Hessians, and BFGS updating, and "EM" denotes the EM algorithm.

It is strongly suggested by Tables 2 and 3 that while the four methods other than the EM algorithm show some deterioration of performance with decreasing mean separation, this deterioration is not nearly as pronounced as that associated with the EM algorithm. We note that the relative ranking of the four algorithms in Table 1 is also seen throughout Tables 2 and 3 except in a few cases. For perspective, we also note that when $\Delta\mu$ is 0.4 or 0.2, the mixture is extremely poorly separated. In these cases, the variances of the MLE's are so large that they are not likely to be useful; in addition, the problem of maximizing $L$ is likely to be very ill-conditioned. (See [25, §3].)

**4. Discussion.** We have observed that the log-likelihood function for maximum-likelihood estimation has special structure similar to that of the residual sum of squares for nonlinear least-squares problems, and we have outlined and given local convergence results for special quasi-Newton methods suggested by this structure which are analogous to well-known nonlinear least-squares methods. Limited testing using an experimental code on mixture estimation problems suggests that these special methods compare very favorably with standard general optimization methods. Both the special methods and standard general optimization methods seem likely to outperform the EM algorithm on mixture estimation problems, especially when the component PDF's are poorly separated.

We indicate two areas in which additional work is needed: (1) Our testing has been limited to significant but simple mixture estimation problems, and extensive additional testing should be done with a broader variety of mixture and other problems. (2) A BFGS-like update analogous to that in Al-Baali and Fletcher [1] should be incorporated in our experimental code and tested along with the DFP-like update which is now being used.

## REFERENCES

[1] M. Al-Baali and R. Fletcher, *Variational methods for non-linear least-squares*, J. Opl. Res. Soc., 36 (1985), pp. 405–421.
[2] D. S. Bunch, *Maximum-likelihood estimation of probabilistic choice models*, SIAM J. Sci. Stat. Comput., 8 (1987), pp. 56–70.
[3] ———, *A comparison of algorithms for maximum-likelihood estimation of choice models*, J. Econometrics, 38 (1988), pp. 147–167.
[4] ———, *A comparison of algorithms for maximum-likelihood estimation of finite mixture distribution models*, to appear.
[5] *Communications in Statistics*, Special Issue on Remote Sensing, Comm. Statist. Theor. Meth., A5 (1976).
[6] J. E. Dennis, Jr., *Algorithms for nonlinear fitting*, in Optimization 1981, M. J. D. Powell, ed., Academic Press, London, 1982, pp. 67–78.
[7] J. E. Dennis, Jr., D. M. Gay, and R. E. Welsch, *An adaptive nonlinear least-squares algorithm*, Trans. Math. Software, 7 (1981), pp. 348–368.
[8] ———, *Algorithm 573 NL2SOL — an adaptive nonlinear least-squares algorithm [E4]*, Trans. Math. Software, 7 (1981), pp. 369–383.
[9] J. E. Dennis, Jr., and R. B. Schnabel, *Numerical Methods for Unconstrained Optimization and Nonlinear Equations*, Prentice-Hall Series in Automatic Computation, Englewood Cliffs, NJ, 1983.
[10] J. E. Dennis, Jr., and H. F. Walker, *Convergence theorems for least-change secant update methods*, SIAM J. Numer. Anal., 18 (1981), pp. 949–987, 19 (1982), p. 443.
[11] B. S. Everitt and D. J. Hand, *Finite Mixture Distributions*, Chapman and Hall, London, 1981.
[12] R. A. Fisher, Ann. Eug., 6 (1935), pp. 187–201.
[13] ———, *A system of scoring linkage data with special reference to pied factors in mice*, Am. Nat., 80 (1946), pp. 568–578.
[14] D. M. Gay, private communication.
[15] D. M. Gay and R. E. Welsch, *Maximum-likelihood and quasi-likelihood for nonlinear exponential family regression models*, J. Amer. Statist. Assoc., 83 (1988), pp. 990–998.
[16] J. D. Gonglewski, *On quasi-Newton methods for maximum-likelihood estimates with applications to the mixture density problem*, Ph.D. dissertation, Department of Mathematics, University of Houston – University Park, Houston, TX, 1986.
[17] R. J. Hathaway, *Constrained maximum-likelihood estimation for a mixture of m univariate normal distributions*, Statistics Tech. Rep. 92, 62F10-2, Univ. of South Carolina, Columbia, SC, 1983.
[18] M. D. Hebden, *An algorithm for minimization using exact second derivatives*, Rept. TP515, A.E.R.E., Harwell, England, 1973.
[19] D. W. Hosmer, Jr., *A comparison of iterative maximum-likelihood estimates of the parameters of a mixture of two normal distributions under three different types of samples*, Biometrics, 29 (1973), pp. 761–770.
[20] M. Loève, *Probability Theory*, Van Nostrand, New York, 1963.
[21] G. J. McLachlan and K. E. Basford, *Mixture Models: Inference and Applications to Clustering*, Statistics: Textbooks and Monographs 84, Marcel Dekker, New York, 1988.
[22] J. J. Moré, *The Levenberg-Marquardt algorithm: implementation and theory*, in Numerical Analysis, G. A. Watson, ed., Lecture Notes in Math. 630, Springer Verlag, Berlin, 1977, pp. 105–116.
[23] K. Pearson, *Contributions to the mathematical theory of evolution*, Phil. Trans. Royal Soc., 185A (1894), pp. 71–110.
[24] R. A. Redner, *Maximum-likelihood estimation for mixture models*, Rep. LEMSCO-14880, Earth Observations Division, Space and Life Sciences Directorate, NASA Johnson Space Center, Houston, TX, 1980.
[25] R. A. Redner and H. F. Walker, *Mixture densities, maximum likelihood, and the EM algorithm*, SIAM Rev., 26 (1984), pp. 195–239.
[26] D. M. Titterington, A. F. M. Smith, and U. E. Makov, *Statistical Analysis of Finite Mixture Distributions*, John Wiley and Sons, Inc., New York 1985.

# CHAPTER 22

## An Approach to Nonlinear $l_\infty$ Approximation

Andrew R. Conn*
Yuying Li†

**Abstract.** Recently we have presented a new approach to nonlinear $l_\infty$ approximation that directly exploits generalisations of the characterisation for the classical best linear Chebyshev approximation.

We are able to produce an algorithm that has the ability to recognise the correct active set more rapidly than the more usual nonlinear programming approaches, which are based on equality quadratic programming methods, while avoiding the inefficiencies typically associated with the several inner iterations normally required by an inequality quadratic programming approach.

In addition to summarising the method, we present details of the line search technique, show that certain degenerate problems give rise to a least squares problem with nonnegativity constraints and include certain technical details, required for example, to avoid the Maratos effect. All the proofs of the theorems are omitted to emphasize the main ideas of the algorithm, their proper references are indicated however.

**Key words.** nonlinear Chebyshev approximation

**AMS(MOS) subject classifications.** 41A50, 65D99, 65F20, 65K05

**1. Introduction.** The underlying problem we wish to consider is to minimize over $\mathbf{R}^n$ the non-smooth function $\psi(x)$ given by the maximum over a finite set, $M = \{1, 2, \cdots, m\}$, of functions $f_i(x) : \mathbf{R}^n \to \mathbf{R}$.

$$(1) \qquad \min_{x \in \Re^n} \max_{i \in M} f_i(x).$$

In this paper, we concentrate on a special case which is the discrete Chebyshev problem, where $\psi(x)$ is given by

$$\psi(x) = \max_{i \in M} |f_i(x)|.$$

---

\* Department of Combinatorics and Optimization, University of Waterloo, Waterloo, Ontario N2L 3G1, Canada. The research of this author was supported in part by NSERC grant A8639.

† Computer Science Department, Cornell University, Upson Hall, Ithaca, NY, 14853. The research of this author was partially supported by the U.S. Army Research Office through the Mathematical Science Institute, Cornell University.

Such problems may have arisen from a discrete approximation to the continuous problem

$$\psi(x) = \max_{t \in T} |f(x,t)|,$$

where $T$ is a compact set.

In any case, the discrete Chebyshev problem,

(2) $$\min_{x \in \Re^n} \max_{i \in M} |f_i(x)|,$$

is the problem of interest in the present article. We are content with finding a *local* minimum of $\psi(x)$ and we assume that the $f_i(x), i \in M$, are twice continuously differentiable.

Most current approaches are based upon the fact that (2) can be transformed into a nonlinear programming problem by adding a single new variable viz.

$$\begin{aligned} \min_{(x,z) \in \Re^{n+1}} \quad & z \\ \text{subject to} \quad & z - f_i(x) \geq 0 \\ & z + f_i(x) \geq 0 \\ & z \geq 0. \end{aligned}$$

Although the structure of this formulation can be exploited to some extent, we are more interested in exploiting directly the structure of an optimal solution to the discrete Chebyshev problem. We are motivated to pursue this latter approach by virtue of the fact that classical Chebyshev theory is able to **characterise** such solutions in the case of continuous linear problems, under certain regularity conditions, and we are able to exploit such a characterisation very successfully in practise (see for example, Barrodale and Phillips [1] and Bartels, Conn and Li [2]).

The basic difficulty is that, in the linear case, there exists a *global* characterisation which is easy for computational exploitation, whereas in the nonlinear case this is not, in general, possible.

In effect, we shall base our algorithm upon *local* attempts at characterisation, which, in the limit, will give the correct characterisation at the solution.

If we consider the one dimensional continuous linear Chebyshev problem to approximate $y(t)$ on the interval $[\alpha, \beta]$ given by

$$\min_{x \in D} \max_{t \in [\alpha, \beta]} |\sum_{i=1}^{n} x_i \phi_i(t) - y(t)|,$$

where $D \subset \mathbf{R}^n$ is a compact set and the $\phi_i$'s are the 'basis functions' for our approximating set, then we determine an approximation to this continuous problem in $t$ by discretising the interval $[\alpha, \beta]$ into $m$ points, say

$$\alpha = t_1 < t_2 < t_3 \ldots < t_m = \beta.$$

The classical theory gives us the following explicit characterisation ( see for example, [17], page 77 ).

THEOREM 1.1 (CHARACTERISATION THEOREM).
Let $\mathcal{L}$ be an $n$ dimensional linear function subspace of $C[\alpha, \beta]$ that satisfies the Haar condition and let $y(t)$ be a continuous function on $[\alpha, \beta]$. Then $\phi^*(t) \stackrel{\text{def}}{=} \phi(x^*, t)$ is the best minimax approximation from $\mathcal{L}$ to $y(t)$ if and only if there exist $n+1$ points $\{t_i\}_{i=0}^n$ such that the conditions:

$$\alpha \leq t_0 < t_1 < \cdots < t_n \leq \beta$$

and

(3) $$|y(t_i) - \phi^*(t_i)| = \|y(t) - \phi^*(t)\|_\infty,$$

and

(4) $$y(t_{i+1}) - \phi^*(t_{i+1}) = -(y(t_i) - \phi^*(t_i)),$$

are satisfied. Such a set of points $\{t_i\}_{i=0}^n$ is often called an **alternant** of $\phi^*(t)$.

There is also an equivalent algebraic characterisation. The following theorem can be found in [17], page 98.

THEOREM 1.2. Let $\mathcal{L}$ be an $n$ dimensional linear function subspace of $C[\alpha, \beta]$ that satisfies the Haar condition. Furthermore, let $\{t_i\}_{i=0}^n$ be a set of reference points from $[\alpha, \beta]$ that are in ascending order:

$$\alpha \leq t_0 < t_1 < \cdots < t_n \leq \beta$$

and let $\{\lambda_i\}_{i=0}^n$ be a set of real multipliers that are not all zeroes, and that satisfy

(5) $$\sum_{i=0}^n \lambda_i \phi(x, t_i) = 0$$

for all functions $\phi(x,t) = \sum_{j=1}^n x_j \phi_j(t)$ (i.e. $\phi \in \mathcal{L}$ with basis functions $\phi_j(t)$). Then every multiplier is nonzero, and their signs alternate.

Equation (5) is called the **characteristic equation**.

Given the characteristic equation (5), with the associated multipliers $\{\lambda_i\}_{i=0}^n$, suppose we are approximating the continuous function $y(t)$ by $\phi(x,t)$, with associated error $f(t) = y(t) - \phi(x,t)$, then we have the following definitions.

DEFINITION 1. The function $\phi(x,t)$ is called a **reference function** with respect to the reference $\{t_i\}_{i=0}^n$ and the function $y(t)$ if and only if:

$$sgn(f_i) = sgn(\lambda_i) \quad \text{for all } i,$$

or

$$sgn(f_i) = -sgn(\lambda_i) \quad \text{for all } i,$$

where $\{\lambda_i\}_{i=0}^n$ is given by the characteristic equation (5).

*If in addition all the $f_i$'s have the same magnitude, called the* **reference deviation**, $\phi(x,t)$ *is a* **levelled reference function**.

Thus, in the linear case we have the following equivalent characterisation.

THEOREM 1.3. *Under the same assumptions of Theorem 1.2, the function of best approximation is the levelled reference function with the maximal reference deviation.*

If we return to algorithms for the linear problem there are two main approaches — dual algorithms, for example Barrodale and Phillips [1] and primal methods, for example Bartels, Conn and Li [2].

The former chooses $n+1$ points, $\{t_i\}_{i=0}^n$ (a reference) and an $x^c$ such that (4) is satisfied. If (3) also holds we are optimal. Otherwise, it is possible to choose a $t_j$ such that the error (value of $|f_j(x^c)|$) is greater than the errors on the reference. We then replace a $t_i$ of the reference by the $t_j$ and iterate.

In contrast, a primal algorithm chooses $n+1$ points and an $x^c$ such that one has $n+1$ activities. If alternation is satisfied, one is optimal. Otherwise, it is possible to find a new $x^+$ such that $n$ of the residuals that determine the $n+1$ activities at $x^c$ remain active but the $n+1^{st}$ residual is less than the maximum residual $\|y(t) - \phi(x^+,t)\|_\infty$. One can then proceed in a direction that maintains the $n$ activities until a new $t_j$ is determined such that $\|y(t)-\phi(x,t)\|_\infty = |y(t_j)-\phi(x,t_j)|$, thus once again satisfying (3), but with a lower maximum absolute residual.

Thus one might remark that dual methods emphasize the alternating sign property whereas primal method emphasize $n+1$ maximal residuals. When both hold optimality is reached.

Some attempts to generalise the concept of alternant to the nonlinear case have been made (see for example Motzkin [15], Rice [20] and Tornheim [21]), but the results are rather restrictive and difficult to exploit computationally since they depend upon properties that are either not possible to predict a priori or, if they do hold globally, are too strong and are rarely satisfied except for very special cases (for example, linear problems under the Haar condition is one useful instance).

This explains why most techniques for nonlinear discrete Chebshev approximation are based upon the nonlinear programming formulation. We wish to do otherwise.

Now, the characteristic equation (5) can be rewritten as

$$\text{(6)} \qquad \sum_{j=0}^n \lambda_i a_i = 0,$$

where $a_i = [\phi_1(t_i), \cdots, \phi_n(t_i)]^T$. This in turn can be rewritten as

$$\text{(7)} \qquad \sum_{j=0}^n \lambda_j \nabla \phi(x,t_j) = 0,$$

which is independent of $x$ in the linear case.

The point is that this form, although then dependent upon $x$, can be generalised to the nonlinear case.

**2. General Theory.** First we require some additional definitions.

DEFINITION 2 ([14]). *At any point $x_0$, the* **linear gradient space** *$J(t)$ of $\phi(x,t)$ refers to*

$$J(t) = span\{\frac{\partial \phi(x_0,t)}{\partial x_1}, \cdots, \frac{\partial \phi(x_0,t)}{\partial x_n}\}$$

*The dimension of this linear function space defined on $t \in [\alpha, \beta]$ is denoted by $d(x_0)$.*

For a linear space $\mathcal{L}$ with the classical Haar condition, $d(x) = n$, for all $x \in \mathbf{R}^n$.

In the discrete case, instead of considering the whole gradient function space $J(t)$, we consider the set of vectors corresponding to the columns of the Jacobian matrix

$$J(t_1, t_2, \cdots, t_m) = [\nabla f_1 \cdots, \nabla f_m].$$

Firstly, we remark that $\nabla f_i(x)$ is equivalent to $\nabla \phi(x, t_i)$ and note that our linear characteristic equation (5) can be written as

(8) $$\sum_{j=0}^{n} \lambda_j \nabla f_j(x) = 0.$$

Note: the Haar condition corresponds to any $n \times n$ submatrix of $J$ being nonsingular. Thus we are led to consider 'minimal' such sets, via the following important concept.

DEFINITION 3. *The vector set $\mathcal{C} = \{\nabla f_{i_0}, \cdots, \nabla f_{i_l}\}$, where the gradients are evaluated at a given fixed point $x$, is called a* **cadre** *if and only if:*
  1. $rank([\nabla f_{i_0}, \cdots, \nabla f_{i_l}]) = l$;
  2. *for any $\{\nabla f_{j_1}, \cdots, \nabla f_{j_l}\} \subset \mathcal{C}$, $rank([\nabla f_{j_1}, \cdots, \nabla f_{j_l}]) = l$.*

Note: the definition is local in that it depends upon $x$.

LEMMA 2.1 ( [8], LEMMA 20 ). *A vector set $\mathcal{C} = \{\nabla f_{i_0}, \cdots, \nabla f_{i_l}\}$ is a cadre if and only if $rank(\mathcal{C}) = l$ and there exists $\{\lambda_j \neq 0\}_{j=0}^{l}$ such that*

$$\sum_{j=0}^{l} \lambda_j \nabla f_{i_j} = 0.$$

DEFINITION 4. *If we take for our $a_{i_j}$, $\nabla f_{i_j}$ and normalise the multipliers $\{\lambda_j\}_{j=0}^{l}$ as follows*

(9) $$\begin{array}{ll} \sum_{j=0}^{l} \lambda_j = 1, & \text{if the sum is nonzero,} \quad \text{(cadres of } \textbf{type 1}) \\ \lambda_0 = 0, & \text{otherwise,} \quad \text{(cadres of } \textbf{type 2}) \end{array}$$

*then such a normalised set is unique and we term $\{\lambda_j\}_{j=0}^{l}$, the* **cadre multipliers** *associated with the cadre, $\mathcal{C}$.*

These multipliers, although asymptotically related to the Lagrange multipliers associated with the underlying minmax problem, are essentially different from

Lagrange multipliers since they are defined for any cadre and are not necessarily based upon maxim al functions.

We also now generalise the idea of a reference.

DEFINITION 5. *For continuous Chebyshev problems*

$$\min_{x \in \mathbf{R}^n} \max_{t \in [\alpha,\beta]} |\phi(x,t) - y(t)|$$

*the set of points* $\{t_{i_j}\}_0^l$ *is called a* **point cadre**, *at* $x_0$, *if and only if* $\{\nabla f_{i_j}(x_0)\}_0^l$ *is a cadre, where* $f_{i_j}(x) = \phi(x, t_{i_j}) - y(t_{i_j})$.

In the special case of the continuous **linear** Chebyshev problem, Descloux [10] called the above point cadre a cadre. In the linear case $\nabla f_{i_j}$ is independent of $x$, but since in the nonlinear case this is not so we use the term point cadre to emphasize the local structure of this generalisation.

It is clear that we can write (2) as the following minimax problem

$$\min_{x \in \mathbf{R}^n} \max_{1 \leq i \leq 2m} f_i(x)$$

where $f_{i+m}(x) = -f_i(x)$. For ease of extension to the general minimax problem, we use the above formulation in this paper.

We are now able to extend the notion of a reference function.

DEFINITION 6. *The set of functions* $\{f_{i_j}\}_{j=0}^l$ *are said to locally form a* reference set *of a minmax problem (1) if* $\mathcal{C} = \{\nabla f_{i_j}\}_{j=0}^l$ *is a cadre such that*

1. *the cadre multipliers* $\{\lambda_j\}_0^l$ *satisfy* $\lambda_j > 0$, $j = 0, \cdots, l$;
2. $\psi(x) f_{i_j}(x) > 0, j = 0, \cdots, l$, *where* $\psi(x) = \max_{0 \leq j \leq l} f_{i_j}(x)$.

*The reference set is further called* a levelled reference set *if the value of each function is the same, viz.,* $f_{i_j}(x) = f_{i_k}(x)$, *for any* $i_j, i_k \in \mathcal{C}$.

Note that, in general, the cadre multipliers may not alternate.

The following is well-known.

THEOREM 2.2 ( FIRST ORDER NECESSARY CONDITIONS ). *If* $x^*$ *is a local minimizer of (1), then there exist multipliers* $\{\lambda_i\}$ *such that*

(10) $$\sum_{i \in \mathcal{A}(x^*,0)} \lambda_i \nabla f_i(x^*) = 0,$$

(11) $$\sum_{i \in \mathcal{A}(x^*,0)} \lambda_i = 1,$$

(12) $$\lambda_i \geq 0, \quad i \in \mathcal{A}(x^*, 0),$$

*where* $\mathcal{A}(x^*, \epsilon) = \{i \mid \psi(x) - f_i(x) \leq \epsilon, i \in M\}$.

In terms of the reference set these first-order optimality conditions can be restated as follows.

THEOREM 2.3. *There exists a set of* $l+1$ *functions* $\{f_{i_j}(x)\}_{j=0}^l$ *which is a levelled reference set at* $x^*$ *on the cadre* $\mathcal{C} = \{\nabla f_{i_j}(x^*)\}_{j=0}^l$ *with the maximum deviation.*

We point out that the cadre corresponding to a reference set is of type 1.

## 3. Main Ideas of the Computational Procedure.

From the previous section, finding a local minimum of the Chebyshev problem is equivalent to locating a levelled reference set including all the active functions.

Thus a natural approach to solving the Chebyshev problem is to
  1. find a cadre;
  2. construct a reference set based on the cadre;
  3. level the reference set.

The algorithm we have developed is a descent algorithm with a line search. In addition to maintaining descent, the algorithm proceeds by recognising the structure of cadres. If a cadre is found, descent directions are defined to construct reference sets which are then levelled.

The following two lemmas help us identifying cadres of type 1 and type 2. Their proofs can be found in [7] ( Lemma 3.2 and Lemma 3.3 ).

LEMMA 3.1 (NECESSARY AND SUFFICIENT CONDITIONS FOR LOCATING A CADRE OF TYPE 2). *Suppose* $A = [\nabla f_{i_0} - \nabla f_{i_1}, \cdots, \nabla f_{i_0} - \nabla f_{i_{l-1}}]$ *is of full rank and that* $Z^T \nabla f_{i_0} \neq 0$, *where the columns of* $Z$ *form a basis for the null space of* $A^T$. *Then, there exists a cadre* $C \subseteq \{\nabla f_{i_0}\}_{j=0}^l$ *with cadre multipliers summing to zero if and only if* $[\nabla f_{i_0} - \nabla f_{i_1}, \cdots, \nabla f_{i_0} - \nabla f_{i_l}]$ *is rank deficient.*

LEMMA 3.2 (NECESSARY AND SUFFICIENT CONDITIONS FOR LOCATING A CADRE OF TYPE 1). *Suppose* $[\nabla f_{i_0} - \nabla f_{i_1}, \cdots, \nabla f_{i_0} - \nabla f_{i_l}]$ *are linearly independent. Then, there exists a cadre* $C \subseteq \{\nabla f_{i_0}\}_{j=0}^l$ *with cadre multipliers summing to one if and only if* $Z^T \nabla f_{i_0} = 0$, *where* $A = [\nabla f_{i_0} - \nabla f_{i_1}, \cdots, \nabla f_{i_0} - \nabla f_{i_l}]$ *and* $Z^T A = 0$.

The cadre structure is monitored through the concept of a **working set**, a collection of indices which are candidates for forming a cadre.

A working set $\mathcal{W} = \{i_0, \cdots, i_l\}$ at a given point, $x^c$, includes preferentially all the $\epsilon$-active functions (i.e. those functions within $\epsilon$ of the maximal functions) but is usually a larger set such that

$$(13) \qquad A = [\nabla f_{i_0} - \nabla f_{i_1}, \cdots, \nabla f_{i_0} - \nabla f_{i_l}]$$

is full-rank.

There are different ways of forming working sets. In our algorithm, a working set $\mathcal{W}$ is made up of the indices of the current $\epsilon$-activities plus a subset (possibly not proper) of the working set from the previous iteration.

The motivation for defining the working set in this manner is that it is from the set of the maximum functions that we expect to determine a levelled reference set.

LEMMA 3.3 ( [8], LEMMA 38 ). *Suppose a cadre of type 1 has been located at the point* $x$ *within the working set* $\mathcal{W} = \{\mu, i_0, \cdots, i_l\}$. *Suppose further, we define*

$v$ as the unique least squares solution to

(14) $$\hat{A}v = -\hat{\Phi}, \quad \text{where}$$
$$\hat{A} = [\nabla f_{i_0} - \sigma_0\sigma_1\nabla f_{i_1}, \cdots, \nabla f_{i_0} - \sigma_0\sigma_l\nabla f_{i_l}],$$
$$\hat{\Phi} = [f_{i_0} - \sigma_0\sigma_1 f_{i_1}, \cdots, f_{i_0} - \sigma_0\sigma_l f_{i_l}],$$

$f_{i_0}$ achieves the current maximum deviation, and $\sigma_j = sgn(f_{i_j})$. Then $v$ is a descent direction for $\psi(x)$.

If (2) is a linear problem, it can be proven that, at $x + v$, $\{f_{k_0}, \cdots, f_{k_l}\}$ form a levelled reference set where

(15) $$k_j = \begin{cases} i_j & \text{if } \lambda_j > 0, \\ i_{j+m} & \text{if } \lambda_j < 0 \text{ and } i_j \leq m, \\ i_{j-m} & \text{if } \lambda_j < 0 \text{ and } i_j > m, \end{cases}$$

and the $\{\lambda_i\}_{i=0}^{l}$ are the cadre multipliers. Thus, $v$ is a desirable direction and the working set is modified to give $\mathcal{W} = \{k_0, \cdots, k_l\}$.

Also, note that the concept of a cadre and cadre multipliers have enabled us, first to consider determining a reference set from an enlarged set and second, to perform multiple dropping — both these concepts are in turn motivated by the generalisation of the characterisation of a solution in the linear case.

We also note that if we have a cadre of type 1 the indices in the working set that correspond to negative cadre multipliers are dropped from the working set. More particularly, if the working set consists uniquely of active functions, indices corresponding to negative cadre multipliers are **automatically** dropped (see Lemma 5.2 of [7]).

Furthermore, if we have a cadre of type 2, we are not in the asymptotic region, as follows directly from the first order conditions for optimality, Theorem 2.2.

On the other hand, if we do not locate a cadre in the working set, we are able to decrease all the active functions and (provided $v$ is not an ascent direction — the usual case) level all the working functions by taking the direction

(16) $$d = \begin{cases} h + v, & \text{if } v \text{ is a descent direction,} \\ h, & \text{otherwise.} \end{cases}$$

Here

(17) $$\begin{aligned} h &= -ZB^{-1}Z^T\nabla f_\mu(x), \\ v &= -A(A^TA)^{-1}\Phi(x), \\ A &= [\nabla f_{i_0} - \nabla f_{i_1}, \cdots, \nabla f_{i_0} - \nabla f_{i_l}], \\ \Phi(x) &= [f_{i_0} - f_{i_1}, \cdots, f_{i_0} - f_{i_l}]^T, \\ Z^TZ &= I_{n-1}, \ A^TZ = 0, \end{aligned}$$

$i_0$ is an index for one of the activities, $B$ is positive definite and the working set, $\mathcal{W}$, is given by $\mathcal{W} = \{i_0, \cdots, i_l\}$.

At a first glance, it seems that the directions $h$ and $v$ depend on the index $i_0$ and thus are not uniquely determined by the current 'structure', i.e., working set. The following theorem shows that this is not the case.

**THEOREM 3.4 ( SEE [13], PAGE 61 ).** *Suppose the working set $\mathcal{W}$ is fixed and $B$ is positive definite. Then, the $h$ and $v$ given above by (16) are independent of the choice of $i_0$.*

If the working set consists only of active functions and we have located a cadre of type 2, all activities change equally with $v$ (up to first order). For a proof, the reader is referred again to Lemma 5.2 of [7]. In addition, for cadres of type 2, all entries in the working set (up to first order) change equally with $h$ (Lemma 5.1 of [7]). Since little is to be gained by levelling (we do not have the correct type of cadre) we discard $v$ and just take $h$ for our search direction.

Before being able to state the algorithm in some detail two major issues remain to be discussed, namely the line search and the definition of and manner in which we handle degeneracy.

Let us begin with the line search.

## 4. The Line Search.
We use a safeguarded line search that is similar to that of [16].

Thus suppose we have a descent direction $d$ and we wish to find an approximation to

$$(18) \qquad \min_{\alpha > 0} \psi(x^k + \alpha d),$$

where $\psi(x) = \max_{i \in M} f_i(x)$.

Define

$$(19) \qquad \nabla \psi_\epsilon^-(x, d) = \max_{i \in \mathcal{A}(x,\epsilon) \cap \nabla f_i^T d < 0} \nabla f_i^T d.$$

The **acceptance criteria** used in our algorithm is the following:

Given any constants $0 < \delta < \beta < 1$ and $0 < \gamma < 1$, we demand that $x^{k+1} = x^k + \alpha^k d^k$ satisfies:

$\boldsymbol{\delta}$ **Condition:** $\psi(x^{k+1}) \leq \psi(x^k) + \delta \alpha^k \nabla \psi_\epsilon^-(x^k, d^k)$

and at least one of the following two:

$\boldsymbol{\beta}$ **Condition:** there exists $i \in \mathcal{A}(x^k, \epsilon)$, $\nabla f_i(x^k)^T d^k < 0$ such that

$$\nabla f_i(x^{k+1})^T d^k \geq \beta \nabla f_i(x^k)^T d^k;$$

$\boldsymbol{\gamma}$ **Condition:** there exists $i \in \mathcal{A}(x^k, \epsilon)$, $\nabla f_i(x^k)^T d^k \geq 0$ or $i \notin \mathcal{A}(x^k, \epsilon)$ such that

$$f_\mu(x^{k+1}) - f_i(x^{k+1}) \leq \gamma[f_\mu(x^k) - f_i(x^k)], \quad \mu \in \mathcal{A}(x^k, 0).$$

We require the following additional assumption:

ASSUMPTION 4.1. Each gradient $\nabla f_i(x)$ satisfies the Lipschitz conditions,

$$|\nabla f_i(z) - \nabla f_i(x)| \leq L \|z - x\|_2 \quad \text{for all } i \in M.$$

The $\delta$ condition ensures that the reduction along each descent direction, $d^k$, has to be at least $\delta \alpha^k \nabla \psi_\epsilon^-(x^k, d^k)$.

The $\beta$ and $\gamma$ conditions essentially enforce that the steplength cannot be smaller than the minimum of $-\zeta \nabla \psi_\epsilon^-(x^k, d^k)$ and $\eta$, where $\zeta$ and $\eta$ are positive constants that depend, in general, on the functions being minimized.

The above conditions are generalisation of the stepsize acceptance criteria for smooth minimisation. Similarly we are able to prove that there always exists an interval of the steplengths satisfying the acceptance criteria.

LEMMA 4.1 ( SEE [13], PAGE 162 ). *Assume that $d^k$ is any descent direction for the maximum function $\psi(x)$ at $x^k$. Then there exists an interval $[\alpha_l, \alpha_r]$ where $\alpha_l < \alpha_r$ such that for all $\alpha \in [\alpha_l, \alpha_r]$, the $\delta$ Condition and either the $\beta$ Condition or the $\gamma$ Condition is satisfied.*

Note that Lemma 4.1 is independent of the definition of the descent direction.

The next lemma shows that if the acceptance criteria are satisfied, the stepsize cannot be too small.

LEMMA 4.2 ( SEE [13], PAGE 160 ). *Assume either the $\beta$ Condition or the $\gamma$ Condition is satisfied with $0 < \beta < 1$ or $0 < \gamma < 1$. Furthermore, assume that the set of descent direction $\{d^k\}$ is bounded and*

$$(20) \quad \nabla f_\mu^T d^k - \nabla f_i^T d^k = -(f_\mu - f_i) \quad \text{for any } i \in \mathcal{A}(x^k, \epsilon) \text{ and } \nabla f_i^T d^k \geq 0.$$

*Then there exist positive constants $\zeta$ and $\eta$ such that*

$$\alpha^k \geq \min\{-\zeta \psi_\epsilon^-(x^k, d^k), \eta\}$$

*is satisfied.*

Since the exact minimum of $\psi(x)$ along $d^k$ could occur only at either an intersection of two or more functions or at a minimum of one of the functions, the following result can easily be established.

LEMMA 4.3 ( SEE [13], PAGE 167 ). *Assume that $d^k$ is any descent direction for the maximum function $\psi(x)$ at $x^k$. Suppose $x^{k+1}$ is the first minimum along the direction $d^k$. Then, at $x^{k+1}$, the $\delta$ Condition and either the $\beta$ Condition or the $\gamma$ Condition are satisfied.*

We are now able to describe the line search procedure.

## Line Search Procedure

**Step 1** [ Initialization ] If newflg = true, $\alpha \leftarrow 1$, $j_0 \leftarrow 0$, Go to Step 2.
Compute the leftmost break point if one exists:

$$-\frac{f_\mu - f_{j_0}}{(\nabla f_\mu - \nabla f_{j_0})^T d} = \min\left\{-\frac{f_\mu - f_j}{(\nabla f_\mu - \nabla f_j)^T d} \bigg| j \notin \mathcal{W}, (\nabla f_\mu - \nabla f_j)^T d < 0\right\}$$

$$\alpha \leftarrow -\frac{f_\mu - f_{j_0}}{(\nabla f_\mu - \nabla f_{j_0})^T d}.$$

Otherwise, set the initial steplength to one.

$$\alpha \leftarrow 1, \quad j_0 \leftarrow 0.$$

**Step 2** [ Evaluation ]
Compute the function values and the gradients at $x^{k+1} = x^k + \alpha d^k$.
If the acceptance criteria are satisfied at $x^{k+1}$, stop.

**Step 3** [ Interpolation ]
(i) Do a cubic interpolation for the function $f_{\mu(x^{k+1})}(x^k + \alpha d^k)$, using both the function values and gradients at the points $x^k$ and $x^{k+1}$. Find its minimum $\alpha_{\mu^{k+1}}$.

(ii) Do a cubic interpolation for $f_{\mu(x^k)}(x)$, using both the function values and gradients at the points $x^k$ and $x^{k+1}$. Find its minimum $\alpha_{\mu^k}$;

(iii) Do two quadratic interpolations for $f_{\mu(x^{k+1})}(x)$ and $f_\mu(x)$, using the function values at the two points $x^k$ and $x^{k+1}$ and the gradients at $x^k$. Find the intersection $\alpha_b$.
If $\alpha_{\mu^k} < \alpha_b$, $\alpha^{k+1} \leftarrow \alpha_{\mu^k}$;
If $\alpha_b < \alpha_{\mu^{k+1}}$, $\alpha^{k+1} \leftarrow \alpha_b$;
Otherwise $\alpha_{k+1} \leftarrow \alpha_{\mu^{k+1}}$.

If $\alpha_{k+1}$ is not in the interval $(0, \alpha^k)$, one step of the bisection method is performed. Otherwise $\alpha \leftarrow \alpha_{k+1}$, go to Step 2.

**5. Degeneracy.** For the discrete Chebyshev problem, degeneracy handling is an important component of any robust algorithm.

DEFINITION 7. *For a general minimax problem (1), the current point $x^c$ is degenerate if and only if there is a cadre $\mathcal{C} = \{\nabla f_\mu, \nabla f_{i_1}, \cdots, \nabla f_{i_l}\}$ such that $\{\mu, i_1, \cdots, i_l\} \subset \mathcal{A}(x^c, 0)$.*

From the construction of the working set, if the problem is degenerate, we have

$$\mathcal{W}^k \subset \mathcal{A}(x^k, 0).$$

Denote

$$\mathcal{W}^k = \{\mu, i_1, \cdots, i_l\},$$

We have already seen that $h$ given by (17) is a descent direction unless the working set includes a cadre of type 1 (Lemma 3.2). Moreover, even in the degenerate case, if $Z^T \nabla f_\mu \neq 0$, there is no difficulty determining descent.

However, if $Z^T \nabla f_\mu = 0$ and there is more than one cadre $\mathcal{C} = \{\nabla f_\mu, \nabla f_{i_1}, \cdots, \nabla f_{i_l}\}$ satisfying $\{\mu, i_1, \cdots, i_l\} \subset \mathcal{A}(x^k, 0)$ it may not be possible to define a search direction such that it decreases the functions in all the cadres, although we know how to define a descending direction on one.

If we consider the cadres which correspond to subsets of active functions, then there can be three types of degenerate points:

**Type i)**    there only exist cadres with cadre multipliers summing to zero;
**Type ii)**    there exists a unique cadre and its cadre multipliers sum to one;
**Type iii)**    there exists more than one cadre and at least one with cadre multipliers summing to one.

For the degenerate points of Type i), there cannot be any reference set consisting of only the active functions. This is because, for any reference set, each of the corresponding cadre multiplier is positive and the sum of them is one. Thus, the current point cannot be optimal. From Lemma 3.3 the vertical direction defined by (14) attempts to construct a levelled reference set from a cadre. Following [8], Lemma 38, for a cadre with cadre multipliers summing to zero, the vertical direction decreases all the functions in the cadre by the same amount. Hence one possible way of constructing a levelled reference set, for degenerate points of Type i), is to decrease all the active functions by the same amount. Note that in this case, the functions in the cadres are all active.

For the degenerate points of Type ii), it is possible that a reference set exists within the active set. If there is such a reference set, then the current point is already a stationary point. Otherwise, a vertical direction can be defined to try to construct a levelled reference set ( See [8], Lemma 37 and equation (14), above ). In fact, at a degenerate point of Type ii), we can still define a descent direction which attempts to construct such a levelled reference set.

For the degenerate points of Type iii), we do not know a direct way of defining a descent direction and we obtain a descent direction by solving a constrained least squares problem.

The following two lemmas are useful in identifying the type of degeneracy.

LEMMA 5.1 ( SEE [13], PAGE 110 ). *Suppose* $\{\nabla f_\mu - \nabla f_{i_1}, \cdots, \nabla f_\mu - \nabla f_{i_{l-1}}\}$ *are linearly independent and* $\mathcal{W} = \{\mu, i_1, \cdots, i_l\}$. *Assume* $\tilde{Z}^T \nabla f_\mu = 0$ *where the columns of* $\tilde{Z}$ *form a basis for the null space of* $\{\nabla f_\mu - \nabla f_{i_1}, \cdots, \nabla f_\mu - \nabla f_{i_{l-1}}\}$. *Assume further that*

$$(\nabla f_\mu - \nabla f_{i_l}) = \sum_{j=1}^{l-1} \hat{\lambda}_j (\nabla f_\mu - \nabla f_{i_j}).$$

Then, there exists a cadre $\mathcal{C} \subseteq \{\nabla f_\mu, \nabla f_{i_1}, \cdots, \nabla f_{i_{l-1}}\}$ with cadre multipliers summing to one and there exists at least another cadre including $\nabla f_{i_l}$.

LEMMA 5.2 ( SEE [13], PAGE 112 ). *Suppose $\mathcal{W} = \{\mu, i_1, \cdots, i_l\}$ is a set of indices and $\{\nabla f_\mu - \nabla f_{i_1}, \cdots, \nabla f_\mu - \nabla f_{i_l}\}$ are linearly independent. Then, there can exist at most one cadre amongst $\{\nabla f_\mu, \nabla f_{i_1}, \cdots, \nabla f_{i_l}\}$.*

Consider the current $\epsilon$-active set $\mathcal{A}(x^k, \epsilon)$. For simplicity of discussion, we assume $\epsilon$ is sufficiently small such that $\mathcal{A}(x^k, \epsilon) = \mathcal{A}(x^k, 0)$. This is no loss of generality since we only have a finite number of functions $f_i(x)$. Denote

$$h_R^k = -Z^k Z^{k^T} \nabla f_\mu,$$

with

$$A^{k^T} Z^k = 0, \quad Z^{k^T} Z^k = I$$

and $\mathcal{W}^k = \mathcal{A}(x^k, \epsilon)$.

It is clear that there exists a cadre $\mathcal{C} = \{\nabla f_\mu, \nabla f_{i_1}, \cdots, \nabla f_{i_l}\}$ such that $\{\mu, i_1, \cdots, i_l\} \subset \mathcal{A}(x^k, 0)$ if and only if some gradients of active functions are linearly dependent.

From the construction of the working set $\mathcal{W}^k$ (details of which are given in [7], page 8), $\mathcal{W}^k = \{\mu, i_1, \cdots, i_l\}$ is chosen such that the corresponding Jacobian matrix $A^k$ is of full rank and the $\epsilon$-active functions are given priority when forming a working set.

Assume $\mathcal{W}^k \subset \mathcal{A}(x^k, \epsilon)$. In this case, there exists $i_{l+1} \in \mathcal{A}(x^k, \epsilon)$ such that

$$(21) \quad (\nabla f_\mu - \nabla f_{i_{l+1}}) = \sum_{j=1}^{l} \hat{\lambda}_j (\nabla f_\mu - \nabla f_{i_j}).$$

**Type i) Degenerate Points:** Since there is no cadre with the cadre multipliers summing to one, from Lemma 3.2, we have $h_R^k \neq 0$. Furthermore, by definition of a degenerate point, there exists a cadre $\mathcal{C}^k$ embedded in the active set $\mathcal{A}(x^k, \epsilon)$ with cadre multipliers summing to zero.

Type i) degeneracy is identified when $\mathcal{W}^k \subset \mathcal{A}(x^k, \epsilon)$ and $\|h_R^k\| > 0$, from (21) and by Lemma 3.1, there exists at least one cadre $\mathcal{C} \subseteq \{\nabla f_\mu, \nabla f_{i_1}, \cdots, \nabla f_{i_l}\}$ with cadre multipliers summing to zero. Following Lemma 3.2, there is no cadre with cadre multipliers summing to one. Hence, if $\{\mu, i_1, \cdots, i_l\} \subset \mathcal{A}(x^k, \epsilon)$, $x^k$ is a degenerate point of Type i).

Thus, moving along the direction which decreases all the functions in the cadre $\mathcal{C}^k$ by the same amount ( whose existence is assured by [7], Lemma 5.2 ) is a constructive way of building up a reference set.

Since $h_R^k \neq 0$, $h^k \neq 0$, assuming $B^k$ is positive definite. The horizontal direction is in the null space of $\{\nabla f_\mu - \nabla f_{i_1}, \cdots, \nabla f_\mu - \nabla f_{i_l}\}$. Furthermore, for any other active function $f_{i_j}$ not in the working set $\mathcal{W}^k$,

$$(\nabla f_\mu - \nabla f_{i_j})^T h^k = 0, \quad i_j \notin \mathcal{W}^k.$$

This comes from the fact that

$$(\nabla f_\mu - \nabla f_{i_j}) = \sum_{\nu \in \mathcal{W}^k} \theta_\nu (\nabla f_\mu - \nabla f_\nu), \quad \text{for any } i_j \notin \mathcal{W}^k.$$

Thus $h^k$ actually decreases all the active functions equally ( up to the first order ) and attempts to build a reference set from $\mathcal{C}^k$.

Hence, in this situation, we just take the horizontal direction as the search direction, i.e., $d^k = h^k$. It is important to realise that if a sequence $\{x^k\}$ converges to a stationary point, then there can only be a finite number of points which are degenerate points of Type i), since, at any stationary point, there exists a cadre with cadre multipliers s umming to one.

**Type ii) Degenerate Points:** If $x^k$ is a degenerate point of Type ii), there exists a unique cadre, based on the current $\epsilon$-active set, with cadre multipliers summing to one. Moreover, $h_R^k = 0$. Since there exists no other cadre, $\mathcal{A}(x^k, \epsilon) \subseteq \mathcal{W}^k$. ( Otherwise, using Lemma 5.1, there exists more than one cadre ).

Assume zero multipliers are detected when the projected gradient $h_R^k = 0$. From Lemma 3.2, there exists a cadre with cadre multipliers summing to one. If $\mathcal{W}^k = \mathcal{A}(x^k, \epsilon)$, following Lemma 5.2, there does not exist any other cadre, based on the current $\epsilon$-active set. Hence, $x^k$ can only be a Type ii) degenerate point.

In this case, all the $\epsilon$-active functions are in the working set $\mathcal{W}^k$. From the proof of Lemma 3.2, one cadre is given by $\mathcal{C}^k = \{ \nabla f_{i_j} \mid \lambda_j^k \neq 0, i_j \in \mathcal{W}^k \}$, where $\lambda^k$ is the least squares solution to

$$(22) \qquad \lambda_0 \nabla f_\mu + \sum_{j=1}^{l} \lambda_j \nabla f_{i_j} = 0, \quad \sum_{j=0}^{l} \lambda_j = 1.$$

If this cadre corresponds to a levelled reference set, then we have found a solution. Otherwise, following a proof similar to Lemma 3.3, it can be shown that the vertical direction, defined by (17), decreases the maximum function $\psi(x)$. Furthermore, this vertical direction attempts to construct a levelled reference set from the cadre $\mathcal{C}^k$.

The multipliers which are the least squares solution to (22) are uniquely defined for this type of degenerate points since $A^k$ has full rank. However, zero multipliers may occur. A zero multiplier in this case indicates that the function does not belong to the cadre which includes the representative function. From (17), the vertical direction will bring the functions with zero multiplers down together if a descending vertical direction is found. However, the functions with zero multiplers are not significant in the definition of the search direction in the sense that whether a descending vertical direction exists or not does not depend on the values of the functions with zero multipliers, since from equation (7.12) of [8],

$$\nabla f_\mu^T v = \sum_{j=0}^{l} \theta_j (f_\mu - \sigma_0 \sigma_j f_{i_j}).$$

Hence, if a zero multiplier occurs, this implies there exist more functions than necessary to form a cadre in the current working set. If $\mathcal{W}^k = \mathcal{A}(x^k, \epsilon)$, then the current point is degenerate. Otherwise, there exists at least one non-$\epsilon$-active function. In this case, it is reasonable to remove a non-$\epsilon$-active function, which is the furtherest away from being active, from the working set, i.e., $\mathcal{W}^k \leftarrow \mathcal{W}^k - I^+$, where

$$I^+ = \begin{cases} \{j_0 \mid f_\mu - f_{j_0} = \max_{j \in \mathcal{W}^k}(f_\mu - f_j)\} & \text{if } \mathcal{A}(x^k, \epsilon) \subset \mathcal{W}^k \\ \emptyset & \text{otherwise.} \end{cases}$$

It is interesting to note that for a degenerate point of Type ii), the definition of the search direction is the same as for a nondegenerate point.

**Type iii) Degenerate Points:** At a degenerate point of Type iii), amongst the gradients of the $\epsilon$-active functions, there exists more than one cadre and at least one with cadre multipliers summing to one.

Type iii) degeneracy is recognised when $\mathcal{W}^k \subset \mathcal{A}(x^k, \epsilon)$ and $\|h_R^k\| = 0$. By Lemma 3.2, there exists a cadre $\mathcal{C} \subseteq \{\nabla f_\mu, \nabla f_{i_1}, \cdots, \nabla f_{i_l}\}$ with cadre multipliers summing to one. Moreover, from (21) and following Lemma 3.1, there exists at least another cadre including $\nabla f_{i_{l+1}}$. Hence, $x^k$ is a degenerate point of Type iii). There is no obvious way of constructing reference sets for this type of degenerate point.

Assume $\mathcal{A}(x^k, \epsilon) = \{i_0, i_1, \cdots, i_l\}$, and $\mu = i_0$. Following a similar approach to [3], we solve the least squares problem given by

(23)
$$\min_{\theta \in \Re^{l+1}} \|\sum_{j=0}^{l} \theta_j \nabla f_{i_j}\|_2$$
subject to
$$\sum_{j=0}^{l} \theta_j = 1$$
$$\theta_j \geq 0, \quad j = 0, \cdots, l.$$

Suppose $\theta^*$ is a solution to (23). Denote

(24)
$$d^k = -\sum_{j=0}^{l} \theta_j^* \nabla f_{i_j}.$$

If the optimum value $\|d^k\| = 0$, the current point $x^k$ is a solution. Otherwise, $d^k$ is the steepest descent direction at the current point in the sense of [9] ( page 64 ), i.e.,

$$\nabla \psi(x^k, d^k) = \min_{\|d\|_2 = 1} \nabla \psi(x^k, d).$$

We modify the working set $\mathcal{W}^k$ as follows:

$$\mathcal{W}^k \leftarrow \{i_j \mid \theta_j^* > 0\}.$$

It is clear that $\mathcal{W}^k \subseteq \mathcal{A}(x^k, \epsilon)$.

The least squares problem with linear constraints (23) can be solved by methods described in [12] ( page 158 ). However, we shall exploit its special structure.

The problem (23) is a least squares problem with both equality and nonnegativity constraints.

We are able to show that we can solve (23) via a nonnegativity constrained least squares problem that handles the single equality implicitly.

Denote

$$e_{n+1}^T = [\underbrace{0, \cdots, 0}_{n}, 1], \quad e^T = [\underbrace{1, \cdots, 1}_{l+1}].$$

$$A = [\nabla f_{i_0}, \cdots, \nabla f_{i_l}], \quad \bar{A}^T = [A^T, e].$$

LEMMA 5.3 ( [13], PAGE 120 ). *Suppose $\lambda^*$ is a solution to the following NNLS (25).*

(25)
$$\min_{\lambda \in \Re^{l+1}} \|\bar{A}\lambda - e_{n+1}\|_2$$
$$\text{subject to}$$
$$\lambda_i \geq 0, \quad i = 0, \cdots, l.$$

*Then*

$$\theta^* = \frac{1}{e^T \lambda^*} \lambda^*$$

*is a solution to (23).*

In the implementation of the algorithm, we directly use the NNLS algorithm from [12].

**6. Maratos Effect.** For nondifferentiable optimization problems, difficulties arise when the iterates have to follow a steep sided groove which is a nonlinear curve across which the function has discontinuous first derivatives (change of sign). If we use a linearization of the discontinuity only, limited progress can be made along this linearization if the merit function is to be reduced. Associated with this difficulty, the Maratos effect that some unit steps fail to reduce the merit function may occur even when the iterates $\{x^k\}$ are arbitrarily close to the solution $x^*$. As a result, it is no longer possible to guarantee superlinear convergence.

In fact, there are examples that indicate that the Maratos effect could occur for the predescribed algorithm where the maximum function $\psi(x)$ has been chosen as the merit function [However, we have not yet seen the Maratos effect *numerically*].

Since in the final iterations of algorithms for nondifferentiable minimization such as a minimax problem, an equivalent nonlinear programming problem is often solved, the Maratos effect is also inevitable unless special strategies are used.

Current available approaches to the Maratos effect include [6], [11] and [4]. The first two are correction methods while [4] use a relaxation technique to allow possible increase of the merit function.

We use the former approach. Thus when close to a stationary point, i.e., a reference set has been found, the horizontal direction is performed first and a vertical direction is conducted afterwards to force the functions to have the same value. This simply amounts to computing the vertical direction by

$$(26) \qquad v = -A(A^T A)^{-1} \Phi(x + h)$$

where $\Phi(x)$ is defined as in (17).

We present the entire algorithm now. A user can request to invoke the process designed to avoid the Maratos effect by setting a flag Mflag = 1.

## ALGORITHM

**Initialization:** Suppose an initial point $x^0$ is given. Set $k \leftarrow 1$, $\mathcal{W}^0 \leftarrow \emptyset$.

*Step 1* [ QR Decomposition ]

Construct the working set (from $\mathcal{A}(x^k, \epsilon) \cup \hat{\mathcal{W}}^{k-1}$ ), Jacobian $A^k$ and its QR decomposition. Assume the columns of $Z^k$ form a basis for the null space of $A^{k^T}$.

If $\mathcal{A}(x^k, \epsilon) \subseteq \mathcal{W}^k$ and $\|Z^{k^T} \nabla f_\mu\| \leq \tau_c^k$, go to Step 2;

If $\mathcal{A}(x^k, \epsilon) \subseteq \mathcal{W}^k$ and $\|Z^{k^T} \nabla f_\mu\| > \tau_c^k$, go to Step 3;

If $\mathcal{A}(x^k, \epsilon) \not\subseteq \mathcal{W}^k$ and $\|Z^{k^T} \nabla f_\mu\| > \tau_c^k$, go to Step 4;

If $\mathcal{A}(x^k, \epsilon) \not\subseteq \mathcal{W}^k$ and $\|Z^{k^T} \nabla f_\mu\| \leq \tau_c^k$, go to Step 5;

*Step 2* [ Cadre "Found" with $\sum_{i \in \mathcal{C}} \lambda_i = 1$ ] If $\mathcal{W}^k$ is a reference set, obtain $B^k = Z^{k^T} G^k Z^k$, where $G^k$ is a positive definite approximation to the Hessian of $\sum_{i \in \mathcal{C}} \lambda_i f_i(x)$ at $x^k$. Compute the horizontal direction $h^k$ from (17); If Mflag = 0, compute the vertical direction $v^k$ from (17); Otherwise, compute the vertical direction from (26). Set the search direction $d^k = h^k + v^k$.

If $\mathcal{W}^k$ is not a reference set, compute the vertical direction according to (14). Set $d^k = v^k$. Go to Step 6.

*Step 3* [ Cadre not Found ]

Obtain $B^k = Z^{k^T} G^k Z^k$, where $G^k$ is a positive definite approximation to the Hessian of $\sum_{i \in \mathcal{C}} \lambda_i f_i(x)$ at $x^k$. Compute the horizontal direction $h^k$ and the vertical direction $v^k$ from (17). If $\nabla f_\mu^T v < 0$, $d^k = h^k + v^k$. Otherwise $d^k = h^k$. Modify $\mathcal{W}^k$ if necessary. Go to Step 6.

*Step 4* [ Cadre "Found" with $\sum_{i \in C} \lambda_i = 0$ ]
Compute $d^k = -Z^k Z^{k^T} \nabla f_\mu^k$. Go to Step 6.

*Step 5* [ More than One Cadre and at Least One with $\sum_{i \in C} \lambda_i = 1$]
Compute the search direction $d^k$ using (24). Set $\tau_c^{k+1} \leftarrow \frac{\tau_c^k}{2}$.

*Step 6* [ Line Search ]
Perform a safeguarded line search. Set $k \leftarrow k+1$. If $\|d^k\|_2 < \tau_s$ and $\mathcal{W}^k$ includes a levelled reference set, stop. Otherwise, go to Step 1. □

**7. Conclusion.** It is well known that a best linear Chebyshev approximation corresponds to a characteristic structure. It is not so well recognised that a solution of a nonlinear Chebyshev problem also possesses a rich structure and characterisation.

Under the classical Haar condition, the best linear Chebyshev approximation, on the real line, is a levelled reference function with the maximum deviation. There exist exactly $n+1$ distinct and ordered points which achieve the maximum deviation and the signs of the residuals on these points alternate.

Our experience with linear Chebyshev problems indicates that it is important for an efficient algorithm to make use of these special properties of a solution. The famous Remez algorithms [19] & [18] and the descent algorithm given in [2] are examples of such algorithms.

For many Chebyshev problems, such as multidimensional problems, nonlinear problems and discrete problems, however, the classical Haar condition does not hold. Thus, whether there exists some significant properties that can be computationally exploited is of some interest.

Nonlinear Chebyshev approximation theory indicates that, theoretically, for certain classes of nonlinear problems at least, useful characterisations still exist. Nonetheless, these theoretical characterisations are not easily computationally constructable or even recognizable.

The first author's experience with the descent algorithm given in [5], suggested that when there exist intermediate points where two or more activities belong to the same peak of the error curve, the efficiency of the method is impeded. We realise, however, that for usual Chebyshev *solutions*, this cannot happen. This suggests that, if we impose the structure of a solution, the algorithm may be improved.

The idea of a cadre has been introduced in [10] to describe a *linear* Chebyshev solution when the classical Haar condition is absent. Starting from this concept, we have been able generalise it to nonlinear Chebyshev problems. Based on the cadre , for nonlinear Chebyshev problems, we have generalised the reference set and levelled reference set concepts which characterise the property of alternating signs for a linear Chebyshev problem with the classical Haar condition.

We have then proceeded to exploit these results computationally.

The global convergence properties of the algorithm have been analysed through establishing the line search acceptance criteria. We point out that, under certain conditions, the algorithm is globally convergent with a two steps superlinear convergence ( see [13], page 191 ).

The characterisation established for a local minimum of the discrete nonlinear Chebyshev problems is a generalisation of the specific properties of the best approximations for continuous linear Chebyshev problems.

Our algorithm has thus been developed to locate a local minimum of a discrete nonlinear Chebyshev problem by attempting to establish its structure and satisfy the characterisation of the local minimum. The algorithm is a method of successive descent on the maximum function with a line search on this function.

The algorithm builds up the structure through construction of the working set which attempts to approximate a reference set. The concept of the working set plays an important role in the algorithm. The functions in the working set are in general not $\epsilon$-active functions, except near a solution. Along with the search directions, the working set attempts to exploit the geometry of the error curve of the solution. The important levelling process is embodied in the vertical directions.

The algorithm has been implemented in a numerically stable way. Initial numerical testing has indicated the efficacy of the method. Details are given in [7].

With suitable modifications, our approach can be applied to general minimax problems. Additional constraints can also be handled using a rather straightforward extension.

# REFERENCES

[1] I. BARRODALE AND C. PHILLIPS, *An improved algorithm for discrete Chebychev linear approximation*, in Proc. 4th Manitoba Conf. on Numer. Math., U. of Manitoba, Winnipeg, Canada, 1974, pp. 177–190.

[2] R. H. BARTELS, A. R. CONN, AND Y. LI, *Primal methods are better than dual methods for solving overdetermined linear systems in the $l_\infty$ sense?*, SIAM J. Numer. Anal., 26 (1989), pp. 693–726.

[3] S. BUSOVAČA, *Handling degeneracy in a nonlinear $l_1$ algorithm*, Tech. Rep. Tech. Rept. CS-85-34, Univ. of Waterloo, Dept. of Computer Science, Univ. of Waterloo, Waterloo, Ontario N2L 3G1, 1985.

[4] R. M. CHAMBERLAIN, C. LEMARECHAL, H. C. PEDERSEN, AND M. J. D. POWELL, *The watchdog technique for forcing convergence in algorithms for constrained optimization*, Math. Prog. Study, 16 (1982), pp. 1–17.

[5] C. CHARALAMBOUS AND A. R. CONN, *An efficient method to solve the minimax problem directly*, SIAM J. Numer. Anal., 15 (1978), pp. 162–187.

[6] T. F. COLEMAN AND A. R. CONN, *Nonlinear programming via an exact penalty function: Asymptotic analysis*, Math. Prog., 24 (1982), pp. 123–136.

[7] A. R. CONN AND Y. LI, *An efficient algorithm for nonlinear minimax problems*, Tech. Rep. CS-88-41, Department of Computer Science, University of Waterloo, Waterloo, Ontario N2L 3G1, Canada, 1988.

[8] A. R. CONN AND Y. LI, *Structure and characterization of discrete chebyshev problems*, Tech. Rep. CS-88-39, Department of Computer Science, University of Waterloo, Waterloo, Ontario N2L 3G1, Canada, 1988.

[9] DEM'YANOV AND MALOZEMOV, *Introduction to Minimax*, Keter Publishing House, Jerusalem, 1974.

[10] J. DESCLOUX, *Dégénéresence dans les approximations de Tschebysheff linéaries et discrètes*, Numerische Mathematik, 3 (1961), pp. 180–187.

[11] R. FLETCHER, *Second order corrections for non-differentiable optimization*, in Lecture Notes in Mathematics 912, G. Watson, ed., Springer Verlag, 1981, pp. 85–114.

[12] C. L. LAWSON AND R. J. HANSON, *Solving Least Square Problems*, Prentice-Hall, 1974.

[13] Y. LI, *An Efficient Algorithm for Nonlinear Minimax Problems*, PhD thesis, University of Waterloo, 1988.

[14] G. MEINARDUS, *Approximation of Functions: Theory and Numerical Methods*, Springer Verlag, 1967. translated by Larry, L. Schumaker.

[15] T. S. MOTZKIN, *Approximation by curves of a unisolvent family*, Bull. Amer. Math. Soc., 55 (1949), pp. 789–793.

[16] W. MURRAY AND M. L. OVERTON, *Steplength algorithms for minimizing a class of nondifferentiable functions*, Computing, 23 (1979), pp. 309–331.

[17] M. J. D. POWELL, *Approximation Theory and Methods*, Cambridge University Press, 1981.

[18] REMEZ, *Sur le calcul effectif des polynomes d'approximation de tchebichef*, Competes Rendues, 199 (1934), pp. 337–340.

[19] ———, *Sur un procédé convergent d'approximations successives pour déterminer les polynômes d'approximation*, Competes Rendues, 198 (1934), pp. 2063–2065.

[20] J. R. RICE, *On the existence and characterisation of best nonlinear Tchebyshev approximation*, Tran. Am. Math. Soc., 110 (1964), pp. 88–97.

[21] L. TORNHEIM, *On n-parameter families of functions and associated convex functions*, Trans. Amer. Math. Soc., 69 (1950), pp. 457–467.